Matrix and finite element displacement analysis of structures

D. J. DAWE

Department of Civil Engineering,
University of Birmingham

CLARENDON PRESS · OXFORD
1984

Oxford University Press, Walton Street, Oxford OX2 6DP

London Glasgow New York Toronto
Delhi Bombay Calcutta Madras Karachi
Kuala Lumpur Singapore Hong Kong Tokyo
Nairobi Dar es Salaam Cape Town
Melbourne Auckland

and associates in
Beirut Berlin Ibadan Mexico City Nicosia

Oxford is a trade mark of Oxford University Press

Published in the United States
by Oxford University Press, New York

British Library Cataloguing in Publication Data

Dawe, D. J.
Matrix and finite element displacement analysis of structures.
—(Oxford engineering science series)
1. Structures, Theory of—Matrix methods
I. Title
624.1'71'015129434   TA642

ISBN 0-19-856211-x
ISBN 0-19-856213-6 Pbk

Library of Congress Cataloging in Publication Data

Dawe, D. J.
Matrix and finite element displacement analysis of structures.

(The Oxford engineering science series)
Bibliography: p.
Includes index.
1. Structures, Theory of—Matrix methods. 2. Finite element method.
I. Title. II. Series.
TA347.D4D38 1983   624.1'7'015129434   83-12101
ISBN 0-19-856211-X
ISBN 0-19-856213-6 Pbk
Filmset and printed in Northern Ireland at The Universities Press (Belfast) Ltd.

To my wife Eileen and daughters Alison and Julia

PREFACE

THE past couple of decades have seen very profound changes take place in the world of structural and stress analysis resulting from the intense development of computer-based methods which can be applied in a very general way to arbitrary structural forms. Chief amongst these methods are the matrix and finite element methods which, though based on fundamental principles of structural mechanics of long standing, awaited the development of adequate digital computer power before becoming practical methods of analysis. With the increasing development of such power the matrix and finite element methods have moved from being the province of a select group of very few research specialists to being of abiding concern not only to innovative research workers but also to many thousands of engineering practitioners who seek numerical solutions to structural problems of great number and variety.

The use of matrix algebra is virtually mandatory in developing procedures which will be suitable for the analysis of complicated structural problems using the digital computer, and the phrase 'matrix analysis of structures' can be interpreted in its broad sense to include any procedure so long as matrices are involved. However, in this book 'matrix analysis of structures' will imply the analysis of skeletal structures (beams, bars, and frames) only, using member (or element) relationships between forces and displacements that are exact within the limits of the background theoretical assumptions. This type of analysis has been taught to students of civil and structural engineering over a relatively long period. Finite element analysis is distinguished herein from matrix structural analysis by the fact that element relationships are usually approximate and are derived in a less basic fashion, commonly through use of work or energy principles. The range of application of the finite element method is vast, both in the type and in the complexity of structures that can be analysed and the method allows numerical solutions to be obtained to many problems that would otherwise be intractable. (The finite element method is even more versatile than this since it is, in fact, a general procedure that is used in other fields of study such as fluid mechanics and heat transfer: the concern in this book is exclusively with solid mechanics, however.) To an ever-increasing extent the structural/stress analyst or designer needs to understand the theoretical basis of the finite element method and to appreciate its potential, and as a consequence the method now often finds a place in the curriculum of undergraduate solid mechanics courses.

There are a number of books in the subject area but these tend to have a high degree of concentration on either matrix analysis of skeletal structures or on finite element continuum analysis, rather than dealing in detail with both. Here detailed and quite separate attention is given to matrix structural analysis, in the early chapters, and then to finite element analysis. In this way it is hoped to provide the reader with largely self-contained and comprehensive descriptions of both topics in the order in which he is likely to meet them in his technical education.

The book is directed at students of solid/structural mechanics in the realms of civil and structural engineering, mechanical engineering, and aeronautical engineering. It is concerned with providing the basic theoretical material which underlies the matrix and finite element displacement approaches, proceeding from simple introductory material to a quite advanced stage at a pace which it is hoped is gentle enough to be assimilated by the newcomer to the subject area. A large part of the material is suitable for senior undergraduate work whilst other material can be used in postgraduate work. It is also hoped that the book will be of value to practising engineers who wish to familiarize themselves with modern methods of structural/stress analysis through a scheme of self-study, and indeed to those already engaged in the subject area. The text is concerned with linear analysis only and it is assumed that the reader has had some exposure to elementary structural mechanics, the theory of elasticity and matrix algebra, although some review of these topics is provided in Chapters 1 and 7 and Appendix A.

As the title indicates, the book is concerned almost exclusively with displacement-type analysis, that is with the type of analysis in which displacements, rather than forces, are the primary variables. Force-type analysis is given very brief attention in Chapter 1 only and is abandoned thereafter. This policy reflects the fact that the displacement (or stiffness) method predominates over the force method in practical structural analysis to a very great extent nowadays.

To aid in the assimilation of material the full worked solutions to a considerable number of examples are provided and there are included also many problems for solution by the reader, with answers provided. The author strongly recommends the reader to attempt a good number and range of these problems since this is seen as a very important aid to the learning process. Those of the worked examples or problems for solution of a numerical nature are such that they involve very few unknowns and can be solved with the aid only of a hand calculator at most. This is designed to clarify the principles of the analysis, which would otherwise be obscured by sheer weight of numbers in large problems, but perhaps it does not always make evident the advantages of the matrix/finite element approach over other approaches. The reader should keep in mind that the systematic procedures described in this book and used in the simple examples detailed herein can readily accommodate very much larger problems (i.e. problems with many more unknowns), which is not the case with other approaches.

The opening chapter of the book is concerned with setting the scene for later developments by presenting a general description of matrix and finite element methods, recapping some very basic considerations of structural analysis, and briefly comparing the displacement and force methods, before abandoning the latter.

The matrix displacement analysis of skeletal structures is described in Chapters 2–4 and this material could be used as a complete and separate course of study. To this end the most basic material is that of Chapter 2 and Sections 3.1–3.4 and 4.1–4.2: this introduces the matrix displacement approach and includes the analysis of bars, beams, and planar frames (pin-jointed and rigid-jointed), taking into account the effect of dead loading acting along members. The remaining sections of Chapters 3 and 4 deal with other kinds of skeletal structure and other specialized effects, and could be studied to the extent appropriate to individual interests and the availability of time.

In Chapters 5 and 6 attention is still restricted to the line elements of skeletal structures but now new procedures of the finite element kind, capable of extension to two- and three dimensional situations, are introduced for the derivation of stiffness relationships. These procedures are based on the use of the Principle of Virtual Displacements or the Principle of Minimum Potential Energy, in conjunction with an assumed displacement field. These principles are interrelated and yield the same solution as one another but in this book priority is given to the Principle of Minimum Potential Energy.

As a further precursor to the finite element analysis of continuum structures, Chapter 7 provides a summary of the equations of three-dimensional elasticity theory and their specialization to the two-dimensional problems of plane stress and strain, of axisymmetric analysis and of classical plate bending. Therein the general finite element displacement formulation for continua is also presented.

Chapter 8 is concerned with the finite element analysis of the plane stress (and plane strain) situation, providing details of the straight-sided triangular and rectangular families of elements, using standard Cartesian co-ordinates. On a similar basis Chapter 9 deals with three-dimensional analysis, introducing tetrahedral and hexahedral elements, and with the analysis of axisymmetric solids. In Chapter 10 attention is given to the more advanced ideas that have formed the basis for the development of two- and three-dimensional elements and the chapter ends with a description of the popular and important category of curvilinear isoparametric elements. Chapter 11 is devoted, in the main, to the finite element analysis of plates in bending based on the use of classical plate theory and some mention is also made of alternative approaches to the plate bending problem and of the analysis of shells.

Chapter 11 concludes the description of the finite element displacement method so far as the analysis of stable, static structures is concerned. For a basic introductory course of instruction on the method it is very important to establish firmly the fundamental procedures and it is recommended that Sections 5.1–5.5, 6.1–6.10, and 7.1–7.4, together with the whole of Chapter 8, be studied in depth. Beyond this introductory level it is a matter of choice as to exactly what topics of those covered in Chapters 9, 10, and 11 are studied.

The main text is concluded in Chapter 12 with a look at the use of the finite element method in two types of solid mechanics problems of the eigenvalue type, namely the calculation of the natural frequencies of free vibration and of the buckling loads of structures.

It has already been made clear that the programming of the matrix and finite element methods for the digital computer is essential if practical problems are to be solved and thus it might be expected that this book would contain the listings of one or more programs. It was in fact rather tempting to do this, but on reflection it was decided that this was not desirable in a book of the nature of this one, partly since space restrictions would not allow other than the most cursory of presentations but also because programs rapidly become outdated.

For those readers with some experience of computer programming it should, in fact, not be too difficult a project to write a simple program for the matrix or finite element analysis of a particular type of structure once the fundamentals of the procedures, described in this text, are thoroughly understood. It is noted that some books are available nowadays which are devoted completely to programming the finite element method (e.g. Hinton, E. and Owen, D. R. J., *Finite element programming*, Academic Press, 1977) and to which the reader wishing to produce programs of reasonable sophistication can refer. It is probable though that the large majority of readers of the present book will be users, rather than writers, of programs and many may well use one or more of the available commercial, general-purpose programs which provide a means of solving problems involving structures of arbitrary type with perhaps thousands of unknowns. These programs make use of a 'library' of dozens of different kinds of elements together with advanced equation-solving techniques and with sophisticated means of inputting and checking data, and of outputting results, often with the aid of automatic graphical interpretation. It is clearly very important that users of such powerful programs are fully aware of the basic theoretical background on which such programs are based and it is hoped that study of the material presented in this book will help them to make sound decisions as to what element types and meshes to employ in a particular problem, how to interpret the results critically, and so on.

Finally, a few words of thanks are appropriate. I am very pleased to record my gratitude to Professor W. H. Wittrick, formerly Beale Professor of Civil Engineering at the University of Birmingham, for his help and encouragement of this project in its early stages. I am very grateful to Mrs Wendy O'Connor for typing the manuscript accurately and with a willing co-operation that considerably eased the task of completing the book. I am much indebted to Mrs Mary Andrews for producing high-quality figures for the book from my pencil sketches.

Birmingham
September, 1983   D.J.D.

CONTENTS

1
GENERAL INTRODUCTION AND STRUCTURAL PRELIMINARIES

1.1. A few historical remarks

THE fundamental concepts of the mathematical theory of elasticity were expounded as long ago as the seventeenth and eighteenth centuries by men such as Hooke, Euler, Lagrange, and the Bernoullis. The contributions of that era concerned the relationships between stresses and strains, the derivation and solution of governing differential equations, and concepts of energy. In the next century the accent was upon the development and application of structural theory based on the mathematical background laid earlier. Important contributions were made by Navier, Clapeyron, Saint-Venant, Airy, Maxwell, Castigliano, and Mohr, amongst others. The work of these pioneering structural investigators continues to form the basis of structural theory since it has the virtues of direct logic and general applicability.

Unfortunately, analytical solutions using the fundamental theories can only be obtained to a limited class of structures of simple geometry, and direct numerical solutions require computing power which was not available until comparatively recently. Consequently in the first half of this century structural engineers turned their attention to developing a variety of indirect methods of solution to particular classes of structural problem. For instance, in civil engineering a class of structure that came in for particular attention, because of its practical importance, was the *skeletal* structure—an assembly of slender bar/beam components connected together at their ends to form a continuous beam or a two- or three-dimensional frame. Here the possible physical idealization of a whole structure into obviously recognizable component parts (or members or elements) presented an obvious avenue of approach.

Using basic principles it is possible, as will be demonstrated later, to develop 'exact' relationships between the moments and lateral forces at the ends of a particular beam member and the corresponding rotations and lateral displacements. (By 'exact' relationships is meant relationships which are exact within the context of an accurate but approximate underlying theory.) These relationships are the well-known slope deflection equations due to G. A. Maney. A skeletal structure can be analysed directly by combining the force–displacement relationships for the individual members subject to satisfying equilibrium (of forces) and compatibility (of displacements) conditions at the joints. In principle the slope deflection method could be used in the analysis of complex skeletal structures but before the advent of the modern digital computer a disadvantage of the method was that it was impossible to solve the resulting set of linear algebraic simultaneous equations for other than very simple problems. Partly because of this the alternative method of moment distribution due to Hardy Cross became popular in design offices; this method avoided the need for a laborious direct solution of a set of simultaneous equations by substituting an iterative procedure more suitable to hand calculations. As is typical of many of the methods of its generation the moment distribution method is a specialized method developed for a particular type of structure, based on certain simplifying assumptions (members are usually assumed to be inextensible for instance), and requiring a degree of physical insight on the part of the structural analyst.

1.2. Matrix methods of analysis of skeletal structures

From the period around 1950 the developments which were taking place in the realm of digital computing presented the structural and stress analyst with ever increasingly powerful aid in the solution of his problems. This progressively led to a change in emphasis away from methods which are simplified and specialized to particular structures, so as to avoid overwhelming manual labour on the part of the analyst, toward more direct and general methods whose basic philosophy is tailored to the digital computer. The availability of computing power meant that a general systematic and repetitive approach was increasingly advantageous and that the solution of any resulting large set of simultaneous equations was no longer a major obstacle. Matrix algebra provided a basis for the efficient organization and manipulation of large quantities of data. Thus, without the need to develop any fundamentally new structural principles, the stage was set for the introduction of what are known as the matrix methods of structural analysis. In the realm of civil engineering this phrase is generally understood (and will be so understood in this work) to mean the analysis of *skeletal* structures.

In Fig. 1.1 are shown a number of typical kinds of skeletal structure whose analysis will be considered in some detail in Chapters 2, 3, and 4 of this text. The basic feature of a skeletal structure is that the individual parts or members or elements of the structure are treated as one-dimensional or line members of the type shown in Fig. 1.1(h). The kinds of skeletal structure shown in Fig. 1.1 are categorized by their geometrical form and by the method of connection of the individual members one to another. Continuous bars and beams are shown in Figs. 1.1(a) and 1.1(b) and are deemed to fall under the heading of skeletal structures

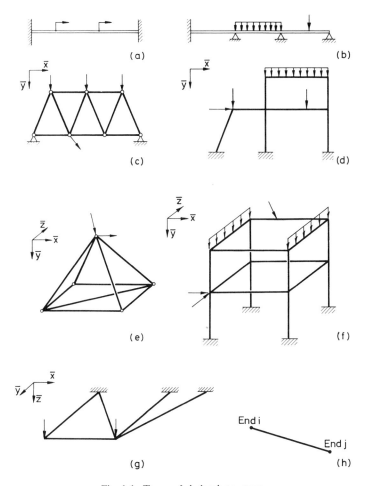

Fig. 1.1. Types of skeletal structure.

since they are assembled from a number of members, although all members are aligned in a common direction. Two-dimensional or planar frames are shown in Figs. 1.1(c) and 1.1(d) and, as the name suggests, all members of the structure lie in a single plane and all applied loads act in this plane. Two types of joint description are commonly used: these are the pin joint (Fig. 1.1(c)) and the rigid joint (Fig. 1.1(d)). In the former joint it is assumed that members meet at a frictionless pin so that all members are free to rotate individually at the joint. Loads are applied at the pin joints only and the members are assumed to carry only an axial force. With the rigid joint the members meeting at the joint are rigidly connected to each other such that any rotation of the joint is common to all members meeting there. The planar rigid jointed frame can carry loads distributed along the lengths of the members as well as at the joints themselves and the members of the frame are subjected to bending moments and shearing forces as well as to axial forces. Three-dimensional or space frames are of unrestricted geometrical form and again two types of joint description are used in direct extension of the two-dimensional situation: Fig. 1.1(e) shows a ball-jointed space frame whilst Fig. 1.1(f) shows a rigid-jointed space frame. The members of the ball-jointed frame are assumed to carry only axial forces, set up by joint loads only, whilst those of the rigid-jointed space frame are subjected to axial and shearing forces, bending moments, and twisting moments due to the application of a general type of loading. The final category of framed structure consi- dered is the grillage, an example of which is shown in Fig. 1.1(g). This is again a two-dimensional rigid-jointed frame but is distinguished from the type of structure shown in Fig. 1.1(d) by the fact that the loads applied to the structure act normal to the plane of the frame rather than in the plane.

Before proceeding, it should be recognized that the skeletal structures described above and shown in Fig. 1.1 are theoretical model structures rather than actual structures. In reality all members and joints of a structure are three-dimensional entities, of course, so the assumption of one-dimensional behaviour of members and of point specification of structure joints is an obvious approximation to the actual situation. However, it is necessary to make such assumptions in order to produce an answer to the structural problem and by and large, provided that the cross-sectional dimensions of the members are small compared with their lengths, the analysis of the theoretical model structure will provide an accurate picture of the overall response of the real structure to loading.

The matrix methods for skeletal structures require the generation of relationships (in matrix form) linking end forces to end displacements for the general line member of such a structure. (Note that the words 'forces' and 'displacements' are used here in the general sense with 'forces' understood to include both translational forces and rotational forces, i.e. moments, and 'displacements' understood to include translational and rotational displacements.) Relationships can be found without undue difficulty for a straight member of uniform cross-section (i.e. a prismatic member) which, within the constraints of the theoretical description of the member as a line member, are exact: that is the relationships represent an exact solution at all points of the differential equations governing the deformation of a line member. Then the relationships for each member of the structure can be combined following certain rules (of equilibrium and compatibility) into a system of simultaneous equations for the whole structure, and the solution of this system of equations effectively gives the exact solution for the theoretical model structure. Two kinds of approach are commonly used. In the matrix *displacement* (or *stiffness* or *equilibrium*) method the member relationships are those for end forces expressed in terms of end displacements by way of a *stiffness matrix* and the unknown quantities in the set of simultaneous equations for the whole structure (i.e. the primary variables) are displacement values at the structure joints. In the matrix *force* (or *flexibility* or *compatibility*) method the member relationships are those for end displacements expressed in terms of end forces by way of a *flexibility matrix* and the primary variables are structure forces. The differences between the displacement and force methods will be discussed later in the chapter, but for the moment it is simply noted that in general the matrix displacement method (MDM) is the more popular method and it is this method which forms the subject of this text.

1.3. The finite-element method

By the end of the first half of this century the design engineers in the aircraft industry had also developed ingenious methods of analysis which were applicable to idealized and specialized structures. In these methods matrix algebra already played some part and with the development of computing power at that time the situation was ripe for a generalization and extension of analysis methods. The need for an improvement in analysis technique was pressing since aircraft structural configurations were changing rapidly and the methods of a few years earlier were insufficiently general to deal with the variety and complexity of the new structural shapes. It was in these circumstances that the finite-element method emerged in its recognizably modern form in the mid 1950s, although it is true to say, at least with the benefit of hindsight, that some much earlier work represent a form of the method. From its origins in the aircraft industry use of the method spread rapidly within the realm of solid mechanics into civil and mechanical engineering (and nowadays applications outside the solid mechanics field are commonplace with a broadening of the mathematical base of the method).

As with the matrix methods of (skeletal) structural analysis the basis of the finite-element method resides in subdividing a complex structure—the theoretical model of the actual structure—into a *finite* number of discrete parts or regions or *elements*. In the matrix analysis of skeletal structures the breakdown of the structure into individual components is physically obvious. With the finite-element method the process is taken much further and is applied to the analysis of two- or three-dimensional continuum structures within which there probably are few, if any, actual physical boundaries between one region and another. It is necessary to subdivide a model continuum structure because, of course, it is generally not possible to obtain a solution satisfying the governing differential equations over the whole theoretical model structure using exact classical methods; the process of sub-division is known as finite-element *discretization* or *idealization*. Structural relationships can be derived for each region of the continuum, i.e. for each finite element, which link force and displacement components at selected reference points on each element. These selected points are termed nodal points or simply *nodes*, and are usually, but not always, located on the boundary of the element. (Just as in the matrix analysis of skeletal structures the reference points are at the member ends.)

The generation of suitable element force–displacement relationships is more difficult than in the matrix displacement method and usually such relationships are approximate, rather than exact, in nature and are popularly based on the use of work or energy principles. The generation of element relationships is obviously a very important part of any finite-element analysis. If suitable (albeit approximate) relationships between forces and displacements can be found the total theoretical model structure can then mentally be reassembled by connecting together the individual finite elements at their nodes subject, again, to the satisfaction of certain conditions of equilibrium and compatibility. The procedure leads to a set of simultaneous algebraic equations for the structure as a whole which can be solved to yield values of the primary variables at the nodes. As with the matrix methods, finite-element methods of the *displacement* and *force* type exist depending on whether nodal displacements or nodal forces are the primary variables. Additionally a finite element *mixed* method exists in which the primary variables are some displacements and some forces. In this volume attention is restricted solely to the finite-element *displacement* method: element properties are therefore characterized by a stiffness matrix. Within the sphere of finite-element structural analysis the displacement approach is much more widely used than are the alternative force or mixed approaches.

The great strength of the finite element method is its versatility, since with the proviso that satisfactory force–displacement relationships can be established at the element level (i.e. an element stiffness matrix can be set up) there is virtually no limit to the type of structure that can be analysed. During the years since the inception of the method many types of finite element have been developed and some of these are shown in Fig. 1.2. In the displacement method the derivation of any element stiffness matrix is based on the assumption that the element deforms in a particular way when the structure is loaded. This deformation pattern is an approximation to the true state of deformation and is usually represented by some polynomial function of the relevant co-ordinate axes. Different finite

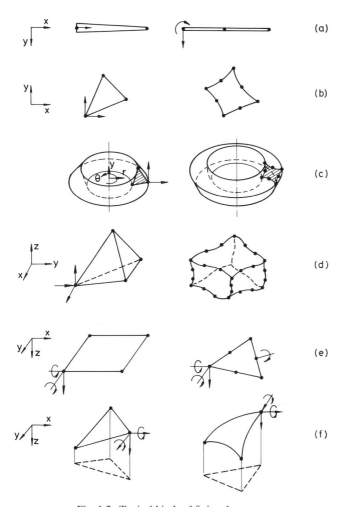

Fig. 1.2. Typical kinds of finite element.

elements are derived by basing the derivation on different deformation patterns and it will be seen later that consequently different numbers of nodal displacement variables are required. The elements shown in Fig. 1.2 fall into a number of basic categories indicated by the letters (a) to (f). The small full circles indicate the nodes. The categories of element shown are as follows:

(a) bar and beam elements which may be non-prismatic.

(b) planar membrane elements used in the solution of plate stretching problems in which loads are applied only in the plane of the element;

(c) axisymmetric elements: these are solid ring-type elements forming part of an axisymmetric structure;

(d) three-dimensional elements for the analysis of general solids;

(e) plate bending elements subjected to loads acting in a direction normal to the plane of the element;

(f) elements for the analysis of curved shells: these elements may themselves be curved or a physical approximation may be introduced by idealizing a curved shell with flat elements which combine the characteristics of planar stretching and plate bending elements.

It is seen that within each classification of element there exist elements of different shapes and elements having different numbers of nodes. Furthermore, at the nodes the displacement quantities which are used as the primary variables of the problem generally differ from element to element. For instance, planar stretching elements may be triangular, rectangular, or quadrilateral in plan form or may have curved sides; a triangular planar element will have at least three nodes but may have considerably more; the nodal displacements for a planar element will probably be the values of the displacement components in the x, y co-ordinate directions but other quantities may also be used. The number of element types that has been developed over the years is very considerable and those shown in Fig. 1.2 should be viewed only as typical elements. Much of the latter part of this book is taken up with discussions of particular element types.

With such a range of available finite elements a corresponding vast range of structures of simple shape or of great complexity can be analysed. A few simple examples are shown in Fig. 1.3. Because of the common basis that exists for the development of element structural properties and for their incorporation into the whole-structure problem, the analyst is not restricted to the use of a single element *type* in an application. Different element types can be used in conjunction with one another provided that they have coincident nodes at their common boundary with identical nodal displacement variables: for example Fig. 1.3(b) shows a reinforced-plate bending problem in which rectangular plate bending elements are used to model the main body of the structure with beam elements used to model the reinforcing members.

It has been mentioned that usually the structural relationships derived for a particular finite element do not exactly represent the behaviour of the part of the theoretical model structure to which it corresponds. However, if element properties are properly formulated, then in general the more elements that are used in representing a structure, the better will be the accuracy of the answer, i.e. with *refinement* of the element mesh *convergence* to the exact solution for the theoretical model structure will occur. Of course there is a price to pay for using more rather than

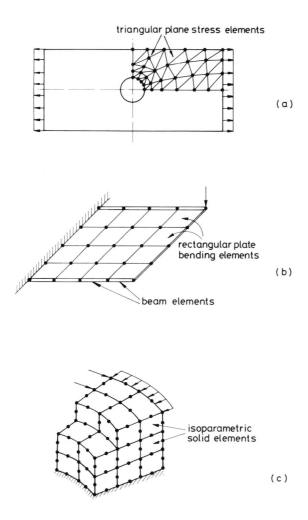

Fig. 1.3. Representation of some simple structures with finite elements: (a) hole-in-plate, plane stress structure; (b) reinforced plate structure; (c) solid structure.

less elements since use of more elements leads to the generation of more simultaneous equations, and more computer storage and time are required to obtain the solution. Thus, in analysing practical continuum structures by the finite-element method there is a balance to be drawn between the conflicting requirements of accuracy and cost of solution. When it has been decided how many elements are going to be used for a particular finite-element analysis of a continuum structure it still has to be decided how these elements should be arranged. As distinct from the situation found in analysing skeletal structures, there does not usually exist a physically obvious, or natural, subdivision of a continuum structure into regions or elements. Different arrangements (meshes) of the same number of elements will give somewhat different answers. In deciding on the arrangement of elements there should, as far as is possible, be taken into account the expected behaviour of the structure under load so that rather more elements are placed in regions of anticipated complicated structural behaviour (e.g. high stress concentration areas such as that around the hole in the problem shown in Fig. 1.3(a)) and rather less elements in regions where simple response is expected.

The finite-element method was originally conceived by engineers in the realm of stress/structural analysis in a rather intuitive manner based on imagining a structure as a physical assembly of component parts, or elements, joined together at certain points (the nodes). Element properties were derived on the basis of elementary structural theory and physical reasoning. The 'physical' approach followed naturally on from the earlier matrix methods of analysis of skeletal structures and the interpretation of the finite-element method in physical terms contributed greatly to its early development.

As the finite-element method evolved it gradually became clear that a more mathematical viewpoint of the method could be taken. This realization led to the establishment by the mid-1960s of a firm and rigorous footing for the method based on the use of certain variational principles of solid mechanics. When displacements are the primary variables the variational principle in question is the principle of minimum potential energy or the associated principle of virtual displacements. Application of the principles in finite-element analysis results in a Rayleigh–Ritz type of procedure with trial displacement fields assumed piecewise over the various regions (elements) of the complete structure. Thus, instead of considering the elements as separate physical pieces of the structure they can be regarded simply as mathematical sub-regions of the whole over which specialized forms of displacement are prescribed. The variational approach provides a formal mathematical verification of the physical approach and illuminates the requirements on which efficient elements should be based. (It also provides a means of extending the scope of the finite-element method to fields other than that of solid mechanics, although this is not considered in this text.)

The variational principles referred to in the preceding paragraph are introduced in Chapter 5 and applied first to the development of properties for simple one-dimensional finite elements in Chapter 6. Extension to two- and three-dimensional analysis, although considerably more complicated, follows naturally and is described in succeeding chapters.

1.4. Commonality of matrix and finite-element methods

It should be clear from what has been written so far that the matrix displacement method of analysis of skeletal structures and the finite-element displacement method have a very great deal in common. Both methods are very systematic and ordered procedures with much of the calculation of a repetitive nature. This makes the methods ideally suited to the modern digital computer and, indeed, the computer is essential for the solution of problems of any real complexity.

The difference between the two methods, apart from the obvious one of the much broader range of application of the finite-element method, might be said to lie in the manner in which the properties of the individual structural component are derived. In the matrix displacement method the structural relationships of the skeletal frame member are exact within the confines of the background behavioural assumptions, whereas the corresponding relationships for a finite element are almost always approximate in nature, being based on assumed displacement patterns. However, it will be seen later that the relationships obtained using the finite-element procedure for the basic bar/beam element (i.e. a uniform one-dimensional element with extensional, torsional, and bending stiffness) are precisely those derived for the skeletal-frame member in the matrix displacement method. Thus, for a skeletal structure formed of prismatic members and carrying only joint loads, the matrix method and the finite-element method will yield precisely the same results. To a very great extent, therefore, the matrix displacement method is just a particular form of the more general finite-element method (although where distributed loads are applied to a skeletal structure there is some difference in philosophy between the two approaches as will be discussed later). Despite this there are advantages in separating the two techniques to some extent, and this has been done in the present work with the matrix displacement method being treated in 'traditional' fashion in Chapters 2, 3, and 4 and the idea of the finite-element, assumed-displacement approach being introduced thereafter.

1.5. Some basic considerations of structural analysis

1.5.1. General remarks

A structure is a physical system whose purpose is to carry a load or a set of loads. Depending upon the nature of the loading the structure may respond in a number of ways: it may, for instance, deform statically in a stable elastic manner or it may yield or it may vibrate or it may buckle. The purpose of the structural analyst is to predict accurately the response of a given structure to a given loading, i.e. to calculate the distribution of internal forces and displacements, or to find the natural frequencies of vibration, or to estimate the level of loading at which instability occurs under the prescribed conditions of geometry and loading. In this text we are mainly concerned with prediction of static behaviour in the elastic range and for the present attention is restricted to static analysis: some consideration is given later to vibration and buckling problems in Chapter 12.

Some of the very basic considerations and conditions on which the static analysis of elastic structures relies are described in this section. Only the background material pertinent to an understanding of the matrix displacement method as presented in Chapters 2, 3, and 4 is given at this stage and the somewhat brief presentation assumes that the reader already has some familarity with the topics covered. Later on in the chapter consideration is given to the different ways in which the basic structural conditions can be used in the analysis of indeterminate structures.

1.5.2. Linear structures: superposition

It will generally be assumed in this work that the structure deflections caused by applied loads are small in comparison with the structure geometry. In other words, there is no significant change of structure geometry under load and it follows that the equilibrium equations can be referred to the unloaded configuration of the structure rather than—as would be the rigorous procedure—to the loaded configuration. It is further assumed that the material from which the structure is made behaves in a linear elastic fashion, i.e. that stress is directly proportional to strain. The combined effect of these two assumptions is that the structure response is *linear*, i.e. the magnitudes of the displacements and internal forces of the structure are directly proportional to the magnitude of the external loading. The assumption that structure response is linear leads, of course, to considerable simplification in analysis and is justified in practice since most real structures respond in an approximately linear fashion to loads within their working range. There are exceptions to this general statement of course: one of these is the behaviour of struts where the deformation of the structure can no longer be considered to be small and the effects of finite deformation must be taken into account.

Assuming that the linear theory of structures holds we can also use the *principle of superposition*. This principle states:

the total effect of a number of loads applied to a structure can be obtained by simple addition of the effects of the individual loads acting separately on the structure.

This statement is illustrated in Fig. 1.4. One important application of the

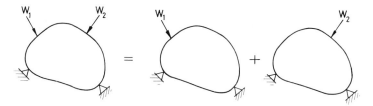

Fig. 1.4. Illustration of the Principle of Superposition.

principle will be seen later in the use of consistent nodal loads in the matrix displacement method to analyse skeletal structures with distributed loadings.

1.5.3. Elasticity, equilibrium, and compatibility

The correct solution to a linear elastic structural problem is one that everywhere satisfies the conditions of (a) elasticity, (b) equilibrium, and (c) compatibiity. These three conditions are of fundamental importance and will now be defined in detail.

(a) *Elasticity condition*

This condition states that, at any point in a structure, the stresses are directly related to the corresponding strains by the particular *stress–strain* or *constitutive* equations which apply for the structure material.

In the general case of a linearly elastic three-dimensional body there are six stress components each of which can be expressed as a linear function of six components of strain. These general constitutive equations will be presented later (in Chapter 7) but for present purposes the interest is in one-dimensional situations. For these latter situations the single constitutive equation is the well-known Hooke's Law which states that deformation (or strain) is directly proportional to the force (or stress) producing it. For direct stress and strain the relationship is

$$\frac{\text{direct stress}}{\text{direct strain}} = \frac{\sigma}{\varepsilon} = E \tag{1.1}$$

whilst for shearing stress and strain it is

$$\frac{\text{shearing stress}}{\text{shearing strain}} = \frac{\tau}{\gamma} = G. \tag{1.2}$$

Here E is Young's modulus of elasticity and G is the shearing modulus or modulus of rigidity for the material.

(b) *Equilibrium condition*

This condition states that the structure as a whole, and all parts thereof, must be in equilibrium under the combined action of applied external and internal forces. Where a structure is envisaged as an assemblage of discrete parts, the equilibrium principle applies at a number of levels. As an illustration consider the simple two-dimensional frame of Fig. 1.5(a). The frame has a pin support at A and a roller support at B.

The equilibrium conditions in two dimensions can be simply written as

$$\Sigma F_x = 0 \qquad \Sigma F_y = 0 \qquad \Sigma M = 0 \qquad\qquad (1.3)$$

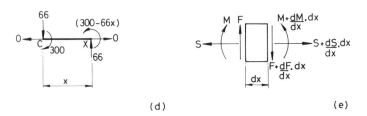

Fig. 1.5. Equilibrium of a two-dimensional frame.

and express the fact that the sum of all forces in any two mutually perpendicular x and y directions and the sum of all moments about any point are equal to zero for any part of or for the whole of the structure.

Applying the equilibrium condition to the whole structure gives the boundary reactions H_A and V_A at the pin A and V_B at the roller B, so that the equilibrium set of forces for the whole structure is as in Fig. 1.5(b). It should be noted that in finding H_A, V_A, and V_B we have to use the boundary conditions on force and moment which are appropriate to the structure; that is the horizontal reaction at B and the moments at A and B are zero.

Applying the equilibrium condition to each complete member and to each joint of the structure gives the equilibrium sets of end and joint forces shown by the *free-body* diagrams in Fig. 1.5(c). (A free-body diagram is a diagram showing the forces acting on a part or parts of a structure.)

Applying the equilibrium condition to *part* of typical member CD, say, gives the equilibrium set of forces shown in Fig. 1.5(d) and hence yields the values of axial force, shearing force, and bending moment at an arbitrary point X within the member CD. Values of these quantities could similarly be found for all points along members AC and BD.

Finally, applying the equilibrium condition to an infinitesimal length of a member between the sections at x and $x+dx$ (see Fig. 1.5(e)) gives

$$\frac{dS}{dx} = 0 \qquad \frac{dF}{dx} = 0 \qquad \frac{dM}{dx} = F$$

where S is the axial force, F is the shearing force, and M is the bending moment at section x, and the positive convention for these quantities in this text is taken to be as shown in Fig. 1.5(e).

(c) *Compatibility condition*

This condition simply states that the displacement of any point of a loaded structure must be compatible (i.e. consistent) with the overall displacement of the whole structure; the displacement must therefore be continuous and single valued everywhere such that the structure still fits together under load. In interpreting this statement we include conditions at the boundary of the structure, i.e. we stipulate that the overall displacement of the whole structure is compatible with the prescribed displacements at the structure boundary.

Where a structure is subdivided, either naturally or artificially, into a number of parts or regions the compatibility condition requires the following.

(i) Within each region the displacement varies smoothly with no discontinuities. For instance, in the portal frame of Fig. 1.5 the axial displacement, normal displacement, and rotation should be single valued along the length of each of the frame members AC, CD, and BD.

(ii) At the boundaries between neighbouring regions the displacements match each other in a manner consistent with the problem under consideration. For the portal frame, for example, joint C is a rigid joint and consequently the translational displacement and the rotation of end C of member AC should be identical with that of end C of member CD. Similar conditions apply at joint D.

(iii) At the boundaries of the whole structure the prescribed displacement boundary conditions are satisfied. For the portal frame this means that at point A—a pin support fixed in position—there should be no horizontal or vertical movement, whilst at point B—a roller support—there should be no vertical movement.

1.5.4. Boundary conditions

The need to satisfy structure boundary conditions has already been mentioned in the last subsection in relation to the particular frame of Fig. 1.5 but a few remarks of a more general nature are appropriate here.

The satisfaction of appropriate boundary conditions is simply a special case of the requirement of satisfying the compatibility and equilibrium conditions over the whole structure. Consequently, two types of boundary condition can be identified.

The first kind of boundary condition pertains to the compatibility condition and requires that prescribed conditions of displacement (the word interpreted in its general sense of including rotation) at the structure boundary be met. Boundary conditions of this type are referred to as *geometric* or *kinematic* or simply *displacement* boundary conditions.

The second kind of boundary condition pertains to the equilibrium condition and requires that prescribed conditions of force (including moment) be satisfied at the structure boundary. Boundary conditions of this kind are referred to as *natural* or *physical* or *force* boundary conditions.

1.5.5. Comments

It has been stated, and bears repeating, that the correct solution to any elastic structural problem is one that fully satisfies the conditions of elasticity, equilibrium, and compatibility. Within the context of linear elasticity it follows that if these three conditions are satisfied the corresponding solution is the unique one for the given loading and no alternative solution is possible.

Of the three conditions the elasticity condition is the most easily accommodated in a solution procedure. The stress–strain relations are obtained from material properties and can be assumed to be precisely known whatever method of solution is used. These relations are used to link the conditions of equilibrium and compatibility and these latter conditions are the principal bases on which the linear analysis of structures is founded.

In dealing with statically loaded skeletal structures (under the heading of the matrix displacement method), it will be shown that it is possible to obtain solutions which satisfy the equilibrium and compatibility conditions everywhere in addition to meeting the requirements of the elasticity condition. Thus, for this type of structure the 'exact' solution to the structural problem can be obtained (that is, the solution is exact within the normal assumptions of the one-dimensional physical nature of the elements and of the engineering theory of beams etc.).

Later in the text it will be seen in analysing other types of structure that it is not always possible to satisfy fully the equilibrium and compatibility conditions. Nevertheless, it will be demonstrated that accurate (but not exact) solutions can be obtained using the finite-element method despite local violations of these conditions. In two- and three-dimensional finite-element displacement analysis the equilibrium condition is usually violated in a local sense both within and across element boundaries (but is satisfied in the overall or macroscopic sense at the nodes), whilst occasionally (in plate and shell analysis) the compatibility condition is also violated at element boundaries.

1.5.6. Work

The basic definition of work done by a force is that it is the product of the force and the distance that it moves in its own direction.

Imagine that a translational force P^* acts through a total distance Δ and assume that in general the magnitude of the force may vary throughout the total movement but that its direction is constant and is the same as the direction in which Δ is measured. During an infinitesimal part of the movement, denoted by $d\delta$, the work dW done is

$$dW = P^* \, d\delta$$

and therefore the work W done during the total movement is

$$W = \int_0^\Delta P^* \, d\delta.$$

Two simple particular cases commonly arise. Firstly, if P^* is constant in magnitude, i.e. $P^* = P$ throughout, then

$$W = P \int_0^\Delta d\delta = P\Delta. \tag{1.4}$$

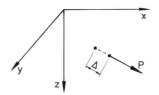

Fig. 1.6. Work of a force.

Secondly, if P^* is increased in linear fashion from zero to the maximum value P during the movement, i.e. $P^* = P\delta/\Delta$, then

$$W = \int_0^\Delta P\frac{\delta}{\Delta}\,\mathrm{d}\delta = \frac{1}{2}\,P\Delta \tag{1.5}$$

and this case corresponds to the application of a gradually increasing load to a linear elastic body.

Although force and displacement are vector quantities (i.e. they have magnitude and direction) work is, of course, a scalar quantity (i.e. it has magnitude only). It is often convenient in structural analysis to express work as the sum of products of components of force and components of displacement. For example, assume that a force P acts (with constant magnitude, say, in this instance) through a distance Δ, measured in its own direction, as shown in Fig. 1.6. If we wish to refer force and displacement quantities to an arbitrary set of mutually perpendicular co-ordinate axes x, y, and z, then let the components of P be P_x, P_y, and P_z and those of Δ be Δ_x, Δ_y, and Δ_z in the x, y, and z directions respectively. Then the work done by P is simply

$$W = P\Delta = P_x\Delta_x + P_y\Delta_y + P_z\Delta_z. \tag{1.6}$$

The above simple statements of work have been concerned with translational forces and displacements but work can also be done by rotational forces (or couples) acting through rotational displacements (or simply rotations). Also, within the context of linear analysis moments and rotations are again vector quantities. Thus the above equations will have their direct counterparts in the case of rotational behaviour with moment M replacing force P and rotation θ replacing displacement Δ in equations (1.4) and (1.5), and with similar replacements for the components in equation (1.6).

1.5.7. Reciprocal theorems

In the linear theory of structures the reciprocal theorems of Betti and Maxwell are often invoked. These theorems will simply be stated here: proofs of the theorems can be quite simply demonstrated using work principles and such proofs are available in many elementary texts.

The reciprocal theorem of Betti states:

the work done by one system of forces F_1 during the deformation caused by a second system of forces F_2 is equal to the work done by the F_2 forces during the deformation caused by the F_1 forces.

The reciprocal theorem of Maxwell is a special form of that of Betti and states:

the displacement of a point 1 due to a force F applied at a point 2 is numerically the same as the displacement of point 2 due to a force F applied at point 1.

In this statement the force F applied separately at points 1 and 2 has a common magnitude but does not necessarily act in the same direction in each instance. The displacements are measured in the same direction as the forces in both systems. Figure 1.7 illustrates the situation and the theorem gives the result

$$\delta_{12} = \delta_{21}.$$

It should be noted that the terms force and displacement are again used in their general sense, i.e. moments (couples) and rotations are included as forces and displacements respectively. It will be seen later that Maxwell's reciprocal theorem requires that stiffness matrices be symmetric.

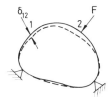 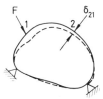

Fig. 1.7. Illustration of the Reciprocal Theorem.

1.6. Determinate and indeterminate structures

Structures carrying a static loading can be categorized as either *statically determinate* or *statically indeterminate*. If all reactions and internal forces in a structure can be found using the equilibrium conditions alone then the structure is statically determinate; if not it is statically indeterminate or hyperstatic or redundant.

Analysis of statically determinate structures is comparatively simple since the internal forces can be calculated completely independently of the physical and material properties (cross-sectional area, second moment of area, Young's modulus of elasticity, etc.) of the various parts of the structure. If required the displacements can then be found in a separate operation from the internal forces by using the elasticity and compatibility conditions to complete the analysis.

In contrast, in order to obtain a solution for a statically indeterminate structure, it is necessary to use directly all three conditions of elasticity, equilibrium, and compatibility so as to obtain either the internal forces or the displacements. Here the internal forces cannot be calculated without prior knowledge of the physical and material properties of the structure parts.

Two quantities that are frequently referred to in structural analysis are the *degree of redundancy* and the *degree of freedom*.

A statically indeterminate structure is said to have a degree of redundancy (or a degree of indeterminacy); this is the number of unknown reaction and internal force components in excess of the number of available equations of static equilibrium. On the other hand any structure—determinate or indeterminate—is said to have a certain number of degrees of freedom. To define a degree of freedom it should be noted that in structural analysis the overall structure behaviour is often artificially characterized by a finite number of reference quantities, i.e. the behaviour of *any* of the infinite number of points throughout the structure is referred to the behaviour at a *limited* number of points by making certain simplifying assumptions. In the matrix and finite-element approaches this comes about by generating expressions which relate certain forces and displacements at certain nodal points based on particular specifications of the behaviour within each member or element as it relates to the behaviour at its nodal points. In the displacement method the nodal quantities are of course displacement-related quantities and these reference displacements are the degrees of freedom of the structure.

To clarify the above points we shall refer back to the simple frame of Fig. 1.5 again. In this case there are three equilibrium equations available (equations (1.3)) which can be applied to the whole or to any part of the structure. Thus the three reactions at A and B can first be found, followed by values of the internal forces at any point, simply by directly using the equilibrium equations. This frame problem is therefore statically determinate. The forces at all points are found without any knowledge of the properties of the members of the frame; this knowledge is only required if we wish to proceed to find the displacements of the frame, in which case the elasticity and compatibility conditions would be used.

In finding the reactions and internal forces of the frame of Fig. 1.5 it might be said that the compatibility condition, although not used explicitly, is satisfied implicitly since, in calculating the reactions for instance, account is indirectly taken of boundary displacement conditions, i.e. we use the facts that only vertical and horizontal reactions act at the pinned end A and that only a vertical reaction acts at the roller support B.

If the support at B were a pin instead of a roller a further unknown (horizontal) reaction would be introduced to the problem. Solution of the modified problem is not now possible by using the three equations of static equilibrium alone; the structure has a single degree of redundancy and direct consideration must be given to the structure displacements to provide an extra equation. If the support at B were fully fixed then the structure has two degrees of redundancy and two extra equations, based on displacements, need be generated.

The number of degrees of freedom of the frame of Fig. 1.5 is easily seen to be nine since there are nine unknown reference displacements comprising three (two translations and one rotation) at joints C and D plus two (horizontal translation and rotation) at B plus one (the rotation) at A. Replacement of the roller support at B by a pin would reduce the degrees of freedom by one; replacement by a fully fixed support would reduce the degrees of freedom by two.

Although some consideration has been given here to the concepts of determinacy and indeterminacy it should be emphasized at this stage that the methods presented in this text do not distinguish between determinate and indeterminate structures. In using matrix and finite-element methods in later chapters we shall generally be dealing with indeterminate rather than determinate structures, but the distinction between the two categories of structure is meaningless and will not enter into consideration. (This is in contrast with the less direct methods of a generation ago when the concept of indeterminacy played an important role.) The two categories have been briefly discussed here only as a prelude to discussing the available methods of analysis of structures that are possible based on the use of the three basic conditions governing the structural problem.

1.7. Methods of analysis

It has been stated that the analysis of redundant structures requires the use of the three conditions of elasticity, equilibrium, and compatibility. The major part of an analysis is completed when we know either the complete distribution of displacements or the complete distribution of internal forces (stresses) throughout the structure; since we are dealing with linear structures the displacements and the internal forces are uniquely related and the one can be obtained from the other with relatively little effort. Clearly if the prime goal of an analysis is the determination of displacements then a different analysis procedure might be followed than in the case where priority is given to determining internal forces. Depending upon the type of structure it may be easier to determine displacements first and internal forces afterwards, or vice versa. Thus it appears logical to set up different methods of analysis for different types of structure, the difference in the methods depending upon the order in which the three basic conditions are applied.

As mentioned earlier there are basically two different types of method available for the systematic analysis of redundant linear structures. These are the *displacement* (or *stiffness* or *equilibrium*) method and the *force* (or *flexibility* or *compatibility*) method. A third, but less used, class of method is the *mixed* method which combines features of the two basic methods. It will be clear by now to the reader that it is the displacement method that will be used throughout the remainder of this text but some brief consideration of the differences between this method and the force method is desirable at this stage.

In the displacement approach it is the structure displacements that are regarded as the primary unknowns, as the name suggests. The elasticity and compatibility conditions are used first to express internal forces in terms of a number n of unknown displacements or degrees of freedom. The equilibrium condition is then applied to yield a set of n simultaneous algebraic equations for these displacements. On solution of this set of equations the displacement values are obtained and then the internal forces can be found by a process of back substitution.

In the force approach the equilibrium condition is applied first to express all internal forces in terms of r unknown redundant forces. The elasticity and compatibility conditions are then employed to yield a set of r simultaneous equations for the redundant forces. Once these redundant forces have been found the values of displacements and other internal forces can be obtained by back substitution.

A simple example will serve to illustrate the two basic approaches.

Example 1.1. The pin-jointed plane framework shown in Fig. E1.1(a) carries horizontal and vertical forces F_u and F_v at the common pin C. A complete analysis requires finding the components of displacement of the common pin (u

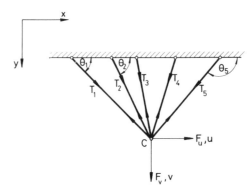

Fig. E1.1(a)

and v) and the forces in all the members of the framework. All frame members have the same value of Young's modulus E. For member j ($j = 1 \ldots 5$) the quantities l_j, A_j, θ_j, T_j, and e_j are the length, cross-sectional area, angle of inclination, tensile force, and extension respectively of member j. (It should be noted that initially all members are assumed to be in tension.)

This problem clearly has two degrees of freedom, i.e. $n = 2$; the degrees of freedom are the displacements u and v. Equally clearly the structure is statically indeterminate. If only two members were present the forces in these two members due to the applied loads could be found using the two available equilibrium equations alone; therefore with five members present the degree of redundancy is $r = 5 - 2 = 3$.

The three basic conditions for linear analysis can be expressed as follows.

(1) *Elasticity.* Using Hooke's Law for each member in turn gives relationships between member tensions and extensions in the form

$$T_j = \frac{A_j E}{l_j} e_j = k_j e_j \qquad \text{for} \qquad j = 1 \ldots 5. \tag{a}$$

The quantity $k_j = A_j E / l_j$ is termed the *extensional stiffness* of member j since it relates force to corresponding displacement.

(2) *Equilibrium.* The common pin at which the external loads are applied is in equilibrium under the action of the applied forces F_u and F_v and the member forces T_j. Thus by resolution of forces

$$F_u = \sum_{j=1}^{5} T_j \cos \theta_j \qquad F_v = \sum_{j=1}^{5} T_j \sin \theta_j. \tag{b}$$

(3) *Compatibility.* The extension e_j (measured along the original direction of member j) of any individual member of the framework is clearly related to the movement of the common pin (if the framework is still to fit together when loaded as required). Fig. E1.1(b) illustrates this relationship which can be expressed as

$$e_j = u \cos \theta_j + v \sin \theta_j \qquad \text{for} \qquad j = 1 \ldots 5. \tag{c}$$

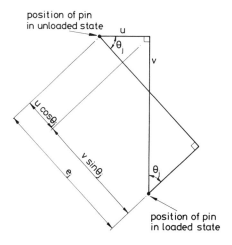

Fig. E1.1(b)

Having established the basic conditions the problem will now be solved using the displacement and force methods in turn.

With the displacement approach the elasticity and compatibility conditions, equations (a) and (b), are first of all combined to give equations for the internal forces in terms of the structure displacements, i.e.

$$T_j = k_j(u \cos \theta_j + v \sin \theta_j) \qquad \text{for} \qquad j = 1 \ldots 5. \tag{d}$$

Now the equilibrium condition is applied by substituting equations (d) into equations (b). This procedure yields expressions for the external loads related to the structure displacements, i.e.

$$
\begin{aligned}
F_u &= u \sum_{j=1}^{5} k_j \cos^2 \theta_j + v \sum_{i=1}^{5} k_j \sin \theta_j \cos \theta_j \\
F_v &= u \sum_{j=1}^{5} k_j \sin \theta_j \cos \theta_j + v \sum_{i=1}^{5} k_j \sin^2 \theta_j.
\end{aligned}
\tag{e}
$$

For given magnitudes of the external loads F_u and F_v these two equations can be solved to yield the two structure displacements, or degrees of freedom, u and v. To complete the analysis the member tensions T_j can be found by back substituting for u and v into equations (d).

With the force method the equilibrium condition is used first. It has been pointed out that the structure has a degree of redundancy of 3. It follows that three of the members of the framework must initially be regarded as redundant (that is the structure could still function as a framework with three members removed) and the equilibrium condition must be used to generate expressions for the tensions in the non-redundant members expressed in terms of the tensions in the remaining members (three in this case) and the applied loads. If members 3, 4, and 5 are chosen to be the redundant members, then using equations (b) we have

$$
\begin{aligned}
F_u &= T_1 \cos \theta_1 + T_2 \cos \theta_2 + \sum_{j=3}^{5} T_j \cos \theta_j \\
F_v &= T_1 \sin \theta_1 + T_2 \sin \theta_2 + \sum_{j=3}^{5} T_j \sin \theta_j.
\end{aligned}
\tag{f}
$$

These equations can be rewritten to express T_1 and T_2 in terms of the redundant forces T_3, T_4, and T_5 and the applied loads as

$$
\begin{aligned}
T_1 &= \frac{F_u \sin \theta_2 - F_v \cos \theta_2 - \sum_{j=3}^{5} T_j \sin(\theta_2 - \theta_j)}{\sin(\theta_2 - \theta_1)} \\
T_2 &= \frac{F_u \sin \theta_1 - F_v \cos \theta_1 - \sum_{j=3}^{5} T_j \sin(\theta_1 - \theta_j)}{\sin(\theta_1 - \theta_2)}.
\end{aligned}
\tag{g}
$$

It should be noted that to generate these equations necessitated a choice of the redundant members; the selection of members 3, 4, and 5 is purely arbitrary and other members could have been chosen in their place.

Proceeding with the force method approach, the elasticity and compatibility conditions are applied next. Substituting equations (a) into equations (c) gives

$$f_j T_j = u \cos \theta_j + v \sin \theta_j \qquad j = 1 \ldots 5 \tag{h}$$

where $f_j = 1/k_j$ is termed the *extensional flexibility* of member j. The first two of these equations yield expressions for the structure displacements u and v in terms of the forces T_1 and T_2, i.e.

$$u = \frac{f_1 T_1 \sin\theta_2 - f_2 T_2 \sin\theta_1}{\sin(\theta_2 - \theta_1)}$$

$$v = \frac{f_1 T_1 \cos\theta_2 - f_2 T_2 \cos\theta_1}{\sin(\theta_1 - \theta_2)}$$

(i)

Substitution of these expressions for the displacements into the remaining three of equations (h) gives

$$f_1 T_1 \sin(\theta_2 - \theta_j) + f_2 T_2 \sin(\theta_j - \theta_1) + f_j T_j \sin(\theta_1 - \theta_2) = 0 \qquad j = 3, 4, 5.$$

(j)

Finally, if the expressions for T_1 and T_2 given by equations (g) are substituted into equations (j) the result is a set of three equations involving three unknown internal forces T_3, T_4, and T_5. This set of equations can be solved to give the values of T_3, T_4, and T_5. Back substitution will then give the remaining internal forces T_1 and T_2 from equations (g) and the two structure displacements from equations (i).

Now that the basic procedures involved in using a displacement or a force approach have been clarified it is opportune to discuss briefly the comparative merits of the two approaches.

It has been seen that for the displacement method the primary unknowns are structure displacements and the number of simultaneous equations that has to be solved is equal to the number of degrees of freedom n, i.e. the number of independent displacement quantities. For the force method the primary unknowns are the structure internal forces, and the number of equations to be solved is equal to the number of redundants r.

One obvious criterion for comparison between the two approaches is to consider the final number of equations involved in solving a problem. This varies with the problem type of course, but generally speaking n is greater than r so that usually the force method has an advantage in this regard. Furthermore, it is usually the internal forces that are of more interest to engineers than are the displacements, so again the force method might appear to be the more suitable.

However, there are other important considerations in which the displacement method has distinct advantages. Even for the simple example given above it is clear that the displacement method is much more straightforward in concept and is easier to apply in a very systematic fashion. Use of the force method involves an arbitrary selection of the redundant quantities and this is a disadvantage in seeking to develop a systematized approach. It is therefore much easier to develop general-purpose computer programs based on the displacement method. Although more equations are usually involved with a displacement method solution it will be seen later that it is a characteristic of the method that it generally produces a set of simultaneous equations which, when represented in matrix form, are of a special 'banded' type. (A banded matrix is a matrix whose only non-zero coefficients are located on or near the leading diagonal.) Special techniques which take advantage of this feature can be used to produce an economic solution of the system of equations. The system of equations obtained using the force method, although often smaller in number, has a less favourable form, being more densely populated and more likely to be ill conditioned.

Thus we see that despite (usually) the generation of a greater number of equations the displacement method applied to the structure as a whole has considerable advantages over the force method. In the view of most structural analysts the advantages of simple logic and straightforwardness of application tilt the balance firmly in favour of the displacement method; this is particularly so when the method is programmed for the computer. Certainly it is the overall displacement approach which is much the more widely used in finite-element work. For these reasons we shall be concerned in succeeding chapters solely with the displacement approach.

1.8. Displacement method: stiffness relationships

It has been pointed out earlier on a number of occasions that the displacement method of structural analysis relies on the establishment of stiffness equations, i.e. equations relating forces to displacements. Referring back to the example of the pin-jointed frame, we see that the final set of simultaneous equations governing the displacement approach (equations (e)) can be written in the form

$$F_u = K_{11}u + K_{12}v$$
$$F_v = K_{21}u + K_{22}v$$

(1.7)

where

$$K_{11} = \sum_{j=1}^{5} k_j \cos^2 \theta_j$$

$$K_{12} = K_{21} = \sum_{j=1}^{5} k_j \sin \theta_j \cos \theta_j$$

$$K_{22} = \sum_{j=1}^{5} k_j \sin^2 \theta_j$$

Equations (1.7) are the *structural stiffness equations* relating the applied forces to the structure displacements; these forces and displacements relate in this case to one particular joint or node of the structure at point A. The quantities K_{11} etc. are termed the structure stiffness coefficients.

At this stage it is convenient to introduce matrix notation† and to rewrite equation (1.7) as

$$\mathbf{F} = \mathbf{KD} \qquad\qquad (1.8)$$

where

$$\mathbf{F} = \begin{Bmatrix} F_u \\ F_v \end{Bmatrix} \qquad \mathbf{D} = \begin{Bmatrix} u \\ v \end{Bmatrix} \qquad \mathbf{K} = \begin{bmatrix} K_{11} & K_{12} \\ K_{21} & K_{22} \end{bmatrix}.$$

In equation (1.8) $\mathbf{F}$ is the column matrix of (known) applied nodal forces, $\mathbf{D}$ is the column matrix of unknown nodal displacements, and $\mathbf{K}$ is the *structure stiffness matrix*. In the matrix and finite-element displacement methods the equations governing the response of the whole structure to an applied static loading are always of the form of equation (1.8). The properties of a stiffness matrix will be discussed at greater length later but for the moment it is noted that the stiffness matrix $\mathbf{K}$ is always a square matrix of course and that in this particular example the stiffness matrix is symmetric, as indicated by the fact that $K_{12} = K_{21}$. In fact, this latter property of the stiffness matrix applies generally and follows directly from Maxwell's reciprocal theorem.

The purpose of Example 1.1 was to show the basic differences in the displacement and force methods. This example shows the form of the final *structure* stiffness equations that is typical of matrix and finite-element displacement methods but it does not reveal the full procedure. The full procedure for the general structure requires the preliminary derivation of stiffness matrices for the individual components or members or elements of a structure and then the assembly of the element stiffness matrices to produce the structure stiffness matrix according to standard rules of compatibility and equilibrium; this procedure is described in the next chapter. The development of suitable element or member stiffness matrices is a major task and a crucial step in finite-element analysis; much of this text is devoted to the task of generating such stiffness matrices.

1.9. Brief remarks on co-ordinate systems used in the book

The development of element and structure stiffness properties in this book will consistently be based on the use of a right-handed set of cartesian co-ordinates. However, it will be seen that particular axes are illustrated as being aligned in different directions in different parts of the book (although having the same direction relative to one another). The

---

† A description of elementary matrix algebra, some knowledge of which is a prerequisite for understanding the rest of this text, is given in Appendix A.

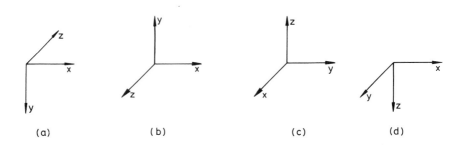

Fig. 1.8. Co-ordinate systems.

purpose of the present remarks is simply to point out the reason for the particular directions chosen at different stages and to make clear that differences in chosen directions have no real effect on the analysis.

In Figs. 1.8(a)–1.8(d) are shown what might appear to be four different co-ordinate systems, but a moment's reflection should convince the reader that all that these figures show are in fact simply four different views of the same thing. Each figure can be obtained from any other merely by changing one's vantage point and therefore all are in reality identical. Hence it does not really matter at all which vantage point is taken except that considerations of convenience or of traditional usage might influence the choice. From a convenience point of view it might be argued that it would be best to use the same vantage point throughout without regard to the type of structure being analysed. On the other hand it is traditional in analysing beams, grillages, and plates to measure deflection positively downwards (because in practice this is the usual sense of the deflection) whereas in three-dimensional continuum analysis it is more usual to take the view represented in Fig. 1.8(c). In this text the policy adopted is influenced by the traditional outlook and hence the illustrations presented in succeeding chapters will reflect the following.†

(1) For skeletal structures, with the exception of the grillage, Fig. 1.8(a) applies. The reason for this choice has its origins in requiring that horizontal beams (lying in the x direction) have positive deflection downwards (in the y direction) with rotation being positive about the z axis in the sense of the right-hand screw rule.

(2) For two-dimensional (planar) continuum structures Fig. 1.8(b) applies with xy being the analysis plane.

(3) For three-dimensional continuum structures Fig. 1.8(c) applies.

(4) For plates and grillages it is again desirable to regard downward

---

† The co-ordinate systems shown in Figs. 1.1 and 1.2 also reflect the statements made here but it should be noted that global cartesian co-ordinates are represented as $\bar{x}$, $\bar{y}$, and $\bar{z}$ in analysing skeletal structures to distinguish them from local co-ordinates.

deflection as positive and so Fig. 1.8(d) applies with the plate or grillage lying in the xy plane and having positive deflection in the z direction.

It is emphasized again, though, that all derived properties presented later apply to *any* right-handed cartesian system and so can be used in conjunction with any of Figs. 1.8(a)–Fig. 1.8(d).

2

THE MATRIX
DISPLACEMENT APPROACH

2.1. Introduction

At the end of Chapter 1 the concept of the stiffness matrix as used in the displacement method was briefly mentioned. In this chapter the aim is to begin the development of the matrix displacement method (MDM) through consideration of very simple structural forms. Attention is focused on one-dimensional bar and beam problems with the aim of firmly establishing the basic procedure before moving on to the consideration of two- and three-dimensional skeletal structures in Chapters 3 and 4.

In the MDM (and in the finite-element method (FEM) generally) a matter of fundamental importance is the development of stiffness relationships for each individual component making up a structure. For the uniform bar and the uniform beam (with concentrated end loads only) it will be seen here that stiffness relationships can be readily established which are exact relationships within the conventional assumptions of bar and beam theory. It follows that the three conditions of elasticity, equilibrium, and compatibility are satisfied throughout the element.

A second matter of fundamental importance is to establish how the stiffness relationships for a whole structure can be obtained once the appropriate relationships for the component parts of the structure are known. Obviously, any valid procedure should be based on satisfying the conditions of equilibrium and compatibility at the points of interconnection of the components and at points on the structure boundary. The most direct procedure will be seen to be effectively a simple process of addition, or superposition, of appropriate component stiffness: the procedure is not restricted to any particular class of structure.

The subject of phraseology should be mentioned at this early stage. In the MDM the individual components making up a skeletal structure are traditionally referred to as *members* and the ends of the members are connected together at *joints*. In the FEM (which embraces the MDM as a special case) the individual components or regions are called *elements* and the connections are at nodal points or, simply, *nodes*. The two types of definition are, of course, interchangeable to a considerable extent in the analysis of skeletal structures. In this and the next two chapters, where we deal with such structures, the FEM terminology—elements and nodes—will largely be used in preference to the traditional MDM terminology so as to retain consistency with later chapters. Use of the word member and joint will by and large be restricted to situations where these words are more appropriate in the sense of physically describing a structure. A line *member* will be regarded as a part of a structure which is clearly distinct in a physical sense from neighbouring parts, e.g. by being inclined at a different angle to that of its neighbour(s) in a frame, or by having different cross-sectional dimensions to its neighbour(s) in a continuous beam or bar; a *joint* then also has a clear physical interpretation as the junction of two or more members. On the other hand an *element* is regarded as simply a piece of a structure which it is convenient for analysis purposes to consider separately, and a *node* is a point where two or more elements are connected. Usually in the analysis of skeletal structures it is convenient to consider a physical line member as a single element so that there is no difference in the number and positioning of the members and elements, or of the joints and nodes. However, it should be realized that it is quite possible for a physical member to be made up of two or more elements and hence for some nodes to be positioned at points other than the physical joints.

2.2. Stiffness matrix of a bar element subjected to axial force

Consider a bar of uniform cross-sectional area A, length l, and Young's modulus E. Let the bar be held at one end and be subjected to a pull of magnitude F in the direction of the axis of the bar at the other end, as shown in Fig. 2.1(a). Let the overall extension of the bar under load be e. We can very easily obtain a relationship between F and e by directly using the conditions of elasticity, equilibrium, and compatibility.

The elasticity condition is that

$$\frac{\text{longitudinal stress}}{\text{longitudinal strain}} = \frac{\sigma_x}{\varepsilon_x} = E$$

The equilibrium condition tells us that the force acting on any section of the bar is equal to the applied force F (as is the reaction at the

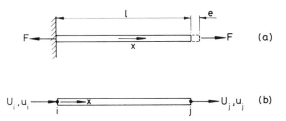

Fig. 2.1. Bar subjected to extension: (a) bar pulled with axial force; (b) bar element.

left-hand end) and we assume in our theoretical model of the bar that this force is distributed evenly over the section so that the axial stress has the uniform value $\sigma_x = F/A$ everywhere.

The compatibility condition requires that the axial displacement varies smoothly and continuously along the bar, with no breaks, from the value zero at the left-hand end to the value e at the right-hand end. For the bar under conditions of uniform stress this implies a uniform value of strain at all sections whose magnitude is $\varepsilon_x = e/l$.

Using these three basic conditions thus yields

$$F = \frac{AEe}{l} \tag{2.1}$$

as the force–displacement relationship for the bar. (This type of relationship has already been used in Example 1.1.)

Now consider both ends of the bar to be capable of movement, as would generally be the case if it was connected at its ends to other parts of a loaded structure. Assume at this stage that the movement of such a bar element still is only possible in the direction of its axis (the x direction) and that the bar element carries only an axial force. Let the ends (nodes) of the bar element now be denoted by the letters i and j as shown in Fig. 2.1(b); forces acting at the ends are denoted by U_i and U_j and end displacements by u_i and u_j (positive in the positive x direction). Relating this situation to that shown in Fig. 2.1(a) it is clear that

$$e \equiv u_j - u_i \tag{2.2}$$
$$F \equiv U_j \qquad\qquad F \equiv -U_i.$$

Combining equations (2.1) and (2.2) gives expressions for the end forces U_i and U_j in terms of the end displacements u_i and u_j of the bar shown in Fig. 2.1(b). These expressions are

$$U_i = \frac{AE}{l}(u_i - u_j)$$

$$U_j = \frac{AE}{l}(-u_i + u_j)$$

In matrix form these expressions can be represented by

$$\mathbf{P} = \mathbf{kd}$$

where

$$\mathbf{P} = \begin{Bmatrix} U_i \\ U_j \end{Bmatrix} \qquad \mathbf{d} = \begin{Bmatrix} u_i \\ u_j \end{Bmatrix} \qquad \mathbf{k} = \begin{bmatrix} k & -k \\ -k & k \end{bmatrix} \tag{2.3}$$

and $k = AE/l$. Here the 2×2 matrix $\mathbf{k}$ linking nodal forces $\mathbf{P}$ and displacements **d** is the *element stiffness matrix* (or ESM) for the axially loaded bar. In developing equation (2.3) the conditions of elasticity, equilibrium, and compatibility have been satisfied throughout the element and so the element stiffness relationships are exact provided that no load acts within the length of the element.

2.3. Bar structure stiffness matrix

The simple stiffness equation (2.3) holds for any bar element having a displacement in the direction of the bar axis only. Now that the stiffness matrix is established for this simple case the next step in the matrix displacement approach is to consider the way in which the stiffness matrix of a structure composed of more than one such element can be assembled. The basis of assembly of the structure stiffness matrix depends on the further use of the conditions of compatibility and equilibrium. (Since the elasticity condition is satisfied for each element it is clearly satisfied for any assemblage of elements). To demonstrate the procedure the simple stepped-bar example shown in Fig. 2.2(a) is now considered. Clearly the structure can be naturally considered to be made up of two distinct regions; thus the stepped bar is *modelled* (or *discretized* or *idealized*) using two elements designated A and B. The points 1, 2, and 3 are the nodes of the structure and forces F_1, F_2, and F_3 are assumed to act at these points. These forces can be either externally applied forces or reaction forces or a combination thereof. (This will depend on the precise nature of the problem and its boundary conditions. Such considerations will be deferred to the next section.) The displacements of the structure at points 1, 2, and 3 are D_1, D_2, and D_3 respectively.

For each element A and B a relationship of the form of equation (2.3) exists between the forces and displacements at the element ends i and j; these forces and displacements are designated U_i^A, u_i^A etc. in Fig. 2.2(b).

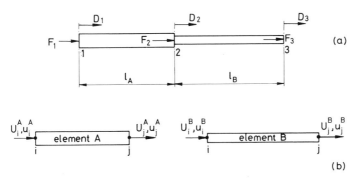

Fig. 2.2. A simple bar extension problem: (a) the structure; (b) the elements.

The element relationships are

$$\begin{Bmatrix} U_i^A \\ U_j^A \end{Bmatrix} = \begin{bmatrix} k_A & -k_A \\ -k_A & k_A \end{bmatrix} \begin{Bmatrix} u_i^A \\ u_j^A \end{Bmatrix} \quad \text{and} \quad \begin{Bmatrix} U_i^B \\ U_j^B \end{Bmatrix} = \begin{bmatrix} k_B & -k_B \\ -k_B & k_B \end{bmatrix} \begin{Bmatrix} u_i^B \\ u_j^B \end{Bmatrix}$$

(2.4)

where

$$k_A = \left(\frac{AE}{l}\right)_A \qquad k_B = \left(\frac{AE}{l}\right)_B$$

These relationships can easily be re-expressed in terms of *structure* quantities; clearly end points i and j of element A are respectively equivalent to points 1 and 2 of the structure, whilst end points i and j of member B are equivalent to points 2 and 3 of the structure. At this stage (before considering the boundary conditions) the structure has three degrees of freedom, namely D_1, D_2 and D_3; the re-expressed relationships for each element can be expanded to order three by simply adding zeros in appropriate locations. This procedure yields

$$\begin{Bmatrix} U_1^A \\ U_2^A \\ 0 \end{Bmatrix} = \begin{bmatrix} k_A & -k_A & 0 \\ -k_A & k_A & 0 \\ 0 & 0 & 0 \end{bmatrix} \begin{Bmatrix} u_1^A \\ u_2^A \\ 0 \end{Bmatrix} \quad \text{and}$$

$$\begin{Bmatrix} 0 \\ U_2^B \\ U_3^B \end{Bmatrix} = \begin{bmatrix} 0 & 0 & 0 \\ 0 & k_B & -k_B \\ 0 & -k_B & k_B \end{bmatrix} \begin{Bmatrix} 0 \\ u_2^B \\ u_3^B \end{Bmatrix}. \quad (2.5)$$

The matrix relationships for the two elements can now be combined together to give the relationships for the whole structure.

The *compatibility* condition for the assembled structure demands that element displacements match the structure displacements at the nodes. Thus

$$u_2^A = u_2^B = D_2 \text{ at the structure internal node}$$
$$u_1^A = D_1, \ u_3^B = D_3 \text{ at the structure boundary nodes.}$$

(2.6)

The *equilibrium* condition for the assembled structure requires that, at any structure node, the external applied force is equal to the algebraic sum of forces acting on all members meeting at that node. Thus

$$F_2 = U_2^A + U_2^B \text{ at the structure internal node}$$

and

(2.7)

$$F_1 = U_1^A, \ F_3 = U_3^B \text{ at the structure boundary nodes.}$$

Now if the relationships (2.6) are substituted into equations (2.5) and then relationships (2.7) are substituted into equations (2.5) the result is

$$\begin{Bmatrix} F_1 \\ F_2 \\ F_3 \end{Bmatrix} = \begin{Bmatrix} U_1^A \\ U_2^A \\ 0 \end{Bmatrix} + \begin{Bmatrix} 0 \\ U_2^B \\ U_3^B \end{Bmatrix} = \begin{bmatrix} k_A & -k_A & 0 \\ -k_A & k_A & 0 \\ 0 & 0 & 0 \end{bmatrix} \begin{Bmatrix} D_1 \\ D_2 \\ 0 \end{Bmatrix} +$$

$$\begin{bmatrix} 0 & 0 & 0 \\ 0 & k_B & -k_B \\ 0 & -k_B & k_B \end{bmatrix} \begin{Bmatrix} 0 \\ D_2 \\ D_3 \end{Bmatrix} \quad (2.8)$$

Performing the matrix addition leads to

$$\begin{Bmatrix} F_1 \\ F_2 \\ F_3 \end{Bmatrix} = \begin{bmatrix} k_A & -k_A & 0 \\ -k_A & (k_A + k_B) & -k_B \\ 0 & -k_B & k_B \end{bmatrix} \begin{Bmatrix} D_1 \\ D_2 \\ D_3 \end{Bmatrix} \quad (2.9)$$

or

F = **KD**

which is the desired stiffness relationship for the whole structure. In this relationship the 3×3 matrix **K** is of course the *structure stiffness matrix* (SSM).

The assembly procedure to produce the structure stiffness equations (2.9) from the element stiffness equations (2.4) has been formally described above and has been shown to depend on the direct application of compatibility and equilibrium conditions. However, observation of the result (equation (2.9)) shows that the structure stiffness matrix **K** is in fact obtained by *simply adding together or superposing the stiffness contributions of the individual elements* provided of course that the element stiffness contributions are properly located in the structure stiffness matrix so as to relate the correct nodal forces and displacements: in equation (2.9) the broken lines within the structure stiffness matrix are included so as to indicate clearly the individual contributions from elements A and B. The reader is advised to study the form of the structure stiffness matrix in relation to the geometry of the structure shown in Fig. 2.2. It should be noted that for this very simple structure only node 2 connects two elements and the combined term $k_A + k_B$ in the structure stiffness matrix relating the force at 2 to the displacement at 2 reflects this fact. In the generation of a structure stiffness matrix by hand the most convenient procedure is by the simple addition or direct superposition of element stiffnesses. Further discussion on and examples of this procedure, which is known as the direct stiffness method, will be given later in this chapter.

2.4. Application of zero-displacement boundary conditions and problem solution

The simple bar problem shown in Fig. 2.2 has not yet been fully defined since no statement has been made as to how the bar is held in space when the external loads are applied. The problem cannot be solved until the bar is held in a proper fashion; for this one-dimensional problem this means that at least one of the axial structure displacements D_1, D_2 and D_3 must be given a specified value to prevent the bar being simply displaced as a rigid body with indeterminate movement. Indeed, until this is done it will be found to be impossible to solve the stiffness equations (2.9) since the structure stiffness matrix is singular (i.e. the corresponding determinant has zero value). Thus the physically impossible structural problem is indicated mathematically by a system of equations which cannot be solved.

Consider that the problem is now made physically reasonable by specifying, for example, that the bar is held against axial movement at both its ends, i.e. that $D_1 = D_3 = 0$, and that the applied force F_2 has some known magnitude. Equation (2.9) now becomes

$$\begin{Bmatrix} F_1 \\ F_2 \\ F_3 \end{Bmatrix} = \begin{bmatrix} k_A & -k_A & 0 \\ -k_A & (k_A + k_B) & -k_B \\ 0 & -k_B & k_B \end{bmatrix} \begin{Bmatrix} 0 \\ D_2 \\ 0 \end{Bmatrix} = \begin{bmatrix} -k_A \\ (k_A + k_B) \\ -k_B \end{bmatrix} D_2 \qquad (2.10)$$

since there is no point in retaining the terms in columns 1 and 3 of the structure stiffness matrix when these terms are only multiplied by zero. Furthermore, the forces F_1 and F_3 are now unknown reactions at the bar ends so that the first and third of these stiffness equations ($F_1 = -k_A D_2$ and $F_3 = -k_B D_2$) relate unknown forces to unknown displacements and are thus of no help in the task of solving for D_2. The only meaningful equation in this task is therefore the second equation

$$F_2 = (k_A + k_B)D_2. \qquad (2.11)$$

Thus it is clear that prescribed boundary conditions of zero displacement are applied simply by crossing out those rows and columns of the structure stiffness matrix which correspond to each and every zero displacement. This is a general rule and does not depend on the type of structure being analysed.

In this simple problem the single non-zero structure degree of freedom is easily obtained from equation (2.11) as

$$D_2 = \frac{F_2}{k_A + k_B} \qquad (2.12)$$

and this is a major part of the solution.

Before proceeding with this problem it should be noted that in the general case more than one degree of freedom will be involved after applying the displacement boundary conditions and the equivalent to equation (2.11) will be a matrix equation of the form

$$\mathbf{F}_r = \mathbf{K}_r \mathbf{D}_r, \qquad (2.13a)$$

where the suffix r indicates a *reduced* matrix, after the displacement boundary conditions have been applied. The solution of these equations for the structure freedoms $\mathbf{D}_r$ can then be expressed as

$$\mathbf{D}_r = \mathbf{K}_r^{-1}\mathbf{F}_r \qquad (2.13b)$$

and this is the equivalent to the single equation (2.12) which arises in the particular problem under consideration. It should be noted that equations (2.13b) should be interpreted as symbolizing the solution of equations (2.13a) rather than being a formal statement of the manner of their solution; in practice equations (2.13a) may be solved in any convenient manner without necessarily involving the inversion of $\mathbf{K}_r$.

Returning to the bar example, if the values of the end reactions are desired these can now be obtained by substituting the known value of D_2 into the first and third of equations (2.10) to give

$$F_1 = -\frac{k_A}{k_A + k_B} F_2 \qquad F_3 = -\frac{k_B}{k_A + k_B} F_2. \qquad (2.14)$$

The minus signs indicate that the reactions are acting from right to left.

The final stage of the displacement method solution is the calculation of element forces by back substituting the now known values of the element end displacements into the element stiffness relationships. In the example this means substituting the values

$$u_i^A = 0 \qquad u_j^A = D_2 = \frac{F_2}{k_A + k_B} \qquad u_i^B = D_2 \qquad u_j^B = 0$$

into equations (2.4). The result is

$$U_i^A = -U_j^A = -\frac{k_A F_2}{k_A + k_B}$$

$$\qquad\qquad (2.15)$$

$$U_i^B = -U_j^B = \frac{k_B F_2}{k_A + k_B}.$$

These are the forces acting *on* the element ends and they have positive value when acting from left to right (see Fig. 2.2(b)). Thus element A is in tension and element B is in compression.

The solution is now complete but before leaving the problem it might be noted that the equilibrium condition can be used to check the results obtained at the following three levels.

(a) Overall equilibrium of the applied force and the reactions:

$$F_1 + F_2 + F_3 = 0.$$

(b) Joint equilibrium:

$$F_2 + U_j^A + U_i^B = 0 \qquad\qquad F_1 = U_i^A \qquad\qquad F_3 = U_j^B.$$

(c) Element equilibrium:

$$U_i^A + U_j^A = 0 \qquad\qquad U_i^B + U_j^B = 0.$$

2.5. Remarks on the solution procedure

The basic procedure in using the matrix displacement method to solve a static problem has been illustrated by analysing a very simple bar problem. The steps involved in the analysis have been seen to be as follows.

(1) Discretize the problem, i.e. decide on the type and arrangement of the elements making up the complete structure.

(2) Establish the stiffness matrix $\mathbf{k}$ for each element of the structure.

(3) Assemble the structure stiffness matrix $\mathbf{K}$ from the element stiffness matrices.

(4) Apply the zero-displacement boundary conditions to obtain the reduced structure matrix $\mathbf{K}_r$. (It should be noted that prescribed displacements need not always be zero displacements (see Section 2.11)).

(5) Solve the reduced set of structure stiffness equations $\mathbf{F}_r = \mathbf{K}_r \mathbf{D}_r$ to yield the structure displacements $\mathbf{D}_r$ (see Appendix A for remarks on equation solving). Here $\mathbf{F}_r$ is the column matrix of *external* forces applied at the structure joints or nodes corresponding in one-to-one fashion with the displacements $\mathbf{D}_r$.

(6) If required, calculate the structure reactions using the structure stiffness equations eliminated previously (in reducing from $\mathbf{K}$ to $\mathbf{K}_r$) in conjunction with the now known structure displacements.

(7) Calculate the element forces using the element stiffness matrices and the known displacements.

These steps apply equally to more complex structural problems analysed using finite elements, as will be seen later. A few remarks of amplification are desirable at this stage.

Step (1) is a straightforward matter in the problem corresponding to Fig. 2.2(a) and in all other applications of this chapter and of Chapters 3 and 4 dealing with bars, beams, and skeletal structures. However, for continuum structures the selection of an element mesh is usually not clear cut.

Step (2) will be a major concern in this text. In the simple example considered above there is no difficulty in deriving the element stiffness matrix but for two- and three-dimensional continuum elements the derivation can become complex. A point not brought out by the above example is that local axes, to which element properties are normally related, often differ from global or structure axes. A transformation is then required between local and global axes as, for example, will be seen to be the case in analysing an arbitrary skeletal frame in the next chapter.

Steps (3) and (4) are often combined in practice since there is little point in assembling the whole **K** matrix and then eliminating specified rows and columns. Instead, only the final structure stiffness matrix **K**$_r$ need be assembled, either by hand or in the computer.

Step (5) is usually the limiting factor in attempting a solution by hand since the manual solution of more than three or four equations becomes very tedious. Similarly, in a computer solution this step usually requires a large proportion of the total computing effort.

Step (6) is often excluded when only **K**$_r$ is assembled directly. In this case the support reactions can be calculated individually if required from the calculated element forces found in step (7).

2.6. Some properties of stiffness matrices

Although only very simple stiffness matrices have been considered so far, these matrices exhibit certain characteristics of a general nature which will apply to all other stiffness matrices developed in this book. Some of the characteristics have already been mentioned earlier but will be included again here for the sake of completeness. The chief properties of a stiffness matrix are now given; these properties apply to both element and (unreduced) structure stiffness matrices.

(1) A stiffness matrix **k** is a square matrix whose typical coefficient k_{ef} is the value of the force P_e required when the displacement d_f is given the value unity and all other displacements have the value zero. (k_{ef} is the coefficient of the stiffness matrix located in row e and column f.) For the bar element whose stiffness matrix is given in equation (2.3) it can be seen that if, for example, we set $u_i = 1$ and $u_i = 0$ the corresponding forces are $U_i = -AE/l$ and $U_j = AE/l$ represented by the values of k_{12} and k_{22}. This situation is illustrated by Fig. 2.3 and is seen to be the correct one.

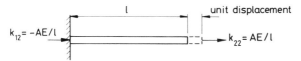

Fig. 2.3. Bar element nodal forces corresponding to unit value of u_j.

(2) The coefficients in any column of the stiffness matrix satisfy the overall equilibrium requirements of the element or structure, as the case may be. This follows from point (1) since the coefficients in any column are the force values acting at the nodes for a given deformed shape and these force values form a self-equilibrating system. For the bar element or for an assemblage of such elements this means that the sum of all coefficients in any column of the stiffness matrix must be zero; this is confirmed on consulting equations (2.3) and (2.9).

(3) The stiffness matrix is symmetrical, i.e. $k_{ef} = k_{fe}$. This property follows from point (1) and use of the reciprocal theorem.

(4) All coefficients on the leading diagonal of a stiffness matrix must be positive since a positive force cannot produce a negative corresponding displacement.

(5) Both the element stiffness matrix and the unreduced structure stiffness matrix are singular, i.e. the determinant of the matrix has the value zero. The physical explanation of this is that until some valid boundary conditions are applied the element or structure is free to move with an arbitrary rigid-body motion in addition to deforming elastically. (It should be noted that the equilibrium conditions are not affected by a rigid-body movement.)

2.7. Stiffness matrix of a bar element subjected to torsion

The behaviour of a prismatic bar when subjected to axial force was considered in Section 2.2 where a very simple stiffness relationship between end forces and end displacements was obtained (equation (2.3)). The problem of the prismatic bar subjected to pure twisting is now considered. This problem is illustrated in Fig. 2.4 and on comparing with Fig. 2.1 it is not difficult to imagine that there is a close correspondence between the types of stiffness relationship pertaining to the axial and torsional cases. In Fig. 2.4 the usual 'double-arrow' convention is used to

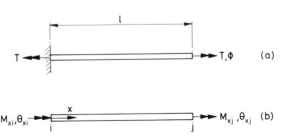

Fig. 2.4. Bar subjected to torsion: (a) bar under the action of twisting moment; (b) bar element.

denote a moment or a rotation, the convention being that the moment and the rotation act about the axis lying along which the arrows are pointing in the clockwise sense when viewed in the direction of the arrows—this is the usual right-hand rule.

The bar shown in Fig. 2.4(a) has one end held whilst a twisting moment T is applied to the free end. If a circular cross-section is assumed it can readily be shown by employment of the three basic conditions of elasticity, equilibrium, and compatibility that

$$T = \frac{GJ}{l} \phi. \tag{2.16}$$

The derivation of this relationship is detailed in elementary texts and assumes that the effect of the twisting moment on the deformation of the bar is simply to rotate any cross-section lying perpendicular to the axis of the bar relative to similar neighbouring cross-sections. The derivation compares directly with the derivation of equation (2.1) for the axial force case. In equation (2.16) G is the modulus of rigidity and J is known as the torsion constant. This latter quantity is a property of the cross-sectional geometry, having dimensions of length to the power 4; for a circular cross-section J is the polar second moment of area. The quantity ϕ is the rotation of the cross-section (or twist) at the free end and for a prismatic bar subjected to a constant torque the twist per unit length ϕ/l is uniform along the bar length.

For the bar element shown in Fig. 2.4(b), whose axis is the x axis and which is subjected to end twisting moments M_{xi} and M_{xj}, it can be seen by direct analogy with the axial force case of Section 2.2 that the stiffness relationship for the torsion case is

$$\begin{Bmatrix} M_{xi} \\ M_{xj} \end{Bmatrix} = \frac{GJ}{l} \begin{bmatrix} 1 & -1 \\ -1 & 1 \end{bmatrix} \begin{Bmatrix} \theta_{xi} \\ \theta_{xj} \end{Bmatrix} \tag{2.17}$$

which has the form $\mathbf{P} = \mathbf{kd}$ again. Here θ_{xi} and θ_{xj} are the rotations about the x axis at ends i and j of the element. This stiffness relationship can be termed exact within the confines of the assumptions noted in the preceding paragraph which lead to our theoretical model of the bar behaviour. It should be noted, however, that a circular cross-section has been specified above and this has been done since only for this shape of cross-section is it true that cross-sections simply rotate under the action of a pure torque. For non-circular cross-sections the application of a torque results in a certain amount of longitudinal deformation—or longitudinal warping—in addition to the rotation. In practice the effect of this is usually small and so the theoretical model represented by equation (2.16), which ignores it completely, will in fact be assumed to apply for any shape of cross-section.

2.8. Stiffness matrix of a beam element

In this section the objective is to derive a stiffness matrix for a beam element, with the derivation based on the so-called simple theory of bending otherwise known as the engineering theory of beams or the Bernoulli–Euler beam theory. It is assumed that the reader is familiar with this theory but some aspects of it are now briefly recalled prior to developing the element properties. (The beam theory is detailed in many elementary texts.)

In the Bernoulli–Euler theory of bending it is assumed that the initially straight beam is subjected to a state of pure bending (i.e. the bending moment has the same value at all sections along the beam) through the application of moments at its ends. Figure 2.5 shows this situation, with the moments taken to be applied in the xy plane and with bending taking place in this plane, or in other words about the z axis where the z axis completes the right-handed set of three cartesian coordinates. This statement implies that the beam cross-section is symmetrical about the xy plane, or in other words that the bending takes place about a principal axis of the section. The longitudinal x axis lies in the neutral surface of the beam where no straining occurs. In the Bernoulli–Euler theory the deformation of the beam is assumed to be such that (i) straight lines which are initially normal to the neutral surface remain straight and normal to this surface in the loaded configuration and (ii) all points lying on a normal to the neutral surface deflect the same amount v in the y direction. These assumptions form the basis of a theoretical model of beam deformation in which the behaviour at any point in the beam can be related to the deflection of the neutral surface and in which the only non-zero stress is the direct stress in the longitudinal x direction; this stress varies linearly through the depth of the beam from a maximum tensile (positive) value at the lower edge to a maximum compressive (negative) value at the upper edge of the beam when the bending moment is in the sense shown in Fig. 2.5. Specifically, for bending about the z axis the well-known relationship between the bending moment M_z at any section, the second moment of area of the section I_z, the longitudinal

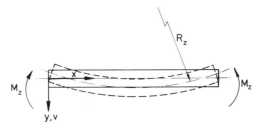

Fig. 2.5. Beam bent in xy plane.

stress σ_x, the distance from the neutral surface y, Young's modulus E and the radius of curvature R_z of the deformed neutral surface is

$$\frac{M_z}{I_z} = \frac{\sigma_x}{y} = \frac{E}{R_z} \tag{2.18}$$

and in developing this relationship the conditions of equilibrium, elasticity, and compatibility are used directly or are implied. With the assumption of small deflection the inverse of the radius of curvature, which is known simply as the curvature χ_z, can be expressed with only small error as

$$\chi_z = \frac{1}{R_z} = -\frac{d^2 v}{dx^2}. \tag{2.19}$$

Thus, from equations (2.18) and (2.19) a relationship between bending moment and curvature at any section along the beam can be obtained in the form

$$M_z = -EI_z \frac{d^2 v}{dx^2} \tag{2.20}$$

and this relationship provides the starting point for the development of a beam element stiffness matrix. The positive sense of bending moment associated with this relationship is that shown in Figs. 2.5 and 1.2(e). (The latter figure also shows the positive sense of shearing force.)

Before proceeding, it bears repeating that the Bernoulli–Euler beam theory symbolized by equations (2.18)–(2.20) is a particular theoretical model of beam behaviour which is based on certain simplifying assumptions which may not always be appropriate for real beams. In particular it assumes pure bending whereas in practice there will almost always be shearing forces present which will produce shear stress across transverse sections of the beam. This obviously invalidates the assumption that the direct stress in the longitudinal direction is the only non-zero stress and also means that initially normal lines will not remain straight and normal to the neutral surface when loading is applied. However, in usual circumstances, provided that the beam is not unduly stocky, these transverse shear effects are small and can be ignored without significant loss in accuracy. Hence it will be assumed in what follows that the Bernoulli–Euler beam theory provides an adequate theoretical model for the bending behaviour of all beams. (A beam theory which takes account of the effects of transverse shear is known as the Timoshenko beam theory.)

Consider now a beam element lying in the xy plane, as shown in Fig. 2.6, with nodes (end points) again labelled i and j. Bending is assumed to take place in the xy plane under the action now of both moments and lateral forces acting at the nodes: at node i a force system acts which

Fig. 2.6. Beam element in xy plane.

comprises a lateral force V_i (positive downwards) and a moment M_{zi} (positive clockwise), and a similar force system acts at node j.† Corresponding to the nodal force system at point i is a nodal displacement system comprising the lateral displacement or deflection v_i and the rotation of the beam $\theta_{zi} = (dv/dx)_i$ with a similar system at point j. The element is assumed to carry no load between its ends. We require to develop relationships between the nodal forces V_i, M_{zi}, V_j, and M_{zj} and the corresponding nodal displacements v_i, θ_{zi}, v_j, and θ_{zj} by the construction of a 4×4 beam element stiffness matrix.

Equilibrium considerations applied to the part of the beam element lying between end i and a general section at distance x along the element show that the bending moment at this section is (see Fig. 2.6)

$$M_z = M_{zi} - V_i x. \tag{2.21}$$

Combining equations (2.20) and (2.21) and then integrating with respect to x twice gives

$$EI_z \frac{d^2 v}{dx^2} = V_i x - M_{zi}$$

$$EI_z \frac{dv}{dx} = V_i \frac{x^2}{2} - M_{zi} x + C_1 \tag{2.22}$$

$$EI_z v = V_i \frac{x^3}{6} - M_{zi} \frac{x^2}{2} + C_1 x + C_2.$$

For the beam bending problem the compatibility condition requires that both the deflection and rotation be continuous throughout the length of the beam. Observation of equations (2.22) shows that this condition is clearly satisfied for the beam element.

---

† It should be noted that the positive senses adopted for the moment (always clockwise) and the lateral force (always downwards) at the nodes differ from the positive senses adopted for the bending moment and the shearing force at a general section and shown in Fig. 1.2(e). The former convention gives unidirectional nodal quantities which are necessary when we come to connect elements together whilst the latter convention is the traditional convention in strength of materials work and is more convenient for the construction of bending moment diagrams etc. The use of the two conventions does not lead to any confusion.

The constants C_1 and C_2 can be evaluated by using the displacement boundary conditions at the left-hand end of the element (node i) that $v = v_i$ and $dv/dx = \theta_{zi}$ at $x = 0$. This gives

$$C_1 = EI_z\theta_{zi} \quad \text{and} \quad C_2 = EI_zv_i \tag{2.23}$$

Further, the displacement boundary conditions at the right-hand end of the element (node j) are $v = v_j$ and $dv/dx = \theta_{zj}$ at $x = l$. Substituting these conditions together with the values of the constants C_1 and C_2 into the last two of equations (2.22) yields the equations

$$EI_z\theta_{zj} = V_i\frac{l^2}{2} - M_{zi}l + EI_z\theta_{zi}$$

and

$$EI_zv_j = V_i\frac{l^3}{6} - M_{zi}\frac{l^2}{2} + EI_z\theta_{zi}l + EI_zv_i.$$

Solving these equations for V_i and M_{zi} gives

$$V_i = \frac{EI}{l^3}(12v_i - 12v_j + 6l\theta_{zi} + 6l\theta_{zj})$$

$$M_{zi} = \frac{EI_z}{l^2}(6v_i - 6v_j + 4l\theta_{zi} + 2l\theta_{zj}) \tag{2.24}$$

When the forces at node i have been found in terms of the end displacements, the corresponding values at node j can easily be obtained by using the equilibrium conditions for the element as a whole, i.e.

$$V_j = -V_i \qquad M_{zj} = V_il - M_{zi}.$$

Therefore

$$V_j = \frac{EI_z}{l^3}(-12v_i + 12v_j - 6l\theta_{zi} + 6l\theta_{zj})$$

$$M_{zj} = \frac{EI_z}{l^2}(6v_i - 6v_j + 2l\theta_{zi} + 4l\theta_{zj}) \tag{2.25}$$

Equations (2.24) and (2.25) are the desired stiffness relationships linking nodal forces and nodal displacements for the beam element. (We note that the expressions for M_{zi} and M_{zj} are the well-known slope–deflection equations.) In matrix form the stiffness relationships can be expressed as

$$\begin{Bmatrix} V_i \\ M_{zi} \\ V_j \\ M_{zj} \end{Bmatrix} = EI_z \begin{bmatrix} 12/l^3 & 6/l^2 & -12/l^3 & 6/l^2 \\ 6/l^2 & 4/l & -6/l^2 & 2/l \\ -12/l^3 & -6/l^2 & 12/l^3 & -6/l^2 \\ 6/l^2 & 2/l & -6/l^2 & 4/l \end{bmatrix} \begin{Bmatrix} v_i \\ \theta_{zi} \\ v_j \\ \theta_{zj} \end{Bmatrix} \tag{2.26}$$

This has the standard form

$\mathbf{P} = \mathbf{kd}$

where $\mathbf{P}$ and $\mathbf{d}$ are the column matrices of nodal forces and nodal displacements respectively and $\mathbf{k}$ is the 4×4 beam element stiffness matrix. The specific forms of $\mathbf{P}$, $\mathbf{d}$, and $\mathbf{k}$ are clear on inspection of equation (2.26) and we note that—as is always the case—there is a one-to-one correspondence between the force quantities of $\mathbf{P}$ and the corresponding displacement quantities of $\mathbf{d}$, i.e. the force V_j for instance is the third quantity in the column matrix $\mathbf{P}$ and the displacement v_j corresponding to this force is the third quantity in $\mathbf{d}$. As with the bar element, the above stiffness matrix for the beam element can be termed 'exact' (within the limitations of the background theoretical model, i.e. the Bernoulli–Euler beam theory, and provided that no loads are carried along the length of the element) since the conditions of elasticity, equilibrium, and compatibility have been satisfied throughout the element.

The matrices in the stiffness equation (2.26) are shown with broken lines subdividing them into a convenient *partitioned* form which can be expressed as

$$\begin{Bmatrix} \mathbf{P}_i \\ \mathbf{P}_j \end{Bmatrix} = \begin{bmatrix} \mathbf{k}_{ii} & \mathbf{k}_{ij} \\ \mathbf{k}_{ji} & \mathbf{k}_{jj} \end{bmatrix} \begin{Bmatrix} \mathbf{d}_i \\ \mathbf{d}_j \end{Bmatrix} \qquad (2.27)$$

where $\mathbf{P}_i$ and $\mathbf{d}_i$ are the sets of forces and displacements respectively at node i, $\mathbf{k}_{ii}$ is the stiffness submatrix relating the forces at node i to the displacements at node i, and similar definitions apply to $\mathbf{P}_j$, $\mathbf{d}_j$, $\mathbf{k}_{ij}$, $\mathbf{k}_{ji}$, and $\mathbf{k}_{jj}$. On comparison with equation (2.26) we see that the submatrices occurring in this equation are

$$\mathbf{P}_i = \begin{Bmatrix} V_i \\ M_{zi} \end{Bmatrix} \qquad\qquad \mathbf{d}_i = \begin{Bmatrix} v_i \\ \theta_{zi} \end{Bmatrix}$$

$$\mathbf{k}_{ii} = \begin{bmatrix} 12EI_z/l^3 & 6EI_z/l^2 \\ 6EI_z/l^2 & 4EI_z/l \end{bmatrix} \qquad (2.28)$$

and so on. We also note that $\mathbf{k}_{ji}^{t} = \mathbf{k}_{ij}$ from the symmetry of the complete element stiffness matrix. The partitioned form of the stiffness relationship shown in equations (2.27) is frequently convenient to use in performing matrix manipulations, whether using beam elements or any other type of element.

The five properties of stiffness matrices which were noted in Section 2.6 apply of course to the beam element stiffness matrix and some elaboration is desirable to provide further illustrations of the meaning of some of the earlier statements.

To illustrate point 1, if $\theta_{zj} = 1$ for example and the other displacements

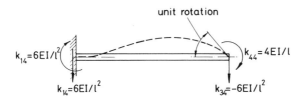

Fig. 2.7. Beam element nodal forces corresponding to unit value of θ_{zj}.

v_i, θ_{zi}, and v_j are all zero then the beam element stiffness relationship predicts that the element nodal forces are

$$\begin{Bmatrix} V_i \\ M_{zi} \\ V_j \\ M_{zj} \end{Bmatrix} = \begin{Bmatrix} k_{14} \\ k_{24} \\ k_{34} \\ k_{44} \end{Bmatrix} = EI_z \begin{Bmatrix} 6/l^2 \\ 2/l \\ -6/l^2 \\ 4/l \end{Bmatrix}.$$

The situation is equivalent to a propped cantilever beam subjected to an applied couple at the prop of magnitude sufficient to induce a corresponding unit rotation at the prop. Figure 2.7 ilustrates this condition and shows the applied couple and the resulting reaction forces and moments. This force system is the correct one of course.

Regarding point (2), the overall equilibrium requirements of the beam element comprise vertical force balance

$$V_i + V_j = 0$$

and moment balance

$$M_{zi} + M_{zj} - V_i l = 0.$$

For the element stiffness matrix this means that for any column f the stiffness coefficients must satisfy the conditions that

$$k_{1f} + k_{3f} = 0$$

and

$$k_{2f} + k_{4f} - lk_{1f} = 0$$

Inspection of the beam element stiffness matrix shows that these conditions are met.

Concerning point (5), the beam element (or any assemblage of beam elements) is capable of both lateral deflection and rotation. Any element or element assemblage must thus be supported in such a way as to prevent both a rigid-body translation (in the lateral direction) and a rigid-body rotation.

Stiffness matrices have now been established in this and preceding sections for three types of element—the bar element loaded axially (see equation (2.3)), the bar element subjected to torsion (equation (2.17)), and the beam element subjected to bending (equation (2.26)). It should be noted that the three types of behaviour have been considered in isolation one from another and are completely uncoupled at the element level such that axial loads, for instance, cause only axial displacements and do not have any effect on bending or twisting deformation. For one-dimensional assemblies of these elements, i.e. continuous bars and beams, this uncoupling of behaviour also applies to the whole structure and means that if a straight bar/beam structure is subjected to a combined loading (e.g. lateral loads and axial loads acting simultaneously) the total solution can be obtained simply by addition of the separate solutions for each type of behaviour. The uncoupling of effects is consistent with the assumptions made in developing the theoretical models of axial, torsional, and bending behaviour used as the basis for development of element properties and is perfectly satisfactory for the great majority of practical applications. (An example of a circumstance in which coupling of behaviour cannot be disregarded is the effect on bending behaviour of the presence of an axial force which is of the order of the buckling load of the component in question (see Chapter 12).)

Before considering numerical examples of the solution of problems represented by assemblies of the elements whose properties have been developed, further consideration is given in the next section to the process of assembly of the structure stiffness matrix which was looked at in preliminary fashion in Section 2.3. However, prior to this a very simple single-element problem will perhaps help to fix the notion of a stiffness matrix.

Example 2.1. Use the element stiffness relationships given by equation (2.26) to calculate the deflection and rotation at the tip of the uniform cantilever beam shown in Fig. E2.1. Also determine the components of reaction at the fixed end.

Since a single element is all that is necessary to represent the beam, we need not distinguish between element and structure quantities and the stiffness equations (2.26) effectively refer to the whole structure.

The boundary conditions of the problem are that the lateral displacement and the rotation have zero value at the built-in end. Thus $v_i = \theta_{zi} = 0$ and the problem degrees of freedom are only v_j and θ_{zj}. The prescribed boundary conditions of zero displacement are applied by eliminating the rows and columns of the stiffness

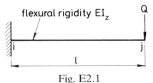

flexural rigidity EI_z

Fig. E2.1

matrix which correspond to v_i and θ_{zi}; this simply leaves the bottom right-hand 2×2 submatrix of equation (2.26). Further, the loads which are applied to the structure nodes and which correspond to the non-zero displacements v_j and θ_{zj}, are the lateral force Q and a zero moment. Thus the structure stiffness equations are

$$\begin{Bmatrix} Q \\ 0 \end{Bmatrix} = EI_z \begin{bmatrix} 12/l^3 & -6/l^2 \\ -6/l^2 & 4/l \end{bmatrix} \begin{Bmatrix} v_j \\ \theta_{zj} \end{Bmatrix}$$

and these, being for the structure after the boundary conditions have been applied, are equations of the form of equation (2.13a).

It is a simple matter to solve these equations for v_j and θ_{zj} in terms of the applied load Q; the result is

$$v_j = \frac{Ql^3}{3EI_z} \qquad \theta_{zj} = \frac{Ql^2}{2EI_z}$$

and these values are the exact ones for tip deflection and rotation within the Bernoulli–Euler theory.

The components of reaction at the fixed end of the beam are equivalent to V_i and M_{zi} and these can be calculated, again using equation (2.26), now that v_j and θ_{zj} have been calculated (and remembering that v_i and θ_{zi} are zero). Thus

$$\begin{Bmatrix} V_i \\ M_{zi} \end{Bmatrix} = EI_z \begin{bmatrix} -12/l^3 & 6/l^2 \\ -6/l^2 & 2/l \end{bmatrix} \begin{Bmatrix} v_j \\ \theta_{zj} \end{Bmatrix} = \begin{Bmatrix} -Q \\ -Ql \end{Bmatrix}$$

This reveals that the vertical reaction at the fixed end is of magnitude Q and acts upwards as evidenced by the negative sign, and the reaction moment is of magnitude Ql acting anticlockwise, again as evidenced by the negative sign. This result is obviously correct from simple static considerations.

2.9. Assembly of the structure stiffness matrix by the direct stiffness method

In the simple bar example it has been seen that formal application of the compatibility and equilibrium conditions in the process of assembling the structure stiffness matrix from element matrices is equivalent to a process of direct superposition of element stiffnesses. This superposition process applies equally to any structure assembled from elements of whatever type and is a consequence of the definition of a stiffness coefficient (as given in point (1) of the properties of stiffness matrices in Section 2.6) and of the principle of superposition.

As far as one-dimensional (or line) elements are concerned it has been demonstrated that the stiffness equation of a bar or a beam element is conveniently expressible in the partitioned form of equation (2.27). It has been seen that in the case of the bar element $\mathbf{k}_{ii}$ etc. are single coefficients and that in the case of the beam $\mathbf{k}_{ii}$ etc. are 2×2 matrices. The same partitioned form also holds for other line elements; it will be shown in the following chapters that for an element of a planar rigid-jointed frame $\mathbf{k}_{ii}$ etc. are 3×3 matrices and for an element of a general three-dimensional rigid-jointed frame $\mathbf{k}_{ii}$ etc. are 6×6 matrices. (For frame structures the element forces and displacements have to be referred to a common global or structure co-ordinate system but the form of the element stiffness equation is still as in equation (2.27)).

Consider that in a structural assembly of line elements an element labelled R connects structure nodes numbered N1 and N2 say (N2>N1), i.e. node i of the element corresponds to node N1 of the structure, node j of the element to node N2 of the structure. The *compatibility* condition applied at the nodes gives

$$\mathbf{d}_i^R = \mathbf{D}_{N1} \qquad \mathbf{d}_j^R = \mathbf{D}_{N2} \tag{2.29}$$

where $\mathbf{d}_i^R$ are the displacements at node i of element R and $\mathbf{D}_{N1}$ are the displacements at node N1 of the structure etc. Thus the stiffness relationships for element R become

$$\mathbf{P}_i^R = \mathbf{k}_{ii}^R \mathbf{D}_{N1} + \mathbf{k}_{ij}^R \mathbf{D}_{N2}$$
$$\mathbf{P}_j^R = \mathbf{k}_{ji}^R \mathbf{D}_{N1} + \mathbf{k}_{jj}^R \mathbf{D}_{N2}. \tag{2.30}$$

The equilibrium condition requires that at any structure node the externally applied forces are balanced by the forces at the ends of elements meeting at that structure node. Thus at nodes N1 and N2 we have

$$\mathbf{F}_{N1} = \mathbf{P}_i^R + \text{similar contributions for all other}$$

$$\text{elements meeting at node N1}$$

$$\mathbf{F}_{N2} = \mathbf{P}_j^R + \text{similar contributions for all other} \tag{2.31}$$

$$\text{elements meeting at node N2}$$

where $\mathbf{F}_{N1}$ and $\mathbf{F}_{N2}$ are the external forces acting at nodes N1 and N2 respectively. Then

$$\mathbf{F}_{N1} = \mathbf{k}_{ii}^R \mathbf{D}_{N1} + \mathbf{k}_{ij}^R \mathbf{D}_{N2} + \text{contributions from other elements}$$
$$\mathbf{F}_{N2} = \mathbf{k}_{ji}^R \mathbf{D}_{N1} + \mathbf{k}_{jj}^R \mathbf{D}_{N2} + \text{contributions from other elements.} \tag{2.32}$$

The structure stiffness matrix $\mathbf{K}$ relates the applied forces $\mathbf{F}$ to the structure displacements $\mathbf{D}$ of course. Thus the contribution of the particular element R to the structure stiffness matrix $\mathbf{K}$ is simply as illustrated schematically in the following:

$$\mathbf{K} = \tag{2.33}$$

The assembly process has been described in terms of stiffness submatrices but the same principles apply to individual stiffness coefficients of course. It is noted that continuum finite elements may have many more than two nodal points, as will be seen later, but the basic procedure of the direct stiffness method remains unaltered.

The following points should be noted.

(1) The submatrix of the structure stiffness matrix lying on the leading diagonal in row and column N_1, say, will be the sum of the submatrices $\mathbf{k}_{ii}$ or $\mathbf{k}_{jj}$ of all elements which meet at joint N1. (Submatrix $\mathbf{k}_{ii}$ applies if node i of the element corresponds to node N1 of the structure; $\mathbf{k}_{jj}$ applies if node j of the element corresponds to node N1 of the structure). Stiffness coefficients on the leading diagonal of a stiffness matrix are often referred to as *direct stiffness* coefficients.

(2) There will be a (non-zero) contribution in an off-diagonal location of the structure matrix such as N1, N2 (i.e. row N1, column N2) only if nodes N1 and N2 are connected by an element; if two such nodes are not directly connected then the associated structure stiffness submatrix is zero. Where there is a contribution it comes from the $\mathbf{k}_{ij}$ or $\mathbf{k}_{ji}(=\mathbf{k}_{ij}^{t})$ element stiffness submatrices; the off-diagonal coefficients are referred to as *indirect* or *cross stiffness* coefficients.

An important point to realize is that the actual process of assembly of a whole structure stiffness matrix from the element stiffness matrices depends solely on the manner in which the individual elements are connected together. The detailed characteristics of the elements are taken account of in deriving the element stiffness matrices but once this is done (and, if necessary, a transformation is made to a common global co-ordinate system) the formation of a structure stiffness matrix by superposition proceeds independently of any knowledge of the structural properties of the elements (except of course that in general situations different elements have different numbers of nodes and different numbers of degrees of freedom at each node). Symbolically the direct superposition of E element stiffnesses $\mathbf{k}_e$ to form the structure stiffness matrix $\mathbf{K}$ can be expressed as

$$\mathbf{K} = \sum_{e=1}^{E} \mathbf{k}_e^{*}$$

where the asterisk associated with $\mathbf{k}_e$ is used to denote that the element stiffness matrices are formally expanded to structure size by the addition of rows and columns of zeros before the summation takes place.

As an example of the assembly procedure by superposition consider the planar rigid-jointed frame structure shown in Fig. 2.8. The stiffness matrix for a general line element of such a structure has not yet been

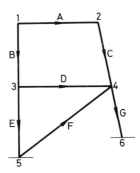

Fig. 2.8. A planar, rigid-jointed frame.

derived (see Chapter 3) but for the moment all that need be known about the typical element stiffness matrix is that it has the form given in equation (2.27). The elements of the frame are referred to by the letters A–G and the structure nodes by the numbers 1–6. (The way in which the nodes are numbered does not matter a great deal in a small problem of this sort but can be important in a more complex problem (see Section 3.8)). The arrows shown on the elements are there to indicate the direction of the local axis of the element, i.e. the arrows run from end i to end j of each element.

The connection between the individual element nodes and the structure nodes is clearly as follows:

$$
\begin{array}{ccc}
\text{Element} & i & j \quad \leftarrow \text{element nodes}\\
A & 1 & 2\\
B & 1 & 3\\
C & 2 & 4\\
D & 3 & 4 \quad \leftarrow \text{structure node numbers.}\\
E & 3 & 5\\
F & 5 & 4\\
G & 4 & 6
\end{array}
$$

Correspondingly the stiffness matrices for the elements A–G can be conveniently expressed in their partitioned forms as

$$
\mathbf{k}^A = \begin{array}{cc} 1 & 2 \end{array}\begin{bmatrix} \mathbf{k}_{ii}^A & \mathbf{k}_{ij}^A \\ \mathbf{k}_{ji}^A & \mathbf{k}_{jj}^A \end{bmatrix}
\qquad
\mathbf{k}^B = \begin{array}{cc} 1 & 3 \end{array}\begin{bmatrix} \mathbf{k}_{ii}^B & \mathbf{k}_{ij}^B \\ \mathbf{k}_{ji}^B & \mathbf{k}_{jj}^B \end{bmatrix}
\qquad
\mathbf{k}^C = \begin{array}{cc} 2 & 4 \end{array}\begin{bmatrix} \mathbf{k}_{ii}^C & \mathbf{k}_{ij}^C \\ \mathbf{k}_{ji}^C & \mathbf{k}_{jj}^C \end{bmatrix}
$$

$$
\mathbf{k}^D = \begin{array}{cc} 3 & 4 \end{array}\begin{bmatrix} \mathbf{k}_{ii}^D & \mathbf{k}_{ij}^D \\ \mathbf{k}_{ji}^D & \mathbf{k}_{jj}^D \end{bmatrix}
\qquad
\mathbf{k}^E = \begin{array}{cc} 3 & 5 \end{array}\begin{bmatrix} \mathbf{k}_{ii}^E & \mathbf{k}_{ij}^E \\ \mathbf{k}_{ji}^E & \mathbf{k}_{jj}^E \end{bmatrix}
$$

$$
\mathbf{k}^F = \begin{array}{cc} 5 & 4 \end{array}\begin{bmatrix} \mathbf{k}_{ii}^F & \mathbf{k}_{ij}^F \\ \mathbf{k}_{ji}^F & \mathbf{k}_{jj}^F \end{bmatrix}
\qquad
\mathbf{k}^G = \begin{array}{cc} 4 & 6 \end{array}\begin{bmatrix} \mathbf{k}_{ii}^G & \mathbf{k}_{ij}^G \\ \mathbf{k}_{ji}^G & \mathbf{k}_{jj}^G \end{bmatrix}
\qquad (2.34)
$$

where the numbers over the columns of the matrices are used to clarify which structure nodes correspond to element nodes i and j.

With this in mind the application of the direct stiffness procedure gives the structure stiffness matrix $\mathbf{K}$ as

$$\mathbf{K} = \begin{matrix} & 1 & 2 & 3 & 4 & 5 & 6 \\ & \begin{bmatrix} (\mathbf{k}_{ii}^{A} + \mathbf{k}_{ii}^{B}) & \mathbf{k}_{ij}^{A} & \mathbf{k}_{ij}^{B} & 0 & 0 & 0 \\ \mathbf{k}_{ji}^{A} & (\mathbf{k}_{jj}^{A} + \mathbf{k}_{ii}^{C}) & 0 & \mathbf{k}_{ij}^{C} & 0 & 0 \\ \mathbf{k}_{ji}^{B} & 0 & (\mathbf{k}_{jj}^{B} + \mathbf{k}_{ii}^{D} + \mathbf{k}_{ii}^{E}) & \mathbf{k}_{ij}^{D} & \mathbf{k}_{ij}^{E} & 0 \\ 0 & \mathbf{k}_{ji}^{C} & \mathbf{k}_{ji}^{D} & \begin{array}{c}(\mathbf{k}_{jj}^{C} + \mathbf{k}_{jj}^{D} \\ + \mathbf{k}_{jj}^{F} + \mathbf{k}_{ii}^{G})\end{array} & \mathbf{k}_{ji}^{F} & \mathbf{k}_{ij}^{G} \\ 0 & 0 & \mathbf{k}_{ji}^{E} & \mathbf{k}_{ij}^{F} & (\mathbf{k}_{jj}^{E} + \mathbf{k}_{ii}^{F}) & 0 \\ 0 & 0 & 0 & \mathbf{k}_{ji}^{G} & 0 & \mathbf{k}_{jj}^{G} \end{bmatrix} \end{matrix} \qquad (2.35)$$

Here $\mathbf{0}$ denotes a null matrix.

The numbers 1–6 at the tops of the columns of the structure stiffness matrix are again there to identify nodes 1–6 of the structure. Since $\mathbf{k}_{ij} = \mathbf{k}_{ji}^{t}$ it is verified that $\mathbf{K}$ is a symmetric matrix.

In manually assembling the structure stiffness matrix of equation (2.35) the procedure starts with drawing a 6×6 gridwork of blank rectangular boxes (and heading the resulting rows and columns 1–6 to distinguish the structure nodes). Then one can proceed by considering each element in turn and progressively adding into the relevant boxes the stiffness contributions of all elements.

It is noticeable that the off-diagonal locations of the frame stiffness matrix contain only a single submatrix or a zero (i.e. null) submatrix. This is typical of an assemblage of line elements since, generally, if two nodal points are connected at all in such a structure it will be by a single element. As has been stated earlier the assembly procedure described here applies also to more complicated continuum structures formed of elements with more than two nodes. Then a non-zero off-diagonal term of a structure stiffness matrix will generally comprise contributions from more than one element stiffness matrix.

No mention has thus far been made of any boundary conditions of the frame of Fig. 2.8. If it is assumed, for instance, that points 5 and 6 are fully fixed, then all displacements at these points are zero. Consequently, rows and columns corresponding to joints 5 and 6 could be eliminated from $\mathbf{K}$ to yield $\mathbf{K}_r$. A better procedure if these boundary conditions apply would be to assemble matrix $\mathbf{K}_r$ directly as a 4×4 assembly of submatrices corresponding to joints 1–4 only.

Three examples are now given of the solution of problems involving simple assemblies of the unidirectional elements. It is perhaps pertinent to recall that the basic properties of the three types of element considered are such that any length of bar or beam structure can be represented structurally in 'exact' fashion by a single element so long as the cross-section is uniform throughout the length and no external loads are applied between the ends of the length. The basic procedure required to obtain a full solution is that summarized in Section 2.5.

Example 2.2. The bar shown in Fig. E2.2(a) is formed of three distinct sections, details of which are given in the figure. Twisting moments of 10 and 15 kN m are applied at the points of changing section. Calculate the rotation of the bar about its axis at points 2 and 3 and hence determine the reaction twisting moments at the ends of the bar and the distribution of torque along the bar ($G = 60 \text{ GN m}^{-2}$).

The discretization (or idealization) of the bar into elements is obvious, with one element used to represent each physical section of the bar. The elements are labelled A, B, and C.

The stiffness matrices for the individual elements are calculated using equation (2.17) and the given problem data. The result, in units of kilonewtons and metres, is

$$\mathbf{k}^A = 10^2 \begin{matrix} 1 & 2 \\ \begin{bmatrix} 12 & -12 \\ -12 & 12 \end{bmatrix} \end{matrix}$$

$$\mathbf{k}^B = 10^2 \begin{matrix} 2 & 3 \\ \begin{bmatrix} 9 & -9 \\ -9 & 9 \end{bmatrix} \end{matrix}$$

$$\mathbf{k}^C = 10^2 \begin{matrix} 3 & 4 \\ \begin{bmatrix} 6 & -6 \\ -6 & 6 \end{bmatrix} \end{matrix}.$$

The connection between the individual element nodes and the structure nodes is

Element   *i*   *j*   ← element nodes

$$\left. \begin{matrix} \text{A} & 1 & 2 \\ \text{B} & 2 & 3 \\ \text{C} & 3 & 4 \end{matrix} \right\} \leftarrow \text{structure node numbers}$$

and as well as being given in this 'connection' table the relationship between element and structure node numbering is also indicated above by providing suitable headings to the columns of the element stiffness matrices. In fact, doing this eliminates the need to record the connection table and it will consequently be dispensed with in the following examples.

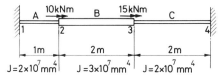

Fig. E2.2(a)

The full structure stiffness matrix $\mathbf{K}$ is easily assembled from $\mathbf{k}^A$, $\mathbf{k}^B$, and $\mathbf{k}^C$ by the direct stiffness method and is

$$\mathbf{K} = 10^2 \begin{array}{cccc} 1 & 2 & 3 & 4 \\ \begin{bmatrix} 12 & -12 & 0 & 0 \\ -12 & 12+9 & -9 & 0 \\ 0 & -9 & 9+6 & -6 \\ 0 & 0 & -6 & 6 \end{bmatrix} \end{array} = 10^2 \begin{array}{cccc} 1 & 2 & 3 & 4 \\ \begin{bmatrix} 12 & -12 & 0 & 0 \\ -12 & 21 & -9 & 0 \\ 0 & -9 & 15 & -6 \\ 0 & 0 & -6 & 6 \end{bmatrix} \end{array}.$$

The boundary conditions of the problem are that $\theta_{x1} = \theta_{x4} = 0$. Applying these conditions is equivalent to striking out the rows and columns of the stiffness matrix which correspond to these freedoms, i.e. rows and columns 1 and 4. This leaves the reduced stiffness matrix $\mathbf{K}_r$, of order 2×2 here, which relates the known externally applied 'forces' (twisting moments in this case) $\mathbf{F}_r$ to the corresponding 'displacements' (rotations in this case) $\mathbf{D}_r$. The column matrix $\mathbf{F}_r$ is clearly equal to $\{10 \quad 15\}$. Thus the general form of the stiffness equations after applying the boundary conditions

$$\mathbf{F}_r = \mathbf{K}_r\, \mathbf{D}_r$$

becomes here

$$\begin{Bmatrix} F_{\theta x2} \\ F_{\theta x3} \end{Bmatrix} = \begin{Bmatrix} 10 \\ 15 \end{Bmatrix} = 10^2 \begin{bmatrix} 21 & -9 \\ -9 & 15 \end{bmatrix} \begin{Bmatrix} \theta_{x2} \\ \theta_{x3} \end{Bmatrix}.$$

(It should be noted that $F_{\theta x2}$ is the externally applied force corresponding to the degree of freedom θ_{x2} etc.)

The solution of this pair of simultaneous equations is

$$\begin{Bmatrix} \theta_{x2} \\ \theta_{x3} \end{Bmatrix} = \frac{1}{1560} \begin{Bmatrix} 19 \\ 27 \end{Bmatrix} \text{rad} = \begin{Bmatrix} 0.01218 \\ 0.01731 \end{Bmatrix} \text{rad}$$

and these are the required values of rotation.

Now that the displacements are known we can back substitute to calculate in turn the reactions at points 1 and 4, using the rows of the structure stiffness matrix which were eliminated in obtaining $\mathbf{K}_r$, and the element forces, using again the individual element stiffness matrices.

The reaction twisting moments are

$$\begin{Bmatrix} F_{\theta x1} \\ F_{\theta x4} \end{Bmatrix} = 10^2 \begin{bmatrix} 12 & -12 & 0 & 0 \\ 0 & 0 & -6 & 6 \end{bmatrix} \begin{Bmatrix} 0 \\ 19/1560 \\ 27/1560 \\ 0 \end{Bmatrix} = \begin{Bmatrix} -190/13 \\ -135/13 \end{Bmatrix} \text{kN m.}$$

The forces (twisting moments) acting *on* the ends i and j of elements A are (see equation (2.17))

$$\mathbf{P}^A = \mathbf{k}^A \mathbf{d}^A$$

or

$$\begin{Bmatrix} M_{xi} \\ M_{xj} \end{Bmatrix}^A = 10^2 \begin{bmatrix} 12 & -12 \\ -12 & 12 \end{bmatrix} \begin{Bmatrix} \theta_{xi} \\ \theta_{xj} \end{Bmatrix}^A = 10^2 \begin{bmatrix} 12 & -12 \\ -12 & 12 \end{bmatrix} \begin{Bmatrix} 0 \\ 19/1560 \end{Bmatrix}$$

$$= \begin{Bmatrix} -190/13 \\ +190/13 \end{Bmatrix} \text{kN m.}$$

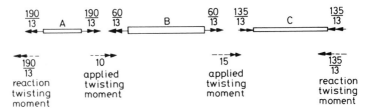

Fig. E2.2(b)

Similarly for elements B and C,

$$\left\{\begin{matrix}M_{xi}\\M_{xj}\end{matrix}\right\}^{B}=10^{2}\begin{bmatrix}9 & -9\\-9 & 9\end{bmatrix}\left\{\begin{matrix}19/1560\\27/1560\end{matrix}\right\}=\left\{\begin{matrix}-60/13\\+60/13\end{matrix}\right\}\text{kN m}$$

$$\left\{\begin{matrix}M_{xi}\\M_{xj}\end{matrix}\right\}^{C}=10^{2}\begin{bmatrix}6 & -6\\-6 & 6\end{bmatrix}\left\{\begin{matrix}27/1560\\0\end{matrix}\right\}=\left\{\begin{matrix}+135/13\\-135/13\end{matrix}\right\}\text{kN m.}$$

This concludes the calculations. For clarity the applied twisting moments and those calculated at the supports and at element ends are shown in Fig. E2.2(b) with their actual sense indicated. A glance at this figure confirms that the equilibrium condition is satisfied for each element, for each node, and for the structure as a whole. It should be noted that it was not really necessary to calculate the reaction twisting moments independently since these are indirectly available from the calculated element end moments. Finally, it is often useful to plot the variation of the force quantity along the structure and this is done here by means of the torque diagram shown in Fig. E2.2(c).

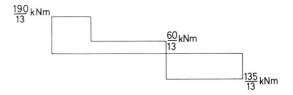

Fig. E2.2(c)

Example 2.3. The beam shown in Fig. E2.3(a) is of uniform cross-section having flexural rigidity $EI_z = 10\,000$ kN m^2. Calculate the displacements (lateral and rotational) at the free end and the rotational displacement at the prop support. Determine the distribution of bending moment along the beam.

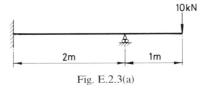

Fig. E.2.3(a)

To represent the beam the two-element idealization shown in Fig. E2.3(b) is used.

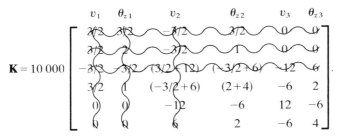

Fig. E.2.3(b)

The stiffness matrices for elements A and B are calculated using the general beam result given in equation (2.26). This gives (in units of kilonewtons and metres)

$$\mathbf{k}^A = 10\,000 \begin{array}{cccc} v_1 & \theta_{z1} & v_2 & \theta_{z2} \\ \begin{bmatrix} 3/2 & 3/2 & -3/2 & 3/2 \\ 3/2 & 2 & -3/2 & 1 \\ -3/2 & -3/2 & 3/2 & -3/2 \\ 3/2 & 1 & -3/2 & 2 \end{bmatrix} \end{array}$$

$$\mathbf{k}^B = 10\,000 \begin{array}{cccc} v_2 & \theta_{z2} & v_3 & \theta_{z3} \\ \begin{bmatrix} 12 & 6 & -12 & 6 \\ 6 & 4 & -6 & 2 \\ -12 & -6 & 12 & -6 \\ 6 & 2 & -6 & 4 \end{bmatrix} \end{array}.$$

It should be noted that the headings over the columns of the element matrices are again an indirect indication of the connection between nodes i and j of the individual elements and nodes 1, 2, and 3 of the structure. (For element A nodes i and j correspond to structure nodes 1 and 2 whilst for element B nodes i and j correspond to structure nodes 2 and 3.) Here, though, the column headings refer to individual degrees of freedom rather than simply node numbers, since in this case there is more than one freedom per node.

The structure stiffness matrix $\mathbf{K}$ is obtained using the direct stiffness method as

$$\mathbf{K} = 10\,000 \begin{array}{cccccc} v_1 & \theta_{z1} & v_2 & \theta_{z2} & v_3 & \theta_{z3} \\ \begin{bmatrix} 3/2 & 3/2 & -3/2 & 3/2 & 0 & 0 \\ 3/2 & 2 & -3/2 & 1 & 0 & 0 \\ -3/2 & -3/2 & (3/2+12) & (-3/2+6) & -12 & 6 \\ 3/2 & 1 & (-3/2+6) & (2+4) & -6 & 2 \\ 0 & 0 & -12 & -6 & 12 & -6 \\ 0 & 0 & 6 & 2 & -6 & 4 \end{bmatrix} \end{array}.$$

The full stiffness matrix is recorded here but it should be noted that since there is no requirement to calculate the structure reactions this is not really necessary. The geometric boundary conditions are that

$$v_1 = \theta_{z1} = v_2 = 0$$

so that the only non-zero degrees of freedom are θ_{z2}, v_3, and θ_{z3}. Consequently the application of the boundary conditions would result in striking out the first three rows and columns of the structure stiffness matrix to obtain $\mathbf{K}_r$. Obviously some effort would be saved by simply setting up only $\mathbf{K}_r$ straight away and ignoring the zero degrees of freedom v_1, θ_{z1}, and v_2 in the structure matrix. In the matrix $\mathbf{K}$ given above this is indicated by the deletion of the unnecessary terms, leaving $\mathbf{K}_r$ as the lower right-hand 3×3 matrix.

The column matrix of external applied loads is†

$$\mathbf{F}_r = \{F_{\theta z 2} \quad F_{v3} \quad F_{\theta z 3}\} = \{0 \quad 10 \quad 0\}$$

where only the force quantities corresponding to non-zero displacements are included.

The structure stiffness equations of the form $\mathbf{F}_r = \mathbf{K}_r \, \mathbf{D}_r$ become

$$\begin{Bmatrix} 0 \\ 10 \\ 0 \end{Bmatrix} = 10\,000 \begin{bmatrix} 6 & -6 & 2 \\ -6 & 12 & -6 \\ 2 & -6 & 4 \end{bmatrix} \begin{Bmatrix} \theta_{z2} \\ v_3 \\ \theta_{z3} \end{Bmatrix}. \tag{a}$$

These equations now have to be solved to give the values of the nodal displacements $\mathbf{D}_r = \{\theta_{z2} \ v_3 \ \theta_{z3}\}$. Symbolically the solution is expressed as

$$\mathbf{D}_r = \mathbf{K}_r^{-1} \mathbf{F}_r$$

but it is quicker to solve the equations directly rather than to proceed via an inversion of the stiffness matrix. The solution is easily found to be

$$\{\theta_{z2} \quad v_3 \quad \theta_{z3}\} = \frac{1}{10\,000} \left\{ 5 \quad \frac{50}{6} \quad 10 \right\}$$

or

$$\theta_{z2} = 0.0005 \text{ rad}, \quad v_3 = 0.000833 \text{ m}, \quad \text{and} \quad \theta_{z3} = 0.001 \text{ rad}.$$

The element end forces can be found by back substituting the now known values of the displacements into the element stiffness relationships typified by equation (2.26). For element A, the stiffness matrix $\mathbf{k}^A$ is recorded above and the nodal displacements are

$$\mathbf{d}^A = \begin{Bmatrix} v_i \\ \theta_{zi} \\ v_j \\ \theta_{zj} \end{Bmatrix}^A = \begin{Bmatrix} v_1 \\ \theta_{z1} \\ v_2 \\ \theta_{z2} \end{Bmatrix} = \frac{1}{10\,000} \begin{Bmatrix} 0 \\ 0 \\ 0 \\ 5 \end{Bmatrix}$$

Therefore element A nodal forces are

$$\begin{Bmatrix} V_i \\ M_{zi} \\ V_j \\ M_{zj} \end{Bmatrix}^A = 10\,000 \begin{bmatrix} 3/2 & 3/2 & -3/2 & 3/2 \\ 3/2 & 2 & -3/2 & 1 \\ -3/2 & -3/2 & 3/2 & -3/2 \\ 3/2 & 1 & -3/2 & 2 \end{bmatrix} \frac{1}{10\,000} \begin{Bmatrix} 0 \\ 0 \\ 0 \\ 5 \end{Bmatrix} = \begin{Bmatrix} 7.5 \text{ kN} \\ 5 \text{ kN m} \\ -7.5 \text{ kN} \\ 10 \text{ kN m} \end{Bmatrix}$$

Similarly, element B nodal forces are

$$\begin{Bmatrix} V_i \\ M_{zi} \\ V_j \\ M_{zj} \end{Bmatrix}^B = 10\,000 \begin{bmatrix} 12 & 6 & -12 & 6 \\ 6 & 4 & -6 & 2 \\ -12 & -6 & 12 & -6 \\ 6 & 2 & -6 & 4 \end{bmatrix} \frac{1}{10\,000} \begin{Bmatrix} 0 \\ 5 \\ 50/6 \\ 10 \end{Bmatrix} = \begin{Bmatrix} -10 \text{ kN} \\ -10 \text{ kN m} \\ 10 \text{ kN} \\ 0 \text{ kN m} \end{Bmatrix}.$$

---

† Note that the braces { } always denote a column matrix but that to save space column matrices are often written across a page.

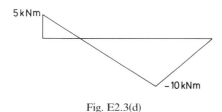

Fig. E2.3(c)

Fig. E2.3(c) shows the two beam elements and the forces, as just calculated, acting on their ends. Each element is easily seen to be in overall equilibrium, as expected, and the calculated values of shearing force and bending moment at the right-hand end of the complete beam match the known conditions at this point of 10 kN applied vertical force and zero bending moment. At the left-hand end element A has a downward vertical force of 7.5 kN and a clockwise moment of 5 kN m acting on it; these act on the beam from the wall and are therefore the values of the reaction force and moment at the built-in support. At the internal node an upward force of 7.5 kN acts on the right-hand end of element A and a force of 10 kN acts in the same sense on the left-hand end of element B. Therefore the reaction force of the prop is 17.5 kN upwards.

The bending moment at any point between the ends of an element can readily be found from simple equilibrium considerations or simply from the knowledge that the bending moment varies linearly along an element length; the bending moment diagram is given in Fig. E2.3(d). It should be noted that if information were required about the beam deflection at other than the nodal points this could be calculated using the expression for v given by equations (2.22) and (2.23).

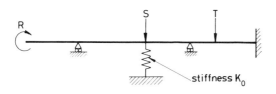

Fig. E2.3(d)

Example 2.4. Assemble the reduced structure stiffness matrix $\mathbf{K}_r$ for the structure shown in Fig. E2.4(a), working symbolically in terms of structure coefficients k_{11} etc. Also write down the column matrix of loads $\mathbf{F}_r$.

Fig. E2.4(a)

The element idealization of the structure is shown in Fig. E2.4(b). The continuous beam is represented by five beam elements, labelled A to E. The spring is represented by element F which has a stiffness relationship of exactly the same form as an axially loaded bar with the stiffness K_0 taking the place of AE/l;

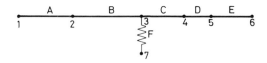

Fig. E2.4(b)

the spring thus simply provides a stiffness K_0 associated with the vertical displacement v_3 and does not contribute to rotational stiffness at point 3.

The problem has eight non-zero degrees of freedom and these are

$$\mathbf{D}_r = \{v_1 \quad \theta_{z1} \quad \theta_{z2} \quad v_3 \quad \theta_{z3} \quad \theta_{z4} \quad v_5 \quad \theta_{z5}\}$$

In assembling $\mathbf{K}_r$ directly—that is without forming $\mathbf{K}$ first and then applying the boundary conditions—the stiffness contributions of the individual elements are as follows. In recording these, only the stiffness coefficients which actually add into $\mathbf{K}_r$ are noted and the headings over the columns again denote structure degrees of freedom and hence effectively specify the location of the element contributions into the structure stiffness matrix. The coefficients k_{11} through to k_{44} obviously represent symbolically the coefficients of a beam element defined explicitly in equation (2.26).

$$
\mathbf{k}^A =
\begin{array}{cccc}
v_1 & \theta_{z1} & v_2 & \theta_{z2} \\
\end{array}
\begin{bmatrix}
k_{11}^A & k_{12}^A & & k_{14}^A \\
k_{21}^A & k_{22}^A & & k_{24}^A \\
 & & & \\
k_{41}^A & k_{42}^A & & k_{44}^A
\end{bmatrix}
\qquad
\mathbf{k}^B =
\begin{array}{cccc}
v_2 & \theta_{z2} & v_3 & \theta_{z3} \\
\end{array}
\begin{bmatrix}
 & & & \\
 & k_{22}^B & k_{23}^B & k_{24}^B \\
 & k_{32}^B & k_{33}^B & k_{34}^B \\
 & k_{42}^B & k_{43}^B & k_{44}^B
\end{bmatrix}
$$

$$
\mathbf{k}^C =
\begin{array}{cccc}
v_3 & \theta_{z3} & v_4 & \theta_{z4} \\
\end{array}
\begin{bmatrix}
k_{11}^C & k_{12}^C & & k_{14}^C \\
k_{21}^C & k_{22}^C & & k_{24}^C \\
 & & & \\
k_{41}^C & k_{42}^C & & k_{44}^C
\end{bmatrix}
\qquad
\mathbf{k}^D =
\begin{array}{cccc}
v_4 & \theta_{z4} & v_5 & \theta_{z5} \\
\end{array}
\begin{bmatrix}
 & & & \\
 & k_{22}^D & k_{23}^D & k_{24}^D \\
 & k_{32}^D & k_{33}^D & k_{34}^D \\
 & k_{42}^D & k_{43}^D & k_{44}^D
\end{bmatrix}
$$

$$
\mathbf{k}^E =
\begin{array}{cccc}
v_5 & \theta_{z5} & v_6 & \theta_{z6} \\
\end{array}
\begin{bmatrix}
k_{11}^E & k_{12}^E & & \\
k_{21}^E & k_{22}^E & & \\
 & & & \\
 & & &
\end{bmatrix}
\qquad
\mathbf{k}^F =
\begin{array}{cc}
v_3 & v_7 \\
\end{array}
\begin{bmatrix}
K_0 & \\
 &
\end{bmatrix}
$$

The stiffness matrix $\mathbf{K}_r$ can be assembled by the direct stiffness procedure and is found to be

v_1	θ_{z1}	θ_{z2}	v_3	θ_{z3}	θ_{z4}	v_5	θ_{z5}
k_{11}^A	k_{12}^A	k_{14}^A	0	0	0	0	0
k_{21}^A	k_{22}^A	k_{24}^A	0	0	0	0	0
k_{41}^A	k_{42}^A	$(k_{44}^A+k_{22}^B)$	k_{23}^B	k_{24}^B	0	0	0
0	0	k_{32}^B	$(k_{33}^B+k_{11}^C+K_0)$	$(k_{34}^B+k_{12}^C)$	k_{14}^C	0	0
0	0	k_{42}^B	$(k_{43}^B+k_{21}^C)$	$(k_{44}^B+k_{22}^C)$	k_{24}^C	0	0
0	0	0	k_{41}^C	k_{42}^C	$(k_{44}^C+k_{22}^D)$	k_{23}^D	k_{24}^D
0	0	0	0	0	k_{32}^D	$(k_{33}^D+k_{11}^E)$	$(k_{34}^D+k_{12}^E)$
0	0	0	0	0	k_{42}^D	$(k_{43}^D+k_{21}^E)$	$(k_{44}^D+k_{22}^E)$

The column matrix of external loads is

$$\mathbf{F}_r = \{0 \quad R \quad 0 \quad S \quad 0 \quad 0 \quad T \quad 0\}$$

2.10. Symmetrical geometry

Symmetrical structures are very common and where they occur consideration should be given to their properties with the aim of reducing the number of degrees of freedom n involved in obtaining a solution. This is of considerable importance, since a high proportion of the effort required to obtain a solution in the displacement method—whether manually or with a computer—is involved in solving a set of simultaneous equations equal in number to n. As far as the computer is concerned the time taken to solve n equations is perhaps proportional to n^3 or n^4.

The most useful and general way of taking advantage of symmetry is to visualize the symmetric portion of a structure as a separate distinct structure and to apply appropriate boundary conditions at the line (or lines) of symmetry. To illustrate the technique consider the simple (statically determinate) beam problem shown in Fig. 2.9(a). Here, not only the structure but also the loading is symmetric and consequently the deflected form of the uniform beam, shown by the broken line in the figure, will be perfectly symmetric. This means that deflections in the right-hand half of the beam equal those in the left-hand half whilst rotations are equal in magnitude but opposite in sign in the two halves; in particular the beam rotation at the centre line is clearly zero. These conditions in the left-hand half of the beam are equivalent to the situation shown in Fig. 2.9(b) and it is only this replacement structure which need be analysed by the displacement method. In the replacement structure the conditions on the displacement quantities at node 3—located at the centre line of the original structure—are that $v_3 \neq 0$, $\theta_{z3} = 0$; this can be viewed physically as a vertical slider support. Once the replacement structure is physically defined the analysis of it proceeds in the same fashion as it if were an

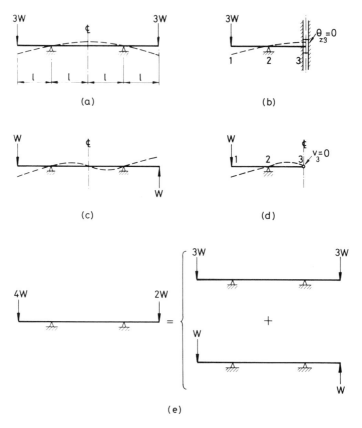

Fig. 2.9. Symmetric and antisymmetric beam problems.

actual structure and only at the end of the analysis need we recall that another half of the beam exists to which the calculated displacements and internal forces can be directly referred.

The application of symmetry conditions in the problem shown in Fig. 2.9(a) reduces the value of n from 6 in the actual structure to 4 in the replacement structure shown in Fig. 2.9(b). These degrees of freedom are $\mathbf{D}_r = \{v_1 \; \theta_{z1} \; \theta_{z2} \; v_3\}$ and the corresponding applied forces are $\mathbf{F}_r = \{3W \; 0 \; 0 \; 0\}$. The assembly of $\mathbf{K}_r$ for the replacement structure proceeds in the usual fashion, as outlined earlier. It should be noted that the magnitude of the reaction moment at point 3 of the replacement structure—which can be calculated by the usual MDM procedure—is of course equal to the value of bending moment at the centre of the actual beam. (The reaction vertical force at point 3 corresponds to the shear force value at the beam centre and is zero since v_3 is not prescribed and no applied load is present.)

In the structure of Fig. 2.9(a) no applied load acts at the axis of symmetry. If such a load was present then it is quite apparent that half of this load should be applied at point 3 of the replacement structure (Fig. 2.9(b)). Further, if there were some sort of structural member lying along the symmetry axis—a vertical spring, say—then half of the member stiffness should be associated with the degree of freedom v_3 of the replacement structure.

So far we have considered only the application of symmetrical loading, but geometrical symmetry can also be exploited for antisymmetrical loading and for general loading.

An example of antisymmetrical loading is shown in Fig. 2.9(c) where the structure itself is the same simple beam just considered. In this situation the deflected form is antisymmetric so that the deflections are equal in magnitude but opposite in sign in the two halves of the beam, and the slopes are equal in the two halves. In particular the deflection at the centre line is zero and also there is a point of contraflexure (i.e. zero bending moment) at the centre. A valid replacement structure is therefore that shown in Fig. 2.9(d) where conditions at point 3 are effectively equivalent to a pin fixed in position; n again is equal to 4, the corresponding freedoms now being $\mathbf{D}_r = \{v_1 \quad \theta_{z1} \quad \theta_{z2} \quad \theta_{z3}\}$ and the applied forces are $\mathbf{F}_r = \{W \quad 0 \quad 0 \quad 0\}$.

A general loading can be accommodated by using the principle of superposition to express the fact that the solution of the general load problem is the summation of the solutions of symmetrical load and antisymmetrical load problems. Fig. 2.9(e) illustrates this for the beam problem. The total solution clearly involves the solution of both the problems illustrated in Fig. 2.9(b) and Fig. 2.9(d). Although the actual problem with six degrees of freedom is now replaced with two problems each having four degrees of freedom, the procedure still has much to recommend it and its advantages would be accentuated for a structural problem having many freedoms.

The basic idea of dealing with symmetric structures has been introduced here with reference to a very simple structure. The philosophy, however, applies to much more complicated structures, including skeletal frames in the MDM and two- and three-dimensional continua in the FEM. Indeed, the use of available geometrical symmetry becomes more important of course as the complexity of the problem—and hence the number of its freedoms—increases. The proportion of an actual structure that need be considered as a replacement structure will obviously depend on the particular structure and its loading; it may be one half, as for the above beam, or a quarter or an eighth or some other proportion. Some further examples of the use of structural symmetry will occur during the text.

Before leaving the topic of structural symmetry, it is noted that an alternative method of taking advantage of structural symmetry is sometimes useful in the *manual* solution of small-order problems. In this second method the symmetry conditions are introduced by directly modifying the stiffness equations pertaining to the whole structure. The method is less general than that outlined above but does sometimes result in a set of equations of a slightly lower order and this obviously can be important when the solution is made manually. The following example will demonstrate this alternative approach as well as the more general one.

Example 2.5. The central portion of the beam shown in Fig. E2.5(a) has a second moment of area of $50 \times 10^6 \text{ mm}^4$ whilst the outer portions have second moments of area of $25 \times 10^6 \text{ mm}^4$. Young's modulus E is 200 GN m^{-2}. Determine the

Fig. E2.5(a)

deflection and rotation of the beam under the applied loads and draw the bending moment diagram. Take advantage of the problem symmetry to obtain the solution by
(a) analysing the symmetric half structure
(b) suitably modifying the stiffness equations of the whole structure.

(a) The symmetric half-structure considered is the left-hand portion of the beam; this portion is idealized with two elements as shown in Fig. E2.5(b) and node 3 corresponds to the centre of the beam so that the length of element B in this figure is, of course, $2m$.

Fig. E2.5(b)

The boundary conditions are that $v_1 = \theta_{z1} = \theta_{z3} = 0$, with the latter condition expressing the symmetry conditions at the beam centre. Correspondingly the problem degrees of freedom and the applied loads are

$$\mathbf{D}_r = \{v_2 \quad \theta_{z2} \quad v_3\} \quad \text{and} \quad \mathbf{F}_r = \{20 \quad 0 \quad 0\}$$

The stiffness matrices for the two elements, based on equation (2.26) and using units of kilonewtons and metres, are

$$
\mathbf{k}^A = \begin{matrix} & v_1 & \theta_{z1} & v_2 & \theta_{z2} \\ \begin{bmatrix} 7500 & 7500 & -7500 & 7500 \\ 7500 & 10000 & -7500 & 5000 \\ -7500 & -7500 & 7500 & -7500 \\ 7500 & 5000 & -7500 & 10000 \end{bmatrix} \end{matrix}
$$

$$
\mathbf{k}^B = \begin{matrix} & v_2 & \theta_{z2} & v_3 & \theta_{z3} \\ \begin{bmatrix} 15000 & 15000 & -15000 & 15000 \\ 15000 & 20000 & -15000 & 10000 \\ -15000 & -15000 & 15000 & -15000 \\ 15000 & 10000 & -15000 & 20000 \end{bmatrix} \end{matrix}.
$$

By the usual direct stiffness procedure the matrix $\mathbf{K}_r$ can be assembled easily and the set of stiffness equations for the structure, after boundary conditions have been applied, becomes

$$
\begin{Bmatrix} 20 \\ 0 \\ 0 \end{Bmatrix} = 7500 \begin{matrix} v_2 & \theta_{z2} & v_3 \\ \begin{bmatrix} 3 & 1 & -2 \\ 1 & 4 & -2 \\ -2 & -2 & 2 \end{bmatrix} \end{matrix} \begin{Bmatrix} v_2 \\ \theta_{z2} \\ v_3 \end{Bmatrix}.
$$

Solving for the nodal displacements gives

$$
\mathbf{D}_r = \{v_2 \quad \theta_{z2} \quad v_3\} = \frac{1}{375}\{2\,\text{m} \quad 1\,\text{rad} \quad 3\,\text{m}\}.
$$

The nodal forces for each element are calculated from

$$
\mathbf{P}^A = \mathbf{k}^A \mathbf{d}^A \qquad \mathbf{P}^B = \mathbf{k}^B \mathbf{d}^B
$$

where

$$
\mathbf{d}^A = \frac{1}{375}\{0 \quad 0 \quad 2 \quad 1\} \qquad \mathbf{d}^B = \frac{1}{375}\{2 \quad 1 \quad 3 \quad 0\}.
$$

The result is

$$
\mathbf{P}^A = \begin{Bmatrix} V_i \\ M_{zi} \\ V_j \\ M_{zj} \end{Bmatrix}^A = \begin{Bmatrix} -20\,\text{kN} \\ -80/3\,\text{kN m} \\ 20\,\text{kN} \\ -40/3\,\text{kN m} \end{Bmatrix} \qquad \mathbf{P}^B = \begin{Bmatrix} V_i \\ M_{zi} \\ V_j \\ M_{zj} \end{Bmatrix}^B = \begin{Bmatrix} 0 \\ 40/3\,\text{kN m} \\ 0 \\ -40/3\,\text{kN m} \end{Bmatrix}
$$

and from this it is a simple matter to construct the bending moment diagram. This diagram is shown for the whole beam in Fig. E2.5(c).

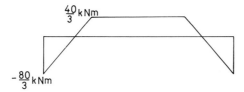

Fig. E2.5(c)

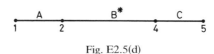

Fig. E2.5(d)

(b) The second way of taking advantage of the symmetry of a problem starts by setting up the stiffness equations for the complete structure. The element idealization is shown in Fig. E2.5(d) and consists of three elements. It should be noted that the element covering the whole of the central portion of the beam is labelled B* simply to avoid confusion with element B used in part (a) which covers only half the central portion; also in Fig. E2.5(d) the nodes are numbered 1, 2, 4, 5 with 3 omitted simply to avoid confusion with node 3 of Fig. E2.5(a) which does not have a counterpart in Fig. E2.5(d).

The stiffness matrix for element A is exactly the same as that given for element A of part (a). Also, the stiffness matrix for element C is equal to that for A, as long as it is borne in mind that the column headings for element C should read v_4, θ_{z4}, v_5, θ_{z5} in order. For element B*, whose length is 4 m,

$$
\mathbf{k}^{B^*} = \begin{array}{c}
\begin{array}{cccc} \quad v_2 \quad & \quad \theta_{z2} \quad & \quad v_4 \quad & \quad \theta_{z4} \end{array} \\
\left[\begin{array}{cccc}
1875 & 3750 & -1875 & 3750 \\
3750 & 10000 & -3750 & 5000 \\
-1875 & -3750 & 1875 & 3750 \\
3750 & 5000 & -3750 & 10000
\end{array} \right]
\end{array}
$$

by use, again, of equation (2.26).

The boundary conditions are that $v_1 = \theta_{z1} = v_5 = \theta_{z5} = 0$ so that the problem degrees of freedom are $\mathbf{D}_r = \{v_2 \; \theta_{z2} \; v_4 \; \theta_{z4}\}$ with the corresponding column matrix of applied loads defined as

$$\mathbf{F}_r = \{20 \quad 0 \quad 20 \quad 0\}.$$

The stiffness matrix $\mathbf{K}_r$ is assembled from $\mathbf{k}^A$, $\mathbf{k}^{B^*}$ and $\mathbf{k}^C$ in the usual way and the set of structure equations $\mathbf{F}_r = \mathbf{K}_r \mathbf{D}_r$ becomes

$$
\begin{Bmatrix} 20 \\ 0 \\ 20 \\ 0 \end{Bmatrix} =
\begin{bmatrix}
9375 & -3750 & -1875 & 3750 \\
-3750 & 20000 & -3750 & 5000 \\
-1875 & -3750 & 9375 & 3750 \\
3750 & 5000 & 3750 & 20000
\end{bmatrix}
\begin{Bmatrix} v_2 \\ \theta_{z2} \\ v_4 \\ \theta_{z4} \end{Bmatrix}.
$$

At this stage the symmetry conditions can be taken into account for it is clear that

$$v_4 = v_2 \quad \text{and} \quad \theta_{z4} = -\theta_{z2}.$$

Bearing in mind these relationships it can also be seen that the third of the system of structure equations is the same as the first and the fourth is the same as the second. Thus only the first two equations need to be considered (and the other two need not have been set up) and effectively only two unknowns are present.

The pertinent equations are then

$$\left\{\begin{matrix} 20 \\ 0 \end{matrix}\right\} = \left[\begin{matrix} 9375-1875 & -3750-3750 \\ -3750-3750 & 20000-5000 \end{matrix}\right]\left\{\begin{matrix} v_2 \\ \theta_{z2} \end{matrix}\right\} = 7500\left[\begin{matrix} 1 & -1 \\ -1 & 2 \end{matrix}\right]\left\{\begin{matrix} v_2 \\ \theta_{z2} \end{matrix}\right\}.$$

(It should be noted that although the stiffness matrix is symmetrical in this case, this is not generally so when using the approach described here.)

The solution is

$$v_2(=v_4) = 2/375 \text{ m} \qquad\qquad \theta_{z2}(=-\theta_{z4}) = 1/375 \text{ rad}$$

which confirms the solution found in (a). The deflection at the centre of the beam could easily be found if desired by using equations (2.22) and (2.23) once the nodal forces of element B* were calculated. It is left to the reader to verify that these nodal forces for element B* are

$$\{V_i \quad M_{zi} \quad V_j \quad M_{zj}\}^{B^*} = \{0 \quad 40/3 \text{ kN m} \quad 0 \quad -40/3 \text{ kN m}\}.$$

The nodal forces for element A are as those for element A of part (a) and the forces for element C follow from those of element A by symmetry. The bending moment diagram given in Fig. E2.5(c) is confirmed as being valid.

In the above problem there are four degrees of freedom if the obvious symmetry is ignored. In solution (a) the problem reduces to one of three degrees of freedom by considering the half structure; this requires the creation of a node at the axis of symmetry (i.e. at node 3 of Fig. E2.5(b)), which would not otherwise be required, so that the correct boundary conditions can be applied. In solution (b) the problem reduces effectively to two degrees of freedom when modifying the stiffness equations of the whole structure. Solution (b) uses less degrees of freedom than does solution (a) in this problem because of the necessity of providing a node at the centre of the beam in solution (a) merely for the purpose of specifying that the slope is zero there. If a node had to be positioned at the centre for any other reason—because a load acted there or because other elements were connected to the beam there—then the same number of freedoms would be involved in solutions (a) and (b).

Solution (a) is an example of the most general and usually most satisfactory approach to the solution of a symmetric probem and is suitable for manual or computer operation. Solution (b) is really only suited to the manual solution of small-order problems, but it can, as in this example, sometimes involve less freedoms than does solution (a) and hence has its uses in this category of problems.

2.11. Further remarks on prescribed displacements

So far the only prescribed displacements considered have been those zero displacements which correspond to unyielding boundary supports. This is far and away the most common type of problem but there do sometimes arise problems in which one or more non-zero displacements are pre-scribed, as for example when a support sinks a given amount, and this situation needs to be considered.

The complete set of structure stiffness equations is

$$\mathbf{F} = \mathbf{KD}.$$

Now out of the total set of structure freedoms $\mathbf{D}$ there will in general be a subset $\mathbf{D}_r$ which are unknown, a subset $\mathbf{D}_s$ which are prescribed non-zero, and a subset $\mathbf{D}_t$ which are prescribed zero. The complete set of equations could be expressed in the partitioned form

$$\begin{Bmatrix} \mathbf{F}_r \\ \mathbf{F}_s \\ \mathbf{F}_t \end{Bmatrix} = \begin{bmatrix} \mathbf{K}_{rr} & \mathbf{K}_{rs} & \mathbf{K}_{rt} \\ \mathbf{K}_{sr} & \mathbf{K}_{ss} & \mathbf{K}_{st} \\ \mathbf{K}_{tr} & \mathbf{K}_{ts} & \mathbf{K}_{tt} \end{bmatrix} \begin{Bmatrix} \mathbf{D}_r \\ \mathbf{D}_s \\ \mathbf{D}_t = \mathbf{0} \end{Bmatrix} \tag{2.36}$$

where, from the symmetry of the matrix, $\mathbf{K}_{sr} = \mathbf{K}_{rs}^t$ etc. Of course some rearrangement of the orders of rows and columns in $\mathbf{K}$ would normally be necessary if this result, with all the displacements of each subset grouped together, were to be achieved. (In fact the rearrangement may not be necessary or desirable in practice but it is convenient to imagine it for the purposes of the following explanation.) Of the forces occuring in equation (2.36) the $\mathbf{F}_r$ forces correspond to the unknown displacements $\mathbf{D}_r$ and will be prescribed, whilst both the $\mathbf{F}_s$ and $\mathbf{F}_t$ forces correspond to prescribed displacements and will be unknown.

If, as has been the case thus far, there are no prescribed non-zero displacements $\mathbf{D}_s$ (or corresponding forces $\mathbf{F}_s$), then from equation (2.36) we obtain the reduced set of stiffness equations

$$\mathbf{F}_r = \mathbf{K}_{rr}\mathbf{D}_r \tag{2.37}$$

which can be solved—as long as sufficient boundary conditions have been applied to make $\mathbf{K}_{rr}$ non-singular—to yield the displacements $\mathbf{D}_r$. In formal terms the solution is expressed as

$$\mathbf{D}_r = \mathbf{K}_{rr}^{-1}\mathbf{F}_r \tag{2.38}$$

and this is the same form as seen earlier with $\mathbf{K}_{rr}$ identical to the usual $\mathbf{K}_r$. If required, the reaction forces $\mathbf{F}_t$ can then be found from

$$\mathbf{F}_t = \mathbf{K}_{tr}\mathbf{D}_r. \tag{2.39}$$

As we have seen, if the reaction forces are not directly required it is only necessary to assemble the reduced stiffness matrix $\mathbf{K}_{rr}$ corresponding to the non-zero freedoms and in practice the solution for $\mathbf{D}_r$ can be obtained by directly attacking equations (2.37) rather than by the formal solution involving the determination of the inverse matrix which is indicated by equation (2.38).

Now, proceeding to the case where there are one or more prescribed non-zero displacements the first set of relationships of equation (2.36) becomes

$$\mathbf{F}_r = \mathbf{K}_{rr}\mathbf{D}_r + \mathbf{K}_{rs}\mathbf{D}_s. \tag{2.40}$$

If the terms on both sides of this equation are premultiplied by $\mathbf{K}_{rr}^{-1}$ to give

$$\mathbf{K}_{rr}^{-1}\mathbf{F}_r = \mathbf{K}_{rr}^{-1}\mathbf{K}_{rr}\mathbf{D}_r + \mathbf{K}_{rr}^{-1}\mathbf{K}_{rs}\mathbf{D}_s$$

then, since the product of a matrix and its inverse is a unit matrix, it can be seen that the unprescribed displacements are given by

$$\mathbf{D}_r = \mathbf{K}_{rr}^{-1}(\mathbf{F}_r - \mathbf{K}_{rs}\mathbf{D}_s). \tag{2.41}$$

With these displacements determined the forces $\mathbf{F}_s$ and $\mathbf{F}_t$ could then be found, if required, from the second and third sets of relationships of equation (2.36) as

$$\mathbf{F}_s = \mathbf{K}_{sr}\mathbf{D}_r + \mathbf{K}_{ss}\mathbf{D}_s \tag{2.42a}$$

$$\mathbf{F}_t = \mathbf{K}_{tr}\mathbf{D}_r + \mathbf{K}_{ts}\mathbf{D}_s. \tag{2.42b}$$

Equation (2.41), like its counterpart equation (2.38) for the conventional problem, is a formal statement of the solution procedure for the unprescribed displacements. Again, though, the actual solution need not proceed strictly in the form indicated and equation (2.41) can be regarded simply as symbolizing the solution of equation (2.40) for the $\mathbf{D}_r$. In a manual solution this means that we set up the system of structure equations corresponding to all the non-zero freedoms (both $\mathbf{D}_r$ and $\mathbf{D}_s$,) extract from this system the equations for the forces $\mathbf{F}_r$ of the form of equation (2.40), and solve these for $\mathbf{D}_r$. The remaining equations for this system will then give the forces corresponding to the prescribed non-zero displacements, i.e. will give $\mathbf{F}_s$ as in equation (2.42a). If the reaction forces $\mathbf{F}_t$ are required these can be determined indirectly from element end forces. The example which follows will illustrate the procedure.

The above description of the way in which conditions of prescribed (zero and/or non-zero) displacements can be accommodated gives a formal viewpoint of the procedure which could be used in obtaining solutions either manually or with a computer. However, it should be noted that the procedure as described is not terribly efficient for computer operation since it involves considerable rearrangement of the full system of equations which is both awkward and time consuming. In practice, therefore, other less formal methods which avoid excessive rearrangement are often adopted for computer use; these methods replace the individual equilibrium equations corresponding to those degrees of freedom which have prescribed values by dummy equations which enforce the required displacement-type boundary conditions.

Example 2.6. For the beam shown in Fig. E2.3(a) remove the 10 kN load and instead apply a displacement of 5/6 mm downwards at the free end. As in Example 2.3 take $EI_z = 10\,000$ kN m^2. Calculate the rotations at the prop and at the free end, and determine the downward force at the free end which corresponds to the prescribed displacement.

The two-element idealization shown in Fig. E2.3(b) is appropriate for this problem. The unprescribed degrees of freedom are θ_{z2} and θ_{z3} and the prescribed non-zero degree of freedom is v_3.

The element stiffness matrices have already been calculated in Example 2.3 as has the stiffness matrix for the structure after the zero-displacement boundary conditions $(v_1 = \theta_{z1} = v_2 = 0)$ are applied. The latter stiffness matrix occurs in equation (a) of Example 2.3 and can be used here. The structure stiffness relationships, corresponding to all non-zero freedoms, are for the present problem

$$\begin{Bmatrix} 0 \\ F_{v3} \\ 0 \end{Bmatrix} = 10^4 \begin{bmatrix} 6 & -6 & 2 \\ -6 & 12 & -6 \\ 2 & -6 & 4 \end{bmatrix} \begin{Bmatrix} \theta_{z2} \\ v_3 = 5/6000 \\ \theta_{z3} \end{Bmatrix} \tag{i}$$

Extracting the first and third of equations (i) we have that

$$\begin{Bmatrix} 0 \\ 0 \end{Bmatrix} = 10^4 \begin{bmatrix} 6 & 2 \\ 2 & 4 \end{bmatrix} \begin{Bmatrix} \theta_{z2} \\ \theta_{z3} \end{Bmatrix} + 10^4 \begin{bmatrix} -6 \\ -6 \end{bmatrix} \{5/6000\}$$

which is of the form

$$\mathbf{F_r = K_{rr}D_r + K_{rs}D_s}$$

as in equation (2.40). The above two equations can be solved easily to give

$$\theta_{z2} = 0.0005 \text{ rad} \qquad\qquad \theta_{z3} = 0.001 \text{ rad}.$$

The remaining equation of equations (i) is

$$F_{v3} = 10^4[-6 \quad -6] \begin{Bmatrix} \theta_{z2} \\ \theta_{z3} \end{Bmatrix} + 10^4[12]\{5/6000\}$$

which is of the form

$$\mathbf{F_s = K_{sr}D_r + K_{ss}D_s}$$

as in equation (2.42a). The result, on substituting the now known values of θ_{z2} and θ_{z3}, is

$$F_{v3} = 10 \text{ kN}$$

On comparing with Example 2.3 it can be seen that these solutions are verified; in Example 2.3 a load of 10 kN was applied at the free end and the calculated displacement under the load was found to be 5/6 mm whereas here it is the displacement that has been prescribed and it is seen that the calculated corresponding load is 10 kN.

If element end forces, bending moment distributions, etc. were now required the procedure would be precisely as described in Example 2.3 of course.

Problems

2.1. Determine the axial displacements at the load points of the stepped bar shown in Fig. P2.1. What are the axial forces in the three sections of the bar? $E = 200 \text{ GN m}^{-2}$.

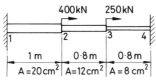

Fig. P2.1

2.2. The uniform beam 1–2–3 shown in Fig. P2.2 has $EI_z = 8 \text{ MN m}^2$. Calculate the rotation at point 2 and hence find all the support reactions. Draw the bending moment diagram.

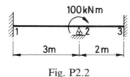

Fig. P2.2

2.3. Determine the vertical displacement and the rotation at the load point of the beam shown in Fig. P2.3. Hence find the vertical reaction and the fixing moment at the left-hand support. Draw the bending moment diagram.

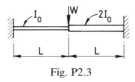

Fig. P2.3

2.4. A uniform horizontal cantilever beam of flexural rigidity EI_z and length L is supported at its tip by a spring which provides a vertical stiffness equal to $3EI_z/L^3$. Determine the tip deflection and rotation, and the moment at the clamped end when a vertical load P acts at the tip.

2.5. Find the rotations at the simple supports of the prismatic beam of Fig. P2.5. Also find the vertical reaction and fixing moment at the clamped end and draw the bending moment diagram. $EI_z = 50 \text{ MN m}^2$.

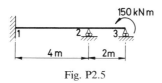

Fig. P2.5

2.6. The stepped shaft shown in Fig. P2.6 is fully restrained against rotation about its axis at its ends and is subjected to the twisting moments shown. The shaft is made from the same material throughout and has solid circular cross-section with outside diameter of 100 mm in regions AB and DE and of 200 mm in region BCD. Determine the distribution of torque along the length of the shaft.

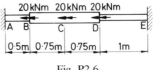

Fig. P2.6

2.7. Consider the torsion problem detailed in Example 2.2. Now replace the fully clamped support at node 4 by a torsional spring which opposes rotation with stiffness 1 MN m rad^{-1}. Calculate the rotations at nodes 2, 3 and 4 and draw a new torque diagram.

2.8. Determine the displacements (vertical, horizontal, and rotational) at the load point of the uniform beam shown in Fig. P2.8. Calculate the values of bending moment at the load point and the two ends of the beam. $E = 200 \text{ GN m}^{-2}$, $A = 40 \text{ cm}^2$, and $I_z = 2500 \text{ cm}^4$.

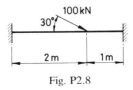

Fig. P2.8

2.9. Fig. P2.9 shows a uniform cantilever beam supported by springs at A and B. The spring stiffnesses are $14 \times 10^5 \text{ N m}^{-1}$ at A and $5 \times 10^5 \text{ N m}^{-1}$ at B. Show that the displacements under the given loading are $\{v_A \ \theta_{zA} \ v_B \ \theta_{zB}\} = 10^{-2}$ $\{3.169 \text{ m} \ 0.2523 \text{ rad} \ 9.941 \text{ m} \ 0.4818 \text{ rad}\}$. Determine the vertical reaction and fixing moment at the clamped end.

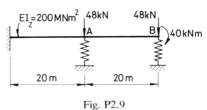

Fig. P2.9

2.10. The beam shown in Fig. P2.10 has symmetric geometry and loading. Taking advantage of this, calculate the displacements at nodes 2 and 3. Hence draw the bending moment diagram for the beam.

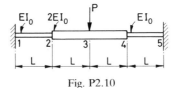

Fig. P2.10

2.11. Determine the central deflection and the rotations at the supports of the continuous beam shown in Fig. P2.11. What are the values of support reaction at the outside supports and of bending moment at the beam centre?. $EI_z =$ 30 MN m^2. (Note the symmetry of the problem).

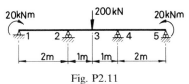

Fig. P2.11

2.12. For the symmetric bar shown in Fig. P2.12 determine the axial displacements at the load points and the axial forces in the various sections of the bar. The extensional rigidity is AE throughout.

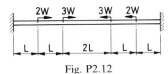

Fig. P2.12

2.13. Consider the beam shown in Fig. P2.10 but now replace the load P at node 3 with the same load at node 2. Calculate the displacements at nodes 2, 3, and 4 and draw the bending moment diagram. Perform the analysis in two stages, as typified by Fig. 2.9(e).

2.14. Repeat Problem 2.1 assuming that on application of the loads the support at point 4 gives axially by an amount 0.8 mm to the right.

2.15. Repeat Problem 2.11 assuming that during application of the loads both the internal supports sink vertically by an amount of 1 mm.

3

MATRIX DISPLACEMENT ANALYSIS OF PLANAR FRAMES WITH JOINT LOADS

3.1. Introduction

In Chapter 2 the basic philosophy of the matrix displacement method has been described and stiffness matrices for bar and beam elements have been developed and used in some simple one-dimensional applications. In this chapter the method is extended to deal with the analysis of planar frames. These skeletal structures are defined physically as two-dimensional assemblies of line members connected together at a number of joints. All structural members lie in a common plane and the applied loading also acts in this plane. (This definition excludes the grillage which also has all its members lying in a common plane but is subjected to a loading acting normal to this plane: the grillage is considered in Chapter 4.)

Two basic types of planar frame can be envisaged depending on the conditions at the physical joints. In a *planar pin-jointed frame* the members are assumed to be connected to each other by frictionless pins and the applied loads act only at the pins. This means that the structural elements of such a frame carry axial forces only and are not subjected to shearing forces and bending moments. On the other hand in a *planar rigid-jointed frame* all the members meeting at a particular physical joint are rigidly connected such that they all rotate an equal amount when loading is applied. In general the individual elements are subject to axial force, shearing force, and bending moment. Aside from the two basic types of planar frame there also exist *non-standard planar frames* in which the connections between members of the frames are not all of one sort, i.e. are not all pinned or all rigid jointed.

The early part of this chapter is concerned with the analysis of the types of frame mentioned in the preceding paragraph. In Chapter 2 the stiffness matrices for the bar and beam elements discussed there are related to a convenient *element*, or *local*, co-ordinate system. This is satisfactory if the structure to be analysed is such that all elements lie in the same direction, as in a continuous beam problem say, but is clearly not going to be satisfactory in analysing a frame structure where different elements are arbitrarily inclined to one another. To apply equilibrium and compatibility conditions conveniently at the structure nodes it is necessary that the nodal forces and displacements of all elements relate to a common *structure*, or *global*, co-ordinate system. This is achieved at the element level by a suitable *transformation* of the bar and beam element stiffness matrices derived in Chapter 2. Once this is done the matrix displacement analysis proceeds as described earlier.

When dealing with rigid-jointed planar frames it is sometimes justifiable to neglect the effect of axial straining of the structural elements. This leads to a simplification in the sense that the number of degrees of freedom of a given problem is reduced, and also leads to solutions which are identical to those obtained using the slope deflection and moment distribution methods. The procedure is discussed in this chapter but it is noted that it is only suitable for use in obtaining solutions by hand. Another topic considered concerns one of a number of complicating features that might arise in analysing planar frames, namely the presence of 'inclined' supports in which the support conditions cannot all be expressed directly in terms of the global axes. The solution is to invoke a transformation between co-ordinate systems. The final topic discussed in this chapter in terms of planar frames, though it is clearly of importance in general matrix displacement and finite-element analysis, concerns the nature of the structure stiffness matrix with regard to the distribution of non-zero terms within it. Such a distribution depends solely on the manner in which the structure nodes are numbered, and in problems with many degrees of freedom it is important to take care with the numbering of nodes so as to produce a 'banded' stiffness matrix with all the non-zero terms located local to the leading diagonal.

The problems considered in this chapter are those in which the loading is applied as concentrated forces and moments acting directly at the physical joints. Other kinds of loading are possible, of course, but consideration of these is deferred to Chapter 4.

3.2. Definitions and conventions for planar frames

In what follows certain quantities are defined for convenience in the context of an analysis of a rigid-jointed plane frame. The definitions remain the same for a pin-jointed frame except that the moments M_z and rotations θ_z can be ignored.

A typical planar frame problem is illustrated in Fig. 3.1(a). The frame nodes (coinciding with the physical joints) have been numbered in some appropriate fashion: later it will be seen that the manner in which the nodes of a structure are numbered is an important practical consideration but for the moment such considerations will be ignored and the node numbering is regarded as a somewhat arbitrary procedure. The elements of the frame can also be identified by a separate (and truly arbitrary)

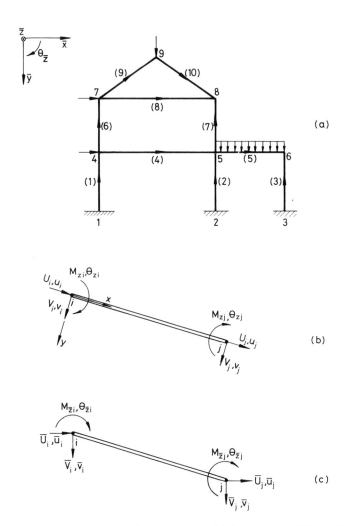

Fig. 3.1. Representation of a plane frame structure: (a) the structure; (b) typical element, showing local quantities; (c) typical element, showing global quantities.

numbering (or lettering) scheme or can be identified by the numbers of the nodes to which they are connected, e.g. element (9) or element 7–9.

The arrows shown on the elements indicate the positive directions chosen for the element longitudinal axes (the x axes), i.e. the arrows run from node i to node j of an element. The decision as to which node of an element is taken as node i and which as node j is not too important, although in a manual solution it is simpler if the structure node corresponding to element node j is numerically larger than the structure node corresponding to element node i. In any event, having decided upon nodes i and j of the element, the decision must not then be altered in proceeding through the full structural analysis.

A typical element of the frame is shown in Fig. 3.1(b). The local axes are x and y with the local axis x running along the element from element node i to node j, as already mentioned, and the local axis y at $90°$ to the x axis (measured clockwise from the x axis). The rotation in the xy plane is θ_z and is positive in the clockwise sense. In the local system the element nodal forces are the values of U, V, and M_z at end points i and j and the corresponding nodal displacements are u, v, and θ_z at these points: these quantities are shown in Fig. 3.1(b). (If the element is part of a pin-jointed frame M_z and θ_z need not be considered.) The element nodal forces and displacements in this local system can be written as column matrices $\mathbf{P}$ and $\mathbf{d}$ respectively, i.e.

$$\mathbf{P} = \left\{ \begin{matrix} \mathbf{P}_i \\ \mathbf{P}_j \end{matrix} \right\} \quad \text{where} \quad \mathbf{P}_i = \left\{ \begin{matrix} U_i \\ V_i \\ M_{zi} \end{matrix} \right\} \quad \text{and} \quad \mathbf{P}_j = \left\{ \begin{matrix} U_j \\ V_j \\ M_{zj} \end{matrix} \right\} \quad (3.1)$$

and

$$\mathbf{d} = \left\{ \begin{matrix} \mathbf{d}_i \\ \mathbf{d}_j \end{matrix} \right\} \quad \text{where} \quad \mathbf{d}_i = \left\{ \begin{matrix} u_i \\ v_i \\ \theta_{zi} \end{matrix} \right\} \quad \text{and} \quad \mathbf{d}_j = \left\{ \begin{matrix} u_j \\ v_j \\ \theta_{zj} \end{matrix} \right\}. \quad (3.2)$$

The global axes are $\bar{x}$ and $\bar{y}$, shown in Fig. 3.1(a). These axes are a fixed set of mutually perpendicular axes whose origin and directions are selected simply for the greatest convenience. The $\bar{x}\bar{y}$ co-ordinate system acts as a reference system for the structure as a whole. The rotation in this system is described by $\theta_{\bar{z}}$ (although clearly the rotations θ_z and $\theta_{\bar{z}}$ in the local and global systems are the same). In the global system the element nodal forces are the values of $\bar{U}$, $\bar{V}$, and $M_{\bar{z}}$ at the end points i and j and the corresponding nodal displacements are $\bar{u}$, $\bar{v}$, and $\theta_{\bar{z}}$ at these points; these quantities are shown in Fig. 3.1(c). Collectively the element forces and displacements in the global system are written as column matrices $\bar{\mathbf{P}}$ and $\bar{\mathbf{d}}$ respectively, i.e.

$$\bar{\mathbf{P}} = \left\{ \begin{matrix} \bar{\mathbf{P}}_i \\ \bar{\mathbf{P}}_j \end{matrix} \right\} \quad \text{where} \quad \bar{\mathbf{P}}_i = \left\{ \begin{matrix} \bar{U}_i \\ \bar{V}_i \\ M_{\bar{z}i} \end{matrix} \right\} \quad \text{etc.} \quad (3.3)$$

and

$$\bar{\mathbf{d}} = \left\{ \begin{matrix} \bar{\mathbf{d}}_i \\ \bar{\mathbf{d}}_j \end{matrix} \right\} \quad \text{where} \quad \bar{\mathbf{d}}_i = \left\{ \begin{matrix} \bar{u}_i \\ \bar{v}_i \\ \theta_{\bar{z}i} \end{matrix} \right\} \quad \text{etc.} \quad (3.4)$$

Having defined the element force and displacement column matrices in both the local and global system we can now define the stiffness matrix of the individual element in both systems. In the local system the element stiffness matrix will be denoted by $\mathbf{k}$ and will link nodal forces $\mathbf{P}$ to displacements $\mathbf{d}$, i.e.

$$\mathbf{P} = \mathbf{kd} \qquad \text{or} \qquad \begin{Bmatrix} \mathbf{P}_i \\ \mathbf{P}_j \end{Bmatrix} = \begin{bmatrix} \mathbf{k}_{ii} & \mathbf{k}_{ij} \\ \mathbf{k}_{ji} & \mathbf{k}_{jj} \end{bmatrix} \begin{Bmatrix} \mathbf{d}_i \\ \mathbf{d}_j \end{Bmatrix}. \tag{3.5}$$

In the global system the element stiffness matrix will be denoted by $\bar{\mathbf{k}}$ such that the stiffness equation is

$$\bar{\mathbf{P}} = \bar{\mathbf{k}}\bar{\mathbf{d}} \qquad \text{or} \qquad \begin{Bmatrix} \bar{\mathbf{P}}_i \\ \bar{\mathbf{P}}_j \end{Bmatrix} = \begin{bmatrix} \bar{\mathbf{k}}_{ii} & \bar{\mathbf{k}}_{ij} \\ \bar{\mathbf{k}}_{ji} & \bar{\mathbf{k}}_{jj} \end{bmatrix} \begin{Bmatrix} \bar{\mathbf{d}}_i \\ \bar{\mathbf{d}}_j \end{Bmatrix}. \tag{3.6}$$

The stiffness matrix $\bar{\mathbf{k}}$ can be obtained from $\mathbf{k}$ by a transformation procedure, as will be seen in the next section.

For the structure as a whole the complete set of nodal displacements will be denoted by column matrix $\mathbf{D}$ where

$$\mathbf{D} = \{\mathbf{D}_1 \ \mathbf{D}_2 \ \mathbf{D}_3 \ \ldots\} \tag{3.7}$$

in which the subscripts $1, 2, 3 \ldots$ are structure node numbers and $\mathbf{D}_1$, say, comprises the displacements $\bar{u}_1$, $\bar{v}_1$, and $\theta_{\bar{z}1}$. (Note that no bar need be placed over $\mathbf{D}$ since the structure displacements are clearly referred to the global system.) The complete set of externally applied loads, acting at the structure nodes and in the directions corresponding to the global system, is denoted by column matrix $\mathbf{F}$ (again the bar is superfluous) where

$$\mathbf{F} = \{\mathbf{F}_1 \ \mathbf{F}_2 \ \mathbf{F}_3 \ \ldots\} \tag{3.8}$$

and $\mathbf{F}_1$, say, comprises external forces $F_{\bar{u}1}$, $F_{\bar{v}1}$, and $F_{\theta\bar{z}1}$ corresponding to the displacements $\bar{u}_1$, $\bar{v}_1$, and $\theta_{\bar{z}1}$ respectively. The structure stiffness matrix is denoted by $\mathbf{K}$ and the set of structure equations is

$$\mathbf{F} = \mathbf{KD}. \tag{3.9a}$$

After application of the zero-displacement boundary conditions the reduced column matrices of nodal displacements and of external loads are denoted by $\mathbf{D}_r$ and $\mathbf{F}_r$ respectively. Correspondingly the reduced structure stiffness matrix is $\mathbf{K}_r$ and the reduced set of equations is

$$\mathbf{F}_r = \mathbf{K}_r\mathbf{D}_r \tag{3.9b}$$

3.3. Planar pin-jointed frames

It has been noted that the typical element of a pin-jointed frame is subjected to a single type of loading only, namely a uniform axial force. Thus the typical element is the bar element whose stiffness relationship in its local rectangular co-ordinate system was shown in Chapter 2 to be

$$\begin{Bmatrix} U_i \\ U_j \end{Bmatrix} = \frac{AE}{l} \begin{bmatrix} 1 & -1 \\ -1 & 1 \end{bmatrix} \begin{Bmatrix} u_i \\ u_j \end{Bmatrix}. \tag{3.10}$$

When such a bar element forms part of a pin-jointed frame with its longitudinal x axis inclined at an angle α, say, from a global axis $\bar{x}$ as in Fig. 3.2(a)† it is clear that a translational force or displacement in the x direction will have components in both the $\bar{x}$ and $\bar{y}$ rectangular co-ordinate directions. Thus the final size of the element stiffness matrix must be 4×4. Correspondingly it is necessary to expand the stiffness matrix of equation (3.10) by the addition of rows and columns corresponding to V forces and v displacements; the extra stiffness coefficients will all be zeros. Equation (3.10) can therefore be rewritten as

$$\begin{Bmatrix} U_i \\ V_i \\ U_j \\ V_j \end{Bmatrix} = \frac{AE}{l} \begin{bmatrix} 1 & 0 & -1 & 0 \\ 0 & 0 & 0 & 0 \\ -1 & 0 & 1 & 0 \\ 0 & 0 & 0 & 0 \end{bmatrix} \begin{Bmatrix} u_i \\ v_i \\ u_j \\ v_j \end{Bmatrix} \qquad \text{or} \qquad \mathbf{P} = \mathbf{kd} \tag{3.11}$$

and still relates to the local system of xy rectangular coordinates.

To transform to the desired relationship in the global system, $\bar{\mathbf{P}} = \bar{\mathbf{k}}\bar{\mathbf{d}}$, it is necessary to consider the relationships that exist between local system displacements $\mathbf{d}$ and global displacements $\bar{\mathbf{d}}$ and similarly between the global $\bar{\mathbf{P}}$ and local $\mathbf{P}$ forces.

Considering the displacements first, we imagine that when the frame, of which the element shown in Fig. 3.2(a) forms part, is loaded, node i of the element moves the amounts $\bar{u}_i$ and $\bar{v}_i$ in the $\bar{x}$ and $\bar{y}$ directions respectively. This movement is illustrated in Fig. 3.2(b) where O represents the initial position of node i and O′ the position to which the node moves after loading. The corresponding displacements in the directions of the local axes are u_i and v_i. From the figure it is clear that the two sets of displacements are related by the equations

$$u_i = \bar{u}_i \cos \alpha + \bar{v}_i \sin \alpha$$

$$v_i = -\bar{u}_i \sin \alpha + \bar{v}_i \cos \alpha.$$

It is noted that α is the angle through which the global axes must be rotated in the clockwise direction to come into coincidence with the local axes. In matrix notation the above equations can be expressed as

$$\begin{Bmatrix} u_i \\ v_i \end{Bmatrix} = \begin{bmatrix} e & f \\ -f & e \end{bmatrix} \begin{Bmatrix} \bar{u}_i \\ \bar{v}_i \end{Bmatrix} \tag{3.12a}$$

---

† The moments and rotations shown in Fig. 3.2(a) should be ignored in this section.

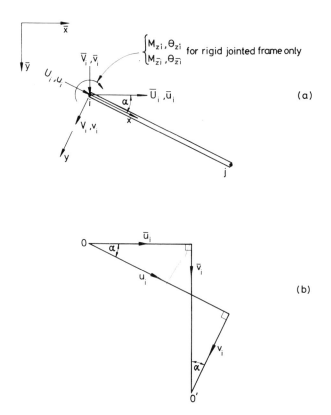

Fig. 3.2. Relationship between local and global quantities for a frame element: (a) the element; (b) displacement transformation.

or

$$\mathbf{d}_i = \mathbf{T}\bar{\mathbf{d}}_i$$

Here $e = \cos \alpha$ and $f = \sin \alpha$. It is sometimes convenient, particularly in programming for the computer, to express these quantities in terms of the co-ordinate positions in the global system of the element nodes. Thus

$$e = \cos \alpha = \frac{\bar{x}_j - \bar{x}_i}{l}, \qquad f = \sin \alpha = \frac{\bar{y}_j - \bar{y}_i}{l} \qquad (3.13)$$

and

$$l = ((\bar{x}_j - \bar{x}_i)^2 + (\bar{y}_j - \bar{y}_i)^2)^{1/2}.$$

In equation (3.12a) the 2×2 matrix $\mathbf{T}$ is the *transformation matrix* or *rotation matrix* for a single node.

Before proceeding it is appropriate to mention, for the purpose of later reference (from Section 4.5), that equation (3.12a) can be expressed more generally as

$$\begin{Bmatrix} u_i \\ v_i \end{Bmatrix} = \begin{bmatrix} \lambda_{x\bar{x}} & \lambda_{x\bar{y}} \\ \lambda_{y\bar{x}} & \lambda_{y\bar{y}} \end{bmatrix} \begin{Bmatrix} \bar{u}_i \\ \bar{v}_i \end{Bmatrix}. \tag{3.12b}$$

Here $\lambda_{x\bar{x}}$ and $\lambda_{x\bar{y}}$ are the direction cosines of the x axis with respect to the $\bar{x}$ and $\bar{y}$ axes, i.e.

$\lambda_{x\bar{x}} = $ cosine of angle between the x and $\bar{x}$ axes $= \cos \alpha$

$\lambda_{x\bar{y}} = $ cosine of angle between x and $\bar{y}$ axes $= \cos(90 - \alpha) = \sin \alpha$

Similarly $\lambda_{y\bar{x}}$ and $\lambda_{y\bar{y}}$ are the direction cosines for the y axis and have the values $\cos(90 + \alpha) = -\sin \alpha$ and $\cos \alpha$ respectively.

Since a similar expression to equation (3.12) applies for node j of the element the complete displacement transformation can be written

$$\mathbf{d} = \tau \bar{\mathbf{d}} \tag{3.14}$$

where

$$\tau = \begin{bmatrix} \mathbf{T} & \mathbf{0} \\ \mathbf{0} & \mathbf{T} \end{bmatrix}$$

is the full 4×4 transformation matrix and

$$\mathbf{d} = \{\mathbf{d}_i \ \mathbf{d}_j\}, \qquad \bar{\mathbf{d}} = \{\bar{\mathbf{d}}_i \ \bar{\mathbf{d}}_j\}.$$

Consider now the relationship between the two systems of forces acting at node i; these forces are shown in Fig. 3.2(a). By simply resolving the U_i and V_i forces into components acting along the $\bar{x}$ and $\bar{y}$ directions and equating the results to the $\bar{U}_i$ and $\bar{V}_i$ forces we obtain

$$\begin{Bmatrix} \bar{U}_i \\ \bar{V}_i \end{Bmatrix} = \begin{bmatrix} e & -f \\ f & e \end{bmatrix} \begin{Bmatrix} U_i \\ V_i \end{Bmatrix}.$$

From the definition of $\mathbf{T}$ in equation (3.12a) we see that this relationship can be expressed

$$\bar{\mathbf{P}}_i = \mathbf{T}^t \mathbf{P}_i. \tag{3.15}$$

Again a similar relationship holds for node j so that the complete force transformation can be written

$$\bar{\mathbf{P}} = \tau^t \mathbf{P} \tag{3.16}$$

where τ is as defined above and $\mathbf{P} = \{\mathbf{P}_i \ \mathbf{P}_j\}$ etc.

At this stage expressions have been established relating displacements and forces in the two co-ordinate systems—the local and global systems—in a manner consistent with the compatibility and equilibrium conditions respectively. To complete development of the planar bar element stiffness matrix in the global system all that is necessary is to substitute equations (3.14) and (3.16) into equation (3.11).

From equation (3.11)

$$\boldsymbol{\tau}^t \mathbf{P} = \boldsymbol{\tau}^t \mathbf{kd}$$

which becomes, using equation (3.14),

$$\boldsymbol{\tau}^t \mathbf{P} = (\boldsymbol{\tau}^t \mathbf{k} \boldsymbol{\tau}) \bar{\mathbf{d}}$$

and then, using equation (3.16),

$$\bar{\mathbf{P}} = \bar{\mathbf{k}} \bar{\mathbf{d}} \tag{3.17}$$

where

$$\bar{\mathbf{k}} = \boldsymbol{\tau}^t \mathbf{k} \boldsymbol{\tau}. \tag{3.18}$$

Equations (3.17) and (3.18) give the desired stiffness relationship in the global system. The column matrices of nodal forces $\bar{\mathbf{P}}$ and displacements $\bar{\mathbf{d}}$ in the system are defined in equations (3.3) and (3.4) with the rotational forces and displacements $\bar{M}_z$ and $\theta_{\bar{z}}$ omitted of course. The element stiffness matrix in the global system, $\bar{\mathbf{k}}$, appears in equation (3.18) in the form of a triple matrix product. This is a standard form of the transformed stiffness which will appear again a number of times in this work for other types of structural element.

It is noted that if the element local stiffness matrix $\mathbf{k}$ is symmetric—as is always the case—then the form of equation (3.18) guarantees that the corresponding element global stiffness matrix $\bar{\mathbf{k}}$ is also symmetric. For the simple bar element the matrix multiplication of equation (3.18) can be easily carried out on substituting for $\mathbf{k}$ and $\boldsymbol{\tau}$ from equations (3.11)– (3.14). The result yields the element stiffness matrix in the global configuration as

$$\bar{\mathbf{k}} = \frac{AE}{l} \begin{array}{c} \\ \left[\begin{array}{cccc} \bar{u}_i & \bar{v}_i & \bar{u}_j & \bar{v}_i \\ e^2 & & & \\ ef & f^2 & \text{symmetric} \\ -e^2 & -ef & e^2 & \\ -ef & -f^2 & ef & f^2 \end{array} \right] \end{array} = \begin{bmatrix} \bar{\mathbf{k}}_{ii} & \bar{\mathbf{k}}_{ij} \\ \bar{\mathbf{k}}_{ji} & \bar{\mathbf{k}}_{jj} \end{bmatrix} \tag{3.19}$$

It should be noted that *for this particular element* $\bar{\mathbf{k}}_{ii} = \bar{\mathbf{k}}_{jj} = -\bar{\mathbf{k}}_{ij} = -\bar{\mathbf{k}}_{ji} = \bar{\mathbf{k}}_0$, say, so that the stiffness matrix can also be written

$$\bar{\mathbf{k}} = \begin{bmatrix} \bar{\mathbf{k}}_0 & -\bar{\mathbf{k}}_0 \\ -\bar{\mathbf{k}}_0 & \bar{\mathbf{k}}_0 \end{bmatrix} \quad \text{where} \quad \bar{\mathbf{k}}_0 = \frac{AE}{l} \begin{bmatrix} e^2 & ef \\ ef & f^2 \end{bmatrix} \tag{3.20}$$

and this simple partitioned form facilitates the manual solution of numerical problems.

Before proceeding further with consideration of the two-dimensional bar element a few remarks of a more general nature concerning the transformation procedure are appropriate.

We have seen that for the bar element the relationships expressing the transformation of displacements (equation (3.14)) and forces (equation (3.16)) each involve the same transformation matrix τ. This is not fortuitous and is not restricted to the particular type of transformation that we have considered here. The pair of transformations given by equation (3.14) and (3.16) reflect what is known as the *Contragradient law*. By using work considerations we can easily see that if the displacements associated with two distinct co-ordinate systems transform as in equation (3.14) then the corresponding forces must transform as in equation (3.16). This follows from the fact that the force components must perform the same amount of work (a scalar quantity) in either system. In general, if the forces and displacements in one co-ordinate system are

$$\mathbf{P} = \{P_1 P_2 P_3 \ldots\} \qquad \text{and} \qquad \mathbf{d} = \{d_1 d_2 d_3 \ldots\}$$

and the forces and displacements in a second co-ordinate system are

$$\bar{\mathbf{P}} = \{\bar{P}_1 \bar{P}_2 \bar{P}_3 \ldots\} \qquad \text{and} \qquad \bar{\mathbf{d}} = \{\bar{d}_1 \bar{d}_2 \bar{d}_3 \ldots\}$$

it follows that

$$\bar{P}_1 \bar{d}_1 + \bar{P}_2 \bar{d}_2 + \bar{P}_3 \bar{d}_3 + \ldots = P_1 d_1 + P_2 d_2 + P_3 d_3 + \ldots$$

or

$$\bar{\mathbf{P}}^t \bar{\mathbf{d}} = \mathbf{P}^t \mathbf{d} \tag{3.21}$$

(or indeed $\bar{\mathbf{d}}^t \bar{\mathbf{P}} = \mathbf{d}^t \mathbf{P}$). When the displacement transformation of the form of equation (3.14) is substituted, the equality expressed in equation (3.21) becomes

$$\bar{\mathbf{P}}^t \bar{\mathbf{d}} = \mathbf{P}^t \tau \bar{\mathbf{d}}$$

and hence it follows that

$$\bar{\mathbf{P}}^t = \mathbf{P}^t \tau.$$

Transposing both sides gives a force transformation of the form of equation (3.16). In transforming from one co-ordinate system to another the Contragradient law ensures that if, say, the relationship between the two displacement systems is known, that between the forces follows in trivial fashion (and vice versa) and then the transformed stiffness matrix must have the form given in equation (3.18).

A further point to note is that the transformation matrix $\boldsymbol{\tau}$ (or $\mathbf{T}$) is an *orthogonal* matrix, i.e. it is a square matrix whose transpose is equal to its inverse, $\boldsymbol{\tau}^t = \boldsymbol{\tau}^{-1}$ (or $\mathbf{T}^t = \mathbf{T}^{-1}$). This can easily be checked for the particular form of $\boldsymbol{\tau}$ given above for the bar element, and it also holds for other transformations from local to global co-ordinate systems given later. It follows that equation (3.16) can be rewritten

$$\bar{\mathbf{P}} = \boldsymbol{\tau}^{-1}\mathbf{P} \qquad \text{or} \qquad \mathbf{P} = \boldsymbol{\tau}\bar{\mathbf{P}} \qquad\qquad (3.22)$$

so that we see that forces in the local system are related to those in the global system in precisely the same way as displacements in the local system are related to those in the global system (see equation (3.14)).

Returning to the specific consideration of the two-dimensional bar element, now that the stiffness matrix $\bar{\mathbf{k}}$ (equation (3.19) or (3.20)) is referred to a common global or structure co-ordinate system it can be used directly in the analysis of arbitrary planar pin-jointed frames; the assembly of a structure stiffness matrix $\mathbf{K}$ proceeds exactly as described in Chapter 2. Boundary conditions are expressed in terms of the global system, of course, as is the applied loading. Solution of the resulting stiffness equation yields the column matrix of nodal displacements for the structure in the global system.

When the structure displacements are known the column matrix $\bar{\mathbf{d}}$ of displacements for any element in the global system can be readily extracted and can be used to calculate element nodal forces. Forces in the global system can be directly calculated using equation (3.17). However, it is usually more useful to calculate the column matrix $\mathbf{P}$ of forces referred to the local co-ordinate axes since this will directly yield the values of axial force in the bar elements. One way of doing this is by using equations (3.11) and (3.14) to obtain

$$\mathbf{P} = (\mathbf{k}\boldsymbol{\tau})\bar{\mathbf{d}}. \qquad\qquad (3.23)$$

Alternatively we could proceed via equations (3.17) and (3.22) to obtain

$$\mathbf{P} = (\boldsymbol{\tau}\bar{\mathbf{k}})\bar{\mathbf{d}}. \qquad\qquad (3.24)$$

Either way, in the case of the pin-jointed frame element the calculation is very simple since the (local) nodal forces V_i and V_j should always be zero and hence only U_i and U_j need be calculated using a reduced form of equation (3.23) (or (3.24)). The result is easily shown to be

$$\begin{Bmatrix} U_i \\ U_j \end{Bmatrix} = \frac{AE}{l} \begin{bmatrix} 1 & -1 \\ -1 & 1 \end{bmatrix} \begin{bmatrix} e\bar{u}_i + f\bar{v}_i \\ e\bar{u}_j + f\bar{v}_j \end{bmatrix}$$

$$= \frac{AE}{l} \begin{bmatrix} e & f & -e & -f \\ -e & -f & e & f \end{bmatrix} \begin{Bmatrix} \bar{u}_i \\ \bar{v}_i \\ \bar{u}_j \\ \bar{v}_j \end{Bmatrix}$$

or

$$U_i = -U_j = \frac{AE}{l}\left(e(\bar{u}_i - \bar{u}_j) + f(\bar{v}_i - \bar{v}_j)\right) \tag{3.25}$$

Example 3.1. The pin-jointed frame shown in Fig. E3.1(a) comprises six members and four nodes. The diagonal members cross at the centre of the frame without being connected there. All members of the frame have the same value of axial rigidity, equal to 150 000 kN.

(a) Form the stiffness matrix for the complete structure in terms of appropriate element stiffness submatrices $\bar{\mathbf{k}}_0$, without regard to any boundary conditions.

(b) Then assume that the frame is held at nodes 1 and 2 against any movement and that two loads of magnitude 100 kN each act at node 4, one in the positive $\bar{x}$ direction and the other in the positive $\bar{y}$ direction. Determine the displacements of nodes 3 and 4, the forces in the frame members, and the reaction forces at nodes 1 and 2.

(a) Each member of the frame will obviously be represented by a single bar element whose stiffness matrix in the global $\bar{x}\bar{y}$ system has the form derived above. The arrows shown on the elements in Fig. E3.1(a) run from element node i to element node j and this indicates the direction of the local axes x. In the last section it was shown that the stiffness matrix for a typical bar element joining element nodes i and j can be expressed in the particular form of equation (3.20). For each of the elements of the frame the submatrix $\bar{\mathbf{k}}_0$ can be easily established. Table E3.1 gives this submatrix for each element in turn (the units are kilonewtons and metres) and also records the connection between element and structure node numbers; the latter information is used in assembling the structure stiffness matrix, of course. There should be no difficulty in deciding on what value of the angle α applies for a particular element if the definition of α given prior to equation (3.12) is borne in mind. (For element F, for instance, α is the positive angle $(360° - \tan^{-1}(\frac{3}{4}))$ or the negative angle $\tan^{-1}(\frac{3}{4})$). However, since it is the sine and cosine of the angle, i.e. the quantities e and f, that we are really concerned with it is perhaps better to calculate these quantities directly using equations

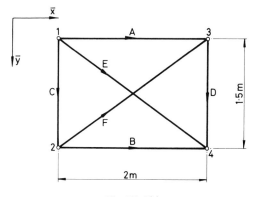

Fig. E3.1(a)

TABLE E3.1

Element	AE/l	α	$e = \cos\alpha$ (equations (3.13))	$f = \sin\alpha$	$\bar{\mathbf{k}}_0$ (eqn (3.20))	Element nodes i and j correspond to structure nodes	
A	7.5×10^4	0	1	0	$10^4 \begin{bmatrix} 7.5 & 0 \\ 0 & 0 \end{bmatrix}$	1	3
B	Details as for A					2	4
C	10×10^4	$90°$	0	1	$10^4 \begin{bmatrix} 0 & 0 \\ 0 & 10 \end{bmatrix}$	1	2
D	Details as for C					3	4
E	6×10^4	$\tan^{-1}(\frac{3}{4})$	0.8	0.6	$10^4 \begin{bmatrix} 3.84 & 2.88 \\ 2.88 & 2.16 \end{bmatrix}$	1	4
F	6×10^4	$-\tan^{-1}(\frac{3}{4})$	0.8	-0.6	$10^4 \begin{bmatrix} 3.84 & -2.88 \\ -2.88 & 2.16 \end{bmatrix}$	2	3

(3.13). Incidentally, in regard to this, it should be realized that the location of the origin of the global co-ordinates $\bar{x}\bar{y}$ is not important; both the origin itself and the directions of the global co-ordinates can be selected simply for greatest convenience.

Having established this basic data the complete SSM can now be set up using the direct stiffness procedure. The result obtained through use of the partitioned form of equation (3.20) for the ESM $\bar{\mathbf{k}}$ and of the node connection data of Table E3.1 is (with the column headings 1–4 corresponding to the numbered nodes of the structure)

$$\bar{\mathbf{K}} = \begin{bmatrix} \bar{\mathbf{k}}_0^A + \bar{\mathbf{k}}_0^C + \bar{\mathbf{k}}_0^E & -\bar{\mathbf{k}}_0^C & -\bar{\mathbf{k}}_0^A & -\bar{\mathbf{k}}_0^E \\ -\bar{\mathbf{k}}_0^C & \bar{\mathbf{k}}_0^B + \bar{\mathbf{k}}_0^C + \bar{\mathbf{k}}_0^F & -\bar{\mathbf{k}}_0^F & -\bar{\mathbf{k}}_0^B \\ -\bar{\mathbf{k}}_0^A & -\bar{\mathbf{k}}_0^F & \bar{\mathbf{k}}_0^A + \bar{\mathbf{k}}_0^D + \bar{\mathbf{k}}_0^F & -\bar{\mathbf{k}}_0^D \\ -\bar{\mathbf{k}}_0^E & -\bar{\mathbf{k}}_0^B & -\bar{\mathbf{k}}_0^D & \bar{\mathbf{k}}_0^B + \bar{\mathbf{k}}_0^D + \bar{\mathbf{k}}_0^E \end{bmatrix}$$

with column headings 1, 2, 3, 4.

(b) The boundary conditions are that

$$\mathbf{D}_1 = \mathbf{D}_2 = \mathbf{0} \quad \text{where} \quad \mathbf{D}_1 = \{\bar{u}_1 \ \bar{v}_1\} \quad \text{etc.}$$

Applying these conditions means that rows and columns corresponding to displacements at nodes 1 and 2 can be deleted from the complete structure stiffness matrix $\mathbf{K}$. The result is

$$\mathbf{K}_r = \begin{bmatrix} \bar{\mathbf{k}}_0^A + \bar{\mathbf{k}}_0^D + \bar{\mathbf{k}}_0^F & -\bar{\mathbf{k}}_0^D \\ -\bar{\mathbf{k}}_0^D & \bar{\mathbf{k}}_0^B + \bar{\mathbf{k}}_0^D + \bar{\mathbf{k}}_0^E \end{bmatrix}.$$

with column headings 3, 4.

(Clearly if the displacement boundary conditions had been assumed at the beginning there would have been no need to assemble the complete matrix $\mathbf{K}$.) Matrix $\mathbf{K}_r$ can now be written in terms of single stiffness coefficients, rather than stiffness submatrices, using the definitions of the $\bar{\mathbf{k}}_0$ submatrices given in Table E3.1. The result is a 4×4 stiffness matrix corresponding to the displacement column matrix $\mathbf{D}_r = \{\bar{u}_3 \ \bar{v}_3 \ \bar{u}_4 \ \bar{v}_4\}$. The external loading column matrix is then

$$\mathbf{F}_r = \{0 \quad 0 \quad 100 \quad 100\}$$

The structure stiffness equations after applying the boundary conditions become (in kilonewtons and metres)

$$\begin{Bmatrix} 0 \\ 0 \\ 100 \\ 100 \end{Bmatrix} = 10^4 \begin{bmatrix} 11.34 & & & \\ -2.88 & 12.16 & \text{Symmetric} & \\ 0 & 0 & 11.34 & \\ 0 & -10 & 2.88 & 12.16 \end{bmatrix} \begin{Bmatrix} \bar{u}_3 \\ \bar{v}_3 \\ \bar{u}_4 \\ \bar{v}_4 \end{Bmatrix} \tag{a}$$

These equations now have to be solved to yield the values of the structure nodal displacements. The result is

$$\begin{Bmatrix} \bar{u}_3 \\ \bar{v}_3 \\ \bar{v}_4 \\ \bar{v}_4 \end{Bmatrix} = \begin{Bmatrix} 0.61893 \\ 2.43704 \\ 0.17449 \\ 2.78519 \end{Bmatrix} \times 10^{-3} \, \text{m.}$$

The end forces for all the elements can be calculated using equations (3.25), once the nodal displacements $\bar{u}_i \dots \bar{v}_j$ of each particular element are extracted from the full list of structure displacements in accordance with the connection details tabulated above. The calculation is quite straightforward and is detailed in Table E3.2. (The quantities AE/l, e, and f used in the calculation are available from Table E3.1. Element C is not included since with both its ends held in position it clearly carries no force.)

The forces U_i and U_j are forces acting from the nodes *on to* the ends i and j of an element and are positive if acting in the same sense as the positive direction of the local x axis. It follows that elements A, B, D, and E are in tension and element F is in compression. The reaction forces in the $\bar{x}\bar{y}$ system at points 1 and 2 can be most readily calculated indirectly from the element forces, simply by summing the horizontal and vertical components of the element forces meeting at a support point. For point 1, for instance, the reaction forces are (see equation (3.15))

$$\mathbf{F}_1 = (\mathbf{T}^t \mathbf{P}_i)^A + (\mathbf{T}^t \mathbf{P}_i)^B$$

TABLE E3.2

Element	Node i		Node j		Element force $U_i \xrightarrow[x]{i} \xleftarrow{j} U_j$ $U_i \ (=-U_j)$
	$10^4 \times \bar{u}_i$	$10^4 \times \bar{v}_i$	$10^4 \times \bar{u}_j$	$10^4 \times \bar{v}_j$	
A	0	0	0.61893	2.43704	−46.420 kN
B	0	0	0.17449	2.78519	−13.087 kN
D	0.61893	2.43704	0.17449	2.78519	−34.815 kN
E	0	0	0.17449	2.78519	−108.642 kN
F	0	0	0.61893	2.43704	58.025 kN

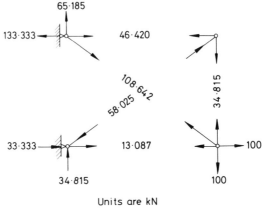

Units are kN

Fig. E3.1(b)

or

$$\begin{Bmatrix} \bar{U}_1 \\ \bar{V}_1 \end{Bmatrix} = \begin{bmatrix} 1 & 0 \\ 0 & 1 \end{bmatrix} \begin{Bmatrix} -46.420 \\ 0 \end{Bmatrix} + \begin{bmatrix} 0.8 & -0.6 \\ 0.6 & 0.8 \end{bmatrix} \begin{Bmatrix} -108.642 \\ 0 \end{Bmatrix} = \begin{Bmatrix} -133.333 \\ -65.185 \end{Bmatrix}$$

and the reaction components at point 2 can be found in a similar fashion. The external loads and the calculated element forces and reaction forces are shown in Fig. E3.1(b). It should particularly be noted that in this figure the arrows giving the senses of the forces are now shown in the traditional manner of indicating the direction of a force acting *on* a joint. This facilitates a desirable check on the validity of the obtained solution by examining equilibrium conditions; it is easily checked that the structure as a whole and all the individual nodes are in equilibrium, as expected.

3.4. Planar rigid-jointed frames

As mentioned earlier the typical element of a rigid-jointed plane frame is subjected to axial force, shearing force, and bending moment. In terms of local geometry it carries U, V, and M_z force components at its ends as shown in Fig. 3.1(b). It can therefore be regarded as an element which combines the functions of the axially loaded bar and the beam elements of Chapter 2. In regarding it as such it is implicitly assumed that axial behaviour is completely uncoupled from bending behaviour; this is valid so long as deformations are sufficiently small that the equilibrium conditions of the deformed structure can be based on the undeformed structure geometry.

In the local or element co-ordinate system the 6×6 stiffness matrix for the element of the rigid-jointed plane frame is therefore obtainable by combining together the stiffness matrices for the bar (equation (2.3)) and the beam (equation (2.26)). The element stiffness relationships in the local system thus become (see Fig. 3.1(b))

$$
\begin{Bmatrix} U_i \\ V_i \\ M_{zi} \\ \hdashline U_j \\ V_j \\ M_{zj} \end{Bmatrix} = \begin{bmatrix} AE/l & & & & \text{Symmetric} & \\ 0 & 12EI_z/l^3 & & & & \\ 0 & 6EI_z/l^2 & 4EI_z/l & & & \\ \hdashline -AE/l & 0 & 0 & AE/l & & \\ 0 & -12EI_z/l^3 & -6EI_z/l^2 & 0 & 12EI_z/l^3 & \\ 0 & 6EI_z/l^2 & 2EI_z/l & 0 & -6EI_z/l^2 & 4EI_z/l \end{bmatrix} \begin{Bmatrix} u_i \\ v_i \\ \theta_{zi} \\ \hdashline u_j \\ v_j \\ \theta_{zj} \end{Bmatrix}
$$

(3.26)

or

$$
\begin{Bmatrix} \mathbf{P}_i \\ \mathbf{P}_j \end{Bmatrix} = \begin{bmatrix} \mathbf{k}_{ii} & \mathbf{k}_{ij} \\ \mathbf{k}_{ji} & \mathbf{k}_{jj} \end{bmatrix} \begin{Bmatrix} \mathbf{d}_i \\ \mathbf{d}_j \end{Bmatrix} \qquad \text{or} \qquad \mathbf{P} = \mathbf{kd}, \quad \text{again.}
$$

As in the case of the element of a pin-jointed frame, we now wish to transform these stiffness equations to relate to the global or $\bar{x}\bar{y}$ co-ordinate system. Figures 3.1 and 3.2 again show the local and global quantities and the transformation procedure is much the same as that described in Section 3.3 except for the presence here of the M_z and θ_z quantities. These local quantities present no extra difficulty since they clearly transform directly into the corresponding global quantities $M_{\bar{z}}$ and $\theta_{\bar{z}}$, i.e. $M_z = M_{\bar{z}}$, $\theta_z = \theta_{\bar{z}}$. The translational forces (U, V) and displacements (u, v) are transformed into corresponding global quantities in exactly the same way as described in Section 3.3. Thus matrix $\mathbf{T}$ becomes here a 3×3 matrix

$$
\mathbf{T} = \begin{bmatrix} e & f & 0 \\ -f & e & 0 \\ 0 & 0 & 1 \end{bmatrix}
$$

(3.27)

and

$$
\boldsymbol{\tau} = \begin{bmatrix} \mathbf{T} & \mathbf{0} \\ \mathbf{0} & \mathbf{T} \end{bmatrix}
$$

again. The element stiffness matrix referred to the global system is then given by an equation of the form of equation (3.18) with this new definition of $\boldsymbol{\tau}$. Here again it is relatively simple to perform the matrix multiplication indicated in equation (3.18) manually. The result of this is the relationship

$$
\bar{\mathbf{P}} = \bar{\mathbf{k}}\bar{\mathbf{d}}
$$

where $\bar{\mathbf{P}}$ and $\bar{\mathbf{d}}$ are as defined in equations (3.3) and (3.4) respectively, and $\bar{\mathbf{k}}$, the ESM in the global reference frame, is given by equation (3.28) below.

$$
\bar{\mathbf{k}} =
\begin{bmatrix}
\dfrac{AE}{l}e^2 + 12\dfrac{EI_z}{l^3}f^2 & & & & & \\[2ex]
\left(\dfrac{AE}{l} - 12\dfrac{EI_z}{l^3}\right)ef & \dfrac{AE}{l}f^2 + 12\dfrac{EI_z}{l^3}e^2 & & \text{Symmetric} & & \\[2ex]
-6\dfrac{EI_z}{l^2}f & 6\dfrac{EI_z}{l^2}e & 4\dfrac{EI_z}{l} & & & \\[2ex]
-\dfrac{AE}{l}e^2 - 12\dfrac{EI_z}{l^3}f^2 & \left(-\dfrac{AE}{l} + 12\dfrac{EI_z}{l^3}\right)ef & 6\dfrac{EI_z}{l^2}f & \dfrac{AE}{l}e^2 + 12\dfrac{EI_z}{l^3}f^2 & & \\[2ex]
\left(-\dfrac{AE}{l} + 12\dfrac{EI_z}{l^3}\right)ef & -\dfrac{AE}{l}f^2 - 12\dfrac{EI_z}{l^3}e^2 & -6\dfrac{EI_z}{l^2}e & \left(\dfrac{AE}{l} - 12\dfrac{EI_z}{l^3}\right)ef & \dfrac{AE}{l}f^2 + 12\dfrac{EI_z}{l^3}e^2 & \\[2ex]
-6\dfrac{EI_z}{l^2}f & 6\dfrac{EI_z}{l^2}e & 2\dfrac{EI_z}{l} & 6\dfrac{EI_z}{l^2}f & -6\dfrac{EI_z}{l^2}e & 4\dfrac{EI_z}{l}
\end{bmatrix}
$$

$$\text{columns: } \bar{u}_i \quad \bar{v}_i \quad \theta_{zi} \quad \bar{u}_j \quad \bar{v}_j \quad \theta_{zj}$$

$$(3.28)$$

Once the structure stiffness equations have been set up and solved the element nodal displacements $\bar{\mathbf{d}}$ in the global system can be extracted for each element in turn and used in the calculation of element nodal forces. Again we can calculate the nodal forces referred to the local or the global axes but usually the former is more convenient since it provides the values of the basic element forces, i.e. the axial force, shearing force, and bending moment. To obtain these local force quantities from the global displacements we can use either equation (3.23) or (3.24), as convenient. The partitioned forms of these equations are sometimes useful in performing the manual calculations; these forms are

$$\mathbf{P}_i = \mathbf{k}_{ii}\mathbf{T}\bar{\mathbf{d}}_i + \mathbf{k}_{ij}\mathbf{T}\bar{\mathbf{d}}_j \qquad \mathbf{P}_j = \mathbf{k}_{ji}\mathbf{T}\bar{\mathbf{d}}_i + \mathbf{k}_{jj}\mathbf{T}\bar{\mathbf{d}}_j \qquad (3.29)$$

or

$$\mathbf{P}_i = \mathbf{T}\bar{\mathbf{k}}_{ii}\bar{\mathbf{d}}_i + \mathbf{T}\bar{\mathbf{k}}_{ij}\bar{\mathbf{d}}_j \qquad \mathbf{P}_j = \mathbf{T}\bar{\mathbf{k}}_{ji}\bar{\mathbf{d}}_i + \mathbf{T}\bar{\mathbf{k}}_{jj}\bar{\mathbf{d}}_j. \qquad (3.30)$$

Example 3.2. Figure E3.2(a) shows a plane frame in which the three members are rigidly connected together at joint 4 and the supports at 1, 2, and 3 are fully fixed. The bending and axial rigidities of the members are as follows: for member 3–4, $EI_z = 10^4$ kN m^2 and $AE = 50 \times 10^4$ kN; for members 1–4 and 2–4, $EI_z = 1.25 \times 10^4$ kN m^2 and $AE = 60 \times 10^4$ kN. Determine the values of the displacements (i.e. horizontal displacement, vertical displacement and rotation) of joint 4 under the given loading. Also, find the forces at the ends of each frame member referred to the local axes and determine the components of reaction at the supports in the $\bar{x}\bar{y}$ reference frame shown in the figure.

In solving this problem obviously each physical member of the frame will be represented by a single element whose stiffness matrix is of the form of equation (3.28). The directions of the local axes of the three elements are arbitrarily assumed to be as indicated in Fig. E3.2(a) by the arrows. The relevant properties

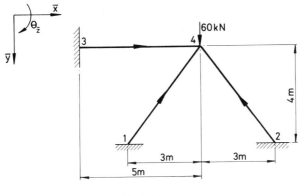

Fig. E3.2(a)

of the individual elements which have not already been noted above are as follows:

element 3–4,   length $l = 5$ m, $e = \cos \alpha = 1$, $f = \sin \alpha = 0$
element 1–4,   length $l = 5$ m, $e = 3/5$ $f = -4/5$
element 2–4,   length $l = 5$ m, $e = -3/5$, $f = -4/5$

(e and f are calculated using equations (3.13)).
  The only non-zero structure degrees of freedom are the three freedoms at point 4, i.e. $\mathbf{D}_r = \{\bar{u}_4 \ \bar{v}_4 \ \theta_{\bar{z}4}\}$. The corresponding applied forces are $\mathbf{F}_r = \{F_{\bar{u}4} \ F_{\bar{v}4} \ F_{\theta\bar{z}4}\} = \{0 \ 60 \ 0\}$. The contributions to the SSM $\mathbf{K}_r$ of each of the three elements are from the submatrix $\bar{\mathbf{k}}_{jj}$ of the ESM (i.e. the lower right-hand 3×3 matrix of $\bar{\mathbf{k}}$) since it is node j of each element which has been chosen to correspond to point 4 of the structure. With these facts in mind the structure stiffness equations, with the boundary conditions already accounted for, are

$$\begin{Bmatrix} 0 \\ 60 \\ 0 \end{Bmatrix} = 10^4 \begin{bmatrix} 10.0 \\ +4.3968 \\ +4.3968 & \text{Symmetric} \\[4pt] 0 & 0.096 \\ -5.7024 & +7.7232 \\ +5.7024 & +7.7232 \\[4pt] 0 & -0.24 & 0.8 \\ -0.24 & -0.18 & +1.0 \\ -0.24 & +0.18 & +1.0 \end{bmatrix} \begin{Bmatrix} \bar{u}_4 \\ \bar{v}_4 \\ \theta_{\bar{z}4} \end{Bmatrix}$$

where each term of the stiffness matrix $\mathbf{K}_r$ is shown as being comprised of the contributions from each of elements 3–4, 1–4, and 2–4, in that order. This becomes

$$\begin{Bmatrix} 0 \\ 60 \\ 0 \end{Bmatrix} = \frac{10^4}{6250} \begin{bmatrix} 117460 & 0 & -3000 \\ 0 & 97140 & -1500 \\ -3000 & -1500 & 17500 \end{bmatrix} \begin{Bmatrix} \bar{u}_4 \\ \bar{v}_4 \\ \theta_{\bar{z}4} \end{Bmatrix}.$$

The solution of these equations gives the structure displacements as

$$\begin{Bmatrix} \bar{u}_4 \\ \bar{v}_4 \\ \theta_{\bar{z}4} \end{Bmatrix} = \begin{Bmatrix} 0.850 \times 10^{-6} \text{ m} \\ 386.555 \times 10^{-6} \text{ m} \\ 33.279 \times 10^{-6} \text{ rad} \end{Bmatrix}.$$

For each element this column matrix of displacements is $\bar{\mathbf{d}}_j$, and furthermore for each element we have $\bar{\mathbf{d}}_i = \mathbf{0}$ which simplifies the calculation of element end forces. These end forces can conveniently be calculated using either equations (3.29) or equations (3.30). If, for instance, the latter equations are used the calculation of the forces at element node j (node 4 of the structure) is of the form.

$$\mathbf{P}_j = \mathbf{T} \bar{\mathbf{k}}_{jj} \bar{\mathbf{d}}_j.$$

For each element $\bar{\mathbf{k}}_{jj}$ can be calculated from equation (3.28) (and this has already been done in setting up the above structure stiffness matrix) and $\mathbf{T}$ is given by equation (3.27). In the case of element 2–4, for example, the forces at element node j are

$$\mathbf{P}_j^{2-4} = \begin{Bmatrix} U_j \\ V_j \\ M_{zj} \end{Bmatrix}^{2-4} = \begin{bmatrix} -0.6 & -0.8 & 0 \\ 0.8 & -0.6 & 0 \\ 0 & 0 & 1 \end{bmatrix} \times 10^4$$

$$\times \begin{bmatrix} 4.3968 & 5.7024 & -0.24 \\ 5.7024 & 7.7232 & 0.18 \\ -0.24 & 0.18 & 1.0 \end{bmatrix} \begin{Bmatrix} 0.850 \\ 386.555 \\ 33.279 \end{Bmatrix} \times 10^{-6}$$

$$= \begin{Bmatrix} -37.1704 \text{ kN} \\ -0.3773 \text{ kN} \\ 1.0265 \text{ kN m} \end{Bmatrix}.$$

By a similar procedure,

$$\mathbf{P}_j^{3-4} = \begin{Bmatrix} 0.0850 \text{ kN} \\ 0.2912 \text{ kN} \\ -0.6615 \text{kN m} \end{Bmatrix} \qquad \mathbf{P}_j^{1-4} = \begin{Bmatrix} -37.0480 \text{ kN} \\ 0.1793 \text{ kN} \\ -0.3650 \text{ kN m} \end{Bmatrix}.$$

The forces at node i of each element could be calculated in a related manner as $\mathbf{P}_i = \mathbf{T} \bar{\mathbf{k}}_{ij} \bar{\mathbf{d}}_j$ but it is easier, when proceeding manually, to determine the forces at i from those at j by using the known relationships linking them. These relationships are, of course, the simple equilibrium equations already incorporated in forming the element stiffness matrix:

$$U_i = -U_j, \qquad V_i = -V_j, \qquad M_{zi} = -M_{zj} + V_j l.$$

Thus it is a simple matter to determine the forces $\mathbf{P}_i^{2-4}$, $\mathbf{P}_i^{3-4}$, and $\mathbf{P}_i^{1-4}$.

Some care is needed in interpreting the signs of the element force components. It is emphasized again that the calculated nodal forces are those acting *on* the element ends and that the positive senses of U, V, and M_z are in the positive x, positive y, and positive θ_z directions respectively. With this in mind the forces acting on the element ends are readily established as being those shown in magnitude and direction in Fig. E3.2(b).

The reactions at the supports, referred to the global system, are the forces

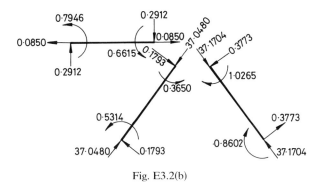

Fig. E3.2(b)

acting on the appropriate ends of elements. For this problem they are the forces $\bar{\mathbf{P}}_i$ for each of the elements of the frame. These forces can be straightforwardly calculated, using equation (3.6) with $\bar{\mathbf{d}}_i = \mathbf{0}$, as

$$\bar{\mathbf{P}}_i = \bar{\mathbf{k}}_{ij}\bar{\mathbf{d}}_j$$

where $\bar{\mathbf{k}}_{ij}$ is obtained from $\bar{\mathbf{k}}$ given by equation (3.28). For element 2–4, for instance, we have that

$$\bar{\mathbf{P}}_i^{2-4} = \left\{\begin{matrix} \bar{U}_i \\ \bar{V}_i \\ \bar{M}_{zi} \end{matrix}\right\}^{2-4} = 10^4 \times \begin{bmatrix} -4.3968 & -5.7024 & 0.24 \\ -5.7024 & -7.7232 & -0.18 \\ -0.24 & 0.18 & 0.5 \end{bmatrix}$$

$$\times \left\{\begin{matrix} 0.8500 \\ 386.5547 \\ 33.2790 \end{matrix}\right\} \times 10^{-6} = \left\{\begin{matrix} -22.0004 \text{ kN} \\ -29.9628 \text{ kN} \\ 0.8602 \text{ kN m} \end{matrix}\right\}$$

and similarly,

$$\bar{\mathbf{P}}_i^{1-4} = \left\{\begin{matrix} 22.0854 \text{ kN} \\ -29.7460 \text{ kN} \\ -0.5314 \text{ kN m} \end{matrix}\right\} \qquad \bar{\mathbf{P}}_i^{3-4} = \left\{\begin{matrix} -0.0850 \text{ kN} \\ -0.2912 \text{ kN} \\ -0.7946 \text{ kN m} \end{matrix}\right\}.$$

It should be noted that the reactions $\bar{\mathbf{P}}_i$ are, of course, directly related to the element forces $\mathbf{P}_i$ by a simple transformation equation $(\bar{\mathbf{P}}_i = \mathbf{T}^t\mathbf{P}_i)$ and this equation could alternatively have been used to find the $\bar{\mathbf{P}}_i$ since the $\mathbf{P}_i$ are already available.

The reaction forces and the applied force acting on the structure are shown in Fig. E3.2(c). It is a simple matter to verify that these forces constitute an equilibrium system, and this serves as a check on the analysis. Apart from overall equilibrium conditions being satisfied we should also find that equilibrium conditions are satisfied for all individual elements and for all joints. It is a good idea to get into the habit of checking overall equilibrium, or alternatively the equilibrium of a free joint, at the conclusion of an analysis as a precaution against numerical error.

Before leaving this example it should be noted that standard structural boundary conditions other than fully fixed supports can be readily accommodated in rigid-jointed frame analysis. If the support at point 2, say, of the frame of Fig. E3.2(a) were a pin then clearly we would have to take account of the extra

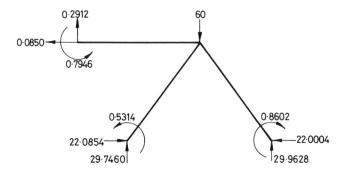

Fig. E3.2(c)

non-zero degree of freedom θ_{z2} and construct a set of four stiffness equations corresponding to $\mathbf{D}_r = (\theta_{z2} \ \bar{u}_4 \ \bar{v}_4 \ \theta_{\bar{z}4})$. On the other hand, if the support at point 2 were a roller support with freedom to rotate and to move horizontally, then two extra freedoms are introduced and $\mathbf{D}_r = \{\bar{u}_2 \ \theta_{z2} \ \bar{u}_4 \ \bar{v}_4 \ \theta_{\bar{z}4}\}$. Apart from the increase in the number of degrees of freedom no extra difficulty arises in accommodating these alternative standard support conditions at the structure boundary.

Example 3.3. Assemble the structure stiffness equations for the rigid jointed frame shown in Fig. E3.3(a) expressing the result in terms of E, A, I_z, and L where E, A, and I_z are the same for all parts of the frame.
For the particular numerical values $AE = 8 \times 10^5$ kN, $EI_z = 4000$ kN m^2, $L = 2$ m and $P = 50$ kN determine the structure displacements and hence construct the bending moment diagram for the frame.

The frame is modelled with elements 1–3, 1–2, and 2–4 whose stiffness matrices can be obtained using equation (3.28).

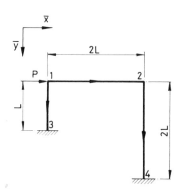

Fig. E3.3(a)

Element 1–3 $e = 0$, $f = 1$

$$\bar{\mathbf{k}}^{1-3} = \begin{bmatrix} & \bar{u}_1 & \bar{v}_1 & \theta_{\bar{z}1} & \bar{u}_3 & \bar{v}_3 & \theta_{\bar{z}3} \\ 12EI_z/L^3 & & & & & \\ 0 & AE/L & & \text{Symmetric} & & \\ -6EI_z/L^2 & 0 & 4EI_z/L & & & \\ -12EI_z/L^3 & 0 & 6EI_z/L^2 & 12EI_z/L^3 & & \\ 0 & -AE/L & 0 & 0 & AE/L & \\ -6EI_z/L^2 & 0 & 2EI_z/L & 6EI_z/L^2 & 0 & 4EI_z/L \end{bmatrix}$$

Element 1–2 $e = 1$, $f = 0$

$$\bar{\mathbf{k}}^{1-2} = \begin{bmatrix} & \bar{u}_1 & \bar{v}_1 & \theta_{\bar{z}1} & \bar{u}_2 & \bar{v}_2 & \theta_{\bar{z}2} \\ AE/2L & & & & & \\ 0 & 3EI_z/2L^3 & & \text{Symmetric} & & \\ 0 & 3EI_z/2L^2 & 2EI_z/L & & & \\ -AE/2L & 0 & 0 & AE/2L & & \\ 0 & -3EI_z/2L^3 & -3EI_z/2L^2 & 0 & 3EI_z/2L^3 & \\ 0 & 3EI_z/2L^2 & EI_z/L & 0 & -3EI_z/2L^2 & 2EI_z/L \end{bmatrix}$$

Element 2–4 $e = 0$, $f = 1$

$$\bar{\mathbf{k}}^{2-4} = \begin{bmatrix} & \bar{u}_2 & \bar{v}_2 & \theta_{\bar{z}2} & \bar{u}_4 & \bar{v}_4 & \theta_{\bar{z}4} \\ 3EI_z/2L^3 & & & & & \\ 0 & AE/2L & & \text{Symmetric} & & \\ -3EI_z/2L^2 & 0 & 2EI_z/L & & & \\ -3EI_z/2L^3 & 0 & 3EI_z/2L^2 & 3EI_z/2L^3 & & \\ 0 & -AE/2L & 0 & 0 & AE/2L & \\ -3EI_z/2L^2 & 0 & EI_z/L & 3EI_z/2L^2 & 0 & 2EI_z/L \end{bmatrix}$$

The boundary conditions are $\mathbf{D}_3 = \mathbf{D}_4 = 0$ and the non-zero structure degrees of freedom are $\mathbf{D}_r = \{\bar{u}_1 \ \bar{v}_1 \ \theta_{\bar{z}1} \ \bar{u}_2 \ \bar{v}_2 \ \theta_{\bar{z}2}\}$. The structure stiffness equations are easily assembled using the direct stiffness procedure and are

$$\begin{Bmatrix} F_{\bar{u}1} \\ F_{\bar{v}1} \\ F_{\theta\bar{z}1} \\ F_{\bar{u}2} \\ F_{\bar{v}2} \\ F_{\theta\bar{z}2} \end{Bmatrix} = \begin{bmatrix} 12EI_z/L^3 + AE/2L & & & & & \\ 0 & AE/L + 3EI_z/2L^3 & & & & \\ -6EI_z/L^2 & 3EI_z/2L^2 & 6EI_z/L & & \text{Symmetric} & \\ -AE/2L & 0 & 0 & AE/2L + 3EI_z/2L^3 & & \\ 0 & -3EI_z/2L^3 & -3EI_z/2L^2 & 0 & 3EI_z/2L^3 + AE/2L & \\ 0 & 3EI_z/2L^2 & EI_z/L & -3EI_z/2L^2 & -3EI_z/2L^2 & 4EI_z/L \end{bmatrix} \begin{Bmatrix} \bar{u}_1 \\ \bar{v}_1 \\ \theta_{\bar{z}1} \\ \bar{u}_2 \\ \bar{v}_2 \\ \theta_{\bar{z}2} \end{Bmatrix}$$

(a)

where for the loading shown in Fig. E3.3(a) only $F_{\bar{u}1} = P$ is non-zero in the column matrix of applied loads.

Now, with the given numerical values of AE, EI_z, L, and P these equations can be solved (on a computer) to yield $\mathbf{D}_r$. The result is

$$\{\bar{u}_1 \quad \bar{v}_1 \quad \theta_{z1} \quad \bar{u}_2 \quad \bar{v}_2 \quad \theta_{z2}\}$$
$$= 10^{-3}\{13.477\,\text{m} \quad -0.028\,\text{m} \quad 6.6015\,\text{rad} \quad 13.433\,\text{m} \quad 0.056\,\text{m} \quad 0.8840\,\text{rad}\}$$

To construct the bending moment diagram for the frame requires only the calculation of moments at element ends (since we know that the bending moment must vary linearly along the length of any element carrying no loads between its ends); this can be achieved by back substituting the nodal displacements into the element stiffness relationships and in this case since the moments are the same in both the local and global system (i.e. $M_{\bar{z}} = M_z$) the easiest way to determine these moments is by using the third and sixth equations of the set of element relationships in the global system $\bar{\mathbf{P}} = \bar{\mathbf{k}}\bar{\mathbf{d}}$. Thus, for element 1–3 for instance

$$\begin{Bmatrix} M_{\bar{z}i} \\ M_{\bar{z}j} \end{Bmatrix}^{1-3} = EI_z \begin{bmatrix} -6/L^2 & 0 & 4/L & 6/L^2 & 0 & 2/L \\ -6/L^2 & 0 & 2/L & 6/L^2 & 0 & 4/L \end{bmatrix} \begin{Bmatrix} \bar{u}_i \\ \bar{v}_i \\ \theta_{\bar{z}i} \\ \bar{u}_i \\ \bar{v}_i \\ \theta_{\bar{z}i} \end{Bmatrix}^{1-3}$$

$$= 4000 \begin{bmatrix} -1.5 & 0 & 2 & 1.5 & 0 & 1 \\ -1.5 & 0 & 1 & 1.5 & 0 & 2 \end{bmatrix} \begin{Bmatrix} 0.013477 \\ -0.000028 \\ 0.066015 \\ 0 \\ 0 \\ 0 \end{Bmatrix}$$

$$= \begin{Bmatrix} -28.048 \\ -54.454 \end{Bmatrix} \text{kN m.}$$

Similarly, for the other two elements

$$\begin{Bmatrix} M_{\bar{z}i} \\ M_{\bar{z}j} \end{Bmatrix}^{1-2} = \begin{Bmatrix} 28.048 \\ 16.613 \end{Bmatrix} \text{kN m} \qquad \begin{Bmatrix} M_{\bar{z}i} \\ M_{\bar{z}j} \end{Bmatrix}^{2-4} = \begin{Bmatrix} -16.613 \\ -18.381 \end{Bmatrix} \text{kN m}$$

The bending moment diagram can now be drawn and is shown in Fig. E3.3(b).

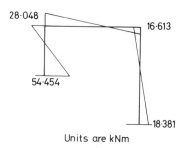

28·048

16·613

54·454

18·381

Units are kNm

Fig. E3.3(b)

3.5. Non-standard planar frames

Two specific types of standard planar frame have been considered so far: in Section 3.3 the stiffness matrix for an element with pin joints at each end was derived and used in the analysis of a frame in which all the joints are pinned, whilst in Section 3.4 the stiffness matrix for an element with rigid joints at each end was derived and used in the analysis of a frame in which all the joints are rigid ones. Frames in which all connections between elements are pinned joints or all are rigid joints are common, of course, but other plane frame structures of what may be described as a non-standard type also occur quite frequently. The word 'non-standard' is intended here to mean that the connections between the elements of a frame are not all of one type which implies that the elements themselves are also not of one single type. (The frames considered in this section are still subjected only to loads acting in their planes; another type of plane frame—the grillage—is considered in Section 4.4 but there the loads act normal to the plane of the frame.) Some simple examples of non-standard plane structures are shown in Fig. 3.3.

In each of Figs. 3.3(a) and 3.3(b) the horizontal and vertical physical members are rigidly jointed together whereas the inclined members are pinned at their ends to the rest of the structure (or to a support). These frames can be modelled with the standard elements discussed in Sections 3.3 and 3.4. Thus the stiffness matrices for the horizontal and vertical

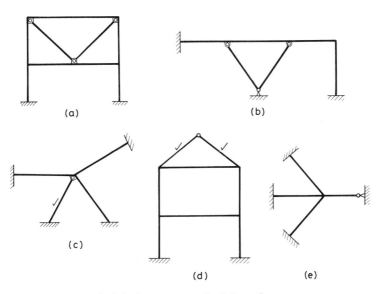

Fig. 3.3. Some non-standard planar frames.

elements are obtained from equation (3.28) whilst those for the inclined pin-ended members could be obtained from equation (3.20) if rows and columns of zeros were added to correspond to the missing rotational displacements and thus to increase the size of $\bar{\mathbf{k}}$ to 6×6. The same result is achieved in more convenient fashion if equation (3.28) is taken to apply to *all* elements, with I_z put equal to zero for the pin-ended elements. Once the element stiffnesses are calculated the assembly process and problem solution then follow in the usual fashion but it should be noted that the calculated $\theta_{\bar{z}}$ rotations apply only to those elements which are rigidly connected together.

In Figs. 3.3(c) and 3.3(d) the structures shown are basically rigid jointed but do include some members, indicated by the symbol $\sqrt{}$, which have one end pinned and the other rigidly fixed to the rest of the structure or to a support. To represent these members requires the generation of stiffness equations for an element of the type shown, in its local reference frame, in Fig. 3.4. It is assumed that end i will be connected to a rigid joint or support whereas end j will be pinned and free of applied moment. The stiffness matrix for this element will now be derived by simply modifying the stiffness equations for the rigid-jointed element (equation (3.26)) by using the condition that the moment at end j has zero value.

From the last of equations (3.26) we have that

$$\frac{M_{zj}}{EI_z} = \frac{6}{l^2} v_i + \frac{2}{l} \theta_{zi} - \frac{6}{l^2} v_j + \frac{4}{l} \theta_{zj} = 0. \tag{3.31}$$

The rotation θ_{zj} can therefore be expressed in terms of the three other nodal displacements as

$$\theta_{zj} = -\frac{3}{2l} v_i - \frac{\theta_{zi}}{2} + \frac{3}{2l} v_j. \tag{3.32}$$

This expression for θ_{zj} can now be substituted into the other five of equations (3.26) to give equations expressing U_i, V_i, M_{zi}, U_j, and V_j in

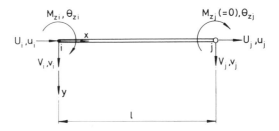

Fig. 3.4. A 'special' frame element with one end pinned.

terms of u_i, v_i, θ_{zi}, u_j, and v_j. Thus the stiffness equations for the special element shown in Fig. 3.4 in its local co-ordinate system are

$$
\begin{Bmatrix} U_i \\ V_i \\ M_{zi} \\ U_j \\ V_j \end{Bmatrix} = \begin{bmatrix} AE/l & & & & \\ 0 & 3EI_z/l^3 & & \text{Symmetric} & \\ 0 & 3EI_z/l^2 & 3EI_z/l & & \\ -AE/l & 0 & 0 & AE/l & \\ 0 & -3EI_z/l^3 & -3EI_z/l^2 & 0 & 3EI_z/l^3 \end{bmatrix} \begin{Bmatrix} u_i \\ v_i \\ \theta_{zt} \\ u_j \\ v_j \end{Bmatrix} \qquad (3.33)
$$

i.e. $\mathbf{P} = \mathbf{kd}$. It should be noted that θ_{zj} is not present in these equations, i.e. it is not used as an independent degree of freedom. This is because it has been turned into a dependent variable (see equation (3.32)) through use of the condition of zero M_{zj}. (The stiffness matrix of equation (3.33) could be increased to size 6×6 to maintain consistency with the size of the standard rigid-jointed frame element matrix by simply adding a sixth row and a sixth column of zeros. This has not been done here because it gives the impression, wrongly, that θ_{zj} is present as a degree of freedom.) On solution of the structure stiffness equations θ_{zj} can be determined, if desired, using equation (3.32). The stiffness for the element in a global co-ordinate system can be readily obtained using the same sort of approach described earlier for the standard pin-jointed and rigid-jointed elements and symbolized by equation (3.18). Here τ has the form

$$
\tau = \begin{bmatrix} \mathbf{T}_1 & \mathbf{0} \\ \mathbf{0} & \mathbf{T}_2 \end{bmatrix} \qquad (3.34)
$$

where $\mathbf{T}_1$ is the 3×3 matrix $\mathbf{T}$ defined by equation (3.27) and $\mathbf{T}_2$ is the 2×2 matrix $\mathbf{T}$ defined by equation (3.12).

In the above the 'special' element has a pin at node j. If the pin were at node i instead then obviously the stiffness matrix of the element is changed but is easily derived by the same general approach. If a pin were present at node i as well as at node j then in addition to equation (3.31) representing $M_{zj} = 0$ there would be a similar condition expressing the fact that $M_{zi} = 0$. Applying this further condition would, of course, lead to the stiffness equations for the usual pin-jointed element discussed in Section 3.3 (see Problem 3.13).

The stiffness equations (3.33) are basically intended, and are necessary, for use where an element of the type illustrated in Fig. 3.4 is connected at its pinned end to the rest of a structure. They can also be used where the pinned end of the special element coincides with a structure support point as in Fig. 3.3(e). Their use in such a circumstance is not really necessary since the support condition can be readily accommodated by using the standard rigid-jointed element (of Section 3.4) and applying the structure boundary conditions at the pin that the translational displacements have zero value but that the rotation has a non-zero value. This would normally be the approach used but it does involve an extra degree of freedom, i.e. $\theta_{\bar{z}}$ at the pin is present as a freedom in the structure stiffness equations, and in a manual solution this might well justify the use of the 'special' element in order to reduce by one the number of structure stiffness equations to be solved.

The 'special' element just described is only one of a number of such elements that might find a use in certain types of problem. For instance a sliding joint may be present in a rigid-jointed frame and a special element for this case could be devised (see Problem 3.14). However, the element shown in Fig. 3.4 is the most common type of special case and no other type will be considered here.

Example 3.4. For the plane structure shown in Fig. E3.4 the supports at points 1 and 4 are fully fixed and that at point 3 is pinned. Members 2–3 and 2–4 are rigidly connected together at point 2 whereas member 1–2 is pinned at 2. For 1–2 and 2–3 $EI_z = 4000 \text{ kN m}^2$, $AE = 10^6 \text{ kN}$; for 2–4 $EI_z = 2000 \text{ kN m}^2$, $AE = 10^5 \text{ kN}$.

For the loading shown determine the structure displacements.

Each of members 1–2 and 2–3 can be represented by a special element of the type described in Section 3.5 (see equation (3.33)) whereas member 2–4 can be represented by a standard rigid-jointed frame element (see equation (3.28)).

Since 2–3 is represented by a special element it follows that $\theta_{\bar{z}3}$ will not appear in the set of structure stiffness equations since the special element already incorporates the fact that end 3 is pinned. (Member 2–3 could, of course, alternatively be represented by a standard rigid-jointed frame element, in which case $\theta_{\bar{z}3}$ would appear in the set of structure stiffness equations; this would increase the number of equations to be solved by one.)

The structure degrees of freedom are therefore $\bar{u}_2$, $\bar{v}_2$, and $\theta_{\bar{z}2}$. It should be noted that $\theta_{\bar{z}2}$ is the rotation of joint 2 *so far as members* 2–3 *and* 2–4 *are concerned*. The rotation of end 2 of member 1–2 has a different value because of the presence of the pin, of course, and this again does not appear in the set of structure stiffness equations since a special element is used to represent this member.

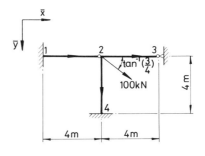

Fig. E3.4

The element stiffness contributions to the reduced structure matrix $\mathbf{K}_r$ are

$$
\begin{array}{cc}
\bar{u}_2 & \bar{v}_2 \\
\begin{bmatrix} 250000 & 0 \\ 0 & 187.5 \end{bmatrix}
\end{array}
\qquad
\begin{array}{ccc}
\bar{u}_2 & \bar{v}_2 & \theta_{\bar{z}2} \\
\begin{bmatrix} 250000 & 0 & 0 \\ 0 & 187.5 & 750 \\ 0 & 750 & 3000 \end{bmatrix}
\end{array}
$$

for elements 1–2 and 2–3 respectively (based on the use of equation (3.33)) and

$$
\begin{array}{ccc}
\bar{u}_2 & \bar{v}_2 & \theta_{\bar{z}2} \\
\begin{bmatrix} 375 & 0 & -750 \\ 0 & 25\,000 & 0 \\ -750 & 0 & 2000 \end{bmatrix}
\end{array}
$$

for element 2–4 (based on the use of equation (3.28) with $e = 0$, $f = 1$).

The applied load of 100 kN breaks down into components of 80 kN and 60 kN acting in the $\bar{x}$ and $\bar{y}$ directions respectively. The structure stiffness equations, of the form $\mathbf{F}_r = \mathbf{K}_r \mathbf{D}_r$, thus become

$$
\begin{Bmatrix} 80 \\ 60 \\ 0 \end{Bmatrix} =
\begin{bmatrix} 500375 & 0 & -750 \\ 0 & 25375 & 750 \\ -750 & 750 & 5000 \end{bmatrix}
\begin{Bmatrix} \bar{u}_2 \\ \bar{v}_2 \\ \theta_{\bar{z}2} \end{Bmatrix}.
$$

The solution of these equations is

$$
\begin{Bmatrix} \bar{u}_z \\ \bar{v}_z \\ \theta_{\bar{z}2} \end{Bmatrix} = 10^{-6}
\begin{Bmatrix} 159.38 \text{ m} \\ 2374.35 \text{ m} \\ -332.25 \text{ rad} \end{Bmatrix}.
$$

Now that the main structure displacements have been calculated the description of the structure displacements can be completed by calculating the rotation at end 2 of 1–2, i.e. $\theta_{\bar{z}2}^{1-2}$, and at the pin support, i.e. $\theta_{\bar{z}3}$. Both these rotations are obtainable using equation (3.32). Thus

$$
\theta_{\bar{z}2}^{1-2} = \frac{3}{2 \times 4} \times 2374.35 \times 10^{-6} = 890.38 \times 10^{-6} \text{ rad}
$$

and

$$
\theta_{\bar{z}3} = -\frac{3}{2 \times 4} \times 2374.35 \times 10^{-6} - \frac{1}{2} \times (-332.25 \times 10^{-6}) = -724.26 \times 10^{-6} \text{ rad}.
$$

It should be noted that if forces and moments were required these could be found in the usual way by back substituting the calculated values of the displacements into the appropriate element stiffness equations.

3.6. Neglect of axial strain in the analysis of planar rigid-jointed frames

In considering the analysis of rigid-jointed frames by the matrix displacement method account has been taken of both bending and axial deformation of the elements representing the frame members. This is, of course, the standard procedure which would normally be used, certainly in obtaining a solution on the computer. The reader will probably be aware though that there exists a large class of practical engineering framed structures in which there is only small error involved if it is assumed that the applied loads are carried wholly by bending of the frame. Such structures include those of the rectangular portal frame type and the more traditional methods of analysis of these structures include the slope deflection and moment distribution methods. The neglect of axial strain involved in the use of these methods means that the members are regarded as being inextensible, i.e. their length does not change whilst they bend under loading. The assumption of inextensibility has the advantage, of course, of reducing the work involved in obtaining a solution by reducing the number of degrees of freedom.

The purpose of the present section is to mention briefly how the assumption of inextensibility can be incorporated in the matrix displacement method. The motive is simply to make clear how solutions can be obtained by the matrix method which will be identical with those obtained by the slope deflection and moment distribution methods, although these latter methods are not described in this text. It is emphasized that it is not generally recommended that the assumption of inextensibility be adopted in the matrix displacement approach other than in the solution by hand of certain types of small-order problem.

In the matrix displacement method the neglect of axial strain in elements means that, in local co-ordinate terms, $u_i = u_j$ and hence the AE/l terms of the element stiffness matrix can be ignored. At the same time care must be taken to ensure that no element is subjected to axial strain where no corresponding axial stiffness is provided. This point is important since it is easy to misapply the inextensibility condition if sufficient care is not taken.

Two simple examples will be used to illustrate the approach.

Example 3.5. Determine the displacements at joint 2 of the rigid-jointed frame shown in Fig. E3.5, both members of which have cross-sectional area A and second moment of area I_z. Obtain solutions by in turn taking account of and neglecting axial deformations.

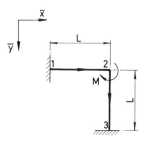

Fig. E3.5

For the full solution in which both bending and axial deformations are included the element stiffness contributions to $\mathbf{K}_r$ are, from equation (3.28),

$$
\begin{array}{ccc}
\bar{u}_2 & \bar{v}_2 & \theta_{\bar{z}2}
\end{array}
$$
$$
\begin{bmatrix}
AE/L & 0 & 0 \\
0 & 12EI_z/L^3 & -6EI_z/L^2 \\
0 & -6EI_z/L^2 & 4EI_z/L
\end{bmatrix}
\quad \text{and} \quad
\begin{array}{ccc}
\bar{u}_2 & \bar{v}_2 & \theta_{\bar{z}2}
\end{array}
\begin{bmatrix}
12EI_z/L^3 & 0 & -6EI_z/L^2 \\
0 & AE/L & 0 \\
-6EI_z/L^2 & 0 & 4EI_z/L
\end{bmatrix}.
$$

for elements 1–2 and 2–3 respectively. The structure stiffness equations are

$$
\begin{Bmatrix} 0 \\ 0 \\ M \end{Bmatrix} =
\begin{bmatrix}
AE/L + 12EI_z/L^3 & 0 & -6EI_z/L^2 \\
0 & 12EI_z/L^3 + AE/L & -6EI_z/L^2 \\
-6EI_z/L^2 & -6EI_z/L^2 & 8EI_z/L
\end{bmatrix}
\begin{Bmatrix} \bar{u}_2 \\ \bar{v}_2 \\ \theta_{\bar{z}2} \end{Bmatrix}.
$$

When a parameter s, where $s^2 = AL^2/I_z$, is introduced the solution of the full set of equations is

$$
\bar{u}_2 = \bar{v}_2 = \frac{3}{4(3+s^2)} \frac{ML^2}{EI_z}
$$

$$
\theta_{\bar{z}2} = \frac{(12+s^2)}{8(3+s^2)} \frac{ML}{EI_z}.
$$

(a)

It should be noted that s^2 will always be a large number for frame members of practical geometry. For a rectangular cross-section of depth d, for instance, $A/I_z = 12d^2$ so that $s^2 = 12(L/d)^2$ and usually $L/d > 10$. Thus, in the limit of very slender geometry when $s^2 \to \infty$, the full solution gives

$$
\bar{u}_2 = \bar{v}_2 \to 0 \qquad \theta_{\bar{z}2} \to ML/8EI_z.
$$

If the axial stiffness AE is now ignored in an attempt to account for element inextensibility a solution can still be obtained which will correspond to putting $s = 0$ in equations (a). This solution is

$$
\bar{u}_2 = \bar{v}_2 = ML^2/4EI_z \qquad \theta_{\bar{z}2} = ML/2EI_z \tag{b}
$$

which might appear plausible if the true solution were not known. However, this solution is clearly not a realistic one, giving as it does a value for the rotation at point 2 which is four times that which should apply for a very slender frame. The point is, of course, that the axial stiffness of the elements has been removed whilst the possibility of their stretching has not been eliminated.

To accommodate the specification of element inextensibility properly in fact requires that $\bar{u}_2$ and $\bar{v}_2$ be set to zero. Then the above set of three structure equations reduces to the single equation

$$
M = \frac{8EI_z}{L} \theta_{\bar{z}2} \qquad \text{or} \qquad \theta_{\bar{z}2} = \frac{ML}{8EI_z}
$$

which is the limiting value of the full solution when $s^2 \to \infty$. This is the value which would be obtained using the slope deflection or moment distribution methods.

Example 3.6. Neglecting axial straining determine the joint displacements of the rigid-jointed frame shown in Fig. E3.3(a) and previously considered in Example 3.3.

The full structure stiffness matrix, with axial straining included, is given in equation (a) of Example 3.3 (with the boundary conditions at points 3 and 4 having been applied). There, a set of six structure equations is involved, corresponding to the degrees of freedom $\bar{u}_1$, $\bar{v}_1$, $\theta_{\bar{z}1}$, $\bar{u}_2$, $\bar{v}_2$, and $\theta_{\bar{z}2}$.

If axial straining is now neglected, i.e. the elements are assumed inextensible, three conditions have to be applied to these degrees of freedom. These are that $\bar{v}_1 = \bar{v}_2 = 0$ and that $\bar{u}_1 = \bar{u}_2$. The conditions for $\bar{v}_1$ and $\bar{v}_2$ can be applied easily in the usual way by eliminating the corresponding rows and columns from the structure stiffness matrix (rows and columns 2 and 5 from the matrix of equation (a) of Example 3.3). The condition that $\bar{u}_1 = \bar{u}_2$ can be accommodated by taking $\bar{u}_1$ as the unknown and simply adding the coefficients in the $\bar{u}_2$ column (the fourth column of the aforementioned matrix) to those in the $\bar{u}_1$ column (the first column). The result of these manoeuvres is the set of equations

$$\begin{Bmatrix} F_{\bar{u}1} \\ F_{\theta\bar{z}1} \\ F_{\bar{u}2} \\ F_{\theta\bar{z}2} \end{Bmatrix} = EI_z \begin{bmatrix} 12/L^3 & -6/L^2 & 0 \\ -6/L^2 & 6/L & 1/L \\ 3/2L^3 & 0 & -3/2L^2 \\ -3/2L^2 & 1/L & 4/L \end{bmatrix} \begin{Bmatrix} \bar{u}_1 \\ \theta_{\bar{z}1} \\ \theta_{\bar{z}2} \end{Bmatrix}.$$

Two things are apparent from this set of equations. Firstly all the stiffness coefficients are proportional to EI_z, with no AE terms appearing. Secondly there are four equations (four applied force components) written in terms of only three unknowns (three displacement components). This situation arises in regard to the $F_{\bar{u}1}$ and $F_{\bar{u}2}$ applied forces. It might appear that these forces should have the prescribed values (see Fig. E3.3(a)) $F_{\bar{u}1} = P$, $F_{\bar{u}2} = 0$, and this is certainly the case in the full solution described in Example 3.3. However, by introducing here the condition that element 1–2 carries no axial strain there is now no capability present for the transfer of part of the applied force P to end 2 of the element. This consideration leads to the conclusion that it is immaterial whether the horizontal load P is imagined to act at node 1, at node 2, or is split between the two nodes. What must be the condition is that $F_{\bar{u}1} + F_{\bar{u}2} = P$; this condition is invoked by adding coefficients in the third row of the above set of equations to those in the first row and, as far as the left-hand column matrix of applied forces is concerned, putting the result equal to P. This leads directly to the following set of three equations in terms of three unknowns:

$$\begin{Bmatrix} F_{\bar{u}1} + F_{\bar{u}2} \\ F_{\theta\bar{z}1} \\ F_{\theta\bar{z}2} \end{Bmatrix} = \begin{Bmatrix} P \\ 0 \\ 0 \end{Bmatrix} = EI_z \begin{bmatrix} 27/2L^3 & -6/L^2 & -3/2L^2 \\ -6/L^2 & 6/L & 1/L \\ -3/2L^2 & 1/L & 4/L \end{bmatrix} \begin{Bmatrix} \bar{u}_1 \\ \theta_{\bar{z}1} \\ \theta_{\bar{z}2} \end{Bmatrix}$$

Solution of these equations gives

$$\{\bar{u}_1 \quad \theta_{\bar{z}1} \quad \theta_{\bar{z}2}\} = \{0.13450L \quad 0.13158 \quad 0.017544\}PL^2/EI_z.$$

For $P = 50$ kN, $L = 2$ m, and $EI_z = 4000$ kN m^2 (the values used in Example 3.3) the result is

$$\{\bar{u}_1(=\bar{u}_2) \quad \theta_{\bar{z}1} \quad \theta_{\bar{z}2}\} = 10^{-3}\{13.450 \text{ m} \quad 6.5789 \text{ rad} \quad 0.8772 \text{ rad}\}$$

and this result compares closely with that of the full solution given in Example 3.3. The bending moments at the supports and the joints can be calculated in the usual way. It is left to the reader to confirm that the magnitudes of the moments at points 1, 2, 3, and 4 for the inextensional solution are 28.070, 16.667, 54.386, and 18.421 kN m respectively; these values are only very slightly different from those found in the full solution. The present result is again identical with what would be obtained using the slope deflection or moment distribution methods.

3.7. Inclined supports

Thus far, all the structures that have been considered have been supported in such a way that the support conditions are readily expressible in terms of components of displacement acting in the directions of the global axes. Whilst this is a common situation it does sometimes happen that the support conditions cannot be expressed in this way because the supports are inclined to the global axes. To see what is meant by this consider the two planar structures shown in Fig. 3.5.

Figure 3.5(a) shows a pin-jointed frame which it is assumed is described in relation to the $\bar{x}\bar{y}$ global axes. However, there is a roller support present at point 4 which allows movement parallel to the inclined surface (in the x' direction) but not perpendicular to the inclined surface (not in the y' direction). Thus the boundary condition to be applied at this support is that $v' = 0$ and the point under discussion here is how this condition can be applied when the problem is basically formulated in terms of global, or $\bar{x}\bar{y}$, quantities. For the structure shown in Fig. 3.5(a) the solution might be simply to choose the $\bar{x}\bar{y}$ axes, whose directions are generally arbitrary, to coincide with the $x'y'$ axes. However, this might be inconvenient for a number of reasons and, in any case, the problem would

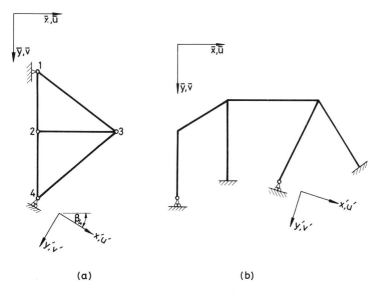

(a)   (b)

Fig. 3.5. Two plane frames with inclined supports.

still arise if more than one roller support were present, as in the case for the rigid-jointed frame shown in Fig. 3.5(b).

The problem of the inclined support can be dealt with in more than one way. Here the method presented will be based on the use of local 'support' co-ordinate systems (the $x'y'$ systems shown in the figures) and will involve transformations between the support system(s) and a global $\bar{x}\bar{y}$ system. To illustrate the method consider the pin-jointed frame shown in Fig. 3.5(a) which has an inclined support at node 4, with the support axes making an angle β_4 with the global axes.

In terms of the $\bar{x}\bar{y}$ global system the complete set of stiffness equations for the given structure, before any boundary conditions are applied, has the form

$$
\begin{Bmatrix} \mathbf{F}_1 \\ \mathbf{F}_2 \\ \mathbf{F}_3 \\ \mathbf{F}_4 \end{Bmatrix} = \begin{bmatrix} \mathbf{K}_{11} & \mathbf{K}_{12} & \mathbf{K}_{13} & \mathbf{0} \\ \mathbf{K}_{21} & \mathbf{K}_{22} & \mathbf{K}_{23} & \mathbf{K}_{24} \\ \mathbf{K}_{31} & \mathbf{K}_{32} & \mathbf{K}_{33} & \mathbf{K}_{34} \\ \mathbf{0} & \mathbf{K}_{42} & \mathbf{K}_{43} & \mathbf{K}_{44} \end{bmatrix} \begin{Bmatrix} \mathbf{D}_1 \\ \mathbf{D}_2 \\ \mathbf{D}_3 \\ \mathbf{D}_4 \end{Bmatrix}. \tag{3.35}
$$

(It is recalled that the convention adopted in this chapter is that $\mathbf{F}$, $\mathbf{K}$, and $\mathbf{D}$ are written without the bar over them but do relate to the $\bar{x}\bar{y}$ global system.)

The boundary condition that is required to be applied at node 4 is that $v'_4 = 0$. To be able to apply this condition we need to replace $\mathbf{D}_4$ by $\mathbf{D}'_4$ and $\mathbf{F}_4$ by $\mathbf{F}'_4$ in equation (3.35): this will, of course, lead to modification within the structure stiffness matrix.

At node 4 the components of displacement in the two co-ordinate systems can be related in the same fashion described earlier when transforming between element and structure co-ordinate systems. Thus, similar to equation (3.12a),

$$
\begin{Bmatrix} u'_4 \\ v'_4 \end{Bmatrix} = \begin{bmatrix} \cos \beta_4 & \sin \beta_4 \\ -\sin \beta_4 & \cos \beta_4 \end{bmatrix} \begin{Bmatrix} \bar{u}_4 \\ \bar{v}_4 \end{Bmatrix} \quad \text{or} \quad \mathbf{D}'_4 = \mathbf{T}_4 \mathbf{D}_4. \tag{3.36}
$$

It follows, given the orthogonal nature of the transformation matrix $\mathbf{T}_4$, that

$$
\mathbf{D}_4 = \mathbf{T}_4^{\mathsf{t}} \mathbf{D}'_4 \tag{3.37}
$$

and that the corresponding force components are related in the same fashion (see Section 3.2) as

$$
\mathbf{F}_4 = \mathbf{T}_4^{\mathsf{t}} \mathbf{F}'_4 \tag{3.38}
$$

where $\mathbf{F}'_4$ are the components of externally applied force at node 4 in the x' and y' directions.

Substituting these latter two relationships into equations (3.35) gives

$$\begin{Bmatrix} \mathbf{F}_1 \\ \mathbf{F}_2 \\ \mathbf{F}_3 \\ \mathbf{T}_4^t\mathbf{F}_4' \end{Bmatrix} = \begin{bmatrix} \mathbf{K}_{11} & \mathbf{K}_{12} & \mathbf{K}_{13} & \mathbf{0} \\ \mathbf{K}_{21} & \mathbf{K}_{22} & \mathbf{K}_{23} & \mathbf{K}_{24} \\ \mathbf{K}_{31} & \mathbf{K}_{32} & \mathbf{K}_{33} & \mathbf{K}_{34} \\ \mathbf{0} & \mathbf{K}_{42} & \mathbf{K}_{43} & \mathbf{K}_{44} \end{bmatrix} \begin{Bmatrix} \mathbf{D}_1 \\ \mathbf{D}_2 \\ \mathbf{D}_3 \\ \mathbf{T}_4^t\mathbf{D}_4' \end{Bmatrix} \qquad (3.39)$$

Now the $\mathbf{T}_4^t$ matrix with which $\mathbf{D}_4'$ is premultiplied in the right-hand column matrix can instead be moved inside the stiffness matrix as a postmultiplier of the terms in the fourth column; also if both sides of the fourth row are premultiplied by $\mathbf{T}_4$ the left-hand column matrix will have $\mathbf{F}_4'$ in its fourth position, since $\mathbf{T}_4\mathbf{T}_4^t = \mathbf{I}$. The result is

$$\begin{Bmatrix} \mathbf{F}_1 \\ \mathbf{F}_2 \\ \mathbf{F}_3 \\ \mathbf{F}_4' \end{Bmatrix} = \begin{bmatrix} \mathbf{K}_{11} & \mathbf{K}_{12} & \mathbf{K}_{13} & \mathbf{0} \\ \mathbf{K}_{21} & \mathbf{K}_{22} & \mathbf{K}_{23} & \mathbf{K}_{24}\mathbf{T}_4^t \\ \mathbf{K}_{31} & \mathbf{K}_{32} & \mathbf{K}_{33} & \mathbf{K}_{34}\mathbf{T}_4^t \\ \mathbf{0} & \mathbf{T}_4\mathbf{K}_{42} & \mathbf{T}_4\mathbf{K}_{43} & \mathbf{T}_4\mathbf{K}_{44}\mathbf{T}_4^t \end{bmatrix} \begin{Bmatrix} \mathbf{D}_1 \\ \mathbf{D}_2 \\ \mathbf{D}_3 \\ \mathbf{D}_4' \end{Bmatrix} \qquad (3.40)$$

and these are the required stiffness equations before the boundary conditions are applied. The latter can now be applied very simply in the usual way by eliminating rows and columns corresponding to $\bar{u}_1$, $\bar{v}_1$, and v_4', and the solution for the remaining structure displacements then proceeds in the normal fashion.

If element forces are desired remember that some of the calculated structure displacements at particular support points are components referred to the support co-ordinates (i.e. $\mathbf{D}_4'$ in the particular case considered here) and will have to be transformed back into components referred to the global system (using equation (3.37) here) before the normal procedure can be applied.

The above description of the way in which inclined supports can be accommodated has been described with regard to plane frames but similar techniques can be used in related circumstances for other kinds of structures; for example inclined supports sometimes occur on parts of the boundary of plane stress continua modelled by finite elements.

Example 3.7. Determine the displacements of the pin-jointed frame shown in Fig. E3.7(a) and hence calculate the forces in the bars. All bars have an axial rigidity of 150 000 kN.

The structure considered here is the same as that of Example 3.1 except for the support conditions at node 2. From that example the stiffness relationships relating applied forces and displacements at nodes 2, 3, and 4 (there are no

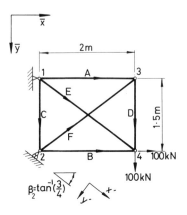

Fig. E3.7(a)

displacements at node 1) are

$$\left\{\begin{array}{c} \mathbf{F}_2 \\ \mathbf{F}_3 \\ \mathbf{F}_4 \end{array}\right\} = \left[\begin{array}{ccc} \bar{\mathbf{k}}_0^B + \bar{\mathbf{k}}_0^C + \bar{\mathbf{k}}_0^F & -\bar{\mathbf{k}}_0^F & -\bar{\mathbf{k}}_0^B \\ -\bar{\mathbf{k}}_0^F & \bar{\mathbf{k}}_0^A + \bar{\mathbf{k}}_0^D + \bar{\mathbf{k}}_0^F & -\bar{\mathbf{k}}_0^D \\ -\bar{\mathbf{k}}_0^B & -\bar{\mathbf{k}}_0^D & \bar{\mathbf{k}}_0^B + \bar{\mathbf{k}}_0^D + \bar{\mathbf{k}}_0^E \end{array}\right] \left\{\begin{array}{c} \mathbf{D}_2 \\ \mathbf{D}_3 \\ \mathbf{D}_4 \end{array}\right\}.$$

The various $\bar{\mathbf{k}}_0$ stiffness submatrices are defined in Example 3.1 and correspond, at all nodes, to the $\bar{x}\bar{y}$ reference system shown in Fig. E3.7. Substituting for the various $\bar{\mathbf{k}}_0$ gives the set of equations (in kilonewtons and metres)

$$\left\{\begin{array}{c} F_{\bar{u}2} \\ F_{\bar{v}2} \\ F_{\bar{u}3} \\ F_{\bar{v}3} \\ F_{\bar{u}4} \\ F_{\bar{v}4} \end{array}\right\} = 10^4 \left[\begin{array}{cc|cc|cc} 11.34 & -2.88 & -3.84 & 2.88 & -7.5 & 0 \\ -2.88 & 12.16 & 2.88 & -2.16 & 0 & 0 \\ \hline -3.84 & 2.88 & 11.34 & -2.88 & 0 & 0 \\ 2.88 & -2.16 & -2.88 & 12.16 & 0 & -10 \\ \hline -7.5 & 0 & 0 & 0 & 11.34 & 2.88 \\ 0 & 0 & 0 & -10 & 2.88 & 12.16 \end{array}\right] \left\{\begin{array}{c} \bar{u}_2 \\ \bar{v}_2 \\ \bar{u}_3 \\ \bar{v}_3 \\ \bar{u}_4 \\ \bar{v}_4 \end{array}\right\}$$

or

$$\left\{\begin{array}{c} \mathbf{F}_2 \\ \mathbf{F}_3 \\ \mathbf{F}_4 \end{array}\right\} = \left[\begin{array}{ccc} \mathbf{K}_{22} & \mathbf{K}_{23} & \mathbf{K}_{24} \\ \mathbf{K}_{32} & \mathbf{K}_{33} & \mathbf{K}_{34} \\ \mathbf{K}_{42} & \mathbf{K}_{43} & \mathbf{K}_{44} \end{array}\right] \left\{\begin{array}{c} \mathbf{D}_2 \\ \mathbf{D}_3 \\ \mathbf{D}_4 \end{array}\right\}.$$

To apply the displacement boundary condition at node 2 corresponding to the indicated inclined roller support, i.e. $v_2' = 0$, requires first setting up a transformation matrix $\mathbf{T}_2$ relating components of displacement in the $\bar{x}\bar{y}$ and $x'y'$ systems by the equation

$$\mathbf{D}_2' = \mathbf{T}_2 \mathbf{D}_2$$

where $\mathbf{D}_2' = \{u_2' \ v_2'\}$ (see equation (3.36)). Here $\mathbf{T}_2$ is

$$\mathbf{T}_2 = \begin{bmatrix} \cos \beta_2 & \sin \beta_2 \\ -\sin \beta_2 & \cos \beta_2 \end{bmatrix} = \begin{bmatrix} 0.8 & 0.6 \\ -0.6 & 0.8 \end{bmatrix}.$$

Following the line of development of equation (3.40) the set of structure stiffness equations which will allow the boundary conditions at node 2 to be applied are

$$\begin{Bmatrix} \mathbf{F}_2' \\ \mathbf{F}_3 \\ \mathbf{F}_4 \end{Bmatrix} = \begin{bmatrix} \mathbf{T}_2\mathbf{K}_{22}\mathbf{T}_2^t & \mathbf{T}_2\mathbf{K}_{23} & \mathbf{T}_2\mathbf{K}_{24} \\ \mathbf{K}_{32}\mathbf{T}_2^t & \mathbf{K}_{23} & \mathbf{K}_{34} \\ \mathbf{K}_{42}\mathbf{T}_2^t & \mathbf{K}_{43} & \mathbf{K}_{44} \end{bmatrix} \begin{Bmatrix} \mathbf{D}_2' \\ \mathbf{D}_3 \\ \mathbf{D}_4 \end{Bmatrix}.$$

The indicated matrix multiplications are easily performed to produce a set of six equations in which the unknowns are u_2', v_2', $\bar{u}_3$, $\bar{v}_3$, $\bar{u}_4$, $\bar{v}_4$. The boundary condition $v_2' = 0$ can now be applied in the usual way simply by eliminating the corresponding row and column. The result is the set of five stiffness equations:

$$\begin{Bmatrix} F_{u'2} \\ F_{\bar{u}3} \\ F_{\bar{v}3} \\ F_{\bar{u}4} \\ F_{\bar{v}4} \end{Bmatrix} = \begin{Bmatrix} 0 \\ 0 \\ 0 \\ 100 \\ 100 \end{Bmatrix} = 10^4 \begin{bmatrix} 8.8704 & & & & \\ -1.344 & 11.34 & & \text{Symmetric} & \\ 1.008 & -2.88 & 12.16 & & \\ -6 & 0 & 0 & 11.34 & \\ 0 & 0 & -10 & 2.88 & 12.16 \end{bmatrix} \begin{Bmatrix} u_2' \\ \bar{u}_3 \\ \bar{v}_3 \\ \bar{u}_4 \\ \bar{v}_4 \end{Bmatrix}.$$

Solving for the nodal displacements gives

$$\{u_2' \quad \bar{u}_3 \quad \bar{v}_3 \quad \bar{u}_4 \quad \bar{v}_4\} = \{-1.5046 \quad 6.2963 \quad 25.4938 \quad 0.6481 \quad 29.0355\}10^{-4}\,\text{m}$$

Calculation of the element (bar) forces can proceed by use of equation (3.25) in the same way as described in Example 3.1 except that it is first necessary at node 2 to transform back from the displacement components u_2' and v_2' to global components $\bar{u}_2$ and $\bar{v}_2$. Thus

$$\mathbf{D}_2 = \mathbf{T}_2^t \mathbf{D}_2' \quad \text{or} \quad \begin{Bmatrix} \bar{u}_2 \\ \bar{v}_2 \end{Bmatrix} = \begin{bmatrix} 0.8 & -0.6 \\ 0.6 & 0.8 \end{bmatrix} \begin{Bmatrix} -1.5046 \times 10^{-4} \\ 0 \end{Bmatrix} = \begin{Bmatrix} -1.2037 \\ -0.9028 \end{Bmatrix} \times 10^{-4}.$$

Then the forces are as detailed in the following table.

Element	AE/l	e	f	Node i		Node j		$U_i = -U_j$ $= \dfrac{AE}{l}(e(\bar{u}_i - \bar{u}_j)$ $+f(\bar{v}_i - \bar{v}_j))$
				$10^4\bar{u}_i$	$10^4\bar{v}_i$	$10^4\bar{u}_j$	$10^4\bar{v}_j$	
A	7.5×10^4	1	0	0	0	6.2963	25.4938	-47.222 (tension)
B	7.5×10^4	1	0	-1.2037	-0.9028	0.6481	29.0355	-13.889 (tension)
C	10×10^4	0	1	0	0	-1.2037	-0.9028	9.028 (compression)
D	10×10^4	0	1	6.2963	25.4938	0.6481	29.0355	-35.417 (tension)
E	6×10^4	0.8	0.6	0	0	0.6481	20.0355	-107.639 (tension)
F	6×10^4	0.8	-0.6	-1.2037	-0.9028	6.2963	25.4938	59.028 (compression)

The forces acting on the nodes from the bars plus the externally applied forces and the reaction forces are shown in Fig. E3.7(b). The reaction forces have not

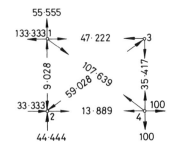

Units are in kN

Fig. E3.7(b)

been calculated here but have been found, very simply, from overall equilibrium conditions. The calculated bar forces form, with the applied forces or reaction forces as appropriate, an equilibrium set at each node.

3.8. Form of structure stiffness matrix: bandwidth

In the matrix displacement analysis of complex structures, and more especially in the finite-element analysis of continua, the number of degrees of freedom is commonly measured in the hundreds and not infrequently in the thousands. The storage of the associated structure stiffness matrix in a digital computer and the subsequent solution of the set of simultaneous equations is obviously a considerable task. For instance, if a problem has 1000 degrees of freedom the full structure stiffness matrix clearly comprises a million numbers and the total number of arithmetic operations involved in solving the equations by an elimination process, say, would be of the order $1000^3/3$. There might also be of the order of twice this number of transfers required between the working and auxiliary stores of the computer.

It is fortunate that the nature of the structure stiffness matrix in the MDM and FEM usually allows very considerable reductions to be made in computer storage and time requirements as compared with the full problem. Firstly, we have seen that the structure stiffness matrix is symmetric and so at most little more than one half of the stiffness coefficients need be stored. Secondly, it is generally possible to number the nodes of a structure so that all the non-zero stiffness coefficients are contained within a comparatively narrow band local to the leading diagonal of the stiffness matrix. Figure 3.6(a) shows diagrammatically such a banded stiffness matrix and very modest examples of banded matrices have appeared earlier in the text, for example as depicted by equation

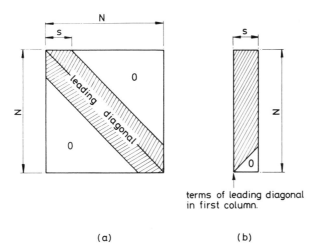

terms of leading diagonal
in first column.

(a)   (b)

Fig. 3.6. The banded nature of a structure stiffness matrix: (a) the stiffness matrix; (b) portion to be stored.

(3.35) and as appearing in Example 2.4. In Fig. 3.6(a) N is the order of the full square structure stiffness matrix and s is the *semi-bandwidth*. The semi-bandwidth is defined as the greatest number of stiffness coefficients in any row of the structure stiffness matrix from the leading diagonal to the outside right-hand non-zero coefficient, inclusive. Instead of storing and operating on N^2 stiffness coefficients we need consider only the Ns coefficients shown diagrammatically in Fig. 3.6(b) since solution techniques are readily available for banded matrices. If it is recognized that in large problems s may be only a few per cent of N it is obvious that very large savings in storage, and in computing time, can be made by taking advantage of the banded nature of the structure stiffness matrix.

The semi-bandwidth depends directly on the way in which the nodes of a structure are numbered and so it is very important when analysing structures with many freedoms to get into the habit of numbering nodes in an efficient manner. The structure-node-numbering scheme which will give the minimum possible semi-bandwidth is that in which the largest difference between node numbers in any single element is as small as possible. For instance, considering skeletal structures, if the structure stiffness matrix $\mathbf{K}$ is imagined, for the moment, as being represented in terms of stiffness submatrices, rather than in its full form in terms of individual coefficients, then a non-zero submatrix $\mathbf{K}_{PQ}$ will be present only if an element joins nodes P and Q in the structure; this submatrix will be distance $|Q-P|$ from the leading diagonal in this representation. Consequently when $\mathbf{K}$ is written in its full form in terms of the coefficients themselves the semi-bandwidth will be the maximum value for any

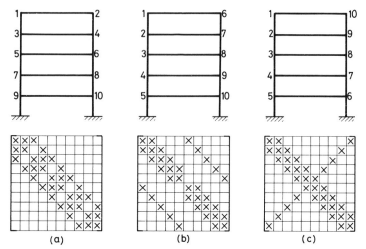

Each X represents a non-zero 3×3 submatrix (for rigid-jointed frame)
Each blank represents a null 3×3 submatrix

Fig. 3.7. Possible node numbering schemes for a simple plane frame.

element of

$$(|Q-P|+1) \times \text{number of degrees of freedom per node.}$$

To clarify the above statements Fig. 3.7 shows three possible node-numbering schemes for a simple rigid-jointed plane framework (three degrees of freedom per node) and the corresponding form of the structure stiffness matrix. Here $N = 30$ and $s = 9$, 18, and 30 for the numbering arrangements shown in Fig. 3.7(a), 3.7(b), and 3.7(c) respectively. Obviously in this simple case it is best to number nodes consecutively 'across' the structure rather than 'along' it, i.e. in the direction where least nodes occur, as in scheme (a). For more-general types of structure the efficient numbering of nodes to provide the minimum semi-bandwidth will not be so easy. No general rules apply and there is quite a skill in complex finite-element analysis in choosing the node numbering so as to keep the semi-bandwidth as small as possible.

Problems

In these problems and in their quoted solutions (see pp. 546–8) the directions of the global coordinates $\bar{x}\bar{y}$ are assumed to be as used in the examples in this chapter, i.e. $\bar{x}$ is directed horizontally to the right and $\bar{y}$ is directed vertically downwards.

3.1. Determine the horizontal displacement at A of the pin-jointed plane structure shown in Fig. P3.1. AE is the same for all bars, as is their length L. Also find the forces in all the bars.

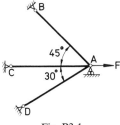

Fig. P3.1

3.2. Find the displacements of point 2 and the forces in the bars of the pin-jointed frame shown in Fig. P3.2. Both bars 1–2 and 2–3 have the same AE values.

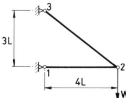

Fig. P3.2

3.3. Calculate the displacements of point B of the pin-jointed frame shown in Fig. P3.3. The cross-sectional area of BC and BE is $\sqrt{2}\,A$ and of all other members is A. Young's modulus is E for all bars. Determine the forces acting in the bars.

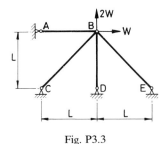

Fig. P3.3

3.4. Assemble the 10×10 structure matrix $\mathbf{K}_r$ for the pin-jointed frame shown in Fig. P3.4. The values of AE are 8 units for horizontal bars, 6 units for vertical bars, and 5 units for inclined bars.

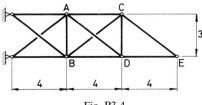

Fig. P3.4

3.5. Each of the five bars of the pin-jointed frame shown in Fig. P3.5 has a cross-sectional area of 20 cm², with $E = 200$ GN m⁻². Form $\mathbf{K_r}$ for the structure and verify that the equations $\mathbf{F_r} = \mathbf{K_r D_r}$ have the solution

$$\mathbf{D_r} = \{0.150 \quad 0.575 \quad 0.198 \quad 0.475\}\,\text{mm}$$

What are the axial forces in the bars?

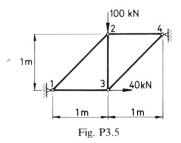

Fig. P3.5

3.6. Determine the nodal displacements of the pin-jointed frame shown in Fig. P3.6. Hence calculate the axial forces in bars 1–3 and 3–5. The extensional rigidity is AE for bars 3–4 and 4–5, and $2AE$ for the other bars.

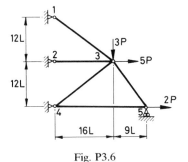

Fig. P3.6

3.7. The plane structure shown in Fig. P3.7 has a rigid joint at point 2 and is fully fixed at points 1, 3, and 4. For members 1–2 and 2–3 $I_z = 4000$ cm⁴ and $A = 40$ cm²; for member 2–4 $I_z = 2000$ cm⁴ and $A = 30$ cm². For all members $E = 200$ GN m⁻². Determine the displacements at point 2 and the end forces for the members. Draw the bending moment diagram.

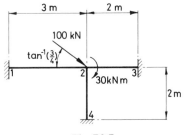

Fig. P3.7

3.8. In a consistent set of units the properties and loading of the plane rigid-jointed frame shown in Fig. P3.8 are

$EI_z = 10^7$, $AE = 10^5$ for both 1–2 and 1–3;

$L = 100$, $W_1 = 32$, $M_1 = 1050$.

Calculate the displacements at node 1 and the end forces for member 1–3.

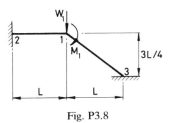

Fig. P3.8

3.9. Taking advantage of the symmetry of the problem determine the nodal displacements of the frame shown in Fig. P3.9. Draw the bending moment diagram for the left-hand half of the structure. Assume $AE = 800$ MN and $EI_z = 5$ MN m^2 for all members.

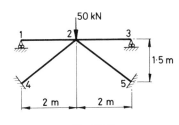

Fig. P3.9

3.10. Figure P3.10 shows a rigid-jointed plane frame. For member 5–6, $EI_z = 1$ MN m^2, $AE = 100$ MN. For all other members $EI_z = 3$ MN m^2, $AE = 500$ MN. Form an appropriate stiffness matrix for the structure, taking note of the symmetry, and calculate the nodal displacements. Hence determine the magnitudes of the axial force, the shearing force, and the bending moment at end 2 of member 2–5.

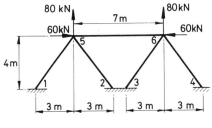

Fig. P3.10

3.11. The planar symmetric structure shown in Fig. P3.11 consists of a fixed-ended beam ACDE supported by cables BC and FD. The cables have an extensional rigidity $AE = EI_z/20$ (metre units) where EI_z is the flexural rigidity of the beam. Find the tension in the cables and the vertical reactions at A and E. Construct the bending moment diagram for the beam.

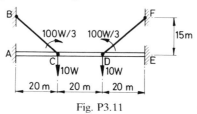

Fig. P3.11

3.12. Repeat Example 3.2, assuming that together with the application of the 60 kN load the support at point 3 moves by amounts of $+2$ mm in the $\bar{x}$ direction and $+3$ mm in the $\bar{y}$ direction.

3.13. Verify that setting $M_{zi} = 0$ in equations (3.33) leads to the stiffness matrix for the standard pin-jointed element.

3.14. Figure P3.14 shows a 'special' element of a frame in its local configuration: the element does not transmit force normal to its axis at its end j, i.e. it has a slider support at j. Derive the stiffness matrix for this element (see Section 3.5).

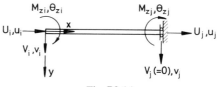

Fig. P3.14

3.15. For the beam shown in Fig. P3.15 calculate the deflection at 2 and the rotation at 3. Hence determine the rotations at 2 in both sections of the beam and find the reaction force and moment at 1. $EI_z = 3$ MN m^2.

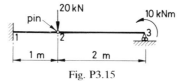

Fig. P3.15

3.16. Figure P3.16 shows a beam ABC to which the vertical member BD is pin-connected. Form matrix $\mathbf{K}_r$ for the structure, assuming that EI_z and AE are the same throughout.

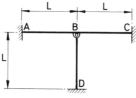

Fig. P3.16

3.17. The two members of a plane structure shown in Fig. P3.17 are connected by a pin at B and are fully clamped at A and C. Form the 2×2 matrix $\mathbf{K}_r$ and solve for the displacements $\bar{u}_B$ and $\bar{v}_B$. Also determine the rotations of the ends of both members at B. $AE = 800$ MN and $EI_z = 6$ MN m^2 throughout.

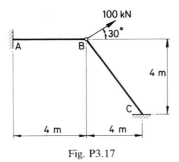

Fig. P3.17

3.18. For the rigid-jointed frame of Fig. P3.18 determine the rotations at points 1 and 2 on the assumption that axial strain can be neglected. EI_z is the same for all members. What is the value of the fixing moment at point 3?

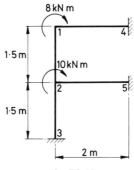

Fig. P3.18

3.19. Neglecting axial strain determine the rotations at points 1 and 2 and the horizontal displacement of member 1–2 of the rigid-jointed frame shown in Fig. P3.19. The flexural rigidity of 1–2 is twice that of 1–3 and 2–4.

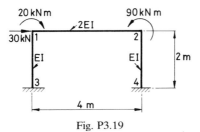

Fig. P3.19

3.20. The plane frame shown in Fig. P3.20 is rigid jointed at point 2 and there is a fully clamped support at point 1. At point 2 there is a support such that only vertical displacement can occur, whilst at point 3 only movement along the indicated plane can occur. Set up the structure stiffness equations (with two degrees of freedom) and solve for the displacements. Draw the bending moment diagram for the frame. $AE = 1000$ MN, $EI_z = 8$ MN m^2 for both members.

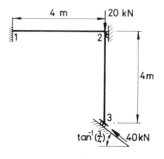

Fig. P3.20

3.21. Figure P3.21 shows a rigid-jointed plane frame for which $AE = 800$ MN and $EI_z = 4$ MN m^2 for the three members. At point 3 the support is fully clamped whilst at point 4 the support is such that rotation and movement along the plane at 30° to the horizontal are permitted. Set up the structure stiffness equations (with eight degrees of freedom) in the $\bar{x}\bar{y}$ co-ordinate system. (It should be noted that this problem differs from that of Example 3.3 only in the support conditions at point 4.)

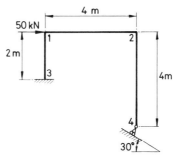

Fig. P3.21

4

OTHER KINDS OF LOADING AND OTHER KINDS OF FRAME

4.1. Introduction

Planar rigid-joined frames are often called upon to carry mechanical loads distributed along the lengths of the individual members as well as, or in place of, concentrated forces or moments at the physical joints. The manner in which the effect of such distributed loadings can be accommodated in the matrix displacement method, without increasing the total number of structure degrees of freedom, is discussed first in this chapter (Section 4.2). The procedure depends upon the use of the principle of superposition and of well-known values of fixed-end forces.

Deformation of a frame, be it pin jointed or rigid jointed, can also occur due to effects other than those of mechanical loading, whether this be concentrated at the joints or distributed. Amongst these effects probably those of temperature change and of lack of fit are of most importance in practice and these two categories are considered in Section 4.3. The effects are accommodated in the matrix displacement method in a manner analogous to the procedure adopted to deal with distributed mechanical loading.

Although the procedures related to distributed loading, temperature change, and lack of fit are introduced with regard to planar frame analysis they are clearly not restricted to such frames. Similarly the concepts described in Chapter 3 for the analysis of planar frames can be directly extended to the analysis of other types of skeletal structure.

The first new type of skeletal structure considered (Section 4.4) is the grillage. This has much in common with the planar rigid-jointed frame of the previous chapter in that it is a planar structure with rigid joints. However, now the loading acts normal to the plane of the structure rather than in the plane and the members of the grillage are subjected to shearing force and bending and twisting moments but not to axial force. Two types of three-dimensional or space frame are considered in Sections 4.5 and 4.6: these are the ball-jointed frame and the rigid-jointed frame. The ball-jointed frame is the three-dimensional equivalent of the pin-jointed frame and so the frame members carry only axial force; this makes the analysis relatively simple. On the other hand the members of a rigid-jointed space frame are generally subjected to axial force, shearing forces, bending moments, and a twisting moment. Consequently the rigid-jointed space frame is, of course, the most complicated skeletal structure to analyse. However, exactly the same principles are involved in the analysis as in the analysis of all other skeletal structures. Again element stiffness matrices expressed in the (three-dimensional) global co-ordinate system can be obtained by appropriate transformation of a local system stiffness matrix.

4.2. Representation of loading between joints

So far it has been assumed that any loading applied to a structure is in the form of concentrated loads acting at the actual physical joints of the structure. This is taken to be so in the case of the pin-jointed frame but clearly a member of a rigid-jointed frame, or a span of a beam, might be called upon to carry loads distributed along its length. In this section a procedure whereby such loads may be accommodated in the standard matrix displacement procedure is described.

One way in which loading between the physical joints can be included is by the introduction of extra nodes along the member lengths. In analysing the frame of Fig. 4.1(a), for example, the point of application of the concentrated load could simply be regarded as a structure node. Similarly the distributed loading acting along the length of the horizontal member could be approximated by a set of concentrated loads acting at suitable points along the member; these points are again extra structure nodes. Obviously the creation of extra nodes has the considerable disadvantage of increasing the size of the SSM. It is desirable to be able to incorporate the effect of loading acting other than at the physical joints without increasing the number of nodal points of a problem, i.e. to have a single element representing each physical member and to have nodes coinciding only with the physical joints. This can be done by using the principle of superposition in conjunction with well-documented values of fixed-end reactions.

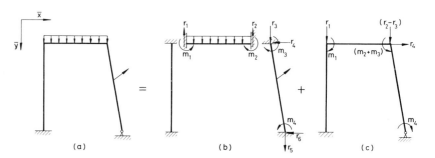

Fig. 4.1. Plane frame subjected to loading between joints: (a) actual problem; (b) set of fixed-end forces; (c) set of consistent nodal loads.

To illustrate the standard procedure consider again the frame of Fig. 4.1(a). Through the use of the superposition principle the actual problem (a) can be regarded as the sum of the problems represented by (b) and (c) of Fig. 4.1.

The load system (b) consists of the actual applied loading plus a system of forces (including, moments of course) acting at the nodes (at the physical joints) which is sufficient to prevent any translation or rotation of any node. These forces are denoted by r for translational forces and m for moments with appropriate suffices. Thus each element (physical member) of the structure is *completely fixed* at its ends and is effectively isolated from the rest of the structure. The forces acting at the ends of the elements are therefore the well-known fixed-end reaction forces. The latter are documented in many texts for a wide range of loading cases; here some of the more common loading cases are recorded in Table 4.1.

TABLE 4.1
Fixed-end reactions due to dead loading

LOADING	MOMENTS	VERTICAL FORCES	HORIZONTAL FORCES
(P, a, b)	$R_1 = -Pab^2/L^2$ $R_2 = Pa^2b/L^2$	$R_3 = Pb(L^2 + ab - a^2)/L^3$ $R_4 = Pa(L^2 + ab - b^2)/L^3$	————
(p uniform)	$R_1 = -pL^2/12$ $R_2 = pL^2/12$	$R_3 = pL/2$ $R_4 = pL/2$	————
(p, a)	$R_1 = \dfrac{-pa^2(6L^2 - 8aL + 3a^2)}{12L^2}$ $R_2 = pa^2(4aL - 3a^2)/12L^2$	$R_3 = pa(2L^3 - 2a^2L + a^3)/2L^3$ $R_4 = pa(2a^2L - a^3)/2L^3$	————
(p triangular)	$R_1 = -pL^2/30$ $R_2 = pL^2/20$	$R_3 = 3pL/20$ $R_4 = 7pL/20$	————
(M, a, b)	$R_1 = Mb(3a-L)/L^2$ $R_2 = Ma(3b-L)/L^2$	$R_3 = -6Mab/L$ $R_4 = 6Mab/L$	————
(P horizontal)	———— ————	———— ————	$R_5 = pL/2$ $R_6 = pL/2$
(P, a, b horizontal)	———— ————	———— ————	$R_5 = Pb/L$ $R_6 = Pa/L$

The load system (c) is a system of nodal forces acting on the structure now *released* in accordance with its actual boundary conditions. The nodal forces are equal in magnitude to the nodal forces of system (b) but are reversed in direction, i.e. load system (c) is the system of negative fixed-end forces. (It should be noted that forces r_5 and r_6 are not shown in system (c) since they correspond to degrees of freedom which are zero in the actual problem.) The structure displacements under the nodal load system (c) can now be found in the usual manner.

Application of the superposition principle tells us the following:

  (i) The displacements of the actual system (a) are equal to the algebraic sum of the displacements of systems (b) and (c). However, the nodal displacements of the actual system (a) are equal to the nodal displacements of system (c) alone, since there are, of course, no nodal displacements of system (b).

 (ii) The internal forces in the actual system (a) are the algebraic sum of the internal forces in systems (b) and (c). A similar statement holds for the reaction forces.

If in addition to distributed loadings there are also actual concentrated loads acting at the nodes these are simply added to the negative fixed-end forces of system (c).

The whole procedure can be itemized as follows.

Stage 1. For all elements (members) express any applied load acting along the length of an element in components referred to the local co-ordinate axes.

Stage 2. Calculate the fixed-end forces at the ends of each element referred to the local axes; standard cases are listed in Table 4.1.

Stage 3. Transform these local fixed-end forces into forces in the global system, if necessary, and apply these latter forces in reversed direction to the nodes (joints) of the structure. This gives the set of nodal loads for the structure which is consistent with (or in other words equivalent to) the actual distributed loading; this set of *consistent* or *equivalent* loads is shown in Fig. 4.1(c).

Stage 4. Calculate the structure displacements and element end forces for the *combined loading* comprising the set of consistent or equivalent nodal loads and any actual nodal loads. This is the standard matrix displacement solution procedure which is carried out in the computer for other than very simple structures. The nodal displacements obtained are those that apply to the actual structure.

Stage 5. Find the true element end forces in the structure by adding together the forces as calculated in stage 4 to the fixed-end forces of stage 2.

Example 4.1   For the uniform cantilever beam shown in Fig. E4.1(a) determine

Fig. E4.1(a)

the deflection and rotation at the free end and the vertical force and bending moment at the fixed end. The flexural rigidity is EI_z.

This is a very simple statically determinate problem, of course, but it will serve as a first illustration of the procedure used in solving a problem involving distributed loading.

The beam is modelled with a single element i-j whose stiffness matrix is given by equation (2.26) since here there is clearly no need to differentiate between local and global co-ordinate systems. Stage 1 of the solution procedure is not needed for the same reason.

Stage 2. The fixed-end forces are obtained from Table 4.1 and are shown in Fig. E4.1(b).

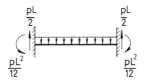

Fig. E4.1(b)

Stage 3. The consistent nodal loads are shown in Fig. E4.1(c).

Stage 4. The degrees of freedom are $\mathbf{D}_r = \{v_j \ \theta_{zj}\}$ and the corresponding applied forces are $\mathbf{F}_r = pL/12\{6 \ -L\}$. The structure stiffness equations, via use of equation (2.26), are

$$\frac{pL}{12}\left\{\begin{array}{c} 6 \\ -L \end{array}\right\} = EI_z \begin{bmatrix} 12/L^3 & -6/L^2 \\ -6/L^2 & 4/L \end{bmatrix}\left\{\begin{array}{c} v_j \\ \theta_{zj} \end{array}\right\}.$$

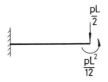

Fig. E4.1(c)

The solution of these equations is

$$v_j = pL^4/8EI_z \qquad \text{and} \qquad \theta_{zj} = pL^3/6EI_z$$

and these values of deflection and rotation at the free end are those that apply to the actual structure. The forces at the element ends, in this stage, are easily found by back substituting the calculated values of v_j and θ_{zj}, together with the values $v_i = \theta_{zi} = 0$, into equation (2.26). Thus

$$
\begin{Bmatrix} V_i \\ M_{zi} \\ V_j \\ M_{zj} \end{Bmatrix} = EI_z
\begin{bmatrix}
12/L^3 & 6/L^2 & -12/L^3 & 6/L^2 \\
6/L^2 & 4/L & -6/L^2 & 2/L \\
-12/L^3 & -6/L^2 & 12/L^3 & -6/L^2 \\
6/L^2 & 2/L & -6/L^2 & 4/L
\end{bmatrix}
\begin{Bmatrix} 0 \\ 0 \\ pL^4/8EI_z \\ pL^3/6EI_z \end{Bmatrix}
$$

$$
= pL \begin{Bmatrix} -1/2 \\ -5L/12 \\ 1/2 \\ -L/12 \end{Bmatrix}
$$

(The values for V_j and M_{zj} agree with the values of the applied forces at node j shown in Fig. E4.1(c) of course.)

Stage 5. The true forces at the element ends are the algebraic summations of the forces of Stage 4 and the fixed-end forces shown in Fig. E4.1(b). Thus the true forces are

$$
\begin{Bmatrix} V_i \\ M_{zi} \\ V_j \\ M_{zj} \end{Bmatrix} = pL \begin{Bmatrix} -1/2 \\ -5L/12 \\ 1/2 \\ -L/12 \end{Bmatrix} + pL \begin{Bmatrix} -1/2 \\ -L/12 \\ -1/2 \\ L/12 \end{Bmatrix} = pL \begin{Bmatrix} -1 \\ -L/2 \\ 0 \\ 0 \end{Bmatrix}.
$$

Therefore a vertical force of magnitude pL acts upwards and a moment of magnitude $pL^2/2$ acts anticlockwise at the fixed end; correspondingly, in the usual convention, the shearing force at this end is $+pL$ and the bending moment is $-pL^2/2$. (The zero values of V_j and M_{zj} agree with the true values of the applied forces at the free end in the actual problem (shown in Fig. E4.1(a)) of course.)

Example 4.2. The plane frame shown in Fig. E4.2(a) has the same dimensions and structural properties as the frame considered in Example 3.2. For the loading

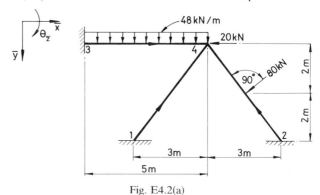

Fig. E4.2(a)

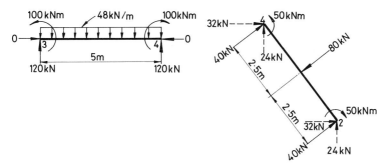

Fig. E4.2(b)

shown in Fig. E4.2(a) determine the values of the displacements at joint 4. Also construct the bending moment diagrams for members 2–4 and 3–4.

Each of the three structural members will be represented again by a single element, of course, and the solution will be described with regard to the five procedural stages noted earlier.

Stage 1. In this problem the applied loading is given in a manner directly referred to the local co-ordinate axes of the loaded elements.

Stage 2. The fixed-end forces referred to the local axes for the loaded elements 3–4 and 2–4 are calculated using the information given in Table 4.1 and are shown in Fig. E4.2(b). For the inclined element 2–4 these forces are the 40 kN forces acting normal to the element and the 50 kN moments.

Stage 3. The only transformation of local fixed-end forces required is of the 40 kN force at the ends of the inclined element; in the global system these resolve into horizontal and vertical components of magnitude 32 kN and 24 kN respectively as shown in Fig. E4.2(b).

In the global system the reversed fixed-end forces are the set of consistent (or equivalent) nodal loads, and only those loads corresponding to non-zero freedoms of the actual structure need be considered. The consistent nodal loads corresponding to freedoms $\bar{u}_4$, $\bar{v}_4$, and $\theta_{\bar{z}4}$ are (see Fig. E4.2(b))

$$\left\{\begin{array}{c} 0 \\ 120 \\ -100 \end{array}\right\} + \left\{\begin{array}{c} -32 \\ 24 \\ 50 \end{array}\right\} = \left\{\begin{array}{c} -32 \\ 144 \\ -50 \end{array}\right\}.$$

Stage 4. In addition to the loads acting along the lengths of the elements there is also an applied horizontal force acting at joint 4 itself. Therefore the final column matrix of combined nodal loads $\mathbf{F}_r$ to be used in the standard equation $\mathbf{F}_r = \mathbf{K}_r \mathbf{D}_r$ is

$$\mathbf{F}_r = \left\{\begin{array}{c} F_{\bar{u}4} \\ F_{\bar{v}4} \\ F_{\theta\bar{z}4} \end{array}\right\} = \left\{\begin{array}{c} -32 \\ 144 \\ -50 \end{array}\right\} + \left\{\begin{array}{c} -20 \\ 0 \\ 0 \end{array}\right\} = \left\{\begin{array}{c} -52 \\ 144 \\ -50 \end{array}\right\}.$$

The stiffness matrix $\mathbf{K}_r$ for the structure has already been set up in Example 3.2. Using this the structure stiffness equations for the calculation of the joint displacements $\mathbf{D}_r$ become

$$\begin{Bmatrix} -52 \\ 144 \\ -50 \end{Bmatrix} = \frac{10^4}{6250} \begin{bmatrix} 117460 & 0 & -3000 \\ 0 & 97140 & -1500 \\ -3000 & -1500 & 17500 \end{bmatrix} \begin{Bmatrix} \bar{u}_4 \\ \bar{v}_4 \\ \theta_{z4} \end{Bmatrix}.$$

The solution of these equations yields the nodal displacements of the actual problem as

$$\begin{Bmatrix} \bar{u}_4 \\ \bar{v}_4 \\ \theta_{z4} \end{Bmatrix} = \begin{Bmatrix} -321.738 \times 10^{-6}\,\text{m} \\ 899.262 \times 10^{-6}\,\text{m} \\ 1763.79 \times 10^{-6}\,\text{rad} \end{Bmatrix}$$

The forces at element nodes in the local system, corresponding to these structural displacements, can be obtained in the same manner as used in Example 3.2. Here we are interested only in elements 2–4 and 3–4, and only in the values of lateral force and bending moment. The element nodal forces of interest can be shown to be as follows: for element 2–4

$$\{V_i \quad M_{zi} \quad V_j \quad M_{zj}\}^{2-4} = \{-4.335 \quad -6.428 \quad 4.335 \quad -15.247\}$$

and for element 3–4

$$\{V_i \quad M_{zi} \quad V_j \quad M_{zj}\}^{3-4} = \{-5.096 \quad -9.213 \quad 5.096 \quad -16.269\}$$

in kilonewtons and metres.

Stage 5. The total nodal forces acting on the elements of the structure due to the actual loading shown in Fig. E4.2(a) are the sum of the nodal forces calculated in Stage 4 and the fixed-end forces shown in Fig. E4.2(b). Thus the total nodal forces (as far as bending behaviour is concerned) are as follows: for element 2–4

$$\begin{Bmatrix} V_i \\ M_{zi} \\ V_j \\ M_{zj} \end{Bmatrix}^{2-4} = \begin{Bmatrix} -4.335 \\ -6.428 \\ 4.335 \\ -15.247 \end{Bmatrix} + \begin{Bmatrix} 40.0 \\ 50.0 \\ 40.0 \\ -50.0 \end{Bmatrix} = \begin{Bmatrix} 35.665 \\ 43.572 \\ 44.335 \\ -65.247 \end{Bmatrix}$$

and for element 3–4

$$\begin{Bmatrix} V_i \\ M_{zi} \\ V_j \\ M_{zj} \end{Bmatrix}^{3-4} = \begin{Bmatrix} -5.096 \\ -9.213 \\ 5.096 \\ -16.269 \end{Bmatrix} + \begin{Bmatrix} -120.0 \\ -100.0 \\ -120.0 \\ 100.0 \end{Bmatrix} = \begin{Bmatrix} -125.096 \\ -109.213 \\ -114.904 \\ 83.731 \end{Bmatrix}.$$

A particular individual structural element is in equilibrium under the action of its nodal forces and of any loading applied along its length. The situation is illustrated for the two elements under consideration in Fig. E4.2(c) (with axial forces omitted). Having established these element force systems it is then a simple matter to determine the bending moment distributions in the elements from rudimentary considerations of equilibrium with which the reader is assumed to be familiar: the distributions are shown in Fig. E4.2(c).

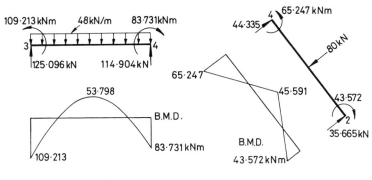

Fig. E4.2(c)

4.3. Effects of temperature change and lack of fit

Structures are quite frequently subjected to variations of temperature whose effects are sufficiently significant that they must be taken into account in the analysis procedure. Such variations may simply be seasonal or may arise as a result of some industrial process when they may be severe and localized. Temperature changes tend to lead, of course, to changes in the lengths of structure members. If such changes in length can occur freely then displacements of the structure will naturally occur but the members will not be stressed by the temperature change. This is the situation that always occurs for statically determinate structures. However, if the changes in length are completely or partially prevented by conditions at supports or by other parts of the structure then stresses will be set up in some or all of the members. This is generally the case for statically indeterminate structures though exceptionally it is possible for such structures to be subjected to temperature change and yet remain unstressed.

Another kind of effect to which a structure may be subjected is that due to a lack of fit of its various parts when it is assembled. It happens quite often, owing to errors made in the fabrication of the individual members and/or in the process of erecting the structure, that the various parts of the structure will not fit together without some force being applied. This again leads to stresses being present in some or all of the members when the structure is forcibly assembled. It is, once more, only for statically indeterminate structures that this can occur. It should be noted that sometimes a lack of fit is due to intention rather than error so that the structure is deliberately prestressed in a way which will allow it to carry an external loading which is subsequently applied more effectively.

The deformation of a structure both through temperature change and from lack of fit usually gives rise then to the situation where the structure

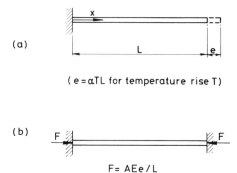

(a)

(e = αTL for temperature rise T)

(b)

F = AEe / L

Fig. 4.2. Effect of temperature change on a bar: (a) free expansion; (b) expansion prevented.

is internally strained and the members are stressed although no externally applied loads act. The two types of problem are handled in identical fashion through a superposition of effects which involves the concept of fixed-end forces just as did the procedure described in Section 4.2 for the inclusion of loads along member lengths.

Consider first the effect of temperature and restrict attention at the moment to the situation shown in Fig. 4.2(a) where a bar of length L is subjected to a temperature rise T which is uniform throughout its volume. If the bar is free to expand its length will increase by $e = \alpha TL$ where α is the coefficient of thermal expansion. If this expansion is prevented completely a compressive axial force will be set up in the bar as shown in Fig. 4.2(b); this force has magnitude $F = AEe/L$†

The procedure for the full solution of a temperature-change problem using the matrix displacement method is shown schematically in Fig. 4.3. Imagine that the frame shown at (a) is free of stress at some datum temperature and that then one or more elements (members) of the frame are subjected to temperature changes (assumed to be increases). These temperature changes are uniform throughout each element but generally may be different from element to element. In fact, for clarity, in Fig. 4.3 only one element—element ①—is assumed to be subjected to a temperature rise and this rise is T. As with the situation described in Section 4.2 the full solution can be thought of as the sum of the two partial solutions shown at (b) and (c). At (b) the ends of each element are completely fixed against any expansion of the element owing to its temperature rise from the datum state and the elements are effectively isolated from each other.

† The total strain ε is that due to stress and to temperature change, i.e. $\varepsilon = \sigma/E + \alpha T$ (See Section 6.12). This total strain is zero in the situation shown in Fig. 4.2(b) $\therefore \sigma = -E\alpha T$ or $F = -AE\alpha T = -AEe/L$ where the negative sign indicates compression.

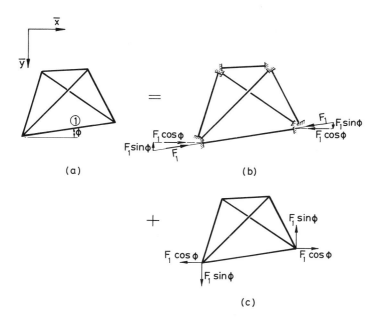

Fig. 4.3. Temperature-change procedure for a plane frame: (a) the frame, with element 1 subjected to temperature rise; (b) fixed-end forces; (c) consistent nodal loads.

The forces required to do this at the element ends are fixed-end forces corresponding to temperature change. They are the forces $F = AEe/L$ (with $e = \alpha TL$), of course, and they act along the axes of the elements but can easily be resolved into components in the directions of the $\bar{x}$ and $\bar{y}$ global axes. Thus, for element ① the global force components are $F_1 \cos \phi$ and $F_1 \sin \phi$. At (c) these force components are reversed in direction and applied to the nodes at the element ends; they can be termed consistent (or equivalent) nodal loads due to temperature change. Where more than one element meeting at a node is subjected to a temperature-change force, the element consistent nodal loads are simply added one to another. The structural problem shown at (c) can now be solved in the usual manner.

The superposition principle leads to the conclusion that the nodal displacements of the actual problem are those calculated at (c) whilst the element end forces of the actual problem are the summations of the forces corresponding to situations (b) and (c).

Before proceeding it should be noted that the frame at (a) may be either pin jointed or rigid jointed and that in both instances the structure is statically indeterminate so that a temperature rise in element ① will lead to the generation of stress. However, if the frame is regarded as pin jointed and if one of the diagonal elements is removed then the structure becomes statically determinate and no stresses will be generated by a temperature change although displacements will occur.

Now consider separately the effect of lack of fit by viewing the quantity e in Fig. 4.2(a) as the amount by which a bar is longer than the space into which it is due to fit. It can be made to fit into this space of length L by applying compressive axial forces F at the bar ends which are once more of magnitude AEe/L; these forces are the fixed-end forces due to excess length. Referring to Fig. 4.3(a) now imagine that element ① is overlong by the amount e. To fit this element into the structure the forces F_1 are applied as in Fig. 4.3(b). As before, these forces are reversed in direction (to become consistent nodal loads due to lack of fit) and applied to the nodes to arrive at the situation shown in Fig. 4.3(c). The procedure is seen to be exactly as that for the temperature-change case and the two types of problem clearly do not differ in principle; to convert from one type of problem to the other only requires equating αTL in the temperature-change problem to the excess length e in the lack-of-fit problem.

In the above only the simplest kinds of temperature-change effect and lack-of-fit effect have been considered in which it is the change of length of the element which leads to fixed-end forces acting in the axial direction. Other situations may arise which will result in different types of fixed-end force and we shall now consider some of these situations which will lead to the occurrence of fixed-end moments.

In problems of temperature change it is quite common for one or more elements of a structure to be subjected to a thermal gradient rather than to a uniform temperature change. Such a situation is shown in Fig. 4.4(a) where a beam lying in the xy plane and having a cross-section which is symmetric about this plane is heated such that the temperature rise of its upper surface is T_2 and that of its lower surface is T_1. (The temperature does not vary along the beam length). A typical beam cross-section is shown to enlarged scale in Fig. 4.4(b); d is the depth of the beam and f is the distance from the top edge to the centroid of the cross-section through which the beam axis passes. The manner of variation of the temperature through the depth is shown in Fig. 4.4(c) where it can be seen that it is assumed that $T_1 > T_2$ and that the variation is linear. If the beam is free to expand, as it is in Fig. 4.4(a), there will in general occur stretching of the centroidal axis plus bending, the latter taking place as a result of the difference in thermal strain through the beam depth. The stretching behaviour (which is not shown in Fig. 4.4) will be prevented if fixed-end axial forces of the type described above, which have magnitude $AE\alpha T$, act. It should be noted that now T is the temperature rise at the centroid, i.e. $T = T_2 + (T_1 - T_2)f/d$. The bending behaviour will be the same at all points along the beam and therefore the beam will bend in a

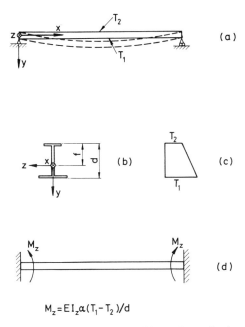

$$M_z = EI_z\alpha(T_1 - T_2)/d$$

Fig. 4.4. Beam subjected to thermal gradient: (a) the beam in the xy plane; (b) beam cross-section; (c) temperature variation through depth; (d) fixed-end moments.

circular arc. By considering the strain at the top or bottom edge which is algebraically additional to the centroidal strain it is quite easy to show using the simple bending formula that the fixed-end moments which would be required to prevent the bending deformation have the magnitude $M_z = EI_z\alpha(T_1 - T_2)/d$. These fixed-end moments are shown in Fig. 4.4(d). It is noted that the magnitude of these moments does not depend on the element length, just as the magnitude of the fixed-end axial force due to a uniform temperature change is independent of element length.

The analogous situation to that just described occurs in lack-of-fit problems when an element has an initial out-of-straightness such that it forms a shallow circular arc. Assume that the curved length of the element L is the correct length for fitting into the rest of the structure, when straightened, but that there is a small out-of-straightness of amount e at the centre of the element. It is simple to show that end moments of magnitude $8EI_z e/L^2$ will be required to straighten the element prior to fitting; these are the fixed-end moments due to initial circular curvature.

In summary, the fixed-end forces corresponding to the effects discussed in this section are presented together in Table 4.2 and the stages in the solution procedure are itemized as follows.

Stage 1. Calculate fixed-end forces due to temperature or lack-of-fit effects for each element in its local co-ordinate system using Table 4.2.

TABLE 4.2

Fixed-end reactions due to temperature change and lack of fit

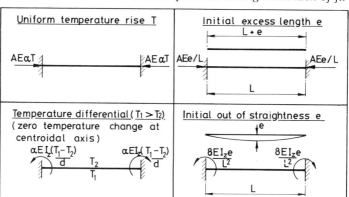

Uniform temperature rise T	Initial excess length e
$AE\alpha T$ $AE\alpha T$	AEe/L AEe/L
Temperature differential ($T_1 > T_2$) (zero temperature change at centroidal axis) $\frac{\alpha E I_z(T_1 - T_2)}{d}$ $\frac{\alpha E I_z(T_1 - T_2)}{d}$	Initial out of straightness e $\frac{8EI_z e}{L^2}$ $\frac{8EI_z e}{L^2}$

Stage 2. Transform these local fixed-end forces into the global system, if necessary, and apply in reversed direction to the nodes of the structure. These latter forces are the set of consistent (or equivalent) nodal loads due to temperature or lack-of-fit effects.

Stage 3. Calculate the structure nodal displacements and the element end forces due to the consistent nodal loads. This is the standard matrix displacement solution procedure, which is usually carried out in the computer. The nodal displacements obtained in this stage are those that apply to the actual structure.

Stage 4. Find the actual element end forces in the structure by adding together the forces as calculated in Stage 3 to the fixed-end forces of Stage 1.

Finally, note that three separate effects have now been considered, each of which gives rise to consistent nodal loads: these are the effects of temperature change and lack of fit considered in this section and the effect of applied dead loads distributed along the lengths of elements considered in Section 4.2. In some problems more than one of these effects may be present and additionally there may be actual loads applied at the nodes. In such circumstances the effects are simply additive, by the superposition principle. Thus the column matrix of combined nodal loads, occurring in the set of equations $\mathbf{F}_r = \mathbf{K}_r\mathbf{D}_r$, is the summation of column matrices of actual nodal loads and of any consistent nodal loads due to any of the effects of distributed loads and/or temperature change and/or lack of fit.

Example 4.3. The five bars of the pin-jointed frame shown in Fig. E4.3(a) all have an axial rigidity of 150 000 kN and a coefficient of thermal expansion of $11 \times 10^{-6}\,°C^{-1}$.

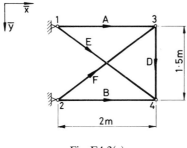

Fig. E4.3(a)

(a) If bars B and D are subjected to a uniform temperature rise of 40 °C and bar E to a uniform rise of 20 °C determine the displacements at points 3 and 4 and the forces in the bars.

(b) If bar F is removed and the same temperature rises as in part (a) are applied determine once more the displacements at 3 and 4 and the bar forces.

(c) Consider again the full frame shown in Fig. E4.3(a) with now no temperature changes involved. Imagine that the bar E is 1 mm shorter than its design length of 2.5 m and hence has to be stretched to fit into its allotted position. What forces are present in the bars of the frame when it is assembled?

(a) It is noted that the frame, and the boundary conditions, are the same as that of Example 3.1.

Stage 1. The fixed-end forces due to the temperature rises are shown in Fig. E4.3(b). These forces are $AE\alpha T = 150 \times 10^3 \ (11 \times 10^{-6})T = 1.65T$ kN. For the inclined bar E the fixed-end forces of 33 kN, shown by the full lines, have horizontal and vertical components of 26.4 kN and 19.8 kN respectively, shown by the broken lines.

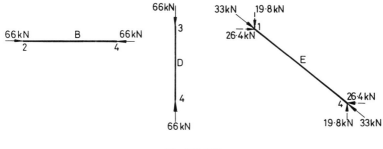

Fig. E4.3(b)

Stage 2. The consistent nodal loads due to the temperature rises are shown in Fig. E4.3(c).

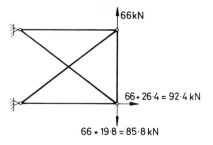

Fig. E4.3(c)

Stage 3. The structure stiffness matrix corresponding to degrees of freedom $\mathbf{D}_r = \{\bar{u}_3 \ \bar{v}_3 \ \bar{u}_4 \ \bar{v}_4\}$ has already been derived in part (b) of Example 3.1 (see equation (a) there) and can be used directly here. With the nodal loads shown in Fig. E4.3(c) the present structure stiffness equations become

$$
\left\{ \begin{array}{c} 0 \\ -66 \\ 92.4 \\ 85.8 \end{array} \right\} = 10^4 \left[\begin{array}{cccc} 11.34 & -2.88 & 0 & 0 \\ -2.88 & 12.16 & 0 & -10 \\ 0 & 0 & 11.34 & 2.88 \\ 0 & -10 & 2.88 & 12.16 \end{array} \right] \left\{ \begin{array}{c} \bar{u}_3 \\ \bar{v}_3 \\ \bar{u}_4 \\ \bar{v}_4 \end{array} \right\}.
$$

The solution gives the following values for the (actual) displacements at points 3 and 4:

$$
\{\bar{u}_3 \ \bar{v}_3 \ \bar{u}_4 \ \bar{v}_4\} = \{-1.0864 \ \ -4.2778 \ \ 7.7136 \ \ 1.7111\}10^{-4} \text{ m}.
$$

The element forces due to these displacements can be found in the same manner as described in Example 3.1 (i.e. by back substitution of the calculated displacements to the stiffness equations of each element in turn). It is left to the reader to verify that the forces $U_i \ (= -U_j)$ for each element (the directions of the local axes are indicated in Fig. E4.3(a)) are as given in the third column of the table that follows in Stage 4.

Stage 4. The actual (or total) element forces are the summations of Stage 1 and Stage 3 forces and are given in the last column of the following table. The inclined bars are in tension; the others are in compression.

	Element force $U_i \ (-U_j)$(kN)		
Element	Stage 1	Stage 3	Total
A	0	8.148	8.148
B	66	−57.852	8.148
D	66	−59.889	6.111
E	33	−43.185	−10.185
F	0	−10.185	−10.185

(b) With bar F removed from the structure the structure stiffness matrix is correspondingly changed by the removal of $\mathbf{k}_0^l$ from the top left-hand quadrant of the structure stiffness matrix used in part (a): $\bar{\mathbf{k}}_0^F$ is detailed in Example 3.1.

The fixed-end forces and consistent nodal loads are unchanged, so Stages 1 and 2 are as for part (a).

Stage 3. The structure stiffness equations become

$$\begin{Bmatrix} 0 \\ -66 \\ 92.4 \\ 85.8 \end{Bmatrix} = 10^4 \begin{bmatrix} 7.5 & 0 & 0 & 0 \\ 0 & 10 & 0 & -10 \\ 0 & 0 & 11.34 & 2.88 \\ 0 & -10 & 2.88 & 12.16 \end{bmatrix} \begin{Bmatrix} \bar{u}_3 \\ \bar{v}_3 \\ \bar{u}_4 \\ \bar{v}_4 \end{Bmatrix}$$

and the solution is

$$\{\bar{u}_3 \quad \bar{v}_3 \quad \bar{u}_4 \quad \bar{v}_4\} = \{0 \quad -9.1667 \quad 8.8000 \quad -2.5667\}10^{-4} \text{ m}.$$

The element forces U_i ($=-U_j$) corresponding to these nodal displacements are found to be 0, -66 kN, -66 kN, and -33 kN for elements A, B, D, and E respectively.

Stage 4. The Stage 1 values of U_i are 0, 66 kN, 66 kN, and 33 kN for elements A, B, D and E respectively. Thus the actual forces, obtained by adding the forces of Stages 1 and 3, are zero for all parts of the structure. This is the expected result since the removal of bar F has resulted in a statically determinate structure; as has been stated in Section 4.3 such a structure will be displaced when the temperature changes but no internal forces will be developed.

(c) *Stage 1.* The fixed-end forces due to lack of fit in element E are $AEe/L = 150 \times 10^3 \times 1 \times 10^{-3}/2.5 = 60$ kN. These forces are shown in Fig. E4.3(d).

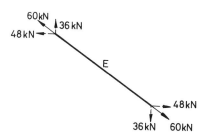

Fig. E4.3(d)

Stage 2. The consistent nodal loads at node 4 are -48 and -36 kN referred to the $\bar{x}\bar{y}$ system.

Stage 3. The structure stiffness equations are

$$\begin{Bmatrix} 0 \\ 0 \\ -48 \\ -36 \end{Bmatrix} = 10^4 \begin{bmatrix} 11.34 & -2.88 & 0 & 0 \\ -2.88 & 12.16 & 0 & -10 \\ 0 & 0 & 11.34 & 2.88 \\ 0 & -10 & 2.88 & 12.16 \end{bmatrix} \begin{Bmatrix} \bar{u}_3 \\ \bar{v}_3 \\ \bar{u}_4 \\ \bar{v}_4 \end{Bmatrix}$$

and the solution is

$$\{\bar{u}_3 \quad \bar{v}_3 \quad \bar{u}_4 \quad \bar{v}_4\} = \{-1.9753 \quad -7.7778 \quad -1.9753 \quad -8.8889\}10^{-4} \text{ m}.$$

The element forces corresponding to these displacements are again found in the manner described in Example 3.1. These forces are given in the third column of the table that follows in Stage 4.

Stage 4. The actual element forces are obtained by adding the Stage 1 and
Stage 3 forces and are given in the last column of the following table.

| | Element force U_i $(=-U_j)$ kN | | |
Element	Stage 1	Stage 3	Total
A	0	14.815	14.815
B	0	14.815	14.815
D	0	11.111	11.111
E	−60	41.481	−18.519
F	0	−18.519	−18.519

Example 4.4. Consider again the rigid-jointed structure of Examples 3.2 and 4.2.
Ignore the loadings present in these earlier examples and instead consider the
situation in which member 2–4 is subjected to a temperature increase of 10 °C on
its inside surface (i.e. its surface nearest to point 1) and 40 °C on its outside
surface, with a uniform gradient in between. The depth of member 2–4 is 0.2 m
and it has a doubly symmetric cross-section. $\alpha = 10^{-5}\,°C^{-1}$. Calculate the displacements at joint 4 and the axial force in member 2–4 due to the temperature change
and construct the bending moment diagram for member 2–4.

Details of the structure and of the elements used to model it are recorded in
Example 3.2.
Stage 1. The stated temperature increase would, if element 2–4 were unre-
strained, lead to both extension and circular bending of the element. Hence the
fixed-end forces will comprise axial forces and moments. The axial forces are
calculated on the basis of a temperature rise of 25 °C at the centroid of the section
and since AE is 60×10^4 kN these forces have magnitude $60 \times 10^4 \times 10^{-5} \times 25 =$
150 kN. With $EI_z = 1.25 \times 10^4$ the end moments are

$$\frac{\alpha EI_z (T_1 - T_2)}{d} = \frac{10^{-5} \times 1.25 \times 10^4 (40 - 10)}{0.2} = 18.75 \text{ kN m.}$$

These fixed-end forces are shown in Fig. E4.4(a).

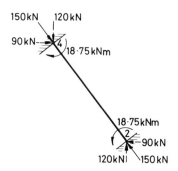

Fig. E4.4(a)

Stage 2. The consistent nodal loads in the global system corresponding to degrees of freedom $\bar{u}_4$, $\bar{v}_4$, and $\theta_{\bar{z}4}$ are clearly -90 kN, -120 kN, and -18.75 kN m respectively.

Stage 3. The structure stiffness matrix has been set up in Example 3.2 and can, of course, be used directly here. The set of stiffness equations becomes

$$
\begin{Bmatrix} F_{\bar{u}4} \\ F_{\bar{v}4} \\ F_{\theta\bar{z}4} \end{Bmatrix} = \begin{Bmatrix} -90 \\ -120 \\ -18.75 \end{Bmatrix} = 1.6 \begin{bmatrix} 117460 & 0 & -3000 \\ 0 & 97140 & -1500 \\ -3000 & -1500 & 17500 \end{bmatrix} \begin{Bmatrix} \bar{u}_4 \\ \bar{v}_4 \\ \theta_{\bar{z}4} \end{Bmatrix}.
\tag{a}
$$

The solution of these equations gives the actual nodal displacements at node 4 as follows:

$$\{\bar{u}_4 \quad \bar{v}_4 \quad \theta_{\bar{z}4}\} = \{-4.9990 \text{ m} \quad -7.8478 \text{ m} \quad -8.2261 \text{ rad}\}10^{-4}.$$

For element 2–4 the nodal forces corresponding to these displacements can now be calculated in the usual manner, exactly as in Example 3.2. The result is (in kilonewtons and metres)

$$\{U_i \quad V_i \quad M_{zi} \quad U_j \quad V_j \quad M_{zj}\}^{2-4}$$
$$= \{-111.332 \quad -2.553 \quad -4.326 \quad 111.332 \quad 2.553 \quad -8.439\}.$$

Stage 4. The actual nodal forces for element 2–4 are obtained by adding the forces of Stage 3 to those of Stage 1 shown in Fig. E4.4(a). Thus

$$
\begin{Bmatrix} U_i \\ V_i \\ M_{zi} \\ U_j \\ V_j \\ M_{zj} \end{Bmatrix} = \begin{Bmatrix} 150 \\ 0 \\ -18.75 \\ -150 \\ 0 \\ 18.75 \end{Bmatrix} + \begin{Bmatrix} -111.332 \\ -2.553 \\ -4.326 \\ 111.332 \\ 2.553 \\ -8.439 \end{Bmatrix} = \begin{Bmatrix} 38.668 \text{ kN} \\ -2.553 \text{ kN} \\ -23.076 \text{ kN m} \\ -38.668 \text{ kN} \\ 2.553 \text{ kN} \\ 10.311 \text{ kN m} \end{Bmatrix}
$$

These forces are shown in Fig. E4.4(b) and from them the bending moment diagram is drawn. Also it is seen that the axial force in element 2–4 is 38.668 kN, compressive.

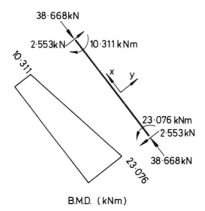

B.M.D. (kNm)

Fig. E4.4(b)

This completes the problem, but before leaving it it is emphasized that there would be little increase in difficulty if the temperature change considered here acted in conjunction with the application of a mechanical, or dead, loading of the sort applied to this same structure in Example 4.2 (see Fig. E4.2(a)). In such a case fixed-end forces could be calculated separately for the distributed dead loading and the temperature change and simply summed to give total fixed-end forces. The column matrix of combined nodal forces for the mechanical loading considered in Example 4.2 (including the load of 20 kN actually acting at node 4) and the temperature effect considered above would then become

$$\begin{Bmatrix} F_{\bar{u}4} \\ F_{\bar{v}4} \\ F_{\theta\bar{z}4} \end{Bmatrix} = \begin{Bmatrix} -52 \\ 144 \\ -50 \end{Bmatrix} + \begin{Bmatrix} -90 \\ -120 \\ -18.75 \end{Bmatrix} = \begin{Bmatrix} -142 \\ 24 \\ -68.75 \end{Bmatrix}.$$

The solution would proceed in the same fashion as when the distributed dead loading or the temperature change were acting alone.

4.4. The grillage

The grillage, or grid, provides the final example of two-dimensional skeletal structural analysis. The grillage comprises an assemblage of line members rigidly connected together where they meet at the joints, with all members and joints lying in a common plane. This description of the structure itself is, of course, identical to that of the planar rigid-jointed frame. The difference between structures designated 'grillages' and those designated 'planar rigid-jointed frames' arises solely in the nature of the loading applied to the structure and the consequent nature of the structural response. With the planar frame it has been seen that all loading is applied in the plane of the frame and it follows that all displacement occurs in that plane; the members of the frame are subjected to bending in the plane of the frame and to extension. With the grillage all loading is applied normal to the plane of the structure and consequent displacement is in that direction; the members of the grillage are subjected to bending out of the plane of the structure and to twisting. Figure 4.5(a) shows an example of a grillage which lies in the $\bar{x}\bar{y}$ plane and carries a lateral load acting in the $\bar{z}$ direction and couples acting about the $\bar{x}$ and $\bar{y}$ axes (i.e. couples whose vectors lie in the $\bar{x}\bar{y}$ plane).

A typical element of the grillage structure combines the functions of the torsionally loaded bar element and the beam element of Chapter 2, with axial displacement assumed not to exist. The element with its local co-ordinate system is shown in Fig. 4.5(b). It is seen that the nodal force quantities are a lateral force W, moment M_y, and twisting moment M_x; corresponding displacement quantities are lateral displacement w and rotations θ_y and θ_x. The xz plane is assumed to be a plane of symmetry for the element cross-section so that bending takes place in the xz plane

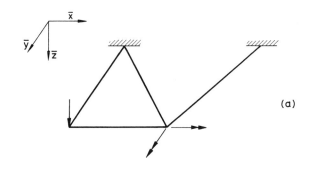

(a)

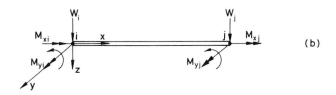

(b)

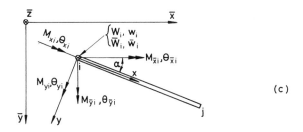

(c)

Fig. 4.5. Representation of a grillage structure: (a) the structure; (b) typical element showing local quantities; (c) typical element lying in the global $\overline{xy}$ plane.

or in other words about the y axis. The relevant second moment of area of the section I_y is thus a principal second moment of area.

One slight complication is apparent from Fig. 4.5 and this relates to the choice of co-ordinate axes and the signs of some of the nodal quantities. This arises since we wish to retain the use of a mutually perpendicular right-handed set of co-ordinate axes x, y, z. With x and z assumed in the directions indicated it follows that the y axis and consequently the M_y and θ_y moment and rotation vectors point in the direction out of the paper, i.e. M_y and θ_y act counterclockwise as viewed in Fig. 4.5(b). This conflicts with the situation shown in Fig. 3.1(b) for bending in the xy plane to which the stiffness equations (3.26) refer. Thus if we wish to use equations (3.26) as the basis for the bending contribution of the present grillage element we have to change the signs of those stiffness coefficients which relate lateral force to rotation or relate moment to lateral displacement. The stiffness equations for the grillage element in its local co-ordinate system, with the torsional contribution also included, are therefore as follows:

$$
\begin{Bmatrix} M_{xi} \\ M_{yi} \\ W_i \\ \hline M_{xj} \\ M_{yj} \\ W_j \end{Bmatrix} = \begin{bmatrix} GJ/l & & & & & \\ 0 & 4EI_y/l & & & \text{Symmetric} & \\ 0 & -6EI_y/l^2 & 12EI_y/l^3 & & & \\ \hline -GJ/l & 0 & 0 & GJ/l & & \\ 0 & 2EI_y/l & -6EI_y/l^2 & 0 & 4EI_y/l & \\ 0 & 6EI_y/l^2 & -12EI_y/l^3 & 0 & 6EI_y/l^2 & 12EI_y/l^3 \end{bmatrix} \begin{Bmatrix} \theta_{xi} \\ \theta_{yi} \\ w_i \\ \hline \theta_{xj} \\ \theta_{yj} \\ w_j \end{Bmatrix}
$$

(4.1)

or

$$
\begin{Bmatrix} \mathbf{P}_i \\ \mathbf{P}_j \end{Bmatrix} = \begin{bmatrix} \mathbf{k}_{ii} & \mathbf{k}_{ij} \\ \mathbf{k}_{ji} & \mathbf{k}_{jj} \end{bmatrix} \begin{Bmatrix} \mathbf{d}_i \\ \mathbf{d}_j \end{Bmatrix} \text{ or } \mathbf{P} = \mathbf{kd} \text{ again.}
$$

The particular order of arrangement of terms in the column matrices $\mathbf{P}$ and $\mathbf{d}$ has been adopted simply to make obvious the close relationship that exists between the grillage and the rigid-jointed plane frame (see equation (3.26)).

We now require to transform the stiffness equations (4.1) to relate to the global $\bar{x}\bar{y}\bar{z}$ co-ordinate system. To this end Fig. 4.5(c) shows a plan view of the grillage element lying in the $\bar{x}\bar{y}$ plane, with α being the angle through which the global $\bar{x}\bar{y}$ axes must be rotated in the clockwise direction about the $\bar{z}$ axis to come into coincidence with the local xy axes. The directions of the global $\bar{z}$ axis and the local z axis are the same, of course, so that there is a one-to-one relationship between lateral forces and between lateral displacements in the two systems; that is $W = \bar{W}$ and $w = \bar{w}$ at both nodes. Further, the transformation from one system to the other of moments and rotations presents no difficulty for these quantities are vector quantities in the same way as are translational forces and displacements, and they can therefore be handled in identical fashion to that described earlier in Sections 3.3 and 3.4 for the translational quantities. It is then clear on comparing Fig. 4.5(c) with Fig. 3.2 and noting the form of equation (3.12) that the displacements in the local and global systems are related in the following way:

$$
\begin{Bmatrix} \theta_{xi} \\ \theta_{yi} \\ w_i \end{Bmatrix} = \begin{bmatrix} e & f & 0 \\ -f & e & 0 \\ 0 & 0 & 1 \end{bmatrix} \begin{Bmatrix} \theta_{\bar{x}i} \\ \theta_{\bar{y}i} \\ \bar{w}_i \end{Bmatrix}
$$

(4.2)

or

$$\mathbf{d}_i = \mathbf{T}\bar{\mathbf{d}}_i$$

Here e and f are defined precisely as in equation (3.13) and the transformation matrix (or rotation matrix) $\mathbf{T}$ is identical to that used in the analysis of planar rigid-jointed frames (see equation (3.27)).

With $\mathbf{T}$ and $\mathbf{k}$ (equation (4.1)) established the 6×6 element stiffness matrix $\bar{\mathbf{k}}$ referred to the global system can be determined explicitly in the same way as described for the planar rigid-jointed frame in Section 3.4. The result is given in equation (4.3). Furthermore the calculation of the grillage element end forces once the structure displacements are known can be made by use of relationships of the form of equations (3.29) and (3.30).

$$
\bar{\mathbf{k}} =
\begin{bmatrix}
& \theta_{\bar{x}i} & \theta_{\bar{y}i} & \bar{w}_i & \theta_{\bar{x}j} & \theta_{\bar{y}j} & \bar{w}_j \\[4pt]
\dfrac{GJe^2}{l}+4\dfrac{EI_y}{l}f^2 & & & & & \\[10pt]
\left(\dfrac{GJ}{l}-4\dfrac{EI_y}{l}\right)ef & \dfrac{GJf^2}{l}+4\dfrac{EI_y}{l}e^2 & & & \text{Symmetric} & \\[10pt]
\dfrac{6EI_y f}{l^2} & -6\dfrac{EI_y}{l^2}e & 12\dfrac{EI_y}{l^3} & & & \\[10pt]
\dfrac{-GJe^2}{l}+2\dfrac{EI_y}{l}f^2 & -\left(\dfrac{GJ}{l}+2\dfrac{EI_y}{l}\right)ef & 6\dfrac{EI_y}{l^2}f & \dfrac{GJe^2}{l}+4\dfrac{EI_y}{l}f^2 & & \\[10pt]
-\left(\dfrac{GJ}{l}+2\dfrac{EI_y}{l}\right)ef & \dfrac{-GJ}{l}f^2+2\dfrac{EI_y}{l}e^2 & -6\dfrac{EI_y}{l^2}e & \left(\dfrac{GJ}{l}-4\dfrac{EI_y}{l}\right)ef & \dfrac{GJ}{l}f^2+4\dfrac{EI_y}{l}e^2 & \\[10pt]
-6\dfrac{EI_y}{l^2}f & 6\dfrac{EI_y}{2}e & -12\dfrac{EI_y}{l^3} & -6\dfrac{EI_y}{l^2}f & 6\dfrac{EI_y}{l^2}e & 12\dfrac{EI_y}{l^3}
\end{bmatrix}
$$

$$(4.3)$$

In this section and in Section 3.4 stiffness relationships have been developed for two separate types of rigid-jointed structure, all the elements of which lie in a common plane. These relationships have a quite similar form and the common features mean that there is little difference between solving a gridwork problem and solving a rigid-jointed plane frame problem, whether manually or by use of the computer. In the former type of problem the planar frame is subjected to wholly out-of-plane loading and in the latter to wholly in-plane loading. It is noted, however, that a general loading could always be considered as a sum of out-of-plane and in-plane components. Thus a solution to the problem of a general loading on the planar structure could be obtained by superposition of separate solutions for out-of-plane and in-plane loadings.

Example 4.5.
(a) Figure E4.5 shows a plan view of a grillage lying in the $\bar{x}\bar{y}$ plane. For members 2–5 and 3–6 the flexural and torsional rigidities are $EI_y = 6.4$ MN m^2 and $GJ = 4$MN m^2. For all other members $EI_y = 12.5$ MN m^2 and $GJ = 10$ MN m^2. Equal loads of magnitude 50 kN are applied at points 5 and 6 in the direction normal to the plane of the grillage, i.e. in the $\bar{z}$ direction. Determine the displacements of the grillage at points 5 and 6.

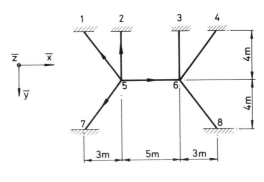

Fig. E4.5

(b) Consider that it is now required to eliminate the rotation of member 5–6 about its longitudinal axis, induced by the two 50 kN loads, by applying appropriate additional couples of equal magnitude at points 5 and 6. What should the magnitudes of these couples be?

(a) The problem is obviously a symmetric one and in a manual solution the 'alternative' method of Section 2.10 will involve the least calculation and will therefore be used here. It is only necessary, then, to set up equations for the forces at node 5 in terms of the displacements at nodes 5 and 6 initially. The local co-ordinates of all elements (members) meeting at structure node 5 are chosen such that end i coincides with node 5. Then for elements 5–2, 5–1, and 5–7 it is only necessary to calculate the submatrices $\mathbf{k}_{ii}$ whilst for element 5–6 we shall need submatrices $\bar{\mathbf{k}}_{ii}$ and $\bar{\mathbf{k}}_{ij}$. It is convenient to tabulate the relevant element properties in the following form.

Element	EI_y (kN m^2)	GJ (kN m^2)	l (m)	$e = (\bar{x}_j - \bar{x}_i)/l$	$f = (\bar{y}_j - \bar{y}_i)/l$
5–2	6 400	4 000	4	0	−1
5–1	12 500	10 000	5	−3/5	−4/5
5–7	12 500	10 000	5	−3/5	4/5
5–6	12 500	10 000	5	1	0

The required element stiffness submatrices can now be calculated using equation (4.3) and are, in kilonewtons and metres.

$$\bar{\mathbf{k}}_{ii}^{5-2} = \begin{bmatrix} 6400 & 0 & -2400 \\ 0 & 1000 & 0 \\ -2400 & 0 & 1200 \end{bmatrix} \qquad \bar{\mathbf{k}}_{ii}^{5-1} = \begin{bmatrix} 7120 & -3840 & -2400 \\ -3840 & 4880 & 1800 \\ -2400 & 1800 & 1200 \end{bmatrix}$$

$$\bar{\mathbf{k}}_{ii}^{5-7} = \begin{bmatrix} 7120 & 3840 & 2400 \\ 3840 & 4880 & 1800 \\ 2400 & 1800 & 1200 \end{bmatrix} \qquad \bar{\mathbf{k}}_{ii}^{5-6} = \begin{bmatrix} 2000 & 0 & 0 \\ 0 & 10\,000 & -3000 \\ 0 & -3\,000 & 1200 \end{bmatrix}$$

$$\bar{\mathbf{k}}_{ij}^{5-6} = \begin{bmatrix} -2000 & 0 & 0 \\ 0 & 5000 & 3000 \\ 0 & -3000 & -1200 \end{bmatrix}.$$

The relevant structure stiffness equations are therefore

$$\begin{Bmatrix} F_{\bar{x}5} \\ F_{\bar{y}5} \\ F_{\bar{w}5} \end{Bmatrix} = \begin{Bmatrix} 0 \\ 0 \\ 50 \end{Bmatrix} = \begin{bmatrix} \theta_{\bar{x}5} & \theta_{\bar{y}5} & \bar{w}_5 & \theta_{\bar{x}6} & \theta_{\bar{y}6} & \bar{w}_6 \\ 22\,640 & 0 & -2400 & -2000 & 0 & 0 \\ 0 & 20\,760 & 600 & 0 & 5000 & 3000 \\ -2400 & 600 & 4800 & 0 & -3000 & -1200 \end{bmatrix} \begin{Bmatrix} \theta_{\bar{x}5} \\ \theta_{\bar{y}5} \\ \vdots \\ \bar{w}_6 \end{Bmatrix}.$$

The conditions of symmetry of the displacements at nodes 5 and 6 are that

$$\theta_{\bar{x}6} = \theta_{\bar{x}5}, \qquad \theta_{\bar{y}6} = -\theta_{\bar{y}5}, \qquad \bar{w}_6 = \bar{w}_5.$$

Thus the structure stiffness equations become

$$\begin{Bmatrix} 0 \\ 0 \\ 50 \end{Bmatrix} = \begin{bmatrix} 20\,640 & 0 & -2400 \\ 0 & 15\,760 & 3600 \\ -2400 & 3600 & 3600 \end{bmatrix} \begin{Bmatrix} \theta_{\bar{x}5} \\ \theta_{\bar{y}5} \\ \bar{w}_5 \end{Bmatrix}. \tag{i}$$

Solving for the displacements gives

$$\begin{Bmatrix} \theta_{\bar{x}5} \\ \theta_{\bar{y}5} \\ \bar{w}_5 \end{Bmatrix} = \begin{Bmatrix} \theta_{\bar{x}6} \\ -\theta_{\bar{y}6} \\ \bar{w}_6 \end{Bmatrix} = 10^{-3} \begin{Bmatrix} 2.3269 \text{ rad} \\ -4.5711 \text{ rad} \\ 20.011 \text{ m} \end{Bmatrix}.$$

(b) The couples required at points 5 and 6 to eliminate rotation of element 5–6 about its axis will be equal to the moment reactions corresponding to the additional restraint that $\theta_{\bar{x}5} = \theta_{\bar{x}6} = 0$. The stiffness equations (i) become

$$\begin{Bmatrix} F_{\theta\bar{x}5} \\ 0 \\ 50 \end{Bmatrix} = \begin{bmatrix} 20\,640 & 0 & -2400 \\ 0 & 15\,760 & 3600 \\ -2400 & 3600 & 3600 \end{bmatrix} \begin{Bmatrix} \theta_{\bar{x}5} = 0 \\ \theta_{\bar{y}5} \\ \bar{w}_5 \end{Bmatrix}.$$

The second and third of these equations can be solved to give

$$\begin{Bmatrix} \theta_{\bar{y}5} \\ \bar{w}_5 \end{Bmatrix} = 10^{-3} \begin{Bmatrix} -4.1118 \text{ rad} \\ 18.001 \text{ m} \end{Bmatrix}$$

and then the first of the equations gives

$$F_{\theta\bar{x}5} = F_{\theta\bar{x}6} = -2400 \times 0.018001 = -43.202 \text{ kN m}$$

with the negative sign indicating that the vectors representing the couples would point in the direction opposite to that of the $\bar{x}$ axis.

Example 4.6. Consider again the grillage shown in Fig. E4.5 and detailed in Example 4.5. Now assume that members 5–1 and 6–4 each carry loads of 48 kN uniformly distributed along their lengths and acting in the $\bar{z}$ direction, whilst member 5–6 similarly carries a load of 60 kN. Determine the set of consistent nodal loads for the structure. Hence calculate the nodal displacements and determine the distributions of bending and twisting moments in member 5–1.

Since distributed loadings are involved the solution procedure described in Section 4.2 is adopted. The loadings are directly related to the local co-ordinate axes so that Stage 1 of the procedure is not needed.
Stage 2. The fixed-end forces for the loaded elements are calculated using the information given in Table 4.1 and are shown in Fig. E4.6(a).

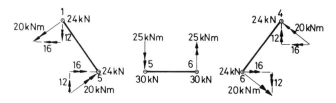

Fig. E4.6(a)

Stage 3. The consistent nodal loads for the structure (i.e. the loads corresponding to the non-zero degrees of freedom) are the reverse of the sums of appropriate element fixed-end forces and are shown in Fig. E4.6(b).
Stage 4. The same symmetry conditions apply as in Example 4.5 and consequently the structure stiffness matrix is identical to that appearing in equation (i) of that example and relates to the degrees of freedom at node 5 only. With the consistent nodal loads shown in Fig. E4.6(b) the structure stiffness equations for

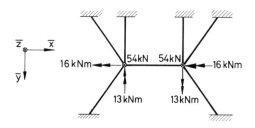

Fig. E4.6(b)

the present example become

$$\begin{Bmatrix} -16 \\ -13 \\ 54 \end{Bmatrix} = \begin{bmatrix} 20\,640 & 0 & -2400 \\ 0 & 15\,760 & 3600 \\ -2400 & 3600 & 3600 \end{bmatrix} \begin{Bmatrix} \theta_{\bar{x}5} \\ \theta_{\bar{y}5} \\ \bar{w}_s \end{Bmatrix}.$$

The solution of these equations yields the actual nodal displacements

$$\begin{Bmatrix} \theta_{\bar{x}5} \\ \theta_{\bar{y}5} \\ \bar{w}_5 \end{Bmatrix} = \begin{Bmatrix} \theta_{\bar{x}6} \\ -\theta_{\bar{y}6} \\ \bar{w}_6 \end{Bmatrix} = 10^{-3} \begin{Bmatrix} 1.7895 \text{ rad} \\ -5.8631 \text{ rad} \\ 22.0560 \text{ m} \end{Bmatrix}.$$

The forces that correspond to these displacements at the ends of element 5–1 can now be calculated using either equation (3.29) or (3.30). The transformation matrix $\mathbf{T}$ for this element is (see equation (4.2)),

$$\mathbf{T}^{5-1} = \begin{bmatrix} e & f & 0 \\ -f & e & 0 \\ 0 & 0 & 1 \end{bmatrix} = \begin{bmatrix} -0.6 & -0.8 & 0 \\ -0.8 & -0.6 & 0 \\ 0 & 0 & 1 \end{bmatrix}.$$

From equation (3.30), bearing in mind that $\bar{\mathbf{d}}_j = \bar{\mathbf{d}}_1 = \mathbf{0}$,

$$\mathbf{P}_i^{5-1} = \mathbf{T}^{5-1} \bar{\mathbf{k}}_{ii}^{5-1} \bar{\mathbf{d}}_5$$

or

$$\begin{Bmatrix} M_{xi} \\ M_{yi} \\ W_i \end{Bmatrix}^{5-1} = \begin{bmatrix} -0.6 & -0.8 & 0 \\ 0.8 & -0.6 & 0 \\ 0 & 0 & 1 \end{bmatrix} \begin{bmatrix} 7120 & -3840 & -2400 \\ -3840 & 4880 & 1800 \\ -2400 & 1800 & 1200 \end{bmatrix}$$

$$\times \begin{Bmatrix} 1.7895 \\ -5.8631 \\ 22.0560 \end{Bmatrix} 10^{-3} = \begin{Bmatrix} 7.2335 \text{ kN m} \\ -16.6741 \text{ kN m} \\ 11.6190 \text{ kN} \end{Bmatrix}$$

Similarly

$$\mathbf{P}_j^{5-1} = \mathbf{T}^{5-1} \bar{\mathbf{k}}_{ji}^{5-1} \bar{\mathbf{d}}_5$$

and it can easily be verified that this becomes

$$\begin{Bmatrix} M_{xj} \\ M_{yj} \\ W_j \end{Bmatrix}^{5-1} = \begin{Bmatrix} -7.2335 \text{ kN m} \\ -41.4211 \text{ kN m} \\ -11.6190 \text{ kN} \end{Bmatrix}.$$

Stage 5. The actual nodal forces for element 5–1 are the summations of the Stage 4 forces and the fixed-end forces shown in Fig. E4.6(a); thus the actual nodal forces are

$$\begin{Bmatrix} M_{xi} \\ M_{yi} \\ W_i \\ M_{xj} \\ M_{yj} \\ W_j \end{Bmatrix}^{5-1} = \begin{Bmatrix} 7.2335 \\ -16.6741 \\ 11.6190 \\ -7.2335 \\ -41.4211 \\ -11.6190 \end{Bmatrix} + \begin{Bmatrix} 0 \\ 20 \\ -24 \\ 0 \\ -20 \\ -24 \end{Bmatrix} = \begin{Bmatrix} 7.2335 \text{ kN m} \\ 3.3259 \text{ kN m} \\ -12.3810 \text{ kN} \\ -7.2335 \text{ kN m} \\ -61.4211 \text{ kN m} \\ -35.6190 \text{ kN} \end{Bmatrix}$$

These forces are shown in Fig. E4.6(c) and from them the bending moment and twisting moment (or torque) diagrams are quite easily constructed and are also shown in the figure.

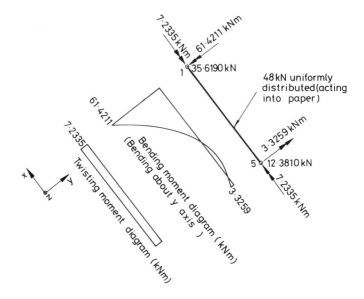

Fig. E4.6(c)

4.5. Ball-jointed space frames

In this and the next section attention is directed to extending the methodology established for two-dimensional structures to three-dimensional or space frames.

First, in this section, the simplest category of space frame is considered, namely the ball-jointed frame, an example of which is shown in Fig. 4.6. In such a frame the members are assumed to be connected to one another by ball joints (or universal hinges in other words) which allow complete and independent freedom of rotation of the ends of all members connected to a joint. At the same time in this idealized frame it is assumed that the only loads are applied translational loads acting at the joints

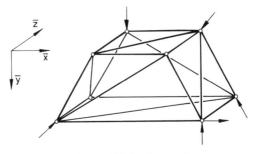

Fig. 4.6. A ball-jointed space frame.

themselves with the members correspondingly not carrying any external loads along their lengths.

The description just given of the ball-jointed frame makes it apparent that this type of frame is the equivalent in three dimensions of the planar pin-jointed frame in two dimensions. Thus, in developing the properties of the typical element of the ball-jointed frame we can draw heavily on the analysis described in Section 3.3.

A typical element of a ball-jointed frame is shown in Fig. 4.7(a) oriented arbitrarily with respect to a fixed cartesian system of global co-ordinates $\bar{x}\bar{y}\bar{z}$. As usual, the local x axis runs along the element from node i to node j. The set of orthogonal local axes is completed by y and z axes which lie in a plane which is perpendicular to the x axis and passes through node i. We need not be more precise than this in the specification of the y and z axes for the ball-jointed frame element since it will be seen

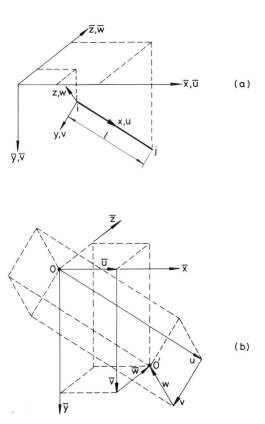

Fig. 4.7. Relationship between local and global quantities for an element of a ball-jointed space frame: (a) the element; (b) displacement transformation.

in what follows that the location and orientation of these axes does not affect the element stiffness matrix.

The development of the element stiffness matrix begins with the following relationships which apply to the local system of xyz orthogonal co-ordinates:

$$\begin{Bmatrix} U_i \\ V_i \\ W_i \\ \hline U_j \\ V_j \\ W_j \end{Bmatrix} = \frac{AE}{l} \left[\begin{array}{ccc|ccc} 1 & 0 & 0 & -1 & 0 & 0 \\ 0 & 0 & 0 & 0 & 0 & 0 \\ 0 & 0 & 0 & 0 & 0 & 0 \\ \hline -1 & 0 & 0 & 1 & 0 & 0 \\ 0 & 0 & 0 & 0 & 0 & 0 \\ 0 & 0 & 0 & 0 & 0 & 0 \end{array} \right] \begin{Bmatrix} u_i \\ v_i \\ w_i \\ \hline u_j \\ v_j \\ w_j \end{Bmatrix} \qquad \text{or} \qquad \mathbf{P = kd} \qquad (4.4)$$

Here u, v, and w are the components of displacement in the x, y, and z directions respectively, and U, V, and W are correspondingly the components of translational force. Equation (4.4) follows naturally on from equations (3.10) and (3.11).

Next, we require the transformation, or rotation, matrix which relates quantities in the local co-ordinate system to those in the global system. For the displacements, for instance, Fig. 4.7(b) represents the movement under loading of node i from initial position 0 to final position 0'. (For two-dimensional analysis the corresponding figure is Fig. 3.2(b).) In the local system the component u is equal to the sum of the projections of $\bar{u}$, $\bar{v}$, and $\bar{w}$ on to the x axis and similar statements apply for v and w. By analogy with equation (3.12b) we have that

$$\begin{Bmatrix} u_i \\ v_i \\ w_i \end{Bmatrix} = \begin{bmatrix} \lambda_{x\bar{x}} & \lambda_{x\bar{y}} & \lambda_{x\bar{z}} \\ \lambda_{y\bar{x}} & \lambda_{y\bar{y}} & \lambda_{y\bar{z}} \\ \lambda_{z\bar{x}} & \lambda_{z\bar{y}} & \lambda_{z\bar{z}} \end{bmatrix} \begin{Bmatrix} \bar{u}_i \\ \bar{v}_i \\ \bar{w}_i \end{Bmatrix} \qquad \text{or} \qquad \mathbf{d}_i = \mathbf{T\bar{d}}_i \qquad (4.5)$$

where the λ values with subscripts are the direction cosines. Matrix $\mathbf{T}$ is the transformation matrix and is again orthogonal.

Of the nine direction cosines appearing in $\mathbf{T}$ attention need be paid only to the three occurring in the first row, as far as the ball-jointed space frame is concerned. This follows from the fact that only the u displacement and U force are of significance in the local system, as indicated by equation (4.4). The three significant terms in the first row of $\mathbf{T}$ are defined in terms of the nodal co-ordinate positions in the global system as

$$\lambda_{x\bar{x}} = \frac{\bar{x}_j - \bar{x}_i}{l} = e \qquad \lambda_{x\bar{y}} = \frac{\bar{y}_j - \bar{y}_i}{l} = f \qquad \lambda_{x\bar{z}} = \frac{\bar{z}_j - \bar{z}_i}{l} = g \qquad (4.6)$$

where

$$l = ((\bar{x}_j - \bar{x}_i)^2 + (\bar{y}_j - \bar{y}_i)^2 + (\bar{z}_j - \bar{z}_i)^2)^{1/2} \qquad (4.7)$$

is the length of the element (see equation (3.13) for the corresponding two-dimensional case). It should be noted that $e^2 + f^2 + g^2 = 1$. The transformation matrix for the ball-jointed frame can thus be written

$$\mathbf{T} = \begin{bmatrix} e & f & g \\ \lambda_{y\bar{x}} & \lambda_{y\bar{y}} & \lambda_{y\bar{z}} \\ \lambda_{z\bar{x}} & \lambda_{z\bar{y}} & \lambda_{z\bar{z}} \end{bmatrix}. \tag{4.8}$$

With matrices $\mathbf{k}$ (equation (4.4)) and $\mathbf{T}$ (equation (4.8)) established the derivation of the element stiffness matrix in the global system $\bar{\mathbf{k}}$ follows the procedure described in Section 3.3. Thus, equations (3.17) and (3.18) apply once more with $\boldsymbol{\tau}$ defined in the form that follows equation (3.14) but now being a 6×6 matrix. The element global stiffness matrix can readily be calculated in explicit form and is

$$\bar{\mathbf{k}} = \frac{AE}{l} \begin{bmatrix} \bar{u}_i & \bar{v}_i & \bar{w}_i & \bar{u}_j & \bar{v}_j & \bar{w}_j \\[4pt] e^2 & & & & \text{Symmetric} & \\ ef & f^2 & & & & \\ eg & fg & g^2 & & & \\ -e^2 & -ef & -eg & e^2 & & \\ -ef & -f^2 & -fg & ef & f^2 & \\ -eg & -fg & -g^2 & eg & fg & g^2 \end{bmatrix} \tag{4.9}$$

As with the pin-jointed planar frame this can be put in the form

$$\bar{\mathbf{k}} = \begin{bmatrix} \bar{\mathbf{k}}_0 & -\bar{\mathbf{k}}_0 \\ -\bar{\mathbf{k}}_0 & \bar{\mathbf{k}}_0 \end{bmatrix} \quad \text{where now} \quad \bar{\mathbf{k}}_0 = \frac{AE}{l} \begin{bmatrix} e^2 & ef & eg \\ ef & f^2 & fg \\ eg & fg & g^2 \end{bmatrix}. \tag{4.10}$$

When the structure displacements are known we may wish to calculate the column matrix of nodal forces referred to the local co-ordinate axes. Equations (3.23) or (3.24) represent the procedure symbolically but again, as with the corresponding planar analysis of Section 3.3, only the axial forces U_i and U_j are of interest. These are given by (see equation (3.25) for the planar case)

$$U_i = -U_j = \frac{AE}{l}(e(\bar{u}_i - \bar{u}_j) + f(\bar{v}_i - \bar{v}_j) + g(\bar{w}_i - \bar{w}_j)). \tag{4.11}$$

Example 4.7. A ball-jointed three-dimensional frame consists of six members, namely AD, BD, CD, DE, AE, and CE. The joints A, B, and C are attached to fixed supports and the co-ordinates (in metres) of the various joints, referred to global axes $\bar{x}\bar{y}\bar{z}$, are as follows:

Joint	$\bar{x}$	$\bar{y}$	$\bar{z}$
A	0	0	0
B	0	1	0
C	1	0	0
D	2	3	4
E	1	0	4

The value of AE/l^3 is $10^5\,\mathrm{N\,m^{-3}}$ for all members.

Determine the 6×6 SSM corresponding to forces and displacements at joints D and E. If external forces are applied at joints D and E corresponding to a column matrix $\mathbf{F}_r$ where

$$\mathbf{F}_r = \{1.7 \quad 4.8 \quad 0 \quad -0.8 \quad -4.8 \quad 4.8\}\,\mathrm{kN}$$

determine the resulting displacements $\mathbf{D}_r$ at these joints.

Calculate the forces in members CE and DE.

Perhaps the first point to note is that the structure is fully defined by the given numerical data and that the analysis can proceed perfectly well without the need of a figure showing the structure geometry.

The global stiffness matrix of the typical element is expressed in equation (4.10) with respect to the 3×3 submatrix $\bar{\mathbf{k}}_0$, the terms of which depend upon the direction cosines e, f, and g which are themselves defined in equation (4.6). Since, in this particular example, it is stated that all elements have a common value of AE/l^3 it is convenient to redefine $\bar{\mathbf{k}}_0$ as

$$\bar{\mathbf{k}}_0 = \frac{AE}{l^3}\begin{bmatrix} (\Delta\bar{x})^2 & \Delta\bar{x}\,\Delta\bar{y} & \Delta\bar{x}\,\Delta\bar{z} \\ \Delta\bar{x}\,\Delta\bar{y} & (\Delta\bar{y})^2 & \Delta\bar{y}\,\Delta\bar{z} \\ \Delta\bar{x}\,\Delta\bar{z} & \Delta\bar{y}\,\Delta\bar{z} & (\Delta\bar{z})^2 \end{bmatrix}$$

where $\Delta\bar{x} = \bar{x}_j - \bar{x}_i$, $\Delta\bar{y} = \bar{y}_j - \bar{y}_i$, and $\Delta\bar{z} = \bar{z}_j - \bar{z}_i$.

The $\bar{\mathbf{k}}_0$ submatrices for the six elements are now easily established using the given nodal co-ordinate data (and taking node i for each element to be the point indicated by the first letter of the element designation) as follows:

Element ij	$\Delta\bar{x}$ (m)	$\Delta\bar{y}$ (m)	$\Delta\bar{z}$ (m)	$\bar{\mathbf{k}}_0$ (kN, m)	Element ij	$\Delta\bar{x}$ (m)	$\Delta\bar{y}$ (m)	$\Delta\bar{z}$ (m)	$\bar{\mathbf{k}}_0$ (kN, m)
AD	2	3	4	$100\begin{bmatrix} 4 & 6 & 8 \\ 6 & 9 & 12 \\ 8 & 12 & 16 \end{bmatrix}$	DE	-1	-3	0	$100\begin{bmatrix} 1 & 3 & 0 \\ 3 & 9 & 0 \\ 0 & 0 & 0 \end{bmatrix}$
BD	2	2	4	$100\begin{bmatrix} 4 & 4 & 8 \\ 4 & 4 & 8 \\ 8 & 8 & 16 \end{bmatrix}$	AE	1	0	4	$100\begin{bmatrix} 1 & 0 & 4 \\ 0 & 0 & 0 \\ 4 & 0 & 16 \end{bmatrix}$
CD	1	3	4	$100\begin{bmatrix} 1 & 3 & 4 \\ 3 & 9 & 12 \\ 4 & 12 & 16 \end{bmatrix}$	CE	0	0	4	$100\begin{bmatrix} 0 & 0 & 0 \\ 0 & 0 & 0 \\ 0 & 0 & 16 \end{bmatrix}$

The stiffness matrix for the structure, corresponding to the non-zero freedoms at nodes D and E, is obtained through the direct stiffness procedure as

$$\mathbf{K}_r = \left[\begin{array}{c:c} \bar{\mathbf{k}}_0^{AD} + \bar{\mathbf{k}}_0^{BD} + \bar{\mathbf{k}}_0^{CD} + \bar{\mathbf{k}}_0^{DE} & -\bar{\mathbf{k}}_0^{DE} \\ \hdashline -\bar{\mathbf{k}}_0^{DE} & \bar{\mathbf{k}}_0^{DE} + \bar{\mathbf{k}}_0^{AE} + \bar{\mathbf{k}}_0^{CE} \end{array} \right].$$

When substitution is made for the various $\bar{\mathbf{k}}_0$ the structure stiffness equations become, using kilonewtons and metres,

$$\begin{Bmatrix} 1.7 \\ 4.8 \\ 0 \\ -0.8 \\ -4.8 \\ 4.8 \end{Bmatrix} = 100 \begin{bmatrix} 10 & & & & & \\ 16 & 31 & & \text{Symmetric} & & \\ 20 & 32 & 48 & & & \\ -1 & -3 & 0 & 2 & & \\ -3 & -9 & 0 & 3 & 9 & \\ 0 & 0 & 0 & 4 & 0 & 32 \end{bmatrix} \begin{Bmatrix} \bar{u}_D \\ \bar{v}_D \\ \bar{w}_D \\ \bar{u}_E \\ \bar{v}_E \\ \bar{w}_E \end{Bmatrix}$$

or

$$\mathbf{F}_r = \mathbf{K}_r \mathbf{D}_r.$$

The solution of these equations is

$$\mathbf{D}_r = \{2 \quad 1 \quad -1.5 \quad 4 \quad -5 \quad 1\} 10^{-3} \, \text{m}$$

The axial forces in the elements (members) can readily be obtained using equation (4.11).

For element CE:

$$l = ((\Delta \bar{x})^2 + (\Delta \bar{y})^2 + (\Delta \bar{z})^2)^{1/2} = (0^2 + 0^2 + 4^2)^{1/2} = 4 \, \text{m}$$

$$e = \frac{\Delta \bar{x}}{l} = 0, \qquad f = \frac{\Delta \bar{y}}{l} = 0, \qquad g = \frac{\Delta \bar{z}}{l} = \frac{4}{4} = 1.$$

Then

$$U_i = -U_j = \left(\frac{AE}{l^3}\right) l^2 g \, (\bar{w}_i - \bar{w}_j) = \left(\frac{AE}{l^3}\right) l^2 g (\bar{w}_C - \bar{w}_E)$$

$$= 100 \times 16 \times 1 \times (0 - 1 \times 10^{-3}) = -1.6 \, \text{kN}$$

i.e. the force in CE is 1.6 kN tensile.
For element DE

$$l = (1^2 + 3^2 + 0^2)^{1/2} = \sqrt{10} \, \text{m}$$

$$e = -1/\sqrt{10}, \qquad f = -3/\sqrt{10}, \qquad g = 0.$$

Then

$$U_i = -U_j = \left(\frac{AE}{l^3}\right) l^2 (e(\bar{u}_D - \bar{u}_E) + f(\bar{v}_D - \bar{v}_E))$$

$$= 100 \times 10 \left(\frac{1}{\sqrt{10}} (2 - 4) 10^{-3} - \frac{3}{\sqrt{10}} (1 + 5) 10^{-3} \right)$$

$$= -\frac{16}{\sqrt{10}} = -1.6\sqrt{10} \, \text{kN}.$$

i.e. the force in DE is $1.6\sqrt{10}$ kN tensile

4.6. Rigid-jointed space frames

This is the most general category of space frame to be considered and is the three-dimensional equivalent of the rigid-jointed planar frames considered earlier. An example of a rigid-jointed space frame is depicted in Fig. 4.8. As the name suggests the members of the frame are assumed to be rigidly connected together at the joints. Loading is applied to the frame at the joints and/or along the members. In addition to the three components of translational displacement which are involved as degrees of freedom in ball-jointed space frame analysis we must now include three components of rotation in the present analysis.

The development of the stiffness matrix of an element of a rigid-jointed space frame begins, as usual, with the setting up of stiffness relationships related to a local co-ordinate system. This system is conveniently specified in much the same way as for the ball-jointed frame in Section 4.5 except that now it is necessary to be precise about the y and z axes, which together with the axial co-ordinate x form an orthogonal set of local coordinates. Here the y and z axes are assumed at this stage to be the principal axes of the cross-section of the element and it is assumed that the cross-section is symmetrical with respect to both principal axes so that the shear centre coincides with the centroid (this uncouples bending and twisting effects).

The element of a rigid-jointed space frame carries axial loading, transverse loadings in the directions of both principal axes, bending moments about the two principal axes, and a twisting moment. Corresponding to these components of force (translational and rotational) are the components of displacement in the local system which are shown, for node i only, in the view of a typical element of a rigid-jointed space frame given by Fig. 4.9. The column matrices of the complete set of 12 force and displacement components are, as usual for a skeletal element, of the form

$$\mathbf{P} = \left\{ \begin{matrix} \mathbf{P}_i \\ \mathbf{P}_j \end{matrix} \right\} \quad \text{and} \quad \mathbf{d} = \left\{ \begin{matrix} \mathbf{d}_i \\ \mathbf{d}_j \end{matrix} \right\}$$

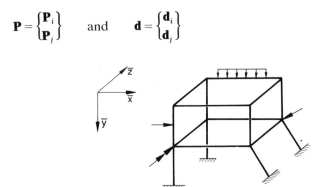

Fig. 4.8. A rigid-jointed space frame.

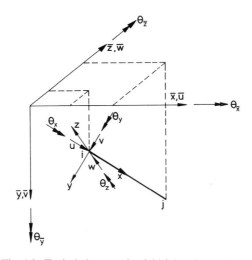

Fig. 4.9. Typical element of a rigid-jointed space frame.

where now

$$\mathbf{P}_i = \{U_i \quad V_i \quad W_i \quad M_{xi} \quad M_{yi} \quad M_{zi}\}$$

$$\mathbf{d}_i = \{u_i \quad v_i \quad w_i \quad \theta_{xi} \quad \theta_{yi} \quad \theta_{zi}\},$$

(4.12)

and $\mathbf{P}_j, \mathbf{d}_j$ are defined in similar fashion with suffix j in place of i of course.

The element of a rigid-jointed space frame combines the functions of the element of a planar rigid-jointed frame (see Section 3.4) and the element of a grillage (see Section 4.4). Consequently the stiffness matrix $\mathbf{k}$ of the space-frame element in its local co-ordinate system can be assembled very simply and directly from the corresponding stiffness matrices of the planar frame and the grillage, i.e. from the stiffness matrices of equations (3.26) and (4.1) with appropriate re-ordering of rows and columns to suit the definitions of $\mathbf{P}$ and $\mathbf{d}$. The result is given in equation (4.13).

$$\mathbf{k} = \begin{bmatrix}
AE/l \\
0 & 12EI_z/l^3 \\
0 & 0 & 12EI_y/l^3 \\
0 & 0 & 0 & GJ/L \\
0 & 0 & -6EI_y/l^2 & 0 & 4EI_y/l \\
0 & 6EI_z/l^2 & 0 & 0 & 0 & 4EI_z/l & & & \text{Symmetric} \\
-AE/l & 0 & 0 & 0 & 0 & 0 & AE/l \\
0 & -12EI_z/l^3 & 0 & 0 & 0 & -6EI_z/l^2 & 0 & 12EI_z/l^3 \\
0 & 0 & -12EI_y/l^3 & 0 & 6EI_y/l^2 & 0 & 0 & 0 & 12EI_y/l^3 \\
0 & 0 & 0 & -GJ/l & 0 & 0 & 0 & 0 & 0 & GJ/l \\
0 & 0 & -6EI_y/l^2 & 0 & 2EI_y/l & 0 & 0 & 0 & 6EI_y/l^2 & 0 & 4EI_y/l \\
0 & 6EI_z/l^2 & 0 & 0 & 0 & 2EI_z/l & 0 & -6EI_z/l^2 & 0 & 0 & 0 & 4EI_z/l
\end{bmatrix}$$

with column headings $u_i \quad v_i \quad w_i \quad \theta_{xi} \quad \theta_{yi} \quad \theta_{zi} \quad u_j \quad v_j \quad w_j \quad \theta_{xj} \quad \theta_{yj} \quad \theta_{zj}$

(4.13)

To transform from the local xyz system to a global $\bar{x}\bar{y}\bar{z}$ system requires the production of a 12×12 transformation matrix. Following on from Section (4.5) this is quite straightforward, in principle at least, if it is realized that the presence of rotations (or moments) does not affect the transformation of translational displacements (or forces) and vice versa; also, as seen with the grillage, the transformation of rotations (or moments) is exactly the same as that of translational displacements (or forces) since all these quantities are vector quantities. Thus, in the present case the transformation matrix is

$$\tau = \begin{bmatrix} \mathbf{T} & \mathbf{0} & \mathbf{0} & \mathbf{0} \\ \mathbf{0} & \mathbf{T} & \mathbf{0} & \mathbf{0} \\ \mathbf{0} & \mathbf{0} & \mathbf{T} & \mathbf{0} \\ \mathbf{0} & \mathbf{0} & \mathbf{0} & \mathbf{T} \end{bmatrix} \tag{4.14}$$

where $\mathbf{T}$ is the 3×3 matrix of direction cosines given in general form in equation (4.5). The global element stiffness matrix can be obtained (numerically rather than explicitly) from the triple matrix product given by equation (3.18) once $\mathbf{T}$ is known. The remainder of this section is concerned with the detailed form of $\mathbf{T}$.

In the case of the ball-jointed space frame only three of the direction cosines occurring in $\mathbf{T}$ were required since the local y and z axes were not precisely defined, and these were easily obtained (see equations (4.6)–(4.8)). For the rigid-jointed space frame the situation is significantly different since the y and z axes have the particular specification, for the moment, that they coincide with the principal axes of the element cross-section. Consequently, all nine direction cosines of $\mathbf{T}$ (see equation (4.5)) have to be evaluated and this is not particularly easy.

The three quantities $\lambda_{x\bar{x}}$, $\lambda_{x\bar{y}}$, and $\lambda_{x\bar{z}}$ in the first row of $\mathbf{T}$ are the direction cosines for the local x axis with respect to the structure axes and will be unchanged from the ball-jointed frame case; thus these quantities are defined in equations (4.6).

To determine the remaining six direction cosines consider, at first, a particular situation in which an element of a rigid-jointed space frame is so oriented that the principal axes of its cross-section lie in horizontal and vertical planes. This will often be the case in practice and corresponds, for example, to an I-section beam whose web lies in a vertical plane; Fig. 4.10 shows this situation where the local z axis is chosen to be horizontal and is also of course perpendicular to the local x axis. The y axis completes the set of orthogonal local co-ordinates and lies in a vertical plane which contains the x axis. Figure 4.11 shows element ij in relation to the local and global co-ordinate systems. In this figure the origins of both systems are shown to coincide. This has been done for convenience;

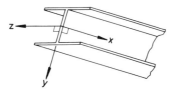

Fig. 4.10. I-section beam with web in vertical plane.

the transformation (or rotation) matrix would be unaffected if the origins did not coincide.

The direction cosines for the y and z axes could be found from direct geometrical considerations but are more easily determined by considering two successive transformations, or rotations, of axes. Remember that the aim is to express displacement or force quantities in the local x,y,z directions in terms of corresponding quantities in the global $\bar{x},\bar{y},\bar{z}$ directions: in the following the reference will be to displacements but could equally well be to forces.

The two successive transformations are as follows.

(i) Between the global $\bar{x}\bar{y}\bar{z}$ system and an intermediate $x^{*}y^{*}z^{*}$ system. This latter system is obtained from the former by rotation through angle ϕ_1 about the $\bar{y}$ axis. Figure 4.12(a) shows the two sets of displacement components when viewed along the negative $\bar{y}$ (or y^{*}) direction. Using the result of equation (3.12), with which Fig. 3.2 is associated, it is clear that the following transformation applies

$$\begin{Bmatrix} u^{*} \\ v^{*} \\ w^{*} \end{Bmatrix} = \begin{bmatrix} \cos\phi_1 & 0 & \sin\phi_1 \\ 0 & 1 & 0 \\ -\sin\phi_1 & 0 & \cos\phi_1 \end{bmatrix} \begin{Bmatrix} \bar{u} \\ \bar{v} \\ \bar{w} \end{Bmatrix}. \tag{4.15}$$

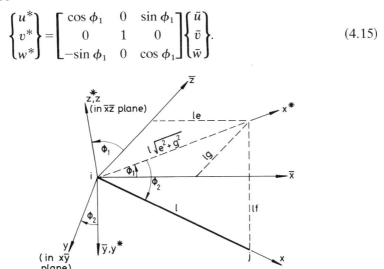

Fig. 4.11. Co-ordinate systems for rigid-jointed space frame analysis.

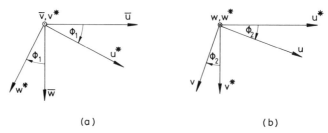

(a)   (b)

Fig. 4.12. Transformation of displacement components in rigid-jointed space frame analysis: (a) between $\overline{xyz}$ and $x^*y^*z^*$ systems; (b) between $x^*y^*z^*$ and xyz systems.

(ii) Between the intermediate $x^*y^*z^*$ system and the local xyz system. The latter system is obtained from the former by rotation through angle ϕ_2 about the z^* (or z) axis. Figure 4.12(b) shows the two sets of displacement components when viewed along the positive z^* direction. Again using equation (3.12) the relationship between the two sets of displacement components is

$$\begin{Bmatrix} u \\ v \\ w \end{Bmatrix} = \begin{bmatrix} \cos\phi_2 & \sin\phi_2 & 0 \\ -\sin\phi_2 & \cos\phi_2 & 0 \\ 0 & 0 & 1 \end{bmatrix} \begin{Bmatrix} u^* \\ v^* \\ w^* \end{Bmatrix}. \tag{4.16}$$

The complete transformation between local and global quantities can be obtained by combining equations (4.15) and (4.16) to give

$$\begin{Bmatrix} u \\ v \\ w \end{Bmatrix} = \begin{bmatrix} \cos\phi_2 & \sin\phi_2 & 0 \\ -\sin\phi_2 & \cos\phi_2 & 0 \\ 0 & 0 & 1 \end{bmatrix} \begin{bmatrix} \cos\phi_1 & 0 & \sin\phi_1 \\ 0 & 1 & 0 \\ -\sin\phi_1 & 0 & \cos\phi_1 \end{bmatrix} \begin{Bmatrix} \bar{u} \\ \bar{v} \\ \bar{w} \end{Bmatrix} = \mathbf{T} \begin{Bmatrix} \bar{u} \\ \bar{v} \\ \bar{w} \end{Bmatrix}. \tag{4.17}$$

Matrix $\mathbf{T}$ is the required 3×3 matrix of direction cosines. This can be evaluated by performing the indicated matrix multiplication and using the following definitions (see Fig. 4.11)

$$\sin\phi_1 = g/(e^2+g^2)^{1/2} \qquad \cos\phi_1 = e/(e^2+g^2)^{1/2}$$
$$\sin\phi_2 = f \qquad\qquad \cos\phi_2 = (e^2+g^2)^{1/2} \tag{4.18}$$

where e, f, and g are given by equations (4.6). The result is

$$\mathbf{T} = \begin{bmatrix} e & f & g \\ \dfrac{-ef}{(e^2+g^2)^{1/2}} & (e^2+g^2)^{1/2} & \dfrac{-fg}{(e^2+g^2)^{1/2}} \\ \dfrac{-g}{(e^2+g^2)^{1/2}} & 0 & \dfrac{e}{(e^2+g^2)^{1/2}} \end{bmatrix} \tag{4.19}$$

The three rows of $\mathbf{T}$ give the direction cosines in turn of the local x, y, and z orthogonal axes as referred to the global $\bar{x}, \bar{y}$, and $\bar{z}$ orthogonal axes. It should be noted that these direction cosines of course obey the general rules for such quantities that

(i) the sum of the squares of the direction cosines for any axis (i.e. for x or for y or for z) is unity, e.g.

$$e^2 + f^2 + g^2 = 1$$

(ii) the sum of the products of corresponding direction cosines for any two axes (i.e. for x and y or for y and z or for x and z) is zero, e.g.

$$e\left(\frac{-ef}{(e^2+g^2)^{1/2}}\right) + f(e^2+g^2)^{1/2} + g\left(\frac{-fg}{(e^2+f^2)^{1/2}}\right) = 0.$$

It will be recalled that we have been considering here an element with the particular specification that the principal axes of the cross-section lie in horizontal and vertical planes and that the z axis is chosen to be both horizontal and perpendicular to the x axis. Now it is clear that these statements do not suffice for an element which is vertical (i.e. whose x axis runs parallel to the $\bar{y}$ axis) since the z axis will then not be uniquely defined by the simple specification that it lies in a horizontal plane and is perpendicular to the x axis of the element. To overcome this difficulty we may specify that for a vertical member the direction of the local z axis coincides with the direction of the global $\bar{z}$ axis. It is then a simple matter to set up $\mathbf{T}$ by inspection for this special case. It should be clear that $\mathbf{T}$ can take one of the following forms

$$\mathbf{T} = \begin{bmatrix} 0 & 1 & 0 \\ -1 & 0 & 0 \\ 0 & 0 & 1 \end{bmatrix} \quad \text{or} \quad \begin{bmatrix} 0 & -1 & 0 \\ 1 & 0 & 0 \\ 0 & 0 & 1 \end{bmatrix} \tag{4.20}$$

depending upon whether the local x axis runs in the positive or negative $\bar{y}$ direction.

Finally, to complete the picture, let us briefly consider the situation that arises if the aforementioned condition that the principal axes of the cross-section lie in horizontal and vertical planes is dispensed with. If this is so an extra complication arises since the principal axes no longer coincide with the y and z axes which were defined above, and the xyz co-ordinate system can no longer, strictly speaking, be regarded as the local system. Rather, a local system now comprises the x_p (or x) axis together with the principal axes denoted by y_p and z_p. Figure 4.13 shows a view looking in the negative x (or x_p) direction which illustrates the relationship of the principal axes co-ordinate system $x_p y_p z_p$ to the earlier defined xyz co-ordinate system. The principal axes system is obtained from the xyz system by a rotation through an angle ϕ_p about the x (or x_p) longitudinal axis of the element. The relationship between the two

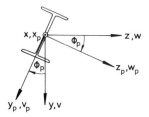

Fig. 4.13. The xyz and $x_p y_p z_p$ co-ordinate systems for a cross-section whose principle axes do not lie in horizontal and vertical planes.

sets of displacement components is

$$\left\{ \begin{matrix} u_p \\ v_p \\ w_p \end{matrix} \right\} = \begin{bmatrix} 1 & 0 & 0 \\ 0 & \cos \phi_p & -\sin \phi_p \\ 0 & \sin \phi_p & \cos \phi_p \end{bmatrix} \left\{ \begin{matrix} u \\ v \\ w \end{matrix} \right\} \tag{4.21}$$

in which the angle ϕ_p would be assumed to be directly specified as part of the structure geometry data. The complete transformation matrix linking displacement components in the principal axes system to those in the global system is obtained by premultiplying matrix **T** given by equation (4.19) by the 3×3 matrix occurring on the right-hand side of equation (4.21) (or, for the special case of a vertical element, premultiplying the appropriate **T** matrix of equations (4.20) by the 3×3 matrix of equation (4.21). It should be noted that the stiffness matrix **k** of equation (4.13) is, of course, to be interpreted as relating to the principal axes co-ordinate system.

Example 4.8. An element of a rigid-jointed space frame runs between the points $(2, 0, 5)$ and $(4, 8, 2)$ in the $\bar{x}\bar{y}\bar{z}$ global co-ordinate system. In the usual manner (see Fig. 4.8) the $\bar{x}\bar{z}$ plane is a horizontal plane whilst the $\bar{y}$ axis runs vertically; also the principal axes of the cross-section lie in horizontal and vertical planes. Determine the matrix **T** which defines the transformation between the local (xyz) and the global systems.

Select node i as the point $(2, 0, 5)$, node j as the point $(4, 8, 2)$.

Element length $l = ((\bar{x}_j - \bar{x}_i)^2 + (\bar{y}_j - \bar{y}_i)^2 + (\bar{z}_j - \bar{z}_i)^2)^{1/2}$
$$= ((4-2)^2 + (8-0)^2 + (2-5)^2)^{1/2} = \sqrt{77} = 8.7750.$$

The direction cosines of the local x axis are

$e = (\bar{x}_j - \bar{x}_i)/l = 2/\sqrt{77} = 0.227\ 921\ 1$
$f = (\bar{y}_j - \bar{y}_i)/l = 8/\sqrt{77} = 0.911\ 684\ 6$
$g = (\bar{z}_j - \bar{z}_i)/l = -3/\sqrt{77} = -0.341\ 881\ 7.$

The matrix $\mathbf{T}$ (equation (4.19)) is then

$$\mathbf{T} = \begin{bmatrix} 0.22792 & 0.91168 & -0.34188 \\ -0.50571 & 0.41089 & 0.75857 \\ 0.83205 & 0 & 0.55470 \end{bmatrix}.$$

Example 4.9. The space frame shown in Fig. E4.9 consists of three mutually perpendicular members, rigidly connected together at point 1 and with fully fixed supports at points 2, 3, and 4. The members are identical, each having a length l of 2 m and a hollow circular cross-section of outside diameter 100 mm and thickness 10 mm. Material properties are $E = 200$ GN m^{-2}, $G = 77$ GN m^{-2}. Assemble the set of stiffness equations for the frame.

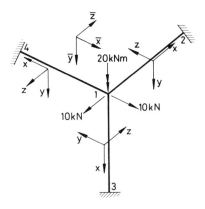

Fig. E4.9

Choose node i for each of the three elements to be at joint 1: hence the local x axis for each element runs outward from 1 along the element. The element stiffness matrix $\mathbf{k}$ in the local system is given in full in equation (4.13) but clearly only the 6×6 submatrix $\mathbf{k}_{ii}$ (in the top left-hand quarter) will contribute to the structure stiffness equations after the boundary conditions are applied.

In this simple space-frame example all elements have identical properties and also the cross-sections are such that

$I_y = I_z = J/2 = I$, say.

Therefore, for each element in its local system

$$\mathbf{k}_{ii} = \begin{bmatrix} AE/l & & & & & \\ 0 & 12EI/l^3 & & & & \\ 0 & 0 & 12EI/l^3 & \text{Symmmetric} & & \\ 0 & 0 & 0 & 2GI/l & & \\ 0 & 0 & -6EI/l^2 & 0 & 4EI/l & \\ 0 & 6EI/l^2 & 0 & 0 & 0 & 4EI/l \end{bmatrix}$$

The transformation matrices for each element are easily established. That for element 1–3 is given by the left-hand one of the two matrices recorded in equation (4.20), i.e.

$$\mathbf{T}^{1-3} = \begin{bmatrix} 0 & 1 & 0 \\ -1 & 0 & 0 \\ 0 & 0 & 1 \end{bmatrix}.$$

Those for elements 1–2 and 1–4 can be evaluated using equation (4.19) or for this particular geometry simply by inspection. The result is

$$\mathbf{T}^{1-2} = \begin{bmatrix} 0 & 0 & 1 \\ 0 & 1 & 0 \\ -1 & 0 & 0 \end{bmatrix} \qquad \mathbf{T}^{1-4} = \begin{bmatrix} -1 & 0 & 0 \\ 0 & 1 & 0 \\ 0 & 0 & -1 \end{bmatrix}.$$

In the global system the pertinent stiffness submatrix for each element has the general form

$$\bar{\mathbf{k}}_{ii} = \begin{bmatrix} \mathbf{T}^t & 0 \\ 0 & \mathbf{T}^t \end{bmatrix} \mathbf{k}_{ii} \begin{bmatrix} \mathbf{T} & 0 \\ 0 & \mathbf{T} \end{bmatrix}.$$

Performing the indicated matrix multiplications yields $\bar{\mathbf{k}}_{ii}^{1-2}$, $\bar{\mathbf{k}}_{ii}^{1-3}$, and $\bar{\mathbf{k}}_{ii}^{1-4}$, and the SSM is then obtained by simple superposition of the element contributions. The result is

$$\mathbf{K}_r = \begin{bmatrix}
\overset{\bar{u}_1}{24EI/l^3 + AE/l} & \overset{\bar{v}_1}{} & \overset{\bar{w}_1}{} & \overset{\theta_{\bar{x}1}}{} & \overset{\theta_{\bar{y}1}}{} & \overset{\theta_{\bar{z}1}}{} \\
0 & 24EI/l^3 + AE/l & & & \text{Symmetric} & \\
0 & 0 & 24EI/l^3 + AE/l & & & \\
0 & -6EI/l^2 & 6EI/l^2 & 8EI/l + 2GI/l & & \\
6EI/l^2 & 0 & 6EI/l^2 & 0 & 8EI/l + 2GI/l & \\
-6EI/l^2 & -6EI/l^2 & 0 & 0 & 0 & 8EI/l + 2GI/l
\end{bmatrix}$$

Now E, G, and l have the prescribed values given above, and the area and second moment of area of the elements cross-sections are

$$A = \frac{\pi}{4}(100^2 - 80^2) \times 10^{-6} = 2.82743 \times 10^{-3} \, \text{m}^2$$

$$I = \frac{\pi}{64}(100^4 - 80^4) \times 10^{-12} = 2.89812 \times 10^{-6} \, \text{m}^4$$

Thus for the loading shown in Fig. E4.9 the stiffness equations for the structure become, using kilonewtons and metres,

$$\begin{Bmatrix} 10 \\ 0 \\ -10 \\ 0 \\ 20 \\ 0 \end{Bmatrix} = \begin{bmatrix}
284482 & & & & & \\
0 & 284482 & & & & \\
0 & 0 & 284482 & & \text{Symmetric} & \\
0 & -869.45 & 869.45 & 2541.7 & & \\
869.45 & 0 & 869.45 & 0 & 2541.7 & \\
-869.45 & -869.45 & 0 & 0 & 0 & 2541.7
\end{bmatrix} \begin{Bmatrix} \bar{u}_1 \\ \bar{v}_1 \\ \bar{w}_1 \\ \theta_{\bar{x}1} \\ \theta_{\bar{y}1} \\ \theta_{\bar{z}1} \end{Bmatrix}.$$

Bibliography

A number of texts are available which deal to varying degrees with the material covered in Chapters 2, 3, and 4 of this book and which can be consulted to broaden the reader's knowledge of the MDM. Amongst these texts are the following.

1 PESTEL, E. and LECKIE, F. A. *Matrix methods in elastomechanics.* McGraw-Hill, New York (1963).
2 MARTIN, H. C. *Introduction to matrix methods of structural analysis.* McGraw-Hill, New York (1966).
3 PRZEMIENIECKI, J. S. *Theory of matrix structural analysis.* McGraw-Hill, New York (1968).
4 WILLEMS, N. and LUCAS, W. *Matrix analysis for structural engineers.* Prentice-Hall, Englewood Cliffs, N.J. (1970).
5 MEEK, J. L. *Matrix structural analysis.* McGraw-Hill, New York (1971).
6 KARDESTUNCER, H. *Elementary matrix analysis of structures.* McGraw-Hill, New York (1974).
7 LIVESLEY, R. K. *Matrix methods of structural analysis* (2nd edn). Pergamon Press, Oxford (1976).
8 McGUIRE, W. and GALLAGHER, R. H. *Matrix structural analysis.* Wiley, New York (1979).
9 WEAVER, W. and GERE, J. M. *Matrix analysis of framed structures* (2nd edn). Van Nostrand, New York (1980).

Problems

4.1. A uniform beam of flexural rigidity EI_z and length L is clamped at one end, simply supported at the other. It carries a uniformly distributed load of p per unit length. Calculate the rotation of the beam at the simple support and the moment at the clamped end.

4.2. Determine the rotations at the three supports of the continuous beam shown in Fig. P4.2. EI_z is uniform throughout. Hence calculate the support reactions and draw the bending moment diagram

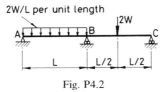

Fig. P4.2

4.3. The continuous beam shown in Fig. P4.3 has uniform value of EI_z throughout. Calculate the rotations at B and C and the vertical force and moment reactions at A and D.

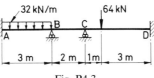

Fig. P4.3

4.4. Consider the beam of Problem 4.1 and imagine that the simple support lifts an amount $pL^4/24EI_z$ at the same time as the distributed loading is applied. What is now the value of the fixing moment at the clamped end?

4.5. Fig. P4.5 shows a uniform clamped beam 1–2–3, with $EI_z = 10$ MN m^2, supported by a spring of stiffness 20 MN m^{-1}. Calculate the deflection and rotation at point 2 and hence construct the bending moment diagram for the beam.

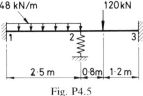

Fig. P4.5

4.6. For each of the plane frames shown in Fig. P4.6 form the column matrix of nodal loads in the $\bar{x}\bar{y}$ system which properly represents the combined loading that needs to be applied in Stage 4 of the solution procedure for frames carrying distributed loading. (List those loads corresponding to the free nodes, in numerical order of the nodes.)

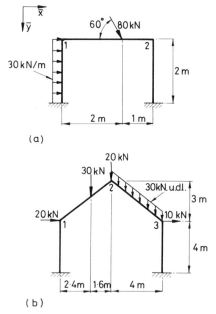

Fig. P4.6

4.7. Determine the column matrix of combined loads which should be applied at nodes 1, 2, and 3 of the rigid-jointed plane frame shown in Fig. P4.7 in order to calculate the nodal displacements. Imagine that these nodal displacements are calculated on the computer and include

$\bar{v}_2 = 4.09$ mm, $\theta_{\bar{z}2} = -0.0029$ rad, $\bar{v}_3 = 1.57$ mm, $\theta_{\bar{z}3} = -0.0071$ rad.

Given that EI_z for member 2–3 is 18 MN m² determine the value of the bending moment at the section where the 50 kN load is acting.

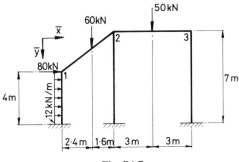

Fig. P4.7

4.8. The plane frame shown in Fig. P4.8 is the same structure as that considered in Problem 3.8 but now the loading is different. Show that the combined loading column matrix for the loading shown in Fig. P4.8 gives the same loading as that specified in Problem 3.8. Determine the end loads for 1–3 under the present system of loading.

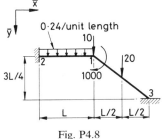

Fig. P4.8

4.9. Find the rotations at B and C of the plane frame shown in Fig. P4.9. Ignore axial strains in the members and assume that EI_z is the same throughout. Calculate the moments at A, B, C, D, and E.

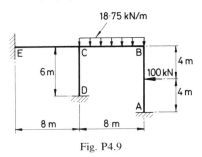

Fig. P4.9

4.10. Consider the beam of Problem 2.5 but now ignore the applied couple at point 3 and instead assume that span 1–2 has a temperature differential applied to it such that the temperature of its lower surface is 50 °C higher than that of its upper surface. The beam section is doubly symmetric, its depth is 0.16 m and

$\alpha = 11 \times 10^{-6} {}^{\circ}\text{C}^{-1}$. Determine the rotations at the simple supports, the vertical reactions at points 1, 2 and 3 and the fixing moment at point 1.

4.11. Consider the pin-jointed frame of Problem 3.3 with the loading indicated in Fig. P3.3 removed. Assume that bar BC has coefficient of expansion α and is heated so that its temperature is increased by T°. Calculate the displacements of point B and hence find the forces acting in all the bars.

4.12. Consider the pin-jointed frame of Problem 3.3 with the loading indicated in Fig. P3.3 removed. Assume that prior to fitting the bar BE is 0.2 per cent longer than its design length. Calculate the forces in all the bars when the frame is assembled.

4.13. Consider the rigid-jointed plane structure of Pro! iem 3.7 and imagine that in addition to the mechanical loading shown in Fig. P3.7 the member 1–2, of doubly symmetric cross-section, is subjected to a temperature change such that the temperature of its upper surface increases by 30 °C and the temperature of its lower surface increases by 90 °C. The depth of this member is 0.2 m and $\alpha = 12 \times 10^{-6} {}^{\circ}\text{C}^{-1}$. Determine the displacements at point 2 and find the forces at the ends of members 1–2 and 2–3.

4.14. The two members of the plane grillage shown in Fig. P4.14 have the same cross section, with $EI_y = 6$ MN m^2 and $GJ = 2$ MN m^2. Determine the displacements at point 1 and find the forces at the ends of both members.

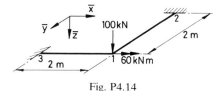

Fig. P4.14

4.15. Fig. P4.15 shows a plan view of a plane grillage. For member 2–1–4, $EI_y = 15$ MN m^2 and $GJ = 4$ MN m^2. For members 1–3 and 1–5, $EI_y = 8$ MN m^2 and $GJ = 2$ MN m^2. Halfway along member 1–3 a load of 100 kN is applied in the $\bar{z}$ direction. Determine the displacements at point 1. Hence find the forces at the ends of member 1–3 and draw the bending moment diagram for this member.

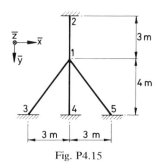

Fig. P4.15

4.16. The members of the plane grillage shown in plan in Fig. P4.16 have the same flexural rigidity ($EI_y = 7.2$ MN m^2) and torsional rigidity ($GJ = 4.5$ MN m^2). Members 1–2, 2–4, and 4–6 each carry uniformly distributed loadings of 36 kN m^{-1} acting normal to the plane of the grillage. Form the SSM for the grillage and determine the nodal displacements.

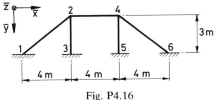

Fig. P4.16

4.17. Figure P4.17 shows a ball-jointed space frame with points 2, 3, 4, and 5 lying at the corners of a square of side 4 m and the free joint 1 being equispaced from these points at a height of 5 m above them. The bars 1–2, 1–3, 1–4, and 1–5 have the same axial rigidity AE of 400 MN. For the loading shown determine the displacements of joint 1 and hence find the forces in the bars.

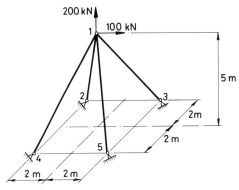

Fig. P4.17

4.18. Consider again the ball-jointed frame of Problem 4.17. Now, in addition to the mechanical loading, imagine that bars 1–3 and 1–5 are both subjected to a temperature increase of 50 °C. If $\alpha = 12 \times 10^{-6}\,°C^{-1}$, determine again the displacement of joint 1 and the forces in the bars.

4.19. Fig. P4.19 shows two views of a ball-jointed space frame. The bars have a cross-sectional area of 10 cm², except for bars AB and BC which have an area of 20 cm². Determine the nodal displacements and hence find the forces in the bars. $E = 200\,GN\,m^{-2}$. (Note the symmetry of the problem.)

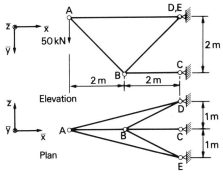

Fig. P4.19

5
VIRTUAL DISPLACEMENTS, POTENTIAL ENERGY, AND THE RAYLEIGH–RITZ METHOD

5.1. Introduction

IN previous chapters the concept of the displacement approach has been firmly established through study of a particular class of structure—the skeletal structure—by means of the method traditionally referred to as the matrix displacement method. It is clear that a crucial requirement of the approach is the establishment of stiffness relationships for the individual elements making up a structure. For the types of structural element so far considered it has been quite easy to establish element stiffness relationships which within the confines of the theoretical model of the real structure, i.e. within the confines of elementary bar and beam theory, are exact. It has been possible to do this by directly satisfying the basic structural conditions of equilibrium, compatibility, and elasticity throughout the element. This is equivalent to satisfying the appropriate (ordinary) differential equation of equilibrium when expressed in terms of displacement. Further, the basic structural conditions are not violated when elements are connected together to form a complete structure.

When we seek to extend the displacement method to the analysis of more complicated two- and three-dimensional continuum structures we shall find, not surprisingly, that we are no longer able to generate exact stiffness relationships. As mentioned in Section 1.3, the earliest attempts at extension adopted a direct approach based on physical reasoning and elementary structural theory but this approach is limited in scope and will not be discussed here. The direct approach was superseded by what may be termed the variational approach in which recourse was made to the principles of work and energy that exist in structural mechanics. In the finite-element displacement approach the relevant principles are the principle of minimum potential energy (PMPE) and the closely related principle of virtual displacements (PVD), and in conjunction with these the Rayleigh–Ritz technique is used to obtain approximate solutions. (Castigliano's first theorem is closely related to both the PMPE and PVD and has also been used to develop element stiffness matrices.) In this chapter these topics are introduced and discussed in detail in relation to simple one-dimensional structural forms, i.e. bars and beams. This provides the basis on which the finite-element analysis of continuum structures can be developed in the chapters that follows. In all circumstances covered in this text the application of the PMPE and the PVD will give identical results. It is largely a matter of personal choice as to which approach is preferred, with certain advantages associated with each; here the PMPE is given precedence.

Before proceeding to describe the variational approach it should be noted that 'weighted residual methods' are sometimes used in finite-element work as an alternative to variational methods. One particular weighted residual method is that of Galerkin in which trial functions are assumed for the pertinent displacement variables of a problem and these functions are used in a manner which gives an approximate solution to the partial (usually) differential equations of the problem in direct fashion. (In the variational approach the governing differential equations are not considered *directly*, as will be seen.) In general it is considered an advantage of the Galerkin method that it operates directly with the governing differential equations of a problem rather than with a functional—which may be difficult to establish—as in the variational approach. However, in the realm of structural analysis with which we are concerned here a well-known suitable functional does exist (the potential energy functional when displacements are primary variables) and the Galerkin method in fact yields results that are identical with those obtained using the variational approach. Since the latter approach has occupied a dominant position in finite-element work it is with this approach that we shall be concerned, and no further consideration will be given to the Galerkin method.

5.2. Strain energy of one-dimensional sytems

5.2.1. Axially loaded bar

Consider the prismatic bar shown in Fig. 5.1(a) of length l and cross-sectional area A which is subjected to an externally applied axial force which is uniformly distributed over the end of the bar and which gradually increases in magnitude through intermediate values F^s to a final value F.

As the bar extends the externally applied force does work in stretching the bar since the force moves through an axial distance which is e^s at an intermediate stage and e finally. The load–elongation curve could generally be of the form shown in Fig. 5.1(b) and the work done by the external force is

$$W_e = \int_0^e F^s \, \mathrm{d}e^s \qquad (5.1)$$

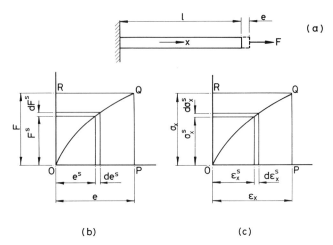

(b)   (c)

Fig. 5.1. Extension of a bar: (a) the bar; (b) load-extension curve; (c) stress–strain curve.

and is the area OPQ under the F–e curve up to the stage of elongation represented by e.

The reader will be aware that in general the bar material behaviour may be elastic (whether linearly elastic or non-linearly elastic) or inelastic; in each case equation (5.1) is valid. If the material behaviour is elastic all the work done in stretching the bar is stored as strain energy which is completely recoverable on removal of the load, in which case the bar then returns to precisely the condition it was in before the load was applied. This constitutes a *conservative* system; the word 'conservative' reflects the fact that the system obeys the principle of conservation of energy which states that the total energy of an isolated system remains constant. In such a system the external and internal forces are termed conservative forces and the work done in moving between two states is independent of the path taken. If, on the other hand, the material behaviour is inelastic, all the work done is not recoverable as strain energy: permanent deformation of the bar would be present on removal of the load and the system is non-conservative. Our concern here is solely with elastic behaviour and conservative systems.

For elastic behaviour, then, the strain energy (or potential energy of deformation) U_p is

$$U_\mathrm{p} = W_e = \int_0^e F^s \, \mathrm{d}e^s. \tag{5.2}$$

It is noted that the strain energy is also equal to the work done *on* the internal forces of the bar and hence is equal to the negative of the work done by the internal forces in the bar, that is

$$U_p = -W_i. \tag{5.3}$$

From elementary considerations the axial stress σ_x and axial strain ε_x are uniform throughout the bar and at any stage s of the deformation are related to the force and extension by the equations

$$F^s = A\sigma_x^s \quad \text{and} \quad e^s = l\varepsilon_x^s. \tag{5.4}$$

Substituting these expressions into equation (5.2) gives

$$U_p = Al\left(\int_0^{\varepsilon_x} \sigma_x^s \, d\varepsilon_x^s\right). \tag{5.5}$$

Here Al is the volume of the bar so that the term in parentheses in this equation is the strain energy per unit volume or, in other words, the strain energy density denoted by ν_p; thus

$$\nu_p = \int_0^{\varepsilon_x} \sigma_x^s \, d\varepsilon_x^s \tag{5.6}$$

and is represented by the area OPQ under the σ_x–ε_x curve in Fig. 5.1(c).

Now attention is further restricted to *linearly* elastic behaviour (and small displacements) in which case the curves of Figs. 5.1(b) and 5.1(c) become straight lines and Hooke's Law applies, that is

$$\sigma_x^s = E\varepsilon_x^s \tag{5.7}$$

at all stages. Thus, from equation (5.6) the strain energy density at load level F becomes

$$\nu_p = \int_0^{\varepsilon_x} E\varepsilon_x^s \, d\varepsilon_x^s = \frac{1}{2}\sigma_x\varepsilon_x = \frac{1}{2}E\varepsilon_x^2 = \frac{1}{2}E\frac{e^2}{l^2}. \tag{5.8}$$

Also, the total strain energy of the bar is

$$U_p = \frac{1}{2}AEl\varepsilon_x^2 = \frac{1}{2}\frac{AE}{l}e^2. \tag{5.9}$$

It is noted that alternative expressions for ν_p and U_p could easily be written in terms of the force F and stress σ_x. However, in the displacement approach the required forms are those given in terms of displacement e or strain ε_x.

In the above we have been concerned with the simple situation in which the bar is of constant cross-section and the strain (and stress) is uniform throughout the whole body; hence the strain energy density ν_p (equation (5.8)) is uniform. In more general situations this will not be so and then the total strain energy U_p is obtained by integrating the strain energy density ν_p throughout the volume of the bar (denoted by vol) as

$$U_p = \int_{vol} \nu_p \, d(vol). \tag{5.10}$$

Later, approximate solutions to structural problems will be obtained through assumptions for the displacements of loaded structures. In the case of the bar this means assuming an expression for the axial displacement as a function of the axial co-ordinate position; that is, using the standard symbols for axial displacement and co-ordinate, we shall assume $u = u(x)$. The strain at any section of the bar is $\varepsilon_x = du/dx$ and consequently, using equations (5.8) and (5.10), the following very useful expression for the bar strain energy is obtained:

$$U_p = \frac{1}{2} \int_0^l AE \left(\frac{du}{dx}\right)^2 dx. \tag{5.11}$$

This expression allows for the situation where the cross-section and/or the axial strain (and stress) varies along the length of the bar.

The above derivation has concerned a particular simple one dimensional system but as well as producing the strain energy expressions for the particular case examined it also helps bring to light some general characteristics of elastic systems. These characteristics are as follows.

(a) Equations (5.8) and (5.9) show that for small-displacement linearly elastic behaviour the strain energy of the bar can be expressed as a *quadratic* function of the strain or as a *quadratic* function of the extension. Within the same behavioural limitations the quadratic nature of the strain energy also applies to all other types of body, be they one, two, or three dimensional.

(b) From equation (5.2) it is apparent that the derivative of strain energy (written as a function of displacement) with respect to extension is the corresponding force. It is also apparent from equation (5.6) that the derivative of strain energy density (written as a function of strain) with respect to strain is the corresponding stress. This result holds whether Hooke's law applies or not. It also applies to systems where more than one stress and strain are involved but there the derivatives are partial; thus typically $\sigma_i = \partial \nu_p / \partial \varepsilon_i$ where the subscript i is used merely to indicate that the stress and strain are particular components corresponding to each other. (ν_p is a function of all relevant components of strain.)

(c) The strain energy and strain energy density are represented by the areas OPQ under the curves in Figs. 5.1(b) and 5.1(c) respectively. Although this text is not concerned with the force method of analysis it should be realized that the areas OQR above the curves in these figures have fundamental significance in the force method. In Fig. 5.1(b) the area OQR represents what is known as the complementary strain energy of the bar $U_c = \int_0^F e^s \, dF^s$, whereas in Fig. 5.1(c) the corresponding area represents the complementary strain energy density $\nu_c = \int_0^{\sigma_x} \varepsilon_x^s \, d\sigma_x^s$. Unlike strain energy, complementary strain energy has no physical significance. Obviously when the material behaviour is linearly elastic the complementary strain energy becomes equal in magnitude to the strain energy. However, the complementary strain energy is none the less properly interpreted as a function of force or stress whereas, as has been mentioned earlier, the strain energy is interpreted as a function of extension or strain. No further consideration will be given to complementary strain energy in this text.

The ideas and results presented above for the axially loaded bar can be readily and directly extended to include other cases of loading, such as torsion and bending; this will now be done in a fairly concise fashion.

5.2.2. Torsionally loaded bar

Consider a uniform circular bar of length l with one end held and the other subjected to an applied twisting moment T which increases gradually in magnitude; Fig. 2.4(a) shows the arrangement. Under the action of the twisting moment the free end of the bar will rotate through an angle ϕ, the total angle of twist. If the twisting moment were plotted against the angle of twist at the free end then the result for non-linear elastic behaviour would be similar to that shown for the axially loaded bar in Fig. 5.1(b), but now T and ϕ replace F and e respectively. The external work done by the twisting moment is stored as strain energy U_p where, restricting attention now to linearly elastic behaviour,

$$U_p = \int_0^\phi T^s \, d\phi^s = \tfrac{1}{2} T\phi. \tag{5.12}$$

As has been mentioned in Section 2.7, under the action of a torque any cross-section of a circular bar simply rotates relative to neighbouring cross-sections. The state of stress set up is one of pure shear and the shear stress τ and corresponding shear strain γ at any point located at radius r are given by elementary torsional theory as

$$\tau = G\gamma = \frac{G\phi r}{l} = \frac{Tr}{J} \tag{5.13}$$

where G and J have their usual meanings as modulus of rigidity and polar second moment of area respectively. Substituting from equation (5.13) into equation (5.12) will give U_p as a function of the end rotation ϕ or of

T and in the former case

$$U_\mathrm{p} = \frac{1}{2} \frac{GJ\phi^2}{l} \tag{5.14}$$

which is analogous to equation (5.9) for the axially loaded bar. It is noted that the strain energy density is

$$\nu_\mathrm{p} = \frac{1}{2} \tau\gamma = \frac{1}{2} G\gamma^2 = \frac{1}{2} G \frac{r^2}{l^2} \phi^2 \tag{5.15}$$

and, unlike the axially loaded bar (see equation (5.8)), is not uniform throughout the bar. For the torsion case the strain energy density represents the area under the curve, or under the straight line in the linearly elastic case, of a τ–γ diagram which would be of similar form to the σ_x–ε_x diagram shown for the axially loaded case in Fig. 5.1(c).

If the bar has a circular cross-section which varies along its length or if the applied twisting moment varies along the bar then the twist per unit length will no longer be constant. In such a case the total strain energy is the sum of the energies of elemental discs of length $\mathrm{d}x$ whose relative angle of twist over the length $\mathrm{d}x$ is $\mathrm{d}\theta_x$. By suitable modification of equation (5.14) we obtain

$$U_\mathrm{p} = \frac{1}{2} \int_0^l GJ \left(\frac{\mathrm{d}\theta_x}{\mathrm{d}x} \right)^2 \mathrm{d}x \tag{5.16}$$

which is of similar form to equation (5.11) for the axially loaded bar.

The above can apply for non-circular cross-sections subject to the reservations expressed in Section 2.7; then J is an appropriate torsion constant.

Because of the close similarity of the equations of the torsionally loaded bar to those of the axially loaded bar the torsion case will not be examined further in this chapter.

5.2.3. Bending of beams

The well-known engineer's theory of bending, or Bernoulli–Euler theory, was outlined in Section 2.8. In this beam theory the only significant stress is the direct stress in the longitudinal or axial direction. In this regard it relates to the axially loaded bar but instead of being uniform over a cross-section the stress (and hence the strain) varies linearly through the beam depth. Despite this the strain energy density at any point can still be represented by equation (5.6) and illustrated by Fig. 5.1(c), although here attention is restricted to linear elasticity. (The equivalent of Fig. 5.1(b) would be a plot of applied moment against cross-sectional rotation.)

The stress, and the corresponding strain, due to bending for a beam lying in the xy plane are (see Section 2.8, particularly equations (2.18)–(2.20) and Fig. 2.5)

$$\sigma_x = E\varepsilon_x = Ey\chi_z = -Ey\frac{\mathrm{d}^2v}{\mathrm{d}x^2}. \tag{5.17}$$

From equation (5.8) the strain energy density for the beam is

$$\nu_\mathrm{p} = \frac{1}{2}E\varepsilon_x^2 = \frac{1}{2}Ey^2\chi_z^2 = \frac{1}{2}Ey^2\left(\frac{\mathrm{d}^2v}{\mathrm{d}x^2}\right)^2. \tag{5.18}$$

The total strain energy for the beam is obtained by integrating ν_p over the beam volume, first by integrating over the cross-sectional area A and then by integrating over the length l. Since, of the terms occurring in the last two expressions of equation (5.18), only y varies over the area and since

$$\int_A y^2\,\mathrm{d}A = I_z$$

then the total strain energy of bending is

$$U_\mathrm{p} = \frac{1}{2}\int_0^l EI_z\chi_z^2\,\mathrm{d}x = \frac{1}{2}\int_0^l EI_z\left(\frac{\mathrm{d}^2v}{\mathrm{d}x^2}\right)^2\,\mathrm{d}x. \tag{5.19}$$

This expression holds, of course, in the general situation when the flexural rigidity EI_z and the curvature χ_z vary along the beam length. It is noted that U_p could, of course, be alternatively expressed as a function of bending moment (since $M_z = EI_z\chi_z$) but that equation (5.19) is the form that is suitable for the displacement approach.

5.3. Principle of virtual displacements

Consider, first of all, the derivation of the principle of virtual displacements as it applies to a single small (rigid) particle. The particle P is shown in Fig. 5.2 and is subjected to a set of N (real) forces F_i. Now we imagine

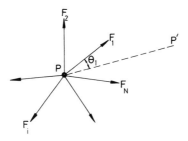

Fig. 5.2. Particle subjected to forces.

that the particle is given a small arbitrary displacement, denoted by δa, in some direction PP'. Such a displacement is referred to as a *virtual displacement*. It is emphasized that the displacement is fictitious and need have nothing whatever to do with any actual motion of the particle that may occur due to the forces acting on the particle. The requirement that the virtual displacement be small is imposed so that the direction of the applied forces can be assumed not to change during the fictitious movement.

The work done by the real forces during the virtual (or imaginary) displacement is called virtual work, denoted by δW, and is the sum of all the products of the virtual displacement and the component of real force in the direction of the virtual displacement, i.e.

$$\delta W = \delta a F_1 \cos \theta_1 + \delta a F_2 \cos \theta_2 + \ldots \delta a F_N \cos \theta_N,$$

or

$$\delta W = \delta a \sum_{i=1}^{N} F_i \cos \theta_i. \tag{5.20}$$

Here θ_i is the angle between the line of action of force F_i and the direction PP'. It should be noted that during the virtual displacement the forces F_i act at their full constant values and hence no factor of $1/2$ is present in the above expression.

Now we shall assume that the particle is, in fact, in equilibrium under the action of the forces F_i. It follows that the resultant of all the forces must vanish and consequently that the component of the resultant in the direction of PP' must equal zero. Since this component is the quantity $\sum_{i=1}^{N} F_i \cos \theta_i$ appearing in equation (5.20) it must be that the work done by the real applied forces F_i in moving through the virtual displacement δa is zero when the particle is in equilibrium under the action of the forces. Since the direction of the displacement δa is completely arbitrary the following statement can be made:

If a particle is in equilibrium under a system of real forces the total work done by the forces during any arbitrary virtual displacement is zero.

This is a statement of the PVD for a particle. Alternatively, the principle can be expressed:

A particle is in equilibrium under a system of real forces if the total work done by the forces during any arbitrary virtual displacement is zero.

In interpreting this statement it is noted that any arbitrary virtual displacement of a particle can be expressed in terms of components in each of the co-ordinate directions relevant to the problem in hand. Thus, for a two-dimensional situation, for instance, the equilibrium condition can be fully examined by considering separate virtual displacements in the directions of each of the two co-ordinate axes.

Fig. 5.3. Set of particles and associated forces.

The above statements allow us to reach the important conclusion that *the principle of virtual displacements is an alternative statement to the conditions of equilibrium.*

The PVD can be readily extended to apply to a system of particles and thence to a solid body of finite dimensions which in general is deformable. To do this we consider a model of the type shown in Fig. 5.3 comprising a set of particles on which acts a system of real forces. The system of forces comprises forces which are external to the particle set, and interparticle forces which are internal to the particle set; the latter forces can be envisaged as being due to weightless springs connecting the particles. The particle set is assumed to be in equilibrium under the action of the force system, that is the set of particles is in overall equilibrium under the action of the external forces whilst individually each particle is in equilibrium under the action of external and internal (interparticle) forces.

The principle of virtual displacements established above can be applied to each and every particle of the set shown in the figure. Because each particle is in equilibrium it follows that the virtual work done by the real forces acting on a particle during a virtual (or imaginary) displacement applied to the set must be zero. Therefore, the virtual work of all the forces acting on all the particles must equal zero. This virtual work can be conveniently viewed as comprising two parts, one corresponding to the virtual work done during the virtual displacement by the external forces, denoted by δW_e, and the other to that done by the internal forces, denoted by δW_i. Obviously the number of particles can be increased without limit, so the above argument applies equally to a finite body. The principle of virtual displacements for a deformable body can thus be expressed:

A deformable body is in equilibrium if the total work done by all the external forces plus the total work done by all the internal forces during any kinematically admissible virtual displacement is zero.

Symbolically the principle is expressed as

$$\delta W_i + \delta W_e = 0. \tag{5.21}$$

In the statement of the PVD for a deformable body it will be noted that the virtual displacement is qualified by the phrase 'kinematically admissible' which implies that such a displacement is not completely arbitrary. In fact the virtual displacement must represent a displaced state that could actually occur physically. This means that the virtual displacement must be such that it maintains continuity between all parts of the structure and is also compatible with the kinematic (or geometric) constraints at the structure boundary. Thus if, say, a rigid reaction is present at some point on the body the virtual displacement should have zero value at that point in the direction of the reaction; correspondingly the reaction force, which can be termed a force of constraint, will do no work during the virtual displacement. As an example, if we are considering a beam clamped at both ends the constraints for the beam are zero displacement and zero rotation at the ends and the virtual displacement should reflect this.

In passing it is noted that in the extreme case of a rigid (rather than deformable) body constraints are imposed on the system of particles such that the distances between the particles remain unchanged during the virtual displacement, and the assumed virtual displacement should reflect this fact. Thus the virtual displacements should be rigid-body motions in which the shape of the body is unchanged; correspondingly there is then no work of internal forces δW_i. The concern in the remainder of this chapter is, however, with deformable rather than rigid bodies.

In the above derivation only the concepts of equilibrium and work have been employed and therefore it can be seen that the validity of the PVD does not depend on any relationship between the real equilibrium force system and the virtual (imaginary) compatible displacement system. The virtual displacements generally need have no connection whatever with any real displacements present in the body in its equilibrium configuration due to the real forces.

In the form given by equation (5.21) the PVD is very general since it holds for non-conservative as well as conservative systems and is thus applicable whether elastic or inelastic deformations are involved and whether displacements are large or small. Our interest is in conservative systems in which deformations are elastic and then the PVD can be specialized and will lead to the PMPE. This is discussed in the next section but before proceeding to this a quite simple example will be given which will illustrate the application of the PVD in circumstances where an exact solution is possible.

Example 5.1. Consider again the truss of Example 1.1 and obtain equations (e) of that example through appliction of the PVD.

The problem is illustrated in Fig. E1.1(a).
The real equilibrium force system comprises the externally applied forces F_u

and F_v and the unknown member tensions in the bars T_j, $j = 1-5$. It might also be said to include the reaction forces at the upper pins of each member but these forces will do no work in operating through an assumed virtual displacement which is compatible with the kinematic constraints of zero displacement at each upper pin. Such a compatible virtual displacement corresponds, for instance, to a movement δu of the common pin C in the x direction. Consistent with this virtual (imaginary) displacement δu the virtual elongation δe_j of the typical member can be found from purely geometrical considerations to be

$$\delta e_j = \delta u \cos \theta_j \qquad j = 1 \ldots 5$$

and this completes the specification of a kinematically admissible virtual displacement system.

During the virtual displacement the external work done by the real applied forces is

$$\delta W_e = F_u \, \delta u$$

since F_v does no work in the selected virtual displacement.

At the same time the internal work done by the real member forces is

$$\delta W_i = - \sum_{j=1}^{5} T_j \, \delta e_j = -\delta u \sum_{j=1}^{5} T_j \cos \theta_j.$$

The minus sign is present here since the internal forces T_j are acting in the opposite sense to the corresponding virtual elongations δe_j, that is work is done *on* rather than *by* the member forces.

Application of the PVD

$$\delta W_i + \delta W_e = 0$$

gives

$$-\delta u \sum_{j=1}^{5} T_j \cos \theta_j + F_u \, \delta u = 0$$

or, cancelling δu,

$$F_u = \sum_{j=1}^{5} T_j \cos \theta_j.$$

As expected this is one of the equilibrium equations of the problem (see equations (b) of Example 1.1). The second equilibrium equation

$$F_v = \sum_{j=1}^{5} T_j \sin \theta_j$$

can easily be generated in like manner by applying a second admissible virtual displacement system corresponding to a displacement δv of the pin C in the y direction.

The above result provides verification in a particular case of the general statement given earlier that the PVD is an alternative statement to the condition of equilibrium. It is noted that the result has been obtained without the use of any constitutive relationships and thus holds for any type of material behaviour, elastic or inelastic.

Equations (e) of Example 1.1 have not yet been derived. To obtain these equations in the present approach it is now necessary to stipulate the type of material behaviour. Consistent with Example 1.1 it is stipulated that the (real) behaviour of the structure is linearly elastic. Thus Hooke's law applies for each member and the relationship between the real force T_j and the real extension e_j is as it was in Example 1.1, that is

$$T_j = \frac{A_j E e_j}{l_j} = k_j e_j \qquad \text{for} \qquad j = 1 \ldots 5.$$

The real extension e_j, which has no connection with the virtual extension δe_j, is related to the real displacement of the pin C under the action of the real applied forces by the equation (see equation (c) of Example 1.1)

$$e_j = u \cos \theta_j + v \sin \theta_j \qquad \text{for} \qquad j = 1 \ldots 5$$

where u and v are the x and y components of the real displacement.

Combining these last two equations and substituting into the above expressions for F_u and F_v gives

$$F_u = u \sum_{j=1}^{5} k_j \cos^2 \theta_j + v \sum_{j=1}^{5} k_j \sin \theta_j \cos \theta_j$$

$$F_v = u \sum_{j=1}^{5} k_j \sin \theta_j \cos \theta_j + v \sum_{j=1}^{5} k_j \sin^2 \theta_j$$

which are equations (e) of Example 1.1. These equations could now be solved to yield the real displacement components u and v and the real member forces T_j could then be readily calculated via the calculation of the extensions e_j.

5.4. Principle of minimum potential energy

To develop the PMPE from the general form of the PVD given by equation (5.21) we now restrict attention to conservative systems in which the deformable body behaves elastically. (The behaviour may in general be linearly elastic or non-linearly elastic but attention here will be restricted to linearly elastic behaviour). It was seen in Section 5.2 that a characteristic of an elastic system is that the negative of the work done during deformation by the internal forces can be equated to the potential energy of deformation or the strain energy. Consequently the internal virtual work δW_i associated with a virtual displacement from an equilibrium position is equal in magnitude and opposite in sign to the change in strain energy δU_p occurring during the virtual displacement. Thus equation (5.21) can be replaced by

$$\delta W_e - \delta U_p = 0 \tag{5.22}$$

which is an algebraic statement of the PVD for elastic bodies. This is a popular form of the PVD which is often used in the development of approximate solutions to structural problems and, in particular, is often used to develop the stiffness matrices of finite elements. The use of this form of the PVD in obtaining approximate solutions will not be explored at the moment since we wish now to proceed directly to a statement of the PMPE; the approach through the PVD will be discussed in Section 5.7.

Since the external forces, like the internal forces, are assumed to be conservative it follows that the work done by these forces during a virtual displacement δW_e can also be expressed in terms of a potential. It is customary to regard the negative of δW_e as a change in potential energy δV_p. Then, from equation (5.22) we obtain

$$\delta U_p + \delta V_p = 0. \tag{5.23}$$

This states that during any admissible virtual displacement from the equilibrium position the sum of the change in strain energy and the change in potential energy of the external forces is zero.

The virtual displacement has been interpreted thus far as an imaginary displacement which might be quite unconnected with the real displacement occurring under load but it can equally well be interpreted as simply a change, or *variation*, of the *real* displacement existing in the equilibrium position. It is this latter interpretation which is adopted in developing the PMPE. The δ sign is now viewed as what is called a variational operator and equation (5.23) is expressed as

$$\delta \Pi_p = 0 \tag{5.24}$$

where

$$\Pi_p = U_p + V_p. \tag{5.25}$$

The quantity Π_p, expressed as a function of the real displacement, is the *total potential energy* of the elastic system and as seen from equation (5.25) is the summation of the strain energy U_p and the potential energy of the external loads V_p.

Equation (5.24) effectively states that if a small kinematically admissible variation of displacement is imposed on an elastic body in its equilibrium position, the resulting change (or first variation) in total potential energy is zero. This means that the total potential energy has a *stationary* value at the equilibrium position. A stationary value could be a maximum value or a minimum value or could correspond to a saddle point, but in fact it can be demonstrated that for stable elastic systems the stationary value is a *minimum* value. Thus we arrive at the principle of minimum potential energy which can be expressed:

Of all possible kinematically admissible displacement configurations that an elastic body can take up the configuration which satisfies equilibrium makes the total potential energy assume a minimum value.

It is reiterated that kinematically admissible displacement configurations are those which satisfy the kinematic boundary conditions and which meet the requirements of compatibility within the body.

The concept of strain energy, or potential energy of deformation U_p, has been discussed for very simple elastic bodies in Section 5.2. There, equations (5.11), (5.16), and (5.19) give expressions for U_p as a quadratic function of the (real) displacements for the axially loaded bar, the torsionally loaded bar, and the beam respectively. The second type of potential energy contributing to the total potential energy Π_p is that of the external loads. For loads unchanging in magnitude, as occurs in static problems, this energy V_p is the negative of the sum of all products of external load and corresponding displacement at the load position measured in the direction that the load acts. Thus, if F_i represents a particular point load on the structure whilst d_i is the corresponding displacement and N is the number of such loads then

$$V_p = - \sum_{i=1}^{N} F_i d_i. \tag{5.26}$$

(Correspondingly $\delta W_e = -\delta V_p = \sum_{i=1}^{N} F_i \, \delta d_i$ as is required for the PVD). If externally applied moments are present a similar expression can be written for their potential energy with moments replacing forces and rotations replacing displacements. If distributed loads are present then V_p will involve integrals of products of load intensity and displacement over that region of the body on which the distributed load acts. A number of examples of the calculations of V_p for different types of loading are included in the examples which are given later in this chapter.

The negative sign associated with the above definition of V_p, (equation (5.26)) reflects the fact that if the unloaded configuration of the elastic body is taken as the datum position then the external loads lose potential for work in moving to the loaded (or deformed) configuration. On the other hand there is an increase in the potential energy of deformation U_p in moving from the unloaded to the loaded configuration.

In truth an elastic body has an infinite number of degrees of freedom and correspondingly there are an infinite number of possible admissible displacement configurations that have to be considered before we can be sure that we have the absolute minimum potential energy value correspondingly precisely to the equilibrium state. When viewed in this light the problem symbolized by equation (5.24) belongs in the realm of the branch of mathematics known as the *calculus of variations*, which is a generalization of the maximum or minimum problem of ordinary differential calculus. The principle of minimum potential energy is referred to as a *variational* principle and the potential energy Π_p is called the potential energy *functional*; a functional is simply a function defined by integrals whose arguments themselves are functions. The variational calculus provides the means of establishing the governing equations of the problem (i.e. the differential equations of equilibrium in the interior of the body and the natural boundary conditions at the exterior of the body) rather than providing the actual solution, which is what we require. We need not consider the topic further in this text, but the interested reader is referred to appropriate mathematical texts or to the last two books of the Bibliography at the end of this chapter which include chapters dealing with variational calculus.

Our concern here is to develop the means of calculating solutions to the structural problem by restricting the degrees of freedom to a finite number and consequently considering a finite restricted class of admissible displacement configurations. By doing this the problem becomes one of differential calculus rather than of variational calculus.

The total potential energy of a system having degrees of freedom d_i ($i = 1 \ldots n$) can be expressed as a function of the degrees of freedom as

$$\Pi_p = \Pi_p(d_i), \qquad i = 1, 2, \ldots, n. \tag{5.27}$$

The condition that the total potential energy be a minimum is met if all the partial derivatives of $\Pi_p(d_i)$ with respect to each degree of freedom in turn are zero. (It is assumed that the partial derivatives are continuous functions.) Thus the minimum potential energy principle, for a system having a finite number of degrees of freedom, can be expressed mathematically as

$$\frac{\partial \Pi_p}{\partial d_1} = \frac{\partial \Pi_p}{\partial d_2} = \ldots = \frac{\partial \Pi_p}{\partial d_n} = 0. \tag{5.28}$$

This procedure will result in a set of n simultaneous algebraic equations which express the equilibrium conditions of the system and which can be solved for the displacements. To illustrate the approach two simple examples are now presented.

Example 5.2. A linear spring of stiffness k carries a load P^* as shown in Fig. E5.2(a). Use the PMPE to determine the deflection Δ of the spring under the load.

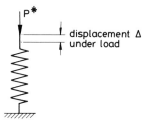

Fig. E5.2(a)

The spring constant k is the equivalent of the extensional stiffness AE/l of a bar, of course, and so the strain energy stored in the spring when it is deflected is, from equation (5.9),

$$U_p = \tfrac{1}{2}k\Delta^2. \tag{a}$$

The potential energy of the external load in the deflected configuration is the negative of the product of P^* and Δ, that is

$$V_p = -P^*\Delta$$

where the negative sign reflects the fact that P^* has lost some of its potential for doing work in moving the distance Δ in the direction in which it acts.

Thus the total potential energy is

$$\Pi_p = \Pi_p(\Delta) = \tfrac{1}{2}k\Delta^2 - P^*\Delta \tag{b}$$

and is a function of the single degree of freedom Δ.

Application of the PMPE $\delta\Pi_p = 0$ is equivalent to the single equation

$$\frac{d\Pi_p}{d\Delta} = 0$$

or

$$k\Delta - P^* = 0. \tag{c}$$

This equation is the expected result, of course, and it effectively states that the system is in equilibrium when the free end of the spring deflects an amount equal to k/P^*. As a simple illustration that the equilibrium position corresponds to a *minimum* value of total potential energy we can readily plot the variation of Π_p (given by equation (b)) with the end displacement Δ. This is shown in Fig. E5.2(b) for specific values of P^* and k of 1 kN and 200 kN m^{-1} respectively. We see that the minimum value of Π_p is -2.5 kN mm and that this occurs when $\Delta = 5$ mm. This value of Δ agrees, of course, with that given by the equilibrium equation (c).

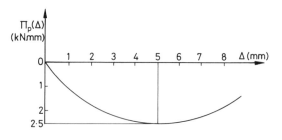

Fig. E5.2(b)

Example 5.3. Consider again the truss of Example 1.1 and obtain equations (e) of that example through application of the PMPE.

For the structure shown in Fig. E1.1(a) the potential energy of the prescribed external loads is

$$V_p = -F_u u - F_v v.$$

The strain energy of the structure is the summation of the strain energies of the individual bars. Using equation (5.9) this becomes

$$U_p = \frac{1}{2} \sum_{j=1}^{5} \left(\frac{AE}{l}\right)_j e_j^2 = \frac{1}{2} \sum_{j=1}^{5} k_j e_j^2.$$

Here, as defined in Example 1.1, the quantity $k_j = (AE/l)_j$ is the extensional stiffness of bar j whilst

$$e_j = u \cos \theta_j + v \sin \theta_j$$

is the (real) extension of bar j under the given loading (see equation (c) of Example 1.1). Using the above three equations the total potential energy of the loaded truss can be expressed as a quadratic function of the two degrees of freedom u and v as

$$
\begin{aligned}
\Pi_p &= \Pi_p(u, v) \\
&= U_p + V_p \\
&= \tfrac{1}{2} \sum_{j=1}^{5} k_j (u^2 \cos^2 \theta_j + 2uv \sin \theta_j \cos \theta_j + v^2 \sin^2 \theta_j) \\
&\quad - F_u u - F_v v. \qquad\qquad (a)
\end{aligned}
$$

Application of the PMPE to this problem is equivalent to using the conditions

$$\frac{\partial \Pi_p}{\partial u} = \frac{\partial \Pi_p}{\partial v} = 0.$$

It follows from the expression for Π_p that

$$\tfrac{1}{2}\left(\sum_{j=1}^{5} k_j (2u \cos^2 \theta_j + 2v \sin \theta_j \cos \theta_j) \right) - F_u = 0$$

and

$$\tfrac{1}{2}\left(\sum_{j=1}^{5} k_j (2u \sin \theta_j \cos \theta_j + 2v \sin^2 \theta_j) \right) - F_v = 0$$

or

$$F_u = u \sum_{j=1}^{5} k_j \cos^2 \theta_j + v \sum_{j=1}^{5} k_j \sin \theta_j \cos \theta_j \qquad\qquad (b)$$

and

$$F_v = u \sum_{j=1}^{5} k_j \sin \theta_j \cos \theta_j + v \sum_{j=1}^{5} k_j \sin^2 \theta_j. \qquad\qquad (c)$$

These equilibrium equations are precisely equations (e) of Example 1.1 which were developed using basic structural principles.

This result verifies again that the equilibrium state of a structure corresponds to a minimum value of the total potential energy.

Before leaving this particular problem it is pertinent to use the results to make some points which have general applicability to structures with any number of degrees of freedom.

(i) If we make the definitions

$$K_{11} = \sum_{j=1}^{5} k_j \cos^2 \theta_j \qquad\qquad K_{12} = \sum_{j=1}^{5} k_j \sin \theta_j \cos \theta_j$$

$$K_{21} = \sum_{j=1}^{5} k_j \sin \theta_j \cos \theta_j \qquad\qquad K_{22} = \sum_{j=1}^{5} k_j \sin^2 \theta_j$$

then the equilibrium equations (b) and (c) can be expressed in matrix form as (see Section 1.8)

$$\begin{Bmatrix} F_u \\ F_v \end{Bmatrix} = \begin{bmatrix} K_{11} & K_{12} \\ K_{21} & K_{22} \end{bmatrix} \begin{Bmatrix} u \\ v \end{Bmatrix} \qquad \text{or} \qquad \mathbf{F} = \mathbf{KD}.$$

Here $\mathbf{K}$ is the structure stiffness matrix and since $K_{12} = K_{21}$ this matrix is symmetric.

(ii) From equation (a) the structure strain energy can be expressed as

$$U_p = \tfrac{1}{2} K_{11} u^2 + K_{12} uv + \tfrac{1}{2} K_{22} v^2$$

or

$$U_p = \tfrac{1}{2} \mathbf{D}^t \mathbf{KD}$$

which is a general matrix expression for the quadratic strain energy.

(iii) The individual coefficient of the stiffness matrix (i.e. K_{ef} where in this example e and f have the values 1 and 2) can be obtained directly from the strain energy expression by the relationship

$$K_{ef} = \frac{\partial^2 U_p}{\partial D_e \, \partial D_f}$$

where D_e and D_f correspond to u and v in this particular example.

5.5. The Rayleigh–Ritz method

In the previous section the PMPE has been described and it has been demonstrated how solutions to simple problems can be obtained. This has served the purpose of introducing the principle and of demonstrating that application of the principle corresponds to satisfaction of the equilibrium condition.

In the two examples considered the solutions obtained using the PMPE, in the manner corresponding to equations (5.27) and (5.28), are exact in the sense that they agree precisely with solutions obtained by direct application of the fundamental conditions of elasticity, equilibrium, and compatibility. It should be realized that this correspondence with the exact solutions occurs because we are implicitly assuming particular restricted classes of admissible displacement configurations which happen to contain the exact solution within them. For instance, in the case of Example 5.2 by using equation (a) as the expression for the strain energy of the spring we imply that the extension per unit length of spring is uniform along the whole spring and hence that the deformation varies linearly from zero at the fixed end to the amount Δ at the free end. Thus we have implicitly chosen a linear displacement configuration (or field) which is the correct shape since the 'exact' solution falls within this category. In this circumstance application of the PMPE only has to provide the magnitude of the free-end deformation in the equilibrium position for the correspondence to be complete. Similar remarks apply to the truss example where again the implied linear variation of displacement along the lengths of the bars used in the application of the PMPE matches the 'exact' situation.

Thus far no real advantage of using the PMPE approach is apparent since for the type of problem considered we can generate solutions rather more easily using other approaches. It is when we turn to considerations of more complex structural problems, for which exact solutions are not known, that the PMPE can be used to great advantage in obtaining approximate solutions. The procedure in these circumstances is to assume or guess the displaced form of the structure under load by expressing the displacement (or displacements) over the structure in the form of a finite admissible expansion of terms which are each products of functions of the space co-ordinates and unknown coefficients. (The number of terms equals the number of degrees of freedom of the model structure.) The assumed displaced form constitutes a restricted class of admissible displacement configurations which generally will *not* contain within it the exact solution of the problem. Application of the PMPE, characterized by equations (5.27) and (5.28), will now provide the best possible solution within the limitations of the restricted class of displacement configurations. The potential energy will be minimized subject to the restriction (or constraint) imposed by the limited assumed form of the structure displacement and generally the true displacement state corresponding to the exact solution exists at a somewhat lower value of total potential energy. Thus the true minimum potential energy state of the structure is not reached and, corresponding to the fact that the assumption of a restricted class of displacement configurations effectively constrains the structural response, the results obtained will indicate a structure somewhat stiffer than the actual structure. Further, since the exact minimum potential energy state of the actual structure is not reached we must expect the equilibrium condition to be violated to a certain degree in this approach; this means that whilst the equilibrium condition of the structure is satisfied in an average sense there will be local violations of the condition within the structure.

The approach described in the preceding paragraph is the well-established Rayleigh–Ritz method. The method has been used extensively in continuum mechanics and, in modified (piecewise) form, it provides the basis for the finite-element displacement formulations described in later chapters. In the Rayleigh–Ritz procedure the assumed displacement fields are popularly of polynomial or trigonometric nature but other types of field are possible. The accuracy of the method depends very largely on how well the assumed displacement field can describe the actual deformed shape of the structure.

Mention has been made above of admissible displacement configurations or functions or fields, and some words of amplification of what this means are appropriate here. The admissibility of a displacement function is linked directly with the satisfaction of the compatibility condition. Firstly, at the boundaries of a structure an admissible function is one which satisfies the geometrical boundary conditions (i.e. those conditions related to prescribed displacements) but need not satisfy the natural boundary conditions (those related to prescribed forces). Secondly, the compatibility condition must be satisfied everywhere in the structure interior and this is equivalent to stating that an assumed displacement field must be continuous to one order less than the highest derivative occurring in the appropriate energy integrand. (This means that the integrand is single valued everywhere.)

To see what this latter statement means physically, consider the simple cases of bar and beam analysis. In the bar problem the highest derivative occurring in the potential energy integrand is du/dx, so here the statement requires simply that u itself be continuous. In the beam problem the corresponding highest derivative is d^2v/dx^2 so the statement requires that any assumed displacement function allows continuity of both v and dv/dx. For the bar and beam these conditions are clearly reasonable in the physical sense.

The fact that an assumed displacement function, or trial function, is admissible is not enough on its own to ensure a close approximation to an exact solution. Care taken in the selection of a sensible trial function will be repaid by improved accuracy of the solution. As has been intimated above, the trial function is often expressed as a finite expansion, or series, of terms since a single term will not usually yield a sufficiently accurate solution. Obviously we would expect that as more terms are progressively included in the trial function the answers obtained would approach progressively closer to the exact solution, that is the answers would *converge* to the true solution, hopefully rapidly. One important factor which affects this is the *completeness* of the series trial function. The formal definition of completeness is a little complicated but in simple terms it effectively means that no appropriate lower-order terms of a series should be omitted whilst higher-order terms are included. Violation of the completeness condition can mean that convergence to the exact solution will not occur, however many series terms are used.

The Rayleigh–Ritz method will now be illustrated by several examples concerning simple one-dimensional structures. These examples can (again) be readily solved in exact fashion by more direct methods but the procedures outlined in the energy approach can be extended to much more complicated structural situations where direct methods are inapplicable.

Example 5.4. Consider the uniform bar of cross-sectional area A and length L shown in Fig. E5.4(a). A distributed axial force of intensity p per unit length (due to gravity perhaps) acts on the bar. Estimate the distribution of displacement and stress in the bar using the Rayleigh–Ritz method.

(i) The first displacement function field with which we shall attempt to obtain a solution is

$$u = C_1 x \tag{a}$$

Here C_1 is called a *generalized displacement* (or generalized co-ordinate) since C_1 is related to a particular point displacement or displacements but is not explicitly expressed as such. (It is obvious in this case that in fact $C_1 = u_L/L$ where u_L is the displacement at $x = L$.)

The displacement u is obviously continuous in the interior of the bar, and the only geometric boundary condition, $u = 0$ at $x = 0$, is satisfied by this displacement function because of the absence of a constant term, C_0 say. Thus the function satisfies the compatibility condition; in other words the function is admissible. The single natural boundary condition $\sigma_x = 0$ at $x = L$ need not be satisfied by the trial function and is not satisfied.

The strain energy or potential energy of deformation of the system is (see equation (5.11))

$$U_p = \frac{AE}{2} \int_{x=0}^{L} \left(\frac{du}{dx}\right)^2 dx = \frac{AE}{2} C_1^2 L.$$

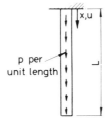

Fig. E5.4(a)

The potential energy of the external loading (the negative of the integrated product of applied load times displacement) is

$$V_p = -\int_{x=0}^{L} u(p\,dx) = -p\int_{x=0}^{L} C_1 x\,dx = -pC_1 L^2/2$$

Thus the potential energy functional is

$$\Pi_p = U_p + V_p = \frac{AE}{2}C_1^2 L - p\frac{L^2 C_1}{2} \tag{b}$$

and the minimization process $\delta\Pi_p = 0$ for this single-degree-of-freedom system is equivalent to

$$\frac{d\Pi_p}{dC_1} = AELC_1 - \frac{pL^2}{2} = 0.$$

Therefore

$$C_1 = \frac{pL}{2AE} \quad \text{and} \quad u = \frac{pLx}{2AE} \tag{c}$$

and it follows that

$$\sigma_x = E\frac{du}{dx} = \frac{pL}{2A}.$$

The value of the total potential energy (equation (b)) associated with the calculated value of C_1 is

$$\Pi_p = -0.125\frac{p^2 L^3}{AE}. \tag{d}$$

To judge the accuracy of the Rayleigh–Ritz approach, based on the displacement function of equation (a), we note that the exact solution to this problem can easily be shown to be that

$$u = \frac{p}{AE}\left(Lx - \frac{x^2}{2}\right) \qquad \sigma_x = \frac{p}{A}(L-x) \qquad \Pi_p = -0.166\frac{p^2 L^3}{AE}. \tag{e}$$

The approximate energy solution and the exact solution are compared diagrammatically for displacement and stress in Fig. E5.4(b). It is clear that the energy

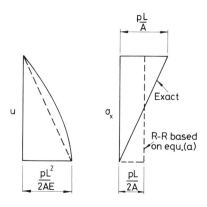

Fig. E5.4(b)

solution is poor in this particular instance. The displacement is approximated rather better than is the stress, and in fact agrees with the exact value at the free end of the bar, but the overall solution is not very accurate, as is further indicated by the considerable disparity in values of approximate and exact Π_p. Clearly the assumed displacement expression represents a restricted class of functions which is not particularly appropriate to the problem under consideration. Given this disadvantage, the Rayleigh–Ritz procedure has produced the best possible solution.

(ii) To try to obtain an improved solution, through the Rayleigh–Ritz approach, we now consider a more realistic admissible trial function by adding the next term in the polynomial series to that of equation (a). Thus we assume

$$u = C_1 x + C_2 x^2. \tag{f}$$

The strain energy now becomes

$$U_p = \frac{AE}{2} \int_{x=0}^{L} (C_1 + 2C_2 x)^2 \, dx$$

and the potential energy of the external load is

$$V_p = -p \int_0^L (C_1 x + C_2 x^2) \, dx.$$

On performing the integrations the total potential energy is

$$\Pi_p = \frac{AE}{2} \left(LC_1^2 + \frac{4}{3} L^3 C_2^2 + 2L^2 C_1 C_2 \right) - p \left(\frac{L^2 C_1}{2} + \frac{L^3 C_2}{3} \right) \tag{g}$$

The minimization process for the two-degree-of-freedom system gives the two algebraic equations

$$\frac{\partial \Pi_p}{\partial C_1} = AE(LC_1 + L^2 C_2) - p \frac{L^2}{2} = 0$$

$$\frac{\partial \Pi_p}{\partial C_2} = AE \left(L^2 C_1 + \frac{4}{3} L^3 C_2 \right) - p \frac{L^3}{3} = 0. \tag{h}$$

Solving these equations for the generalized displacements gives

$$C_1 = \frac{pL}{AE}, \qquad C_2 = -\frac{p}{2AE}. \tag{i}$$

Corresponding to these values, the displacement u (equation (f)), the stress $\sigma_x (= E \, du/dx)$, and the potential energy Π_p (equation (g)) are found to be as given in equation (e); thus the exact solution of the problem has been found in the energy approach based on the two-term displacement field. The attainment of the exact solution through the use of the assumed displacement field of equation (f) could have been foreseen of course. The assumed field happens to contain the exact field within it, and this being so the energy approach ensures that the calculated values of generalized displacements are those that correspond to the true minimum energy level, i.e. the true equilibrium state.

It is worth noting that the remarks made at the end of Example 5.3 apply equally to a problem (such as the present one) in which we are working in terms of generalized displacements rather than actual point displacements. Thus equations (h) of the present problem are stiffness equations

$$\begin{Bmatrix} pL^2/2 \\ -pL^3/3 \end{Bmatrix} = AE \begin{bmatrix} L & L^2 \\ L^2 & \frac{4}{3}L^3 \end{bmatrix} \begin{Bmatrix} C_1 \\ C_2 \end{Bmatrix}$$

relating generalized forces to generalized displacements through a symmetric stiffness matrix. The coefficients of this stiffness matrix are appropriate partial second differentials of the strain energy, e.g. $K_{12} = \partial^2 U_p/\partial C_1 \, \partial C_2$.

(iii) Thus far we have considered two levels of a polynomial series trial function. The trial functions of parts (i) and (ii) have been both admissible and complete in each case. With regard to the completeness of the series, equation (a) represents the lowest-order term of a polynomial series which is appropriate to the problem, since any constant term C_0 which might otherwise be expected to be present in the series would have the value zero in this case owing to the boundary condition and the choice of co-ordinate origin. Similarly, equation (f) represents the two lowest-order polynomial terms appropriate to the problem. It has been seen that convergence does occur with this sort of approach and, in fact, occurs so rapidly that the exact solution is achieved with the two-term series.

Now, simply to illustrate the meaning and importance of completeness, consider the same problem and take the trial function to be

$$u = C_2 x^2. \tag{j}$$

By the same approach as in part (i) it is easy to obtain

$$\Pi_p = \tfrac{2}{3}AEL^3 C_2^2 - \tfrac{1}{3}pL^3 C_2. \tag{k}$$

Thence the minimization process gives

$$\frac{d\Pi_p}{dC_2} = 0 = \frac{4}{3} AEL^3 C_2 - \frac{pL^2}{3}.$$

Therefore

$$C_2 = \frac{p}{4AE} \qquad u = \frac{px^2}{4AE}$$

$$\sigma_x = \frac{px}{2A} \qquad \Pi_p = -0.04166\,\frac{p^2 L^3}{AE} \tag{l}$$

The displacement u and stress σ_x obtained from this equation are shown in Fig. E5.4(c) along with the exact solution. Obviously the result obtained is very poor.

The trial function (j) is admissible since it is a smooth continuous function and it satisfies the geometric boundary condition $u = 0$ at $x = 0$. However, the function is an incomplete polynomial since the x^2 term is present but the x term has been omitted, and this accounts for its unsatisfactory performance. If higher-order terms such as $C_3 x^3$, $C_4 x^4$, etc. were added to the trial function (j) the resulting function would still be inappropriate and convergence to the exact solution would not occur however many terms were used. The lack of completeness because of the absence of an x term means that the stress and strain at the support point $x = 0$ will be zero and this is clearly incorrect.

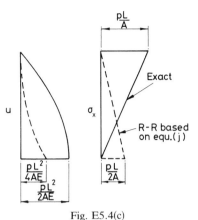

Fig. E5.4(c)

Example 5.5. Consider the application of the Rayleigh–Ritz method to the solution of a uniform simply supported beam carrying a central point load, as shown in Fig. E5.5(a).

To satisfy the admissibility requirements the assumed expression for the lateral deflection v as a function of x must match the displacement boundary conditions $v(0) = v(L) = 0$ in addition to satisfying compatibility within the length of the beam. (The deflection v and the slope, dv/dx must be single valued and continuous.)

The first admissible displacement field we shall try is a half sine wave

$$v = C_1 \sin \frac{\pi x}{L}. \tag{a}$$

Apart from being admissible and appearing a reasonable shape this field has the advantage of properly representing the obvious symmetry of the problem. The natural (or force) boundary conditions of the problem are

$$\frac{d^2 v}{dx^2}(0) = \frac{d^2 v}{dx^2}(L) = 0$$

(i.e. zero moment at the beam ends) and these are also met by the field, but it is emphasized again that this is not a necessary requirement of the assumed displacement field; normally satisfaction of the natural boundary conditions is achieved in an approximate sense as a result of applying the PMPE.

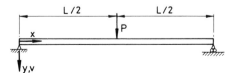

Fig. E5.5(a)

At $x = L/2$, $v = C_1$ and therefore the potential energy of the external load is

$$V_\mathrm{p} = -PC_1.$$

The strain energy is (see equation (5.19))

$$U_\mathrm{p} = \frac{EI_z}{2} \int_0^L \left(\frac{\mathrm{d}^2 v}{\mathrm{d}x^2}\right)^2 \mathrm{d}x = \frac{EI_z}{2} \int_0^L \left(-C_1 \frac{\pi^2}{L^2} \sin \frac{\pi x}{L}\right)^2 \mathrm{d}x = \frac{EI_z \pi^4}{4L^3} C_1^2.$$

The total potential energy is therefore

$$\Pi_\mathrm{p} = U_\mathrm{p} + V_\mathrm{p} = \frac{EI_z}{4} \frac{\pi^4}{L^3} C_1^2 - PC_1$$

and the minimization process $\delta \Pi_\mathrm{p} = 0$ becomes for this single degree-of-freedom system

$$\frac{\mathrm{d}\Pi_\mathrm{p}}{\mathrm{d}C_1} = \frac{EI_z \pi^4}{2L^3} C_1 - P = 0 \quad \text{and thus} \quad C_1 = \frac{2PL^3}{EI_z \pi^4}.$$

The solution for the displacement field is therefore

$$v = \frac{2PL^3}{EI_z \pi^4} \sin \frac{\pi x}{L}$$

and the maximum deflection at the beam centre is $v_\mathrm{c} = PL^3/48.70EI_z$. This latter value compares with the exact solution of $v_\mathrm{c} = PL^3/48EI_z$ and thus is in error by only about 1.5 per cent. (The Rayleigh–Ritz solution for the displacement is smaller than the true displacement, corresponding to an over-stiff solution as expected). Before looking at this solution in any more detail we shall consider the assumption of a second admissible displacement field by adding the second symmetric sine function to equation (a) to give

$$v = C_1 \sin \frac{\pi x}{L} + C_2 \sin \frac{3\pi x}{L}. \tag{b}$$

For this two-degree-of-freedom problem we have

$$V_\mathrm{p} = -(PC_1 - PC_2)$$

and

$$U_\mathrm{p} = \frac{EI_z}{2} \int_0^L \left(-C_1 \frac{\pi^2}{L^2} \sin \frac{\pi x}{L} - C_2 \frac{9\pi^2}{L^2} \sin \frac{3\pi x}{L}\right)^2 \mathrm{d}x.$$

In performing the integration use is made of the standard result that

$$\int_0^L \sin \frac{m\pi x}{L} \sin \frac{n\pi x}{L} \mathrm{d}x = \tfrac{1}{2} \text{ for } m = n$$

$$= 0 \text{ for } m \neq n.$$

Then

$$\Pi_\mathrm{p} = U_\mathrm{p} + V_\mathrm{p} = \frac{EI_z \pi^4}{4L^3} (C_1^2 + 81C_2^2) - PC_1 + PC_2$$

and the minimization procedure gives

$$\frac{\partial \Pi_p}{\partial C_1} = \frac{EI_z \pi^4}{2L^3} C_1 - P = 0$$

and

$$\frac{\partial \Pi_p}{\partial C_2} = \frac{81 EI_z \pi^4}{2L^3} C_2 + P = 0.$$

It should be noted that in this particular case, because of the nature of the assumed displacement field, the algebraic equations are uncoupled, i.e. C_1 can be determined from the first equation alone and C_2 can be separately determined from the second equation.† The result is

$$C_1 = \frac{2PL^3}{\pi^4 EI_z} \qquad C_2 = -\frac{2PL^3}{81 \pi^4 EI_z}$$

and consequently

$$v = \frac{2PL^3}{\pi^4 EI_z} \left(\sin \frac{\pi x}{L} - \frac{1}{81} \sin \frac{3\pi x}{L} \right)$$

At the beam centre $v_c = PL^3/48.11 EI_z$ with an error of only 0.23 per cent.

We have seen that comparison of point values of displacement does not necessarily give a good idea of the overall accuracy of an energy method solution. To obtain a fuller picture of the solution accuracy we now need to compare values of the total potential energy and to look at distributions along the beam of relevant quantities. It is noted that the exact solution to the problem for displacement can easily be shown to be

$$v = \frac{PL^3}{EI_z} \left(\frac{1}{16} \frac{x}{L} - \frac{1}{12} \frac{x^3}{L^3} \right)$$

and that the accuracy of the Rayleigh–Ritz displacements at any general point along the beam is of much the same order as that quoted above for displacements at a particular point, the beam centre. The comparison of energy method and exact distributions of shearing force $-EI_z \, d^3v/dx^3$ and bending moment $-EI_z \, d^2v/dx^2$ is shown in Fig. E5.5(b). These quantities are determined with substantially less accuracy in the energy approach than is the displacement and it is clear once again that point values of force quantities can be significantly in error. Finally the calculated total potential energy values are $-1.02660 \times 10^{-2} P^2 L^3/EI_z$ for the one-term solution, $-1.03927 \times 10^{-2} P^2 L^3/EI_z$ for the two-term solution, and $-1.04167 \times 10^{-2} P^2 L^3/EI_z$ for the exact solution.

In this problem the restricted classes of displacement functions that have been used are of a sinusoidal nature and do not contain the exact solution (a cubic polynomial) within them. Because of this the true solution has not been found by the energy approach. However, if the number of terms in the sine series were to

---

† It should also be noted that for the sine function series (but *not* generally) the value of any generalized displacement is unchanged by the addition of extra terms in the series. Thus C_1 has the same value when the two-term series (equation (b)) is used as when the single-term (equation (a)) is used.

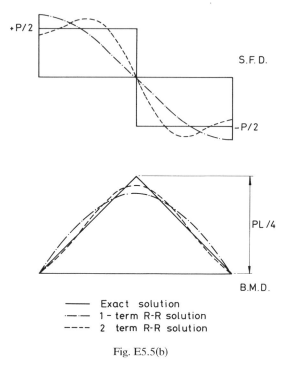

S.F.D.

B.M.D.

——— Exact solution
—·— 1- term R-R solution
---- 2 term R-R solution

Fig. E5.5(b)

be progressively increased we should expect to obtain a solution agreeing more and more closely with the exact solution.

Example 5.6. Examine the solution of the problem illustrated in Fig. E5.6(a), a cantilever beam carrying a uniformly distributed load, by assuming a displacement field of the form

$$v = C_2 x^2 + C_3 x^3. \tag{a}$$

It is clear, first of all, that the assumed displacement field is admissible since compatibility is satisfied within the beam and the geometric boundary conditions $v = dv/dx = 0$ at $x = 0$ are complied with. The latter has been achieved simply by omitting from the field any constant term C_0 and any linear term $C_1 x$; having

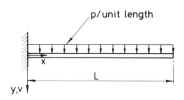

Fig. E5.6(a)

done this the two lowest-order available polynomial terms are those given in the displacement expression and so the displacement field is complete. The natural boundary conditions

$$\frac{d^2v}{dx^2}(L) = \frac{d^3v}{dx^3}(L) = 0$$

corresponding to zero bending moment and shearing force at the free end are not explicitly satisfied for non-zero C_2 and C_3.

The strain energy is

$$U_p = \frac{EI_z}{2} \int_0^L \left(\frac{d^2v}{dx^2}\right)^2 dx = \frac{EI_z}{2} \int_0^L (2C_2 + 6C_3x)^2 \, dx$$

and the potential energy of the external loading is

$$V_p = -\int_0^L vp \, dx = -p \int_0^L (C_2x^2 + C_3x^3) \, dx.$$

Then

$$\Pi_p = U_p + V_p = \frac{EI_z}{2}(4LC_2^2 + 12L^3C_3^2 + 12C_2C_3L^2) - p\left(C_2\frac{L^3}{3} + C_3\frac{L^4}{4}\right)$$

and the minimization procedure gives the coupled pair of algebraic equations:

$$4LC_2 + 6L^2C_3 - \frac{pL^3}{3EI_z} = 0$$

$$12L^3C_3 + 6L^2C_2 - \frac{pL^4}{4EI_z} = 0.$$

The solution of these simultaneous equations is

$$C_2 = \frac{5L^2p}{24EI_z} \qquad C_3 = -\frac{pL}{12EI_z}$$

so that

$$v = \frac{pL^4}{24EI_z}\left(\frac{5x^2}{L^2} - \frac{2x^3}{L^3}\right)$$

and the deflection at the free end is $v_L = pL^4/8EI_z$. The value of the total potential energy corresponding to this solution is $\Pi_p = -0.02431L^5p^2/EI_z$.

The exact solution to this problem can be readily generated from elementary principles (also see Example 4.1). The exact displaced form is

$$v = \frac{pL^4}{24EI_z}\left(\frac{6x^2}{L^2} - \frac{4x^3}{L^3} + \frac{x^4}{L^4}\right)$$

and at the free end $v_L = pL^4/8EI_z$. The corresponding true minimum value of total potential energy is $\Pi_p = -0.025L^5p^2/EI_z$.

Again we have the situation here that the assumed displacement form (equation (a)) does not contain within it the true displaced form. Thus the true minimum energy state, or true equilibrium state, cannot be generated. Nevertheless the

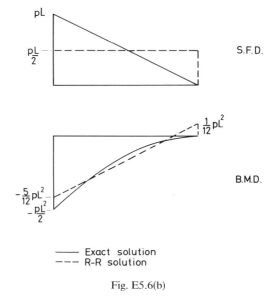

Fig. E5.6(b)

Rayleigh–Ritz procedure ensures the best possible solution, given the restricted nature of the assumed displacement field. The displacement is accurately predicted along the whole length of the beam—and is exact at the free end—and the corresponding minimum energy value is very close to the true value. However, along the beam the derived quantities of shearing force and bending moment do not compare nearly as well with the exact values, as is indicated in Fig. E5.6(b).

We would expect to generate the exact solution through the energy approach by augmenting the displacement field of equation (a) by a quartic term $C_4 x^4$ (see Problem 5.3.)

Example 5.7. The symmetric stepped beam shown in Fig. E5.7(a) has fully clamped ends. Assume a polynomial expression for the displacement v in the left-hand half of the beam and hence develop a solution using the Rayleigh–Ritz method.

The selection of the trial function is not so straightforward in this example. In the left-hand half of the beam there are three geometric boundary conditions that have to be satisfied. These are that $v = 0$ at $x = 0$, $dv/dx = 0$ at $x = 0$ and $dv/dx = 0$ at $x = 3l$. The lowest-order polynomial that will satisfy these conditions and give non-zero values for displacement and bending moment along the beam is a cubic polynomial. Thus in terms of generalized displacements we assume a complete

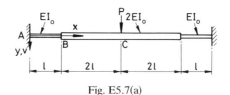

Fig. E5.7(a)

cubic polynomial of the form

$$v = C_0 + C_1 x + C_2 x^2 + C_3 x^3.$$

The generalized displacements C_0 and C_1 must, however, be zero to satisfy the first two of the stated geometric boundary conditions. The third such condition, i.e. $dv/dx = 0$ at $x = 3l$, gives a relationship between C_2 and C_3 such that $C_3 = -2C_2/9l$. Therefore the assumed trial function that satisfies all the geometric boundary conditions is

$$v = C_2\left(x^2 - \frac{2x^3}{9l}\right) \qquad \text{for} \qquad 0 \leqslant x \leqslant 3l.$$

The strain energy for the whole beam is

$$U_p = 2\frac{EI_0}{2}\int_0^l \left(\frac{d^2v}{dx^2}\right)^2 dx + 2\left(\frac{2EI_0}{2}\right)\int_l^{3l}\left(\frac{d^2v}{dx^2}\right)^2 dx$$

and on substituting for v this becomes

$$U_p = \frac{164EI_0 l C_2^2}{27}.$$

The potential energy of the external loading is, with v_c denoting the central deflection,

$$V_p = -Pv_c = -3C_2 l^2 P.$$

The total potential energy $\Pi_p = U_p + V_p$ is thus available as a function of the single variable C_2. The minimization procedure $d\Pi_p/dC_2 = 0$ yields

$$C_2 = \frac{81Pl}{328EI_0}$$

and it follows that

$$v = \frac{81Pl}{328EI_0}\left(x^2 - \frac{2x^3}{9l}\right) \quad (0 \leqslant x \leqslant 3l) \tag{a}$$

is the expression for the beam displacement corresponding to the state of minimum total potential energy. At the beam centre $(x = 3l)$ $v_c = 243Pl^3/328EI_0 = 0.74085Pl^3/EI_0$.

Now the exact solution of this problem can be readily generated using, for example, the MDM and gives the value $v_c = 0.77083Pl^3/EI_0$ so that the error in the Rayleigh–Ritz approach is only 3.9 per cent as far as the central deflection is concerned. However, we shall obtain a better idea of the accuracy of the Rayleigh–Ritz solution by examining the distribution of bending moment and shearing force along the beam.

For v defined by equation (a), and bearing in mind that the flexural rigidity changes abruptly at $x = l$ from EI_0 to $2EI_0$, we have

$$M_z = -EI_0\frac{d^2v}{dx^2} = -\frac{81}{164}Pl\left(1 - \frac{2}{3}\frac{x}{l}\right) \qquad \text{for} \qquad 0 \leqslant x \leqslant l$$

and

$$M_z = -2EI_0\frac{d^2v}{dx^2} = -\frac{162}{164}Pl\left(1 - \frac{2}{3}\frac{x}{l}\right) \qquad \text{for} \qquad l \leqslant x \leqslant 3l.$$

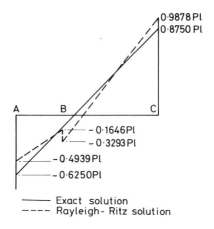

Fig. E5.7(b)

This distribution of bending moment is shown in Fig. E5.7(b) along with the exact distribution (obtained using the MDM, say). It is clear that there is something basically wrong with our Rayleigh–Ritz approach in this problem, since a discontinuity in bending moment occurs at $x = l$, whereas obviously the moment values on each side of the change of section (the joint) should be equal in magnitude to satisfy joint equilibrium. The distribution of shearing force shown in Fig. E5.7(c) also indicates a basic error in the adopted approach.

The difficulty with this problem is that we have assumed a displacement field which is smooth and continuous up to the third derivative. This means that the bending moment $-EI_z \, d^2v/dx^2$ and the shearing force $-EI_z \, d^3v/dx^3$ must be discontinuous if EI_z changes abruptly, and this is at variance with the true situation. The step changes in bending moment and shearing force would occur at $x = l$ when using the Rayleigh–Ritz approach whatever *single* displacement field were assumed, and in these circumstances the exact solution could never be achieved. The displacement field may be said to have excessive continuity at $x = l$. What is needed is a displacement field that allows continuity of v and dv/dx at the section $x = l$ but discontinuity of d^2v/dx^2 and higher derivatives. This can only be achieved with the use of a *separate* displacement field for *each* of the appropriate parts of the beam (i.e. for $0 \leqslant x \leqslant l$ and for $l \leqslant x \leqslant 3l$). This leads to the idea of a

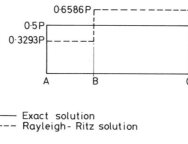

Fig. E5.7(c)

piecewise Rayleigh–Ritz procedure in which separate displacement fields are assumed for a finite number of separate regions of a structure. If the regions are referred to as elements then we are very close here to the concept of the finite-element method.

Comments on Examples 5.4 to 5.7. The above bar and beam examples have served to illustrate the basic features of the approximate Rayleigh–Ritz procedure. They make clear the great influence that the choice of the admissible displacement field has on the accuracy of solution. In some simple problems it is an easy matter to select a displacement field, written in terms of generalized displacements, which will lead through the minimization process to the exact solution. This is not typical of more complex situations, though; in such situations exact solutions will not be available and the energy method plays a valuable role in providing accurate approximate solutions.

The examples have shown that the potential energy method generally provides accurate estimates of structure displacements, even where the assumed displacement field is not particularly appropriate. However, the derived force-type quantities are predicted with substantially less accuracy than is the displacement. We have seen that point values of moment and force can be very significantly in error, though in an integrated average sense over the whole structure the errors involved are usually not great. This corresponds to the fact that equilibrium is satisfied in an average sense but is not satisfied at each and every point in the structure when a restricted class of admissible displacement functions is used. The accuracy of a derived quantity decreases with differentiation; for example in beam problems we expect the energy approach to provide a more accurate solution for bending moments (proportional to the second derivative of displacement) than for shearing forces (proportional to third derivatives) and this is borne out by the above examples. We note that application of the energy method involves integration over the region of the structure and this operation tends to smooth out any local inaccuracies in an assumed displacement function. It is this smoothing process which leads to a quite accurate satisfaction of conditions of equilibrium in the average sense, although such conditions may be badly violated locally. The subsequent differentiation to obtain force quantities exposes the local errors again.

The conventional Rayleigh–Ritz method has been frequently used in the solution of structural problems of a two- or three-dimensional continuum nature. Usually, however, when this has been done the structure geometry and loading are relatively uncomplicated and this facilitates the selection of approximate assumed displacement fields. It is clear that such selection will be very difficult, if not impossible, to make for structures of any real complexity if a single displacement field is to apply to the whole structure. This was seen to be so in Example 5.7 even for a quite simple beam problem. As a way out of this difficulty we might instead imagine applying the Rayleigh–Ritz method in a piecewise fashion based on the assumption of a number of relatively simple *local* displacement fields, each applying only to a particular limited region or element of the whole structure. Of course there will need to be rules governing the linking up of the local displacement fields at the boundaries of the patchwork of regions. Basically these rules will be such that the total structure displacement field obtained by combining all the local fields should still be admissible by satisfying relevant compatibility conditions. A simple example will be used to illustrate this philosophy. (The beam problem of Example 5.7 is considered again in Problem 5.11).

Example 5.8. Consider again the uniform bar of Example 5.4, but this time seek a solution based on the concept of using local displacement fields in the Rayleigh–Ritz procedure.

In Example 5.4 we saw that the assumption of a linear displacement field over the whole structure did not yield a very accurate solution. One way of obtaining an improved solution was to increase the order of the whole-structure displacement field to quadratic (and this happened to give the exact solution). Here we explore an alternative approach to obtaining an improved solution by assuming a linear displacement field in each of two or more regions of the structure.

In Fig. E5.8(a) the bar is shown artificially divided into two regions, A and B, of length $L/2$. (The regions need not be of the same length.) It is assumed that under load the axial displacements of each region can be represented by complete linear functions, i.e.

$$u = a_0 + a_1 x_1 \text{ for region A}$$
$$u = b_0 + b_1 x_2 \text{ for region B.} \tag{a}$$

These two regional fields must be combined together such that the overall structure field is admissible, i.e. the geometric boundary condition at $x_1 = 0$ must be satisfied and the displacement u must be continuous (single valued) over the whole length of the bar.

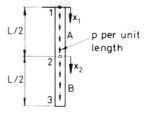

Fig. E5.8(a)

The first of the admissibility requirements is met by simply setting $a_0 = 0$. The second will be met provided that u is made continuous at the boundary of the two regions, i.e. provided that the value of u for region A at point 2 (where $x_1 = L/2$) equals that of u for region B at point 2 (where $x_2 = 0$). From equations (a) this requirement means that $a_1 L/2 = b_0$ and thus the assumed displacement field can be expressed in terms of two generalized displacements only, b_0 and b_1 say. We could proceed to evaluate the potential energy—expressed in terms of the generalized displacements—for each of the two regions in turn and to obtain the total potential energy of the whole structure by straightforward addition of energies. Then the solution could be obtained by minimizing this energy to yield the values of the two generalized displacements in the manner of Examples 5.4–5.7. However, we shall not proceed in this way here since it is more appropriate when using local displacement fields to work in terms of the values of actual displacements at particular points rather than in terms of generalized displacements. This is of direct help in visualizing the connection between adjacent regions.

The local displacement fields (equation (a)) can easily be re-expressed in terms of convenient point displacements by considering 'boundary' conditions at the ends of each region. For region A (ignoring for the moment the fact that the displacement $u_1 = 0$) we have that

$$u = u_1 \quad \text{at} \quad x_1 = 0$$
$$u = u_2 \quad \text{at} \quad x_1 = L/2.$$

Using these conditions in the expresion for u in region A (the first of equations (a)) gives the following relationships between the generalized displacements a_0 and a_1 and the point displacements u_1 and u_2:

$$a_0 = u_1 \qquad a_1 = \frac{2}{L}(u_2 - u_1).$$

It follows that in region A

$$u = u_1\left(1 - \frac{2}{L}x_1\right) + u_2\left(\frac{2}{L}x_1\right). \tag{b}$$

Similarly in region B

$$u = u_2\left(1 - \frac{2}{L}x_2\right) + u_3\left(\frac{2}{L}x_2\right). \tag{c}$$

Having settled the form of the assumed displacements we can now determine the potential energy for each region, and by summation for the whole bar, as a function of the point displacements.

For region A

$$u = u_2\left(\frac{2}{L}x_1\right) \text{ from equation (b) with } u_1 = 0, \text{ and } \varepsilon_x = \frac{2u_2}{L}$$

$$U_p^A = \frac{AE}{2}\int_0^{L/2}\left(\frac{2}{L}u_2\right)^2 dx_1$$

$$V_p^A = -\int_0^{L/2} u(p\,dx_1) = -\int_0^{L/2} u_2\frac{2}{L}x_1 p\,dx_1$$

and therefore

$$\Pi_p^A = \frac{AE}{L} u_2^2 - \frac{pL}{4} u_2.$$

For region B

u is given by equation (c) and $\varepsilon_x = \frac{2}{L}(u_3 - u_2)$

$$U_p^B = \frac{AE}{2} \int_0^{L/2} \left(\frac{2}{L}(u_3 - u_2)\right)^2 dx_2$$

$$V_p^B = -\int_0^{L/2} \left(u_2\left(1 - \frac{2}{L}x_2\right) + u_3\left(\frac{2}{L}x_2\right)\right) p \, dx_2$$

and therefore

$$\Pi_p^B = \frac{AE}{L}(u_2^2 + u_3^2 - 2u_2 u_3) - p\frac{L}{4}(u_2 + u_3).$$

For the complete bar the total potential energy in terms of the point displacements is

$$\Pi_p = \Pi_p^A + \Pi_p^B$$

or

$$\Pi_p = \frac{AE}{L}(2u_2^2 + u_3^2 - 2u_2 u_3) - p\frac{L}{4}(2u_2 + u_3). \qquad (d)$$

The minimization procedure gives

$$\frac{\partial \Pi_p}{\partial u_2} = \frac{AE}{L}(4u_2 - 2u_3) - \frac{pL}{2} = 0$$

and

$$\frac{\partial \Pi_p}{\partial u_3} = \frac{AE}{L}(-2u_2 + 2u_3) - \frac{pL}{4} = 0.$$

The solution of this pair of simultaneous equations gives the displacements of points 2 and 3 as

$$u_2 = \frac{3}{8}\frac{pL^2}{AE} \qquad u_3 = \frac{pL^2}{2AE}$$

and both values coincide with the exact solution. However, this does not mean that we have found the exact solution for the bar as a whole. To examine the overall accuracy of the above solution we substitute the calculated point displacements back into the expressions for total potential energy (equation (d)) and for the displacement and stress in each region. The value of Π_p is in fact $-0.15625p^2L^3/AE$ which is much closer to the exact value $(-0.166p^2L^3/AE)$ than was the value corresponding to the use of the single linear field in Example 5.4 $(-0.125p^2L^3/AE)$. The distributions of displacement and stress for the two-region solution are shown in Fig. E5.8(b) and compared with the exact

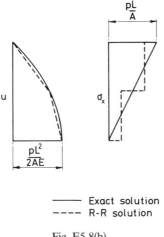

$\frac{pL}{A}$

u

σ_x

$\frac{pL^2}{2AE}$

——— Exact solution
– – – R-R solution

Fig. E5.8(b)

solution. The expected result is achieved in that subject to the restricted nature of the assumed displacement field—linear displacement and uniform stress in each region—the energy method has produced the best possible solution wherein the value of the uniform stress in each region is equal to the average true value of stress in the region. It should be noted that in this solution the displacement u is single valued at the boundary of the two regions but that the strain ($\varepsilon_x = du/dx$) and stress ($\sigma_x = E\varepsilon_x$) are not, and need not be continuous. The stress values are exact at the centres of the regions.

Clearly there is nothing to prevent the extension of the above procedure by subdividing the bar into three regions or four regions etc., although the work involved in obtaining a solution increases progressively of course. We would expect that the more regions that we use the better will be the accuracy of the solution and this generally is true. (To be precise a more accurate solution, i.e. a lower value of total potential energy, is only *guaranteed* when using a greater number of regions as compared with a lesser number if the point displacements corresponding to the lesser number of regions are all included as point displacements in the analysis using the great number of regions.)

The above procedure of artificially subdividing a complete structure into convenient regions and assuming a displacement field for each region can, as mentioned earlier, be termed the *piecewise* Rayleigh–Ritz procedure. The procedure forms the basis of the variational viewpoint of the finite-element displacement method. In the finite-element method, as we shall see, the piecewise Rayleigh–Ritz procedure is made systematic by the use of matrices. The regions become elements and the points at which reference displacement values are used are nodal points.

5.6. Approximate solutions by the principle of virtual displacements

The main aim of this chapter has been to introduce the principle of minimum potential energy and to show how this principle can be used in conjunction with the Rayleigh–Ritz procedure to obtain approximate solutions. The advantages of operating within the PMPE are that the requirements guiding the selection of assumed displacements are clarified, that a formal bound exists on the calculated value of the total potential energy, and that operating within the variational principle opens the way to non-structural applications, although we are not concerned with the latter in this text.

The PMPE was derived earlier in this chapter via the more general PVD and it is therefore apparent that the PVD can also be used to generate approximate solutions to structural problems, just as was the PMPE in Examples 5.4–5.8. In finite-element work the derivation of element properties is often made through the PVD rather than through the PMPE and therefore some description of the PVD approach is desirable here. It is emphasized though, that results obtained through the PVD approach in linear elastic problems will be identical with those obtained through the PMPE approach (provided of course that identical assumptions are made for the displacement fields).

To consider the PVD further let us return to the algebraic expression of the principle for elastic bodies given by equation (5.22), that is

$$\delta W_e - \delta U_p = 0. \tag{5.22}$$

In this equation δW_e is the work done by the external forces during a kinematically admissible virtual displacement and δU_p $(= -\delta W_i)$ is the change in strain energy occurring during the virtual displacement.

In Section 5.2 we considered amongst other one-dimensional systems a bar subjected to an axial force, as in Fig. 5.1(a), which increases progressively in magnitude from zero to a value F. Correspondingly the bar extension increases to a value e, the stress to a value σ_x, and the strain to a value ε_x; graphs of F versus e and σ_x versus ε_x are shown in Figs. 5.1(b) and 5.1(c). All the quantities F, e, σ_x, and ε_x shown in Fig. 5.1 are real quantities and at load level F the bar is in an equilibrium configuration.

To apply the PVD (equation (5.22)) to the simple bar problem consider now the imposition on this equilibrium state of a virtual displacement δe of the end of the bar. Correspondingly there will be a strain $\delta \varepsilon_x$ in the bar. The force F and stress σ_x remain constant during the virtual displacement. The effect of these changes is shown on the F–e and σ_x–ε_x diagrams in Fig. 5.4.

The work done by the external force F during the virtual displacement is

$$\delta W_e = F \delta e \tag{5.29}$$

and is represented by the hatched area in Fig. 5.4(a). The change in strain

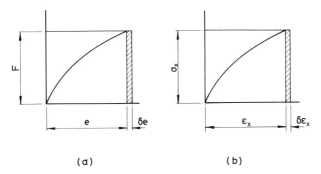

Fig. 5.4. Effect of a virtual displacement on a stretched bar: (a) load-extension curve, (b) stress–strain curve.

energy density occurring during the virtual displacement is

$$\delta v_{\mathrm{p}} = \sigma_x \, \delta\varepsilon_x \qquad (5.30)$$

and is represented by the hatched area in Fig. 5.4(b).

Now the real stress σ_x in the bar in the equilibrium state can be expressed in terms of the corresponding real strain ε_x once we decide on what kind of material behaviour to allow. This behaviour has already been assumed to be elastic and now it is further stipulated that it be linear elastic, in line with the rest of this text. Thus Hooke's law applies and equation (5.30) becomes

$$\delta v_{\mathrm{p}} = \delta\varepsilon_x E \varepsilon_x. \qquad (5.31)$$

Correspondingly the change in strain energy due to the virtual displacement for the whole bar is

$$\delta U_{\mathrm{p}} = \int_0^l \delta\varepsilon_x A E \varepsilon_x \, \mathrm{d}x. \qquad (5.32)$$

We are interested in obtaining approximate solutions based on the assumption of displacement forms and so need to re-express δU_{p} in terms of displacements. Now the real strain ε_x and the real displacement u at any co-ordinate position x are related by $\varepsilon_x = \mathrm{d}u/\mathrm{d}x$. In the same way the virtual strain $\delta\varepsilon_x$ and virtual displacement δu are related by $\delta\varepsilon_x = \mathrm{d}(\delta u)/\mathrm{d}x$. Therefore δU_{p} becomes

$$\delta U_{\mathrm{p}} = \int_0^l \left(\frac{\mathrm{d}}{\mathrm{d}x}(\delta u) \right) A E \frac{\mathrm{d}u}{\mathrm{d}x} \, \mathrm{d}x \qquad (5.33)$$

and this expression applies for any axially loaded bar, whether or not the cross section or the strain varies along the bar length.

It is emphasized that in equation (5.32) the strain $\delta\varepsilon_x$ occuring due to the virtual displacement need have no connection whatever with the real strain ε_x present in the bar in the equilibrium position. It follows that the virtual displacement δu need not be related to the real displacement u in equation (5.33). However, it is very much the usual practice in seeking approximate solutions, whether based on the conventional or the piece-wise form of the Rayleigh–Ritz procedure, to assume that the distribution of the virtual displacement is of the same form as is the distribution of the real displacement. (Example 5.9 which follows adopts this approach.)

Before ending consideration of the axially loaded bar it is noted that the expression for δW_e given by equation (5.29) pertains to the situation where a single-point load acts at one end of the bar. Loading of a more general nature can be accounted for quite easily; for instance if N concentrated axial loads F_i act at points along the bar then

$$\delta W_e = \sum_{i=1}^{N} F_i\, \delta d_i \qquad\qquad (5.34)$$

where δd_i is the virtual displacement corresponding to load F_i.

Expressions for the change in strain energy δU_p (and the external virtual work δW_e) can readily be established for the torsionally loaded bar and the Bernouilli–Euler beam by similar reasoning to that just described for the axially loaded bar. In terms of appropriate displacements the results for δU_p are

$$\delta U_p = \int_0^l \left(\frac{\mathrm{d}(\delta\theta_x)}{\mathrm{d}x}\right) GJ \frac{\mathrm{d}\theta_x}{\mathrm{d}x}\, \mathrm{d}x \qquad\qquad (5.35)$$

for the torsionally loaded bar and

$$\delta U_p = \int_0^l \left(\frac{\mathrm{d}^2(\delta v)}{\mathrm{d}x^2}\right) EI_z \frac{\mathrm{d}^2 v}{\mathrm{d}x^2}\, \mathrm{d}x \qquad\qquad (5.36)$$

for the beam.

To illustrate the application of the PVD in obtaining approximate solutions we shall resolve one of the problems dealt with earlier using the PMPE and the Rayleigh–Ritz method; Example 5.6 is selected for this purpose.

Example 5.9. Resolve Example 5.6 using the PVD.

We assume that the beam shown in Fig. E5.6(a) is in equilibrium under the action of the given loading when its (real admissible) displacement field is defined by

$$v = C_2 x^2 + C_3 x^3. \qquad\qquad (a)$$

This displacement field is the same as that assumed in Example 5.6.

We also assume that the virtual displacement field has the same form as the real displacement field; thus

$$\delta v = x^2 \, \delta C_2 + x^3 \, \delta C_3.$$

Then from equation (5.36), the change in strain energy occurring due to the virtual displacement is

$$\delta U_p = \int_0^L (2 \, \delta C_2 + 6x \, \delta C_3) EI_z (2C_2 + 6xC_3) \, dx$$

$$= EI_z (4LC_2 \, \delta C_2 + 6L^2 C_3 \, \delta C_2 + 6L^2 C_2 \, \delta C_3 + 12L^3 C_3 \, \delta C_3).$$

The work done by the external loading during the virtual displacement is

$$\delta W_e = \int_0^L p \, \delta v \, dx = p \int_0^L (x^2 \, \delta C_2 + x^3 \, \delta C_3) \, dx = p \frac{L^3}{3} \delta C_2 + p \frac{L^4}{4} \delta C_3.$$

Therefore application of the PVD as expressed by equation (5.22) gives

$$\delta W_e - \delta U_p = \delta C_2 \left(p \frac{L^3}{3} - EI_z (4LC_2 + 6L^2 C_3) \right)$$

$$+ \delta C_3 \left(p \frac{L^4}{4} - EI_z (6L^2 C_2 + 12L^3 C_3) \right) = 0.$$

Now the virtual quantities δC_2 and δC_3 are each of arbitrary magnitude and so the only way in which this equation can be satisfied is if the multipliers of each virtual quantity are individually zero, that is if

$$4LC_2 + 6L^2 C_3 = \frac{pL^3}{3EI_z} \qquad\qquad 6L^2 C_2 + 12L^3 C_3 = \frac{pL^4}{4EI_z}.$$

As expected these simultaneous equations, whose solution yields the values of the generalized displacements C_2 and C_3, are precisely the equations generated in Example 5.6 by using the Rayleigh–Ritz procedure and the PMPE. On solving for C_2 and C_3 it is a simple matter to determine the distributions of displacement, bending moment, and shearing force along the beam and these distributions will coincide with those presented in Example 5.6 of course. The one significant quantity that is not determined in the PVD approach, as compared with the PMPE approach, is the value of the total potential energy Π_p.

In applying the PVD here we have assumed that a particular displacement field (equation (a)) is appropriate to an equilibrium state. We know that in fact this initial assumption is incorrect and that consequently, in a rigorous sense, we are not correctly applying the PVD since we are not considering the external work and strain energy associated with a virtual displacement *from an equilibrium position*. We assume that the nearer our initial assumed displacement corresponds to the equilibrium configuration, the more accurate will be our ultimate solution. However, we do not have a formal convincing proof of this such as can be provided for the alternative approach through the minimization procedure of the PMPE (although the numerical results obtained from the two approaches are identical).

Bibliography

1   CHARLTON, T. M. *Energy principles in applied statics.* Blackie, London (1959).
2   AU, T. Elementary structural mechanics, Chapter 8. Prentice-Hall, Englewood Cliffs, NJ (1963).
3   TIMOSHENKO, S. P. and GERE, J. M. *Mechanics of materials,* Chapter 11. Van Nostrand–Reinhold, New York (1973).
4   DYM, C. L. and SHAMES, I. H. *Solid mechanics: A variational approach.* McGraw-Hill, New York (1973).
5   RICHARDS, T. H. *Energy methods in stress analysis,* Ellis Horwood, Chichester (1977).

Problems

In Problems 5.1–5.13 it is intended that solutions be obtained using the PMPE in conjunction with the Rayleigh–Ritz procedure.

5.1. Refer to Example 5.5 and obtain further solutions to the problem described there based on the use of the following displacement fields in turn:

(a) $v = C_1 \sin \dfrac{\pi x}{L} + C_2 \sin \dfrac{3\pi x}{L} + C_3 \sin \dfrac{5\pi x}{L}$

(b) $v = C_1 \sin \dfrac{\pi x}{L} + C_2 \sin \dfrac{3\pi x}{L} + C_3 \sin \dfrac{5\pi x}{L} + C_4 \sin \dfrac{7\pi x}{L}.$

In each case determine the values of central deflection and total potential energy and sketch the shearing force and bending moment diagrams. Collect results together for the various displacement fields used here and in Example 5.5 and study the manner of convergence of the results.

5.2. For the beam considered in Example 5.5 obtain a solution by taking advantage of the problem symmetry and assuming the displacement field

$$v = C(3L^2 x - 4x^3)$$

for the left-hand half of the beam only. Note that this field satisfies the geometric boundary conditions that $v = 0$ at $x = 0$ and $(dv/dx) = 0$ at $x = L/2$. Show that the solution obtained is exact.

5.3. Refer to Example 5.6 and obtain a further solution to the problem by assuming the displacement field

$$v = C_2 x^2 + C_3 x^3 + C_4 x^4.$$

Show that the exact result is obtained. What would happen if an extra term $C_5 x^5$ were added to the displacement field?

5.4. Refer to Example 5.6 but now replace the uniformly distributed load with a distributed load whose intensity varies linearly from zero at the free end to q per unit length at the clamped end. Examine solutions obtained by using in turn a cubic polynomial and a quartic polynomial to represent v: calculate the free-end deflection and the total potential energy and sketch the shearing force and bending moment diagrams. What would the order of the polynomial displacement field need to be to obtain the exact solution?

5.5. Repeat Example 5.6 but now include the presence of a spring of stiffness $k = 4EI_z/L^3$ which provides some restraint in the vertical direction at the free end. (EI_z is the flexural rigidity of the beam.)

5.6. In Example 5.6 the expression assumed for v is written in terms of two independent constants C_2 and C_3. These two constants could be linked together by requiring that the bending moment at the free end be zero (i.e. by explicitly satisfying one of the natural boundary conditions of the problem). Examine the effect of this on the accuracy of solution as compared with that demonstrated in Example 5.6.

5.7. A uniform simply supported beam of length L carries a distributed loading whose intensity varies from zero at each end to a maximum of p per unit length at the beam centre. Use in turn the two displacement fields of Example 5.5 to obtain solutions for the deflection at the centre and the total potential energy. Draw the shearing force and bending moment diagrams corresponding to each assumed displacement field and to the exact solution.

5.8. A uniform beam of length L is clamped at both ends and carries a load of total magnitude W which is uniformly distributed over the central half of the span. Assume the displacement field

$$v = C_2(x^2 - 4x^3/3L) \qquad \text{for} \qquad 0 \leqslant x \leqslant L/2$$

where x is measured from one end. Determine the deflection at the beam centre and the value of the total potential energy. Draw the bending moment diagram and compare it with the exact solution (which can be obtained using the MDM).

5.9. A bar of length L has a cross-sectional area which varies linearly from value $2A$ at end 1 to A at end 2. End 1 is held against any movement whilst the bar is stretched by an axial force W applied at end 2. Obtain solutions for the axial displacement and axial stress distributions, and the value of the total potential energy, based on the use of the following displacement fields in turn:

(a) $u = C_1 x$
(b) $u = C_1 x + C_2 x^2$.

Compare the results obtained with the exact solution.

5.10. Refer to Example 5.8 and obtain a further solution based on the use of linear local displacement fields in three regions of length $L/3$. Compare the results for displacement, total potential energy, and stress with the exact solution and with the Rayleigh–Ritz procedure solutions detailed in Examples 5.4 and 5.8.

5.11. Refer to Example 5.7 and consider the use of two distinct local displacement fields, one in region AB and another in region BC. Assume that each local field is basically a cubic polynomial but note that some terms will be eliminated since the following geometric conditions must apply:
(i) $v = dv/dx = 0$ at A;
(ii) $dv/dx = 0$ at C;

(iii) v and dv/dx must be continuous across the inter-region interface at B. Relate generalized displacements to the (nodal) point displacements v_B, θ_B, and v_C, and hence express Π_p as a function of these point displacements. Proceed to show that the exact result is obtained; this result is recorded in Example 5.7.

5.12. The stepped bar shown in Fig. P5.12, of the same material throughout, is held against axial movement at both ends and is subjected to an axial force W at the point of change of cross-section. The exact solution of this problem is readily obtainable using the MDM and is given, for general geometry, in Sections 2.3 and 2.4. Obtain a full solution (for u_2, Π_p, and the distribution of stress) based on the single displacement field

$$u = C(x - x^2/L)$$

where x is measured from the left-hand end. Comment on the deficiencies of this approach.

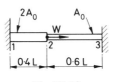

Fig. P5.12

5.13. Obtain a full solution to Problem 5.12 based on the use of local linear displacement fields in each of regions 1–2 and 2–3 in the manner of Example 5.8. Verify that this approach yields the exact solution.

5.14–5.18. Solve Examples 5.4–5.8 using the PVD. In all cases assume the same real displacement field as used in a particular example and assume a similar form for the virtual displacement. Verify that in each case the same solution is obtained as was obtained using the PMPE.

6

ONE-DIMENSIONAL FINITE ELEMENTS

6.1. Introduction

HAVING introduced the PMPE and the Rayleigh–Ritz procedure in Chapter 5 we are now in a position to consider their use in the development of stiffness relationships both for individual elements (or regions of the whole) and for element assemblages. The mode of procedure has already been strongly hinted at in Example 5.8 where local displacement fields were employed in specific regions in a piecewise Rayleigh–Ritz procedure for the solution of a bar problem. In this chapter this type of procedure is formalized as the finite-element displacement method. Attention is restricted to one-dimensional elements (or 'line' elements) but the procedures are of a general nature which can later be standardized for all types of finite element.

Although the PMPE is really applied to the structure as a whole, that is to the assemblage of elements or regions, we can develop individual element properties using the principle without giving consideration to the place of the element in the overall structure (just as was done in a direct way in the MDM approach). This is possible because displacement fields are separately assumed for each element or region and any integral over the assemblage, occurring in the expression for potential energy, can be considered as the sum of the integrals over the various regions.

The line finite elements considered in this chapter are straight bar and beam elements. The most basic of these, considered first in Sections 6.2–6.4, are shown to correspond precisely to the 'exact' axially loaded bar and the beam developed and used in the MDM work; there will be seen to be a straightforward explanation as to why this is so. (The torsionally loaded bar is not specifically considered as its analysis is very similar to that of the axially loaded bar.) The procedure by which a distributed loading is accommodated in a consistent fashion within the FEM is developed in Section 6.5. The result is the column matrix of the consistent or equivalent nodal loads; for the basic elements this corresponds to the system of reversed fixed-end forces in the MDM. The assembly process within the context of the PMPE is then discussed in Section 6.6 and it is shown that the end result is identical in practice with the direct stiffness procedure. Thus far in the chapter all the results obtained by the energy approach are the same as those obtained by the direct approach of the MDM. Stress calculation is considered next, in

Section 6.7, and this is straightforward in principle through the use of an element stress matrix. However, it is noted that conventional finite-element stress calculation for the simple line elements yields different results to that of the MDM approach when distributed loads are present. Sections 6.8–6.10 are concerned with a more detailed look at the nature of and requirements for element displacement fields and with a description of some refined one-dimensional finite elements. The technique of static condensation, which is sometimes useful in producing a reduced set of stiffness equations by removing certain degrees of freedom from consideration, is described in Section 6.11. Finally, in Section 6.12 it is demonstrated how the effects of initial strain—due to temperature change or lack of fit say—can be incorporated into the FEM approach.

On a point of detail, it should be noted that from this chapter on element nodes are denoted by numbers rather than by the letters which were used in the MDM work. The lettering system was used in the early part of the text for greater clarity in distinguishing element nodes from the numbered structure nodes. By this stage there should be no confusion in the reader's mind between the element and structure labelling schemes, and element node numbering is the more convenient when dealing later with elements having a relatively large number of nodes.

In this and succeeding chapters the development of the FEM is described through use of the PMPE. It is noted again, though, that identical results would be obtained if instead the PVD were used; this is shown to be so in Appendix B.

6.2. Stiffness matrix for the basic bar element

By the 'basic' bar element is meant the uniform bar element, previously met with in Section 2.2, which has nodes at its ends only and consequently has two degrees of freedom. The element is illustrated in Fig. 6.1 and it is assumed here that the only forces acting on the element are point axial forces U_1 and U_2 acting at the ends. The two degrees of freedom of the element are the axial displacements u_1 and u_2 of the nodes.

In developing the element stiffness matrix the first (and crucial) consideration concerns the selection of the assumed displacement field. We need to be able to express the displacement field either indirectly or directly in terms of the element nodal displacements. Since only two

Fig. 6.1. The basic bar finite element.

nodal displacements are involved, this means that if a displacement field is written in terms of generalized displacements it must contain only two such displacements. Consequently for the basic bar it is assumed that the element deformation can be represented by the linear field

$$u = A_0 + A_1 x.$$

In matrix form this can be written

$$u \equiv \mathbf{u} = [1 \quad \mathbf{x}] \begin{Bmatrix} A_0 \\ A_1 \end{Bmatrix}$$

or

$$\mathbf{u} = \boldsymbol{\alpha} \mathbf{A} \tag{6.1}$$

It should be noted that in equation (6.1) the matrix symbolism $\mathbf{u}$ is used in denoting the single displacement u for this element so as to maintain generality with later developments in which $\mathbf{u}$ becomes a column matrix containing two or three displacement components in continuum applications.

The origin of the x co-ordinate has been taken to be at node 1 but it is emphasized that it could be located at some other convenient point along the element. The generalized displacements can be related to the nodal displacements by using the element 'boundary conditions' that

$$u = u_1 \quad \text{at} \quad x = 0.$$

$$u = u_2 \quad \text{at} \quad x = l.$$

Thus, using equation (6.1) the nodal displacements are expressed in terms of the generalized displacements as

$$\begin{Bmatrix} u_1 \\ u_2 \end{Bmatrix} = \begin{bmatrix} 1 & 0 \\ 1 & l \end{bmatrix} \begin{Bmatrix} A_0 \\ A_1 \end{Bmatrix}$$

or

$$\mathbf{d} = \mathbf{CA}. \tag{6.2}$$

These equations can easily be solved to give the inverse relationship

$$\begin{Bmatrix} A_0 \\ A_1 \end{Bmatrix} = \begin{bmatrix} 1 & 0 \\ -1/l & 1/l \end{bmatrix} \begin{Bmatrix} u_1 \\ u_2 \end{Bmatrix}$$

or

$$\mathbf{A} = \mathbf{C}^{-1} \mathbf{d}. \tag{6.3}$$

From equations (6.1) and (6.3) the displacement at any point can be expressed directly in terms of the nodal displacements as

$$\mathbf{u} = (\boldsymbol{\alpha}\mathbf{C}^{-1})\mathbf{d} = \mathbf{N}\mathbf{d} \tag{6.4}$$

where

$$\mathbf{N} = \boldsymbol{\alpha}\mathbf{C}^{-1},$$

i.e.

$$\mathbf{N} = [1 \quad x]\begin{bmatrix} 1 & 0 \\ -1/l & 1/l \end{bmatrix} = \left[\left(1 - \frac{x}{l}\right) \quad \frac{x}{l}\right]. \tag{6.5}$$

$\mathbf{N}$ is referred to as the 'matrix of shape functions' or the 'shape function matrix'. We shall defer consideration of shape functions until Section 6.9 but it should be noted here that if appropriate shape functions are known the development of element properties can commence with the equation $\mathbf{u} = \mathbf{N}\mathbf{d}$ rather than with equation (6.1).

In the bar problem there exists only one strain, ε_x and one corresponding stress, σ_x, acting in the axial direction. The strain can be obtained using equations (6.4) and (6.5) as

$$\varepsilon_x = \frac{\mathrm{d}u}{\mathrm{d}x} = \left[\frac{\mathrm{d}\mathbf{N}}{\mathrm{d}x}\right]\mathbf{d} = [-1/l \quad 1/l]\begin{Bmatrix} u_1 \\ u_2 \end{Bmatrix}$$

or

$$\varepsilon_x = \mathbf{B}\mathbf{d}. \tag{6.6}$$

The rectangular matrix $\mathbf{B}$ has been obtained from $\mathbf{N}$ by appropriate differentiation. For the basic bar element the individual components of $\mathbf{B}$ are constant terms, since with the assumption of the linear displacement field the corresponding strain must be uniform over the element length.

The stretching strain energy of the bar element of volume V_e is (from Section 5.2)

$$U_p = \frac{1}{2}\int_{V_e} \sigma_x^t \varepsilon_x \, \mathrm{d}V_e = \frac{1}{2}\int_{V_e} \varepsilon_x^t E \varepsilon_x \, \mathrm{d}V_e \tag{6.7}$$

where E is Young's modulus which relates the stress σ_x to the strain through the single constitutive equation

$$\sigma_x = E\varepsilon_x. \tag{6.8}$$

The presence of the terms σ_x^t and ε_x^t in equation (6.7) requires some explanation. For the present one-dimensional problem the stress σ_x and strain ε_x are single terms (i.e. are scalar as distinct from matrix quantities). However, the strain is expressed in equation (6.6) in the form of a matrix product and the arrangement of terms shown in equation (6.7) must be adopted so that U_p becomes a scalar quantity, as it should be of course. This is in line with the expression for work as $\mathbf{P^t d}$ or $\mathbf{d^t P}$ used in Section 3.3 and below (equation (6.14)).

Substituting equation (6.6) into equation (6.7) the element strain energy becomes

$$U_p = \tfrac{1}{2}\mathbf{d^t}\left(\int_{V_e} \mathbf{B^t EB}\,dV_e\right)\mathbf{d}$$

For (6.9)

$$\mathbf{k} = \int_{V_e} \mathbf{B^t EB}\,dV_e$$

this has the general form

$$U_p = \tfrac{1}{2}\mathbf{d^t kd}. \tag{6.10}$$

If A denotes the cross-sectional area $\mathbf{k}$ becomes

$$\mathbf{k} = \int_0^l \mathbf{B^t}(AE)\mathbf{B}\,dx. \tag{6.11}$$

Using the expression for $\mathbf{B}$ given in equation (6.6) and taking A and E to be uniform along the element length gives

$$\mathbf{k} = AE\int_0^l \begin{Bmatrix} -1/l \\ 1/l \end{Bmatrix}[-1/l \quad 1/l]\,dx = \frac{AE}{l}\begin{bmatrix} 1 & -1 \\ -1 & 1 \end{bmatrix}. \tag{6.12}$$

Further, from equation (6.10),

$$U_p = \frac{1}{2}[u_1 \quad u_2]AE\begin{bmatrix} 1 & -1 \\ -1 & 1 \end{bmatrix}\begin{Bmatrix} u_1 \\ u_2 \end{Bmatrix} \tag{6.13}$$

and it is clear that U_p is a quadratic function of the nodal displacements u_1 and u_2.

The potential energy of the nodal forces, denoted by V_{pn} (with the subscript n included so as to distinguish this potential energy from that of distributed forces, presented later) is

$$V_{pn} = -U_1 u_1 - U_2 u_2$$

or in matrix form

$$V_{pn} = -\mathbf{d^t P}$$

where (6.14)

$$\mathbf{P} = \{U_1 \quad U_2\}$$

is the column matrix of nodal forces.

The total potential energy of the element is

$$\Pi_p = U_p + V_{pn}$$

or, using equations (6.10) and (6.14),

$$\Pi_p = \tfrac{1}{2}\mathbf{d}^t\mathbf{k}\mathbf{d} - \mathbf{d}^t\mathbf{P}. \qquad (6.15)$$

From Chapter 5 we know that application of the PMPE corresponds to the system of equations

$$\frac{\partial \Pi_p}{\partial d_i} = \frac{\partial U_p}{\partial d_i} + \frac{\partial V_{pn}}{\partial d_i} = 0 \qquad i = 1, 2, \dots . \qquad (6.16)$$

where, for the bar element, $d_1 \equiv u_1, d_2 \equiv u_2$ of course. Performing the differential operation indicated in equation (6.16) on the potential energy given by equation (6.15) for each displacement in turn gives the result (the reader can readily check this general result for the particular case of the simple bar element considered here),

$$\mathbf{k}\mathbf{d} - \mathbf{P} = \mathbf{0}$$

or   (6.17)

$$\mathbf{P} = \mathbf{k}\mathbf{d}$$

Equation (6.17) is immediately recognizable as the set of stiffness equations for the element, since it relates nodal forces to nodal displacements, and hence $\mathbf{k}$, given by equations (6.12), must be the element stiffness matrix. Referring to equation (6.12) it can be seen that the bar element stiffness matrix derived here via assumed displacements is precisely the same matrix as that derived by direct analysis in Section 2.2 and hence is 'exact'. There is a simple explanation as to why this is so (see Section 6.4), but for the moment it is merely emphasized that usually the element stiffness matrix derived in the manner described in this section is approximate.

One further point of a general nature should be mentioned. We have seen that the element strain energy U_p is a quadratic function of the nodal displacements (see equations (6.10) and (6.13)) and that partial differentiation of U_p with respect to each displacement d_i in turn yields $\mathbf{k}\mathbf{d}$. If we now differentiate again with respect to each displacement d_j the result is

$$\frac{\partial^2 U_p}{\partial d_i\, \partial d_j} = k_{ij} \qquad (6.18)$$

where k_{ij} is the individual stiffness coefficient in the ith row and jth column of $\mathbf{k}$. Equation (6.18) gives a neat analytical expression for the stiffness coefficient of any finite element. The fact that the order of differentiation does not alter the value of a second derivative indicates that the stiffness matrix must be symmetrical, i.e. $k_{ij} = k_{ji}$.

6.3. Stiffness matrix for the basic beam element

Here we consider through assumed displacements and the PMPE the stiffness properties of the 'basic' beam element having end nodes only, with the values of lateral displacement v and rotation θ_z ($= dv/dx$) as its freedoms at these nodes. The element is of uniform cross-section and is illustrated in Fig. 6.2; it corresponds to the beam element considered in Section 2.8 using a direct approach.

The derivation of the beam element stiffness matrix starts with the assumption of a suitable expression to represent the lateral displacement v of the deformed element. Since four nodal freedoms are involved we assume a cubic field, written in terms of four generalized displacements. Thus

$$\mathbf{u} \equiv v = [1 \quad x \quad x^2 \quad x^3]\{A_0 \quad A_1 \quad A_2 \quad A_3\}$$

or

$$\mathbf{u} = \boldsymbol{\alpha}\mathbf{A} \tag{6.19}$$

Corresponding to this assumption for v the rotation at any point along the beam is given by

$$\theta_z \equiv \frac{dv}{dx} = [0 \quad 1 \quad 2x \quad 3x^2]\mathbf{A}.$$

The relevant element 'boundary conditions' are that $v = v_1$ and $\theta_z = \theta_{z1}$ at $x = 0$, and that $v = v_2$ and $\theta_z = \theta_{z2}$ at $x = l$. Applying these conditions gives

$$\begin{Bmatrix} v_1 \\ \theta_{z1} \\ v_2 \\ \theta_{z2} \end{Bmatrix} = \begin{bmatrix} 1 & 0 & 0 & 0 \\ 0 & 1 & 0 & 0 \\ 1 & l & l^2 & l^3 \\ 0 & 1 & 2l & 3l^2 \end{bmatrix} \begin{Bmatrix} A_0 \\ A_1 \\ A_2 \\ A_3 \end{Bmatrix} \qquad \text{or} \qquad \mathbf{d} = \mathbf{CA} \tag{6.20}$$

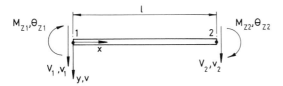

Fig. 6.2. The basic beam finite element.

Solving for the generalized displacements gives

$$\mathbf{A} = \mathbf{C}^{-1}\mathbf{d} \tag{6.21}$$

where

$$\mathbf{C}^{-1} = \begin{bmatrix} 1 & 0 & 0 & 0 \\ 0 & 1 & 0 & 0 \\ -3/l^2 & -2/l & 3/l^2 & -1/l \\ 2/l^3 & 1/l^2 & -2/l^3 & 1/l^2 \end{bmatrix}. \tag{6.22}$$

It follows that

$$\mathbf{u} \equiv v = (\boldsymbol{\alpha}\mathbf{C}^{-1})\mathbf{d} = \mathbf{N}\mathbf{d} \tag{6.23}$$

where now the shape function matrix is

$$\mathbf{N} = \begin{bmatrix} 1 - \dfrac{3x^2}{l^2} + \dfrac{2x^3}{l^3} & x - \dfrac{2x^2}{l} + \dfrac{x^3}{l^2} & \dfrac{3x^2}{l^2} - \dfrac{2x^3}{l^3} & -\dfrac{x^2}{l} + \dfrac{x^3}{l^2} \end{bmatrix}. \tag{6.24}$$

In Engineers' Theory of Bending (with bending taking place in the xy plane) the only significant strain component is the axial strain which varies linearly through the beam depth in the form

$$\varepsilon_x = -y\frac{\mathrm{d}^2 v}{\mathrm{d}x^2} \tag{6.25}$$

where y is the distance from the neutral axis. The corresponding stress σ_x is related to this strain by the simple Hookean relation, as in equation (6.8). The bending strain energy can obviously be expressed as a quadratic function of the strain in the form of equation (6.7) of the last section. It would then be a straightforward matter to express the beam strain energy in the form of equations (6.9) and (6.10) with $\mathbf{B}$ being the matrix relating the true strain to the nodal displacements. Rather than work in terms of true strain and stress, though, it is convenient and traditional in beam analysis to work in terms of curvature χ_z and bending moment M_z which can be regarded as a 'pseudo'-strain and a 'pseudo'-stress respectively. As has been seen earlier the curvature and the bending moment are defined as

$$\chi_z = \frac{\varepsilon_x}{y} = -\frac{\mathrm{d}^2 v}{\mathrm{d}x^2} \quad \text{and} \quad M_z = EI_z\chi_z. \tag{6.26}$$

An important point to note is that the product of curvature and moment is equal to the integral over the beam cross-section of the product of the true strain ε_x and the true stress σ_x. Thus, as an alternative to the expression for element strain energy given by equation (6.7), we can write (see Section 5.3)

$$U_{\mathrm{p}} = \tfrac{1}{2} \int_0^l M_z^{\mathrm{t}} \chi_z \, \mathrm{d}z = \tfrac{1}{2} \int_0^l \chi_z^{\mathrm{t}} E I_z \chi_z \, \mathrm{d}x. \tag{6.27}$$

The curvature, or pseudo-strain, at a point along the beam element can be written directly in terms of the nodal displacements using equations (6.23) and (6.24) as

$$\chi_z = -\frac{\mathrm{d}^2 v}{\mathrm{d}x^2} = -\left[\frac{\mathrm{d}^2}{\mathrm{d}x^2} \mathbf{N}\right]\mathbf{d} \quad \text{or} \quad \chi_z = \mathbf{B}\mathbf{d} \tag{6.28}$$

where

$$\mathbf{B} = \left[\frac{6}{l^2} - \frac{12x}{l^3} \quad \frac{4}{l} - \frac{6x}{l^3} \quad \frac{-6}{l^2} + \frac{12x}{l^3} \quad \frac{2}{l} - \frac{6x}{l^2}\right]. \tag{6.29}$$

It is emphasized that, for the particular case of the beam, the matrix $\mathbf{B}$ given here relates the pseudo-strain, and not the true strain, to the nodal displacements. Substituting this expression for the pseudo-strain into equation (6.27) yields the expression for the element strain energy as

$$U_{\mathrm{p}} = \tfrac{1}{2}\mathbf{d}^{\mathrm{T}}\left(\int_{x=0}^l \mathbf{B}^{\mathrm{t}}(EI_z)\mathbf{B}\,\mathrm{d}x\right)\mathbf{d}$$

or

$$U_{\mathrm{p}} = \tfrac{1}{2}\mathbf{d}^{\mathrm{t}}\mathbf{k}\mathbf{d}$$

where

$$\mathbf{k} = \int_0^l \mathbf{B}^{\mathrm{t}}(EI_z)\mathbf{B}\,\mathrm{d}x. \tag{6.30}$$

The potential energy of the nodal forces can be written in the form of equation (6.14) of the last section, where now the column matrix of nodal forces for the beam element is

$$\mathbf{P} = \{V_1 \quad M_{z1} \quad V_2 \quad M_{z2}\} \tag{6.31}$$

By precisely the same procedure as adopted in the last section it can easily be verified that the matrix $\mathbf{k}$ defined by equation (6.30) is the stiffness matrix of the basic beam element. Using the definition of $\mathbf{B}$ given in equation (6.29) the stiffness matrix for a *uniform* beam element becomes, on carrying out the appropriate matrix multiplication and subsequent integration,

$$\mathbf{k} = EI_z \begin{array}{c} \quad v_1 \qquad \theta_{z1} \qquad v_2 \qquad \theta_{z2} \\ \begin{bmatrix} 12/l^3 & & & \\ 6/l^2 & 4/l & \text{Symmetric} & \\ -12/l^3 & -6/l^2 & 12/l^3 & \\ 6/l^2 & 2/l & -6/l^2 & 4/l \end{bmatrix} \end{array} \tag{6.32}$$

This stiffness matrix will be recognized as being the same as that derived in Section 2.8 so that again, as with the bar element, the 'exact' stiffness matrix has been derived through the FEM displacement approach. The reason why this is so is discussed in the next section.

Before leaving the beam element it should be noted that it is possible to re-express the stiffness matrix $\mathbf{k}$ given by equation (6.30) in a form that is perhaps more convenient for manual generation of the stiffness matrix. In equation (6.30) matrix $\mathbf{B}$ links the (pseudo) strain directly to the element nodal freedoms (see equation (6.28)) and is obtained by appropriate differentiation of matrix $\mathbf{N}$. However, the displacement field was originally expressed in terms of generalized displacements by equation (6.19) and we could use the form (equation (6.23))

$$v = (\boldsymbol{\alpha}\mathbf{C}^{-1})\mathbf{d}$$

to generate an expression for the pseudo-strain χ_z which corresponds to that of equation (6.28). Thus

$$\chi_z = -\frac{\mathrm{d}^2 v}{\mathrm{d}x^2} = (\boldsymbol{\beta}\mathbf{C}^{-1})\mathbf{d} \tag{6.33}$$

where $\boldsymbol{\beta}$ is obtained from $\boldsymbol{\alpha}$ by appropriate differentiation. Clearly, for the basic beam element

$$\boldsymbol{\beta} = [0 \quad 0 \quad 2 \quad 6x] \tag{6.34}$$

and equally clearly

$$\mathbf{B} = \boldsymbol{\beta}\mathbf{C}^{-1}. \tag{6.35}$$

It follows that equation (6.30) could be re-expressed as

$$\mathbf{k} = (\mathbf{C}^{-1})^{\mathrm{t}} \left(\int_0^l \boldsymbol{\beta}^{\mathrm{t}}(EI_z)\boldsymbol{\beta} \, \mathrm{d}x \right) \mathbf{C}^{-1}. \tag{6.36}$$

The advantage of this form, so far as manual operations are concerned, is that the required matrix multiplication and subsequent integration of terms represented within the outer brackets of equation (6.36) is very simple. Even with the necessary post-multiplication by $\mathbf{C}^{-1}$ and pre-multiplication by its transpose the approach represented by equation (6.36) often has advantages over that represented by equation (6.30). Clearly, this type of approach could also have been used for the bar element but there the matrix operations involved in generating $\mathbf{k}$ (according to equation (6.11)) are trivial in any case. The approach can also be used in generating the stiffness matrix of any general finite element if we start with the expression for the displacement field written in terms of generalized displacements and provided that it is possible to calculate $\mathbf{C}^{-1}$. (It can happen that matrix $\mathbf{C}$ is singular on rare occasions for some types of element. This reflects a lack of independence of the individual nodal freedoms, one from another, but will not occur for any of the elements discussed in this text.)

6.4. Comments on the basic bar and beam elements

6.4.1. Summary of the results of Sections 6.2 and 6.3

In deriving the bar and beam element stiffness we have seen that the same basic procedure is adopted in both cases. The stiffness matrix is derived from the element strain energy and has the form of a volume integral

$$\mathbf{k} = \int_{V_e} \mathbf{B}^t E \mathbf{B} \, dV_e \tag{6.37a}$$

or alternatively

$$\mathbf{k} = (\mathbf{C}^{-1})^t \left(\int_{V_e} \boldsymbol{\beta}^t E \boldsymbol{\beta} \, dV_e \right) \mathbf{C}^{-1}. \tag{6.37b}$$

In these forms, where the integrations are taken over the element volume, the matrices $\mathbf{B}$ and $\boldsymbol{\beta}$ relate the *true* strain to the nodal displacements and to the generalized displacements respectively. For the beam element this means that the $\mathbf{B}$ and $\boldsymbol{\beta}$ matrices in these equations are the matrices given by equations (6.29) and (6.34) respectively *multiplied by* y.

Although we have yet to consider two- and three-dimensional situations, it can be foreseen that the expressions for the element stiffness matrix will have much the same forms as those given above in these extended situations. In fact it will be seen in the next chapter that these expressions apply for general finite elements except that the unidirectional elastic modulus E is replaced by an elasticity matrix $\mathbf{E}$ when more than one dimension (and hence more than one stress and strain) is involved. Of course in multi-dimensional situations the complexity of the analysis increases but the great advantage of the finite-element displace-ment approach is that the same systematic procedure is adopted for the derivation of element properties whether the individual element be simple or very complicated.

We now return to consideration of the one-dimensional situation. The stiffness matrix for the bar and beam elements can be simplified to the following alternative forms involving integration over the length only:

$$\mathbf{k} = \int_0^l (\mathbf{B}^t E^1 \mathbf{B}) \, dx \qquad (6.38a)$$

or

$$\mathbf{k} = (\mathbf{C}^{-1})^t \left(\int_0^l \boldsymbol{\beta}^t E^1 \boldsymbol{\beta} \, dx \right) \mathbf{C}^{-1}. \qquad (6.38b)$$

Here E^1 represents a scalar quantity which is the axial rigidity AE for a bar element or the flexural rigidity EI_z for a beam element. It should be noted here, though, that for the beam element the matrices $\mathbf{B}$ and $\boldsymbol{\beta}$ in these expressions are now those that occur in the expression for *pseudo*-strain, or curvature, and are identically defined by equations (6.29) and (6.34).

Equations (6.38a) or (6.38b) can be used to generate the stiffness matrix for any straight bar or beam finite element. With regard to this it will be seen later in this chapter that refined elements can be based on higher-order displacement fields. Further, variable cross-section can be accommodated by appropriate expression of E^1 as a function of x in either of equations (6.38a) or (6.38b).

6.4.2. The exact nature of the basic bar and beam elements

It has been demonstrated that the stiffness matrices derived for the basic bar and beam finite elements through the PMPE are exactly the same as those developed earlier (in Chapter 2) using an 'exact' approach. This clearly means that for these simple elements the three basic conditions of elasticity, compatibility, and equilibrium are satisfied throughout the elements.

The *elasticity* condition is satisfied directly by the use of the proper constitutive equation (equation (6.8)).

The *compatibility* condition is satisfied throughout the element since the assumed displacement field (equation (6.1) for the bar and equation (6.19) for the beam) is smooth and continuous over the element length, i.e. the deformed element can have no artificial breaks or kinks along its length.

The *equilibrium* condition is identically satisfied at all points within the element, as can readily be seen from consideration of the governing differential equations of equilibrium of the bar and beam problems. For a uniform bar not carrying any distributed loading along its length the governing differential equation is

$$\frac{\mathrm{d}^2 u}{\mathrm{d}x^2} = 0 \tag{6.39}$$

and this is clearly satisfied at all points by the assumed bar element displacement field of equation (6.1). For a uniform beam, again not carrying a distributed loading, the governing differential equation is

$$\frac{\mathrm{d}^4 v}{\mathrm{d}x^4} = 0 \tag{6.40}$$

and this is satisfied by the assumption of equation (6.19).

In finite-element analysis complete satisfaction of all three basic conditions within an element is not at all usual. It is simple to meet the requirements of the elasticity and compatibility conditions within an element. (There are difficulties, however, in meeting compatibility requirements in an assembly of elements for some classes of problem, as will be discussed in later chapters). Usually the equilibrium condition is not satisfied at all points within elements and the derivation of (approximate) element stiffness equations proceeds without any direct reference to the equilibrium condition. Even in the simple world of bars and beams this local violation of equilibrium would occur if the element were based on the displacement fields of equations (6.1) or (6.19) and either carried a distributed loading along its length or was of varying cross-section.

6.5. Representation of distributed loading

If a finite element is subjected to any form of mechanical loading other than direct nodal loading the influence of such loading can be properly accommodated within the PMPE approach by including the potential energy of the loading in the expression for the total potential energy of the element. In solid mechanics it is usual to speak of two kinds of distributed forces which may act upon a body; these are termed *body* forces and *surface* forces. Body forces act over the *volume* of the body and may arise from the effects of gravity, acceleration, magnetic attraction, etc. Surface forces—or surface 'tractions'—act over part or whole of the *surface area* of the body and arise very commonly as applied static loadings. In restricting attention in the present chapter to one-dimensional elements it is consistent with the underlying assumptions of one-dimensional theory to work in terms of body and surface forces per unit length of the element under consideration. Examples of the way in which such forces are accommodated in the conventional Rayleigh–Ritz approach have already been considered for simple bar and beam problems in Chapter 5. Here, a very similar procedure is adopted with regard to the one-dimensional finite element.

Consider the bar element discussed in Section 6.2 and illustrated in Fig. 6.1. In Section 6.2 it was assumed that the only forces acting on the element are concentrated nodal forces at points 1 and 2. It followed that the potential energy of the forces acting on the element had the simple form of equation (6.14) and that correspondingly the total potential energy for the element was given by equation (6.15). If, now, we assume that distributed loads act over part or whole of the element length it is clear that the potential energy of these loads has to be included in the analysis.

Let us assume that the bar element is subjected to the distributed loading shown in Fig. 6.3 (*in addition* to the nodal forces which are not shown). The loading comprises a body force $r(x)$ per unit length acting over the complete bar length and a surface force or traction $t(x)$ per unit length acting over part of the element length $\bar{x}$. Both loadings act in the positive sense of the axial displacement. Denoting the potential energy of the distributed loading by V_{pd} we have that

$$V_{pd} = -\int_x ur\,dx - \int_{\bar{x}} ut\,dx. \tag{6.41}$$

It should be noted that for the first term on the right-hand side of this equation the integration extends over the whole element (that is for the whole range of x), whilst for the second term the integration is taken only over that portion of the element where the surface traction is prescribed (that is over the range $\bar{x}$).

The element displacement field has the form (as in equation (6.4))

$$u = \mathbf{Nd} = \mathbf{d}^t\mathbf{N}^t. \tag{6.42}$$

Substituting this expression for u into equation (6.41) gives

$$V_{pd} = -\mathbf{d}^t\int_x \mathbf{N}^t r\,dx - \mathbf{d}^t\int_{\bar{x}} \mathbf{N}^t t\,dx. \tag{6.43}$$

This can be expressed as

$$V_{pd} = -\mathbf{d}^t\mathbf{Q} \tag{6.44}$$

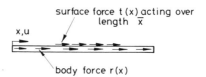

Fig. 6.3. Bar element subjected to body and surface forces.

where

$$\mathbf{Q} = \int_x \mathbf{N}^t r \, dx + \int_{\bar{x}} \mathbf{N}^t t \, dx \qquad (6.45)$$

is a column matrix.

The total potential energy Π_p of the element carrying a distributed loading is obtained by adding V_{pd} to the strain energy U_p and the potential energy of nodal forces V_{pn}. Thus (see equation (6.15))

$$\Pi_p = \tfrac{1}{2}\mathbf{d}^t\mathbf{kd} - \mathbf{d}^t\mathbf{P} - \mathbf{d}^t\mathbf{Q}. \qquad (6.46)$$

Minimizing the total potential energy in the usual way gives

$$\mathbf{P} + \mathbf{Q} = \mathbf{kd} \qquad (6.47)$$

which is the set of element stiffness equations where distributed loads are present.

The column matrix $\mathbf{Q}$ provides a set of loads at the element nodes which replaces in a consistent fashion the effect of any loadings distributed along the length of the element. $\mathbf{Q}$ is known as the 'kinematically consistent' or 'kinematically equivalent' column matrix of nodal loads or simply as the 'consistent' or 'equivalent' load matrix. Often in finite-element work distributed loads are simply 'lumped' at the nodes in proportions that appear reasonable in a direct physical sense. This procedure is often quite effective when dealing with simple elements and loadings but in general the consistent procedure outlined above is much to be preferred.

Concentrated loads acting at points along the element, other than at the nodes, can be accounted for in consistent fashion by treating them as specialized distributed tractions falling within the category of terms symbolized by the second term on the right-hand side of equation (6.45). Thus if, for example, a concentrated load T_c acts (in the positive sense of the displacement) at the location $x = x_c$ this second term is augmented by the contribution

$$\mathbf{N}_c^t T_c$$

where $\mathbf{N}_c$ is the shape function matrix evaluated at $x = x_c$. Obviously any number of concentrated loads can be accommodated in this way. The expression for $\mathbf{Q}$ in equation (6.45) should therefore be interpreted as including the possible presence of concentrated load effects in the second term on the right-hand side.

Where the displacement field is initially expressed in terms of generalized displacements we have that $\mathbf{N} = \boldsymbol{\alpha}\mathbf{C}^{-1}$ and the column matrix

Q given by equation (6.45) can be alternatively expressed as

$$\mathbf{Q} = (\mathbf{C}^{-1})^t \int_x \boldsymbol{\alpha}^t r \, dx + (\mathbf{C}^{-1})^t \int_{\bar{x}} \boldsymbol{\alpha}^t t \, dx. \tag{6.48}$$

The above derivation has been made with reference initially to the bar element of Section 6.2. However, it should be clear that the results obtained—which are represented by equations (6.45), (6.47), and (6.48)—are applicable to any bar or beam element (and can be extended to include two- and three-dimensional elements as will be seen later). For a beam element the forces $r(x)$ and $t(x)$ act normally to the element axis, of course, rather than along it. Also, for a beam concentrated loads between nodes may include both normal loads T_c and couples M_c acting at a point or points. For the point couple case, with the couple located at $x = x_c$, the contribution to **Q** is of the form

$$\frac{d}{dx} (\mathbf{N}^t)_c M_c.$$

Consider, now, two simple examples of the calculation of the column matrix **Q**.

Example 6.1. A uniform basic bar element (shown in Fig. 6.1) carries a uniformly distributed loading along its whole length of intensity p per unit length. What is the system of nodal forces which is kinematically consistent with this distributed loading?

For this trivial problem we have, using equation (6.45) with **N** defined in equation (6.5) and $t(x) = p$, that

$$\mathbf{Q} = \int_{\bar{x}} \mathbf{N}^t t \, dx = p \int_0^l \begin{Bmatrix} 1 - x/l \\ x/l \end{Bmatrix} dx = pl \begin{Bmatrix} \frac{1}{2} \\ \frac{1}{2} \end{Bmatrix}$$

Thus half of the total distributed load is allocated to each end node. This is the expected result and is the assumption that would have been made in a simple lumping procedure.

Example 6.2. A uniform basic beam element (shown in Fig. 6.2) carries a concentrated load together with the linearly varying distributed surface loading shown in Fig. E6.2. Determine **Q** for this loading.

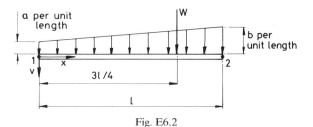

Fig. E6.2

Matrices $\boldsymbol{\alpha}$, $\mathbf{C}^{-1}$, and $\mathbf{N}$ are available for this element in Section 6.3 (equations (6.19), (6.22), and (6.24)) and so $\mathbf{Q}$ can be determined using either of equations (6.45) or (6.48). If the latter equation is used we have that

$$\mathbf{Q} = (\mathbf{C}^{-1})^t \int_0^l \boldsymbol{\alpha}^t t \, dx + (\mathbf{C}^{-1})^t \boldsymbol{\alpha}_c^t W.$$

The linearly varying loading can be expressed as

$$t = a + gx/l$$

where $g = b - a$. It is then easy to show that

$$\int_0^l \boldsymbol{\alpha}^t t \, dx = \frac{l}{60} \{60a + 30g \quad l(30a + 20g) \quad 20a + 15g \quad l(15a + 12g)\}$$

Associated with the concentrated load is the matrix $\boldsymbol{\alpha}_c^t$ which is obtained from $\boldsymbol{\alpha}^t$ simply by putting $x = 3l/4$. Thus

$$\boldsymbol{\alpha}_c^t = \{1 \quad 3l/4 \quad 9l^2/16 \quad 27l^3/64\}.$$

Premultiplication of $\int_0^l \boldsymbol{\alpha}^t t \, dx$ and $\boldsymbol{\alpha}_c^t$ by $(\mathbf{C}^{-1})^t$ gives the contributions to $\mathbf{Q}$ of the linearly varying distributed loading and of the concentrated load respectively. The result is

$$\mathbf{Q} = \frac{l}{60} \begin{Bmatrix} 30a + 9g \\ l(5a + 2g) \\ 30a + 21g \\ -l(5a + 3g) \end{Bmatrix} + \frac{W}{64} \begin{Bmatrix} 10 \\ 3l \\ 54 \\ -9l \end{Bmatrix}.$$

Clearly this result would have been unlikely to have been obtained using a crude lumping procedure.

The above examples have been concerned with the basic bar and beam elements, and we know that the exact stiffness matrices for these elements can be obtained through the PMPE for a uniform element carrying only end loads because the displacement field happens to exactly represent this situation. When distributed loading is present on these basic elements we might similarly expect that the system of consistent nodal loads calculated through the PMPE bears some direct relationship to the system of fixed-end forces used in the MDM (see Table 4.1). This is indeed so, for the consistent nodal loads are the reverse of the fixed-end forces.

6.6. The assembly process within the PMPE approach

Thus far in this chapter our concern has been with the development of *element* relationships through the use of assumed displacements and the PMPE. The element has been treated as a complete entity in the sense that no consideration has yet been given to the way in which the element fits into the structural assembly. The element may carry a distributed loading whose effect is represented by the consistent loads $\mathbf{Q}$ and it has forces $\mathbf{P}$ acting at its nodes which in general represent the supportive effect of the rest of the structure on the element. We have seen that the set of stiffness equations for the element is of the form of equation (6.47). Now we wish to assemble the set of stiffness equations for the complete structure from those of the individual elements by considerations of potential energy. This assembly procedure, which applies to any type of finite element, will, not surprisingly, turn out to be effectively identical to the direct stiffness assembly procedure of the MDM described in earlier chapters. However, the more mathematical viewpoint adopted here of the FEM as a piecewise Rayleigh–Ritz procedure has some important advantages over the physical viewpoint of earlier chapters. This is especially so when we come to consider two- and three-dimensional structures in later chapters.

In the mathematical, or variational, viewpoint of the FEM the individual element is considered simply as a finite-size region of the whole body which it is convenient for analysis purposes to consider separately from the rest of the structure. For each such region a localized displacement field is assumed which represents the deformed state of the region when the structure is loaded. This localized field is restricted to the region and is expressed (either directly or indirectly) in terms of certain reference displacements at selected points (the nodes) located on the boundary of, and sometimes within, the region. Based on the assumed local displacement field the potential energy of each region can be evaluated in terms of the reference displacements of that region (only) by performing appropriate integrations over the region (as has already been seen for the basic bar and beam elements). Then, provided that certain conditions are met at the boundaries between neighbouring regions, the potential energy of the whole loaded structure can be obtained as the summation of the potential energies of all the individual regions and of the potential energy of any point loads acting directly at the structure nodes. (The certain conditions mentioned here relate to the satisfaction of the compatibility condition for the structure as a whole. This will be discussed later.) The minimization of the total potential energy then yields the system of stiffness (or equilibrium) equations for the structure. The procedure may be termed a 'piecewise Rayleigh–Ritz procedure' for obvious reasons.

It has been seen that the potential energy of a region (element), identified now by the appellation e, *when considered in isolation* is

$$\Pi_p^e = U_p^e + V_{pd}^e + V_{pn}^e. \tag{6.49}$$

Here

$$U_p^e = \tfrac{1}{2}\mathbf{d}_e^t \mathbf{k}_e \mathbf{d}_e \tag{6.50}$$

is the strain energy or potential energy of deformation of the element e,

$$V_{\mathrm{pd}}^e = -\mathbf{d}_e^t \mathbf{Q}_e \qquad (6.51)$$

is the potential energy of distributed loads acting on the element (i.e. of all externally applied loadings acting anywhere other than directly at the nodes), and

$$V_{\mathrm{pn}}^e = -\mathbf{d}_e^t \mathbf{P}_e \qquad (6.52)$$

is the potential energy of forces acting on the element from its nodes.

When consideration is given to the *complete region* or *whole structure* it becomes clear that the V_{pn}^e contribution is no longer appropriate. The typical element is 'connected' at its boundary to neighbouring elements and the column matrix $\mathbf{P}_e$ contains forces at the interelement boundaries which are *external* to the individual element but are *internal* so far as the structure as a whole is concerned (just as, for example, in Fig. 2.2 the forces U_j^A and U_i^B at structure node 2 are external to elements A and B respectively but are internal to the structure as a whole; for the structure the actual external force at node 2 is F_2). Thus, for the whole structure the V_{pn}^e contributions from each element should be replaced by the potential energy of any concentrated forces $\mathbf{F}$ which are *externally* applied directly at the structure node points. This potential energy is simply

$$-\mathbf{D}^t \mathbf{F} \qquad (6.53)$$

where, as before, $\mathbf{D}$ is the column matrix of all structure nodal freedoms.

The total potential energy of an assemblage of E elements, having n structure degrees of freedom, is denoted $\bar{\bar{\Pi}}_{\mathrm{p}}$ and can now be written

$$\bar{\bar{\Pi}}_{\mathrm{p}} = \bar{\bar{U}}_{\mathrm{p}} + \bar{\bar{V}}_{\mathrm{pd}} - \mathbf{D}^t \mathbf{F}. \qquad (6.54)$$

Here $\bar{\bar{U}}_{\mathrm{p}}$ and $\bar{\bar{V}}_{\mathrm{pd}}$ are the strain energy and the potential energy of distributed loads of the complete structure. These quantities are direct summations of contributions from each of the individual elements; that is

$$\bar{\bar{U}}_{\mathrm{p}} = \sum_{e=1}^{E} U_{\mathrm{p}}^e \quad \text{and} \quad \bar{\bar{V}}_{\mathrm{pd}} = \sum_{e=1}^{E} V_{\mathrm{pd}}^e. \qquad (6.55)$$

Now, U_{p}^e and V_{pd}^e are defined in equations (6.50) and (6.51), and therein $\mathbf{k}_e$ is the stiffness matrix of element e and $\mathbf{Q}_e$ is the column matrix of consistent loads corresponding to any distributed loading acting on element e. The column matrix $\mathbf{d}_e$ is the list of nodal displacements for element e and, of course, all the components of $\mathbf{d}_e$ for each element of the assemblage will occur somewhere in the column matrix $\mathbf{D}$ of structure nodal displacements. Consequently we can re-express equations (6.50) and (6.51) symbolically as

$$U_p^e = \tfrac{1}{2}\mathbf{D}^t\mathbf{k}_e^*\mathbf{D} \qquad \text{and} \qquad V_{pd}^e = -\mathbf{D}^t\mathbf{Q}_e^* \tag{6.56}$$

where $\mathbf{Q}_e^*$ and $\mathbf{k}_e^*$ are the element consistent load and stiffness matrices enlarged to 'structure size'. This means that $\mathbf{Q}_e^*$ is of size $n \times 1$ and contains the individual components of $\mathbf{Q}_e$ in those locations corresponding to the locations in $\mathbf{D}$ of the components of $\mathbf{d}_e$ with zeros added in all other locations. Similarly $\mathbf{k}_e^*$, of size $n \times n$, is obtained from $\mathbf{k}_e$ by placing the individual components of $\mathbf{k}_e$ in row and column locations corresponding to the positions of the components of $\mathbf{d}_e$ in $\mathbf{D}$ and adding zeros in all other locations.

From equations (6.54), (6.55), and (6.56) the total potential energy of the assemblage is

$$\bar{\bar{\Pi}}_p = \tfrac{1}{2}\sum_{e=1}^{E} \mathbf{D}^t\mathbf{k}_e^*\mathbf{D} - \sum_{e=1}^{E} \mathbf{D}^t\mathbf{Q}_e^* - \mathbf{D}^t\mathbf{F}$$

$$= \tfrac{1}{2}\mathbf{D}^t\left(\sum_{e=1}^{E} \mathbf{k}_e^*\right)\mathbf{D} - \mathbf{D}^t\sum_{e=1}^{E} \mathbf{Q}_e^* - \mathbf{D}^t\mathbf{F} \tag{6.57}$$

Minimizing this potential energy, using the relevant conditions that $\partial\bar{\bar{\Pi}}_p/\partial D_i = 0$ $(i = 1, 2, \dots n)$, gives the n equilibrium equations

$$\left(\sum_{e=1}^{E} \mathbf{k}_e^*\right)\mathbf{D} - \sum_{e=1}^{E} \mathbf{Q}_e^* - \mathbf{F} = 0$$

This can be expressed

$$\mathbf{F} + \mathbf{F}_Q = \mathbf{KD} \tag{6.58a}$$

where

$$\mathbf{K} = \sum_{e=1}^{E} \mathbf{k}_e^* \tag{6.59}$$

is the assembled structure stiffness matrix and

$$\mathbf{F}_Q = \sum_{e=1}^{E} \mathbf{Q}_e^* \tag{6.60}$$

is the column matrix of consistent loads for the assembled structure.

The result expressed by equations (6.58)–(6.60) and obtained through the PMPE approach is identical with that established by direct methods in the earlier chapters dealing with the MDM. *It is thus confirmed that the method of direct superposition (i.e. the direct stiffness method) can be used to assemble structure stiffness equations in general finite-element applications and it is this method which is almost always used in practice.*

The above analysis has made no mention of the geometric boundary conditions of the structure. Clearly, though, if some structure freedoms are prescribed the analysis could be readily adjusted to accommodate this. The resulting procedures for accommodating the prescribed displacements would be identical with those already discussed for the MDM method and the corresponding set of structure equations would be

$$\mathbf{F}_r + \mathbf{F}_{Or} = \mathbf{K}_r \mathbf{D}_r \qquad (6.58b)$$

where as before r denotes a 'reduced' matrix.

The above analysis has also implied that for each and every element the local co-ordinate axes—to which the element stiffness and consistent load matrices are referred—coincide with a common global system of axes. This is often the case, but if it is not it is clear that for all affected elements a transformation procedure has to be adopted prior to the assembly process. The form of the transformation is in fact identical with that given in Chapter 3 but it is perhaps appropriate now to re-derive it within the context of the PMPE approach. We assume that the nodal displacements of a particular element in the local system are $\mathbf{d}_e$ and that these can be related to the nodal displacements $\bar{\mathbf{d}}_e$ in the global system by a transformation of the form

$$\mathbf{d}_e = \tau \bar{\mathbf{d}}_e \qquad (6.61)$$

(from equation (3.14)). The potential energy of the element expressed in the local system terms is given by equations (6.49)–(6.52) and it is assumed that $\mathbf{k}_e$ and $\mathbf{Q}_e$, the stiffness matrix and consistent load matrix in the local system, are known. Substituting equation (6.61) into equations (6.49)–(6.52) gives the potential energy of the element expressed in global system terms as

$$\Pi^e = \tfrac{1}{2}\bar{\mathbf{d}}_e^t \bar{\mathbf{k}}_e \bar{\mathbf{d}}_e - \bar{\mathbf{d}}_e^t \bar{\mathbf{P}}_e - \bar{\mathbf{d}}_e^t \bar{\mathbf{Q}}_e \qquad (6.62)$$

where

$$\bar{\mathbf{k}}_e = \tau^t \mathbf{k}_e \tau, \qquad \bar{\mathbf{P}}_e = \tau^t \mathbf{P}_e, \qquad \bar{\mathbf{Q}}_e = \tau^t \mathbf{Q}_e. \qquad (6.63)$$

Here $\bar{\mathbf{k}}_e$ and $\bar{\mathbf{Q}}_e$ are the required element stiffness and consistent load matrices referred to the common co-ordinate system. These matrices can now be added into the structure matrices $\mathbf{K}$ and $\mathbf{F}_O$ in the usual fashion, as described above.

In summary, it has been shown in this section that application of the PMPE yields the same results as did the procedures used earlier in the MDM so far as the assembly process (of element properties into structure properties) is concerned and so far as transformation of element properties from a local to a global co-ordinate system is concerned.

6.7. Element stresses

On solution of the system of simultaneous equations for the whole structure the structure nodal displacements are available and from these can be extracted the column matrix **d** of local nodal freedoms for each individual finite element. For the one-dimensional element it is then a simple matter to calculate the single stress component arising as a result of the nodal displacements. For a bar element we have, using equations (6.6) and (6.8), that

$$\sigma_x = E\varepsilon_x = E\mathbf{Bd}. \tag{6.64}$$

Usually the stress varies within an element and matrix **B** needs to be evaluated at any particular point or points where the stress value is required; for the basic bar, though, the terms of **B** are constants. For a beam element an expression of the form of equation (6.64) could also apply but if we work in terms of the more convenient moment (or pseudo-stress) value we have, using equations (6.26) and (6.28), the modified form

$$M_z = EI_z\mathbf{Bd}. \tag{6.65}$$

In general we can express the stress (or pseudo-stress) in the form

$$\sigma_x \text{ (or } M_z) = \mathbf{Sd} \tag{6.66}$$

where **S** is a *stress matrix* whose form is apparent from equations (6.64) and (6.65). It should be noted that it is sometimes convenient to substitute $(\boldsymbol{\beta}\mathbf{C}^{-1})$ for **B** in calculating the stress matrix.

Only on rare occasions will the stresses calculated from the nodal displacements using the FEM be the 'exact' ones throughout a structure. From our experience of the conventional Rayleigh–Ritz procedure in Chapter 5 we expect that stress-type quantities will often reveal considerable local errors in a FEM analysis (i.e. a piecewise Rayleigh–Ritz analysis) even when calculated nodal displacements are very accurate. The reasons for this have already been discussed quite fully in Chapter 5 but it bears repetition that this is due to the fact that equilibrium conditions are not satisfied within the individual finite element other than in an overall sense. We expect to approximate the equilibrium conditions more and more closely throughout a structure as we employ more and more finite elements in an analysis and hence to achieve progressively a more and more accurate prediction of the exact stress distribution.

For the *basic* bar and beam finite elements it has been demonstrated that the element stiffness matrices are identical with the 'exact' ones used in the MDM and thus there will be no difference in the calculated results (for displacements or stresses) obtained using the FEM or the MDM so long as any actual loading is concentrated at the nodes. However, where distributed loads are present there does arise a significant difference in the calculation of stress-type quantities between the two approaches, although calculated nodal displacements are still identical in the FEM and the MDM. It has been seen, in Section 4.2, that in the MDM a distributed loading is properly and precisely accommodated by invoking the principle of superposition so that an actual problem is considered effectively as the sum of two problems, as shown in Fig. 4.1. In the first of these problems (Fig. 4.1(b)) each element of the structure has completely fixed ends (no nodal displacements) and is in equilibrium under the action of the distributed loading applied to it and the fixed-end forces. In the second (Fig. 4.1(c)) the nodal freedoms are released and overall deformation occurs under the action of a set of nodal loads corresponding to the reversed fixed-end forces. Now, in the conventional FEM the calculation of stresses is based on an equation of the form of equation (6.66) and hence the stresses are those due only to the nodal displacements resulting from the action of the set of nodal loads which is kinematically consistent with the actual distributed loading. For the basic one-dimensional elements this set of consistent nodal loads is the same as the set of reversed fixed-end forces used in the MDM. Hence we can see that the lack of satisfaction of equilibrium in the interior of a basic element in the FEM corresponds to neglect of the stresses produced by the distributed loading acting on the fully restrained element. In conventional finite-element stress calculation of structures assembled from *basic* bar and beam elements we are effectively considering the situation illustrated by Fig. 4.5(c) only. To clarify these remarks consider the very simple statically determinate example of a uniformly loaded cantilever beam which was first considered by the MDM in Example 4.1.

Example 6.3. The uniformly loaded cantilever is illustrated in Fig. E6.3(a). Determine the distribution of bending moment along the beam by the FEM.

We shall consider first of all the use of a single basic beam element to represent the beam.

Using the methods of Section 6.5 (see Example 6.2) it is very simple to determine the consistent nodal loads corresponding to the distributed loading. Only two non-zero nodal freedoms are involved, v_2 and θ_{z2}, and the consistent

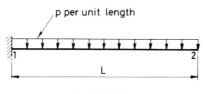

Fig. E6.3(a)

loads for the structure are thus

$$\mathbf{F}_{Qr} = \{F_{v2} \quad F_{\theta z 2}\} = \frac{pL}{12}\{6 \quad -L\}.$$

These loads are the same as the reversed fixed-end forces of Example 4.1 shown in Fig. E4.1(c). As in the MDM the structure stiffness equations in the FEM approach are

$$\frac{pL}{12}\left\{\begin{array}{c} 6 \\ -L \end{array}\right\} = EI_z \left[\begin{array}{cc} 12/L^3 & -6/L^2 \\ -6/L^2 & 4/L \end{array}\right]\left\{\begin{array}{c} v_2 \\ \theta_{z2} \end{array}\right\}$$

and the solution for the nodal displacements is the exact one, that is

$$v_2 = \frac{pL^4}{8EI_z} \qquad \theta_{z2} = \frac{pL^3}{6EI_z}$$

To calculate the distribution of the bending moment by the FEM we use the relationship $M_z = \mathbf{Sd}$ where the stress matrix $\mathbf{S} = EI_z\mathbf{B}$ and where $\mathbf{B}$ is given for the basic beam element by equation (6.29). Thus, bearing in mind that only v_2 and θ_{z2} are non-zero, we have that the bending moment at any point x along the element is

$$M_z(x) = EI_z\left[\frac{6}{L^2} - \frac{12x}{L^3} \quad \frac{4}{L} - \frac{6x}{L^3} \quad \frac{-6}{L^2} + \frac{12x}{L^3} \quad \frac{2}{L} - \frac{6x}{L^2}\right]\left\{\begin{array}{c} 0 \\ 0 \\ pL^4/8EI_z \\ pL^3/6EI_z \end{array}\right\}.$$

From this the values of the moment at the beam ends are

$$M_z(0) = -\frac{5}{12}pL^2 \qquad M_z(L) = +\frac{pL^2}{12}$$

and $M_z(x)$ varies linearly between the ends, of course. This distribution of bending moment is shown together with the exact result in Fig. E6.3(c). As mentioned above, this standard procedure of bending moment calculation in the FEM, when using the basic beam element, corresponds to that part of the MDM solution represented by Fig. E4.1(c) and effectively neglects the moment distribution contributed by the situation shown in Fig. E4.1(b). (It is noted that the single-element FEM solution of this uniformly loaded cantilever beam problem is precisely the same as that obtained by the conventional Rayleigh–Ritz procedure in Example 5.6. This is to be expected since the Rayleigh–Ritz solution was based on the assumption of a cubic polynomial displacement field.)

If we wish to improve the prediction of the bending moment distribution using the FEM and basic beam elements we must use more elements to model the beam. (Alternatively we could use a single 'refined' element based on a higher-order displacement field (see Section 6.10).) When two equal-length basic elements are used, for instance, as in Fig. E6.3(b) the column matrix of non-zero

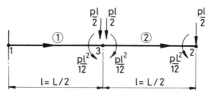

Fig. E6.3(b)

structure freedoms is

$\mathbf{D}_r = \{v_2 \quad \theta_{z2} \quad v_3 \quad \theta_{z3}\}.$

The corresponding kinematically consistent set of loads comprising contributions from both elements is (with $l = L/2$)

$$F_{Qr} = \frac{pL}{24}\begin{Bmatrix} 0 \\ 0 \\ 6 \\ -L/2 \end{Bmatrix} + \frac{pL}{24}\begin{Bmatrix} 6 \\ -L/2 \\ 6 \\ L/2 \end{Bmatrix} = \frac{pL}{24}\begin{Bmatrix} 6 \\ -L/2 \\ 12 \\ 0 \end{Bmatrix}$$

$$\quad\quad\quad\quad\text{element 1}\quad\quad\quad\text{element 2}$$

Using the element stiffness matrix given by equation (6.32) the structure stiffness equations, assembled using the direct stiffness procedure, become

$$\frac{pL}{24}\begin{Bmatrix} 6 \\ -L/2 \\ 12 \\ 0 \end{Bmatrix} = EI_z \begin{bmatrix} 96/L^3 & & \text{Symmetric} & \\ -24/L^2 & 8/L & & \\ -96/L^3 & 24/L^2 & 192/L^3 & \\ -24/L^2 & 4/L & 0 & 16/L \end{bmatrix} \mathbf{D}_r$$

and the solution is

$$\mathbf{D}_r = \frac{pL^4}{384EI_z}\{48 \quad 64/L \quad 17 \quad 56/L\}$$

which agrees with the exact solution, as expected. The bending moment distribution in each element is then calculated in the same manner as for the single-element case, noting that in matrix $\mathbf{B}$ the co-ordinate origin of x is at the left-hand end of the individual element in question. For element 1,

$$\mathbf{d} = \frac{pL^4}{384EI_z}\{0 \quad 0 \quad 17 \quad 56/L\}$$

and

$$M_z(0) = -\frac{23}{48}pL^2 \quad\quad\quad M_z(l) = -\frac{5}{48}pL^2.$$

For element 2,

$$\mathbf{d} = \frac{pL^4}{384EI_z}\{17 \quad 56/L \quad 48 \quad 64/L\}$$

and

$$M_z(0) = -\frac{5}{48}pL^2 \quad\quad\quad M_z(l) = +\frac{1}{48}pL^2.$$

The FEM bending moment distribution for the two-element idealization is also shown in Fig. E6.3(c) and considerable improvement as compared with the single-element result is evident. It is noted from this figure that for the FEM results there is a non-zero bending moment value at the free end of the beam; this is typical of the method, but the residual moment would become vanishingly small if a large number of elements were used. It is further noted that there is not a discontinuity at the beam centre in the bending moment distribution forecast by

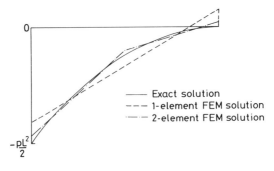

Fig. E6.3(c)

the two-element solution; this is untypical, as such discontinuities are common-place but reduce in magnitude with increase in the number of elements used.

6.8. Requirements for element displacement fields

In the finite-element displacement method the critical step is the choice of the element displacement field. On this choice will directly depend the accuracy achieved in the solution of a structural problem wherein the structure is idealized as an assemblage of elements. Clearly the aim is to ensure that the exact solution is progressively approached as more elements are used—that is that *convergence* to the exact solution occurs in the limit as element size decreases or, in other words, as the element mesh is refined—and that the rate of approach (or rate of convergence) is reasonably rapid.

Since in the variational approach the FEM is viewed as a piecewise Rayleigh–Ritz procedure it is clear that the requirements for element displacement fields will bear a distinct relationship to the requirements for overall displacement fields in the conventional Rayleigh–Ritz procedure. In the latter it has been seen, in Chapter 5, that the trial fields need to meet the criteria of completeness and of compatibility (or admissibility); if this is done convergence to the true solution takes place with increase in the number of terms used and the total potential energy of the energy method solution is always greater than or (in the limit) equal to the true potential energy, i.e. an upper-bound solution to the energy is obtained. In the FEM the following four conditions related to the element displacement fields are postulated as being necessary to ensure convergence rigorously whilst maintaining the upper-bound nature of the solution; these conditions apply to all types of finite element.

(1) The element displacement field should include the proper representation of any possible rigid body motion of the element.

A rigid body motion is a motion in which no straining occurs. The statement means that if an element, representing a part of a whole structure, happens to have values of nodal degrees of freedom which are consistent with a rigid body motion, then no strains should be developed within the element. This is clearly desirable.

(2) The element displacement field should include representation of any possible states of constant strain.

States of constant strain are, like rigid body motion states, very basic states. (In fact rigid body motion states can be regarded as special cases of constant strain states in which the strains have zero value.) The philosophy lying behind the adoption of this criterion is that in the limit, if the size of elements is sufficiently small, the strains within each element will clearly be very nearly uniform. This makes the representation of constant strain states necessary in the limit of infinitesimal size elements to ensure ultimate convergence to the true solution, but accurate representation of constant strain states in 'large' elements is also desirable to help give a rapid rate of convergence.

The above two criteria concerning simple basic states of strain in the individual finite element are akin to the completeness condition for trial functions used in the conventional Rayleigh–Ritz procedure. The two remaining criteria (of the four mentioned above) relate to the admissibility, or compatibility, of the whole structure displacement field when this is assembled from the piecewise contributions of the displacement fields of the individual finite elements.

(3) Within each element the displacement field should be smooth and continuous.
(4) At the boundaries between adjacent elements the compatibility condition should be satisfied.

These two conditions mean that when an assemblage of elements is loaded no discontinuities of displacement (or of the displacement derivative in bending problems) should occur anywhere in the assemblage, i.e. no gaps or overlaps (or kinks) should occur within elements or at element boundaries. This can be expressed more formally by considering the nature of the total potential energy functional. The potential energy is a function of displacements and displacement derivatives of up to the mth order. To evaluate the potential energy uniquely therefore requires that derivatives of up to the mth level exist and are finite everywhere. This is assured if continuity of derivatives up to the $(m-1)$th level is achieved. Thus we conclude that unique evaluation of the total potential energy of the whole structure—and hence the complete satisfaction of compatibility requirements—is obtained if continuity of derivatives of one order less than the highest-order derivative appearing in the energy expression is maintained. In finite-element terminology elements whose displacement fields satisfy all compatibility requirements are called *conforming* elements, whilst elements whose fields violate the compatibility requirements (at interelement boundaries) are *non-conforming*.

It is easy to show that the displacement fields of the only elements we have been concerned with so far—the basic bar and beam elements— satisfy the above four criteria exactly.

For the bar we have, dealing with the criteria in order (with A_0, A_1 and A_2 as constant terms), the following.

(i) The only rigid body motion is the simple translation in the x direction

$$u = A_0$$

and this is contained in the assumed displacement field (equation (6.1)).

(ii) The state of constant strain is

$$\varepsilon_x = \mathrm{d}u/\mathrm{d}x = A_1$$

and again the required term A_1x is included in the expression for u.

(iii) and (iv) The highest derivative occuring in the potential energy functional is of order 1 (i.e. the derivative $\mathrm{d}u/\mathrm{d}x$). Therefore continuity of the displacement u itself over an element assemblage is all that is required to meet compatibility conditions. Such continuity is clearly available in the element interior, from the smooth form of equation (6.1), and is also available at element boundaries—simply the node *points* for the one-dimensional elements—since the values of the displacement are matched at the nodes. It is noted that where the compatibility condition requires continuity of displacement(s) alone this is referred to as a C_0 continuity requirement.

For the beam we have the following.

(i) There are two rigid body motions to consider, one a translational motion

$$v = A_0$$

and the other a rotational motion

$$\theta_z = \frac{\mathrm{d}v}{\mathrm{d}x} = A_1 \qquad \text{so that} \qquad v = A_1x.$$

Both these motions are contained in the element displacement field (equation (6.19)).

(ii) The state of constant 'strain' is

$$-\chi_z = \frac{\mathrm{d}^2 v}{\mathrm{d}x^2} = A_2$$

and the required term to give this result, namely $A_2 x^2$, appears in the element displacement field.

(iii) and (iv) The highest derivative in the potential energy functional is of order 2 (i.e. the second derivative $\mathrm{d}^2 v/\mathrm{d}x^2$). Therefore continuity of v and $\mathrm{d}v/\mathrm{d}x$ is required over the element assemblage to meet compatibility conditions. This continuity is available both in the element interior and at the element boundaries where values of v and $\mathrm{d}v/\mathrm{d}x$ in the adjacent elements are matched. Where continuity of displacement(s) and first derivative(s) is required this is referred to as a C_1 continuity requirement.

Of course, straight one-dimensional elements are quite trivial and we would expect more difficulty in satisfying the above criteria when we consider two- and three-dimensional situations. It is particularly noted that in such situations the condition of interelement compatibility (condition (4)) applies at the whole of common boundaries and not just at nodal *points*.

It will be seen later that all the criteria can be satisfied quite readily in plane stress or strain and three-dimensional elasticity applications where only C_0 continuity is required. However, in plate bending applications simultaneous satisfaction of conditions (2) and (4) poses a problem since in elements based on classical plate theory the requirement, as in the case of the beam, is for C_1 continuity. In curved shell analysis the situation is even more complicated, with difficulties arising in satisfying conditions (1), (2), and (4). It is noted that condition (3) is easily satisfied in all classes of problem when using polynomial displacement fields.

As mentioned earlier, in circumstances where the above four conditions are satisfied we can be assured that the finite-element solution will yield an upper bound on the true total potential energy and that convergence to the true level will occur with mesh refinement. Two points should be noted regarding this statement. Firstly, the bound is only on the total potential energy, and whilst this means that the solution will be overstiff in a general or average sense it does not mean that the displacement at any particular point is necessarily less than the corresponding true displacement (except in the case where a single point load acts on a structure when the FEM displacement at the load point will necessarily not exceed the corresponding true displacement.) Nor does it mean that any statement can be made as to whether calculated FEM stress values are greater or smaller than the exact stresses at particular points. Secondly, whilst we would normally expect to obtain a more accurate solution when using a relatively large number of elements as compared with that obtained when using a lesser number of elements, this is only absolutely *guaranteed* in certain circumstances. These circumstances are that all the nodal freedoms of the coarser element assemblage are included as nodal freedoms in the finer element assemblage. When this condition applies for a series of element assemblages—each succeeding assemblage being a refinement of the previous one—we can expect the results to demonstrate *monotonic* convergence to the true potential energy (or stiffness) from above. (By monotonic convergence is meant convergence such that the accuracy of the solutions obtained from the various assemblages improves progressively with mesh refinement.)

It should be made clear, at this early stage, that in practice convergence to the true solution can be achieved even when an element displacement field does not exactly satisfy all the above four conditions. There exist quite a number of types of successful finite element in which one or more of the conditions is violated: most of these elements are thin-plate or shell elements where it has been noted that the C_1 continuity requirement at element boundaries is difficult to meet. Non-conforming elements are often used in plate and shell work with the non-conformity usually arising as discontinuities in slope, or kinks, at interelement boundaries. Of course, consequent lack of satisfaction of the compatibility condition in the element assemblage destroys the rigour of the analysis and means that the calculated total potential energy is no longer necessarily an upper bound to the true value. Effectively the presence of the kinks at interelement boundaries corresponds to infinite values of curvature and these infinities are not accounted for in the expression for the whole structure potential energy as a summation of the element energies and the potential of nodal loads. Nevertheless, conformity can be progressively restored as more and more elements are used, and if at the same time the constant strain condition is satisfied, convergence will occur to the true solution. Such convergence may not be monotonic and may be from above or below, but the actual errors involved may well be less than if a conforming element is used. This and the other points mentioned above with regard to two- and three-dimensional elements will be discussed further where appropriate later in the text.

6.9. Shape functions and interpolation polynomials

For the finite elements considered so far, the generation of element properties has commenced with the assumption of a polynomial displacement field written in terms of a number of generalized displacements as in equations (6.1) and (6.19). Subsequently the relevant displacement at any point in the element is expressed in terms of the nodal freedoms of that element via application of the element 'boundary conditions' and inversion of the $\mathbf{C}$ matrix. This procedure was the one used in the early development of the FEM and is still commonly adopted—as it will be to a considerable extent in this text—because it is usually very straightforward to implement. It does, however, have the disadvantage of requiring the calculation of $\mathbf{C}^{-1}$ (and in extreme cases this inverse does not exist). It has been seen that the procedure leads to an equation of the form

$$\mathbf{u} = \alpha\mathbf{C}^{-1}\mathbf{d} = \mathbf{Nd} \qquad\qquad (6.4), (6.23)$$

linking the displacement at a general point in an element to the nodal displacements. Here, matrix $\mathbf{N}$ for a one-dimensional element is a function of the single co-ordinate x and this matrix has already been referred to as the shape function matrix. Clearly, if we can somehow derive or deduce the shape function matrix directly we can then avoid the preliminary stage involving α and $\mathbf{C}$. Before considering how to do this let us examine the nature of the shape function matrix, with particular regard to the elements that we have considered so far.

For the one-dimensional element the shape function matrix $\mathbf{N}$ can be expressed in the form (see equations (6.5) and (6.24))

$$\mathbf{N}(x) = [N_1(x) \quad N_2(x)\ldots] \qquad\qquad (6.67)$$

where the $N_i(x)$ $(i = 1, 2, \ldots)$ are polynomial functions of x (each being of the same order as that of $\mathbf{N}$) and are the individual shape functions that correspond on a one-to-one basis with the nodal displacements d_i. The expanded form of equation (6.4) (or of equation (6.23)) is thus

$$u \text{ (or } v) = N_1(x)d_1 + N_2(x)d_2 + \ldots. \qquad\qquad (6.68)$$

It is clear that the total expression for displacement is simply a superposition of terms associated with each nodal freedom in turn.

Considering the basic bar element (Fig. 6.1) first, we have that $\{d_1 \ d_2\} \equiv \{u_1 \ u_2\}$ and, from equation (6.5) that

$$N_1(x) = 1 - \bar{x} \qquad\qquad N_2(x) = \bar{x} \qquad\qquad (6.69)$$

where $\bar{x}$ is used for convenience to represent x/l.

These two linear shape functions, which each apply over the whole element length, are shown in Fig. 6.4. It can be seen that $N_1(0) = 1$, $N_1(l) = 0$, $N_2(0) = 0$ and $N_2(l) = 1$. It will be clear that these conditions must apply when reference is made to equation (6.68) in the light of the knowledge that u must equal u_1 $(\equiv d_1)$ at $x = 0$ and u must equal u_2 $(\equiv d_2)$ at $x = l$. If elements with a greater number of nodes are considered in which only displacements (and not any displacement derivatives) are used as nodal freedoms it follows that a property of the typical shape function

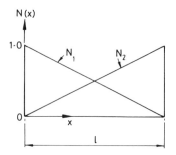

Fig. 6.4. Linear shape functions for the basic bar element.

N_i is that it has unit value at the nodal point where the displacement is d_i and zero value at all other nodal points.

Turning to the basic beam element (Fig. 6.2) there are four nodal freedoms which are

$$\{d_1 \quad d_2 \quad d_3 \quad d_4\} \equiv \{v_1 \quad \theta_{z1} \quad v_2 \quad \theta_{z2}\}$$

where $\theta_z = dv/dx$. From equation (6.24) the corresponding four cubic shape functions are

$$N_1(x) = 1 - 3\bar{x}^2 + 2\bar{x}^3 \qquad N_2(x) = l(\bar{x} - 2\bar{x}^2 + \bar{x}^3)$$
$$N_3(x) = 3\bar{x}^2 - 2\bar{x}^3 \qquad N_4(x) = l(-\bar{x}^2 + \bar{x}^3) \tag{6.70}$$

These shape functions are illustrated in Fig. 6.5. N_1 (associated with v_1) has unit value at node 1, zero value at node 2, and zero slope (first derivative, dv/dx) at nodes 1 and 2. N_2 (associated with θ_{z1}) has zero value at nodes 1 and 2, unit slope at node 1, and zero slope at node 2. The properties of shape functions N_3 and N_4 can be similarly defined.

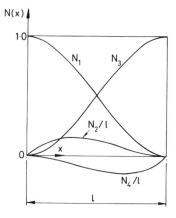

Fig. 6.5. Cubic shape functions for the basic beam element.

The simple shape functions defined in equations (6.69) and (6.70) could have been deduced *directly* from particular *interpolation formulae.* Such formulae occur in the realm of numerical analysis and enable the expression of a variable at any point throughout a certain region in terms of the values of the variable (and perhaps its derivatives) at a number of particular points in the region. For the one-dimensional elements the representation of the displacement in shape function terms corresponds precisely to the use of two particular interpolation formulae, namely the Lagrangian formula (for the bar) and the Hermitian formula (for the beam).

In one-dimensional *Lagrangian* interpolation a variable, y say, is assumed to have known values $y_1, y_2 \ldots y_p$ at points $1, 2 \ldots p$ in a region of interest (see Fig. 6.6). The points need not be spaced at uniform intervals; they are located at co-ordinate positions $x_1, x_2 \ldots x_p$. The interpolation formula fits a polynomial curve of order $p-1$ through all these particular points. The formula expresses the value of the variable y at any location x in the region as

$$y = l_1 y_1 + l_2 y_2 + \ldots l_p y_p = \sum_{i=1}^{p} l_i y_i \tag{6.71}$$

where the l_i are polynomial functions of x and are called Lagrange interpolation polynomials, each of which is of order $p-1$. The polynomials are defined as

$$l_1 = \frac{(x-x_2)(x-x_3)\ldots(x-x_p)}{(x_1-x_2)(x_1-x_3)\ldots(x_1-x_p)}$$

$$l_2 = \frac{(x-x_1)(x-x_3)\ldots(x-x_p)}{(x_2-x_1)(x_2-x_3)\ldots(x_2-x_p)} \tag{6.72}$$

$$\vdots$$

$$l_p = \frac{(x-x_1)(x-x_2)\ldots(x-x_{p-1})}{(x_p-x_1)(x_p-x_2)\ldots(x_p-x_{p-1})}.$$

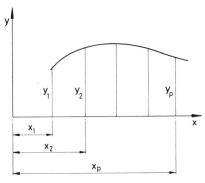

Fig. 6.6. One-dimensional Lagrangian interpolation.

It can be seen that a Lagrange interpolation polynomial l_i has the value unity at $x = x_i$ and the value zero at all other of the p particular points. It is thus defined in precisely the same fashion as is the bar-type shape function discussed above and we conclude that Lagrange interpolation polynomials can be directly used as shape functions in that category of finite elements where displacement values alone are used as nodal freedoms, i.e. in problems requiring only C_0 continuity. The p particular points correspond to the element nodes, y corresponds to the displacement, and the l_i correspond to the shape functions. For the particular case when $p = 2$, when the origin of x is at point 1 and when the interval between points 1 and 2 is l, the expressions for l_1 and l_2 from equation (6.72) are identical with those given in equations (6.69). For other cases, when $p > 2$, the Lagrange interpolation polynomial given in equation (6.72) provide ready made shape functions for the generation of the stiffness properties of refined bar elements (see Section 6.10). Further, products of the polynomials can also be used in two- and three-dimensional finite-element analysis, as will be seen in later chapters.

In one-dimensional *Hermitian* interpolation the variable y *and one or more of its derivatives* $(\mathrm{d}y/\mathrm{d}x, \mathrm{d}^2y/\mathrm{d}x^2 \ldots)$ are assumed to have known values at p particular points (the nodes), located at co-ordinate positions $x_1, x_2 \ldots x_p$, in the region of interest (the element). If values of the first r derivatives of y are known at the points, the variable y is expressed as

$$y = \sum_{i=1}^{p} \left(H_{0i}^r y_i + H_{1i}^r \left(\frac{\mathrm{d}y}{\mathrm{d}x} \right)_i + \ldots H_{ri}^r \left(\frac{\mathrm{d}^r y}{\mathrm{d}x^r} \right)_i \right). \tag{6.73}$$

Here the H_{ji}^r functions (of x) are Hermite polynomials. These polynomials are referred to as rth-level Hermite polynomials; thus H_{ji}^1 is a first-level Hermite polynomial because values of the variable and its first derivative are involved, H_{ji}^2 is called a second-level Hermite polynomial because values of the second derivative of the variable are also involved, etc. The first of the two subscripts associated with each H polynomial, that is j, gives the order of the derivative of y associated with that polynomial whilst the second, that is i, indicates the point number. The total number of terms involved in the expression for y is $p(r+1)$ and correspondingly y is a polynomial function of order $p(r+1)-1$. The Hermite polynomials have the property that

$$\frac{\mathrm{d}^k (H_{ji}^r)}{\mathrm{d}x^k} = 1 \qquad \text{for} \qquad k = j \text{ and } x = x_i,$$

$$\frac{\mathrm{d}^k (H_{ji}^r)}{\mathrm{d}x^k} = 0 \qquad \text{for} \qquad k \neq j \text{ or } x \neq x_i. \tag{6.74}$$

There are many different combinations of values of p and r that could be considered, and rather complicated formulae exist from which the corresponding definitions of the Hermite polynomials can be obtained. Here, though, only the results for two cases will be mentioned. In the first of these, if $p = 2$ and $r = 1$ we have an interpolation scheme suitable for an element with two (end) nodes using values of the variable and its first derivative at each node. The corresponding first-level Hermite polynomials are designated H_{01}^1, H_{11}^1, H_{02}^1, and H_{12}^1 and are then precisely the cubic functions N_1, N_2, N_3, and N_4 respectively, given by equations (6.70) for the basic beam element. In the second, if $p = 2$ and $r = 2$ we again have only two nodes, but now include the second derivative of the variable (or the negative of the curvature in the beam problem) at the nodes. The relevant second-level Hermite polynomials are quintic functions of x, defined as (for element length l, x co-ordinate origin at left-hand end)

$$H_{01}^2 = 1 - 10\bar{x}^3 + 15\bar{x}^4 - 6\bar{x}^5$$
$$H_{11}^2 = l(\bar{x} - 6\bar{x}^3 + 8\bar{x}^4 - 3\bar{x}^5)$$
$$H_{21}^2 = \frac{l^2}{2}(\bar{x}^2 - 3\bar{x}^3 + 3\bar{x}^4 - \bar{x}^5)$$
$$H_{02}^2 = 10\bar{x}^3 - 15\bar{x}^4 + 6\bar{x}^5 \qquad (6.75)$$
$$H_{12}^2 = l(-4\bar{x}^3 + 7\bar{x}^4 - 3\bar{x}^5)$$
$$H_{22}^2 = \frac{l^2}{2}(\bar{x}^3 - 2\bar{x}^4 + \bar{x}^5).$$

These polynomials are illustrated in Fig. 6.7. The Hermite polynomials

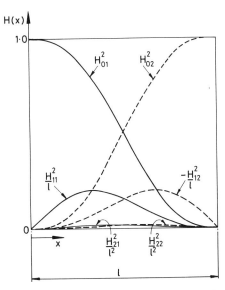

Fig. 6.7. Second-level Hermite polynomial functions.

provide shape functions which are appropriate for use in C_1 continuity problems (i.e. beam and plate problems) but can also be used in certain circumstances in C_0 continuity problems, as will be seen in the next section.

As well as using interpolation formulae to write down element shape functions directly, it should be emphasized that such shape functions can also be devised directly by insight, intuition, or trial. The direct shape function approach has considerable advantages in some important areas of two- and three-dimensional finite-element analysis.

6.10. Refined one-dimensional elements

The formulations of the basic bar and beam elements presented earlier in this chapter are 'exact' so long as the elements are of uniform cross-section and so long as no distributed loading is present. The elements are termed 'basic' because their displacement fields and their associated nodal freedoms are of the simplest form which is consistent with the satisfaction of the conditions governing the selection of element displacement fields. There is nothing to prevent expanding these basic displacement fields by adding extra polynomial terms so as to increase the level of sophistication of the representation of displacement and stress within the element. Elements in which this is done will be referred to as 'refined' or 'high-order' elements and will be of use in dealing with non-uniform geometry and distributed loading. Of course, if the number of polynomial terms in an assumed displacement field is increased then the number of element degrees of freedom must be correspondingly increased. This is a disadvantage but the expectation would be that an assemblage of few refined elements would yield more accurate results than an assemblage of a greater number of basic elements, when the total number of structure degrees of freedom involved is the same. Once the choice of refined displacement field is made and the nodal freedoms are allocated, the derivation of element stiffness and consistent load matrices follows the standard procedure indicated by equations (6.38a) or (6.38b) for stiffness and equations (6.45) or (6.48) for consistent load.

As a first level of refinement for a bar element we consider adopting a quadratic displacement field (and correspondingly linear strain variation), i.e. in terms of generalized displacements we assume the displacement field

$$u = A_0 + A_1 x + A_2 x^2. \tag{6.76}$$

In addition to using values of u at the bar ends (i.e. at the *external* nodes) as element degrees of freedom, as in the basic element, a third degree of freedom must now be specified. The obvious way to do this is to

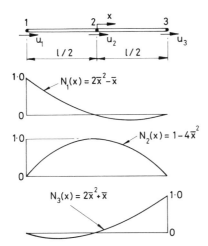

Fig. 6.8. The three-node bar element and its quadratic shape functions.

introduce a third node in the interior of the bar to which the value of u is the nodal freedom. This *internal* node can be located anywhere between the bar ends but the mid-point is the logical choice and thus the three-degrees-of-freedom refined bar element is as shown in Fig. 6.8. To express the displacement at any point in terms of the nodal displacements we can proceed to derive the $\mathbf{C}$ matrix and hence obtain the form $u = \boldsymbol{\alpha}\mathbf{C}^{-1}\mathbf{d}$. Alternatively, the Lagrange interpolation formula, equations (6.71) and (6.72), allows the direct specification of the appropriate shape functions occurring in the form $u = \mathbf{Nd}$. These shape functions are shown in Fig. 6.8. (It should be noted that here the co-ordinate origin is chosen to be at the beam centre but could be elsewhere, of course.)

The next level of refinement for the bar is based on the assumption of a cubic displacement field (and quadratic strain) which can be expressed in terms of four generalized displacements as $u = A_0 + A_1 x + A_2 x^2 + A_3 x^3$. Two extra degrees of freedom are now involved, as compared with the basic bar element. Here there are two convenient ways in which the four degrees of freedom of the cubic element can be arranged, each incorporating the values of u at the external element nodes of course. These alternative arrangements are illustrated in Fig. 6.9 and comprise

    (a) a model with four (equispaced) nodes, with values of u alone as degrees of freedom

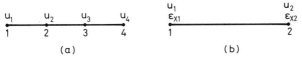

Fig. 6.9. Two versions of the cubic-displacement bar element: (a) with four nodes; (b) with two nodes.

(b) a model with end nodes only, with values of u and du/dx (i.e. the strain ε_x) as degrees of freedom.

Proceeding either indirectly via the use of generalized displacements or directly from interpolation formulae each displacement field can be expressed in terms of the nodal freedoms. In the direct or shape function approach, the functions for model (a) are again given by the Lagrange interpolation formula whilst those for model (b)—involving the displacement and its first derivative—will be recognized as being the Hermite polynomials already employed in generating the properties of the basic beam element (see Fig. 6.5). A word of caution is appropriate here concerning model (b). In this refined element the strain du/dx is used as a degree of freedom at the external nodes and normally at interelement junctions all nodal freedoms of adjoining elements are connected. Thus the axial strain values of neighbouring elements are constrained to be equal at their common node. If the bar problem under consideration is one in which the axial strain varies smoothly and continuously there is no difficulty. However, applications can readily be envisaged in which step changes in strain occur quite naturally and properly, as when a concentrated load is applied at some point along the bar, say, or when a step change in cross-section occurs. In these circumstances, if such a point is the junction between two model (b) elements there is an obvious conflict with the physical requirements of the problem if the strain values of the two elements are equated at this point. The elements are said to be 'over-connected' or 'connected with excess continuity'. The situation only arises, of course, because in this model we are using as degrees of freedom a higher displacement derivative than is strictly needed to satisfy the requirements of the potential energy formulation.

The process of refinement for bar elements could be continued to embrace higher-order polynomials but, in practice, the cubic displacement level is probably as far as we would go.

Turning to the consideration of beam elements we recall that the basic element has a cubic displacement field and that the compatibility condition requires the matching of displacement and its first derivative at interelement junctions. Various refined beam elements are possible but here only two models will be mentioned, both of which are based on the assumption of the quintic displacement field (and hence cubic moment variation)

$$v = A_0 + A_1 x + A_2 x^2 + A_3 x^3 + A_4 x^4 + A_5 x^5. \tag{6.77}$$

As with the cubic bar element, two versions of the quintic beam element come readily to mind. These are illustrated in Fig. 6.10 and each has six

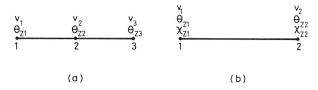

Fig. 6.10. Two versions of the qunitic-displacement beam element: (a) with three nodes; (b) with two nodes.

degrees of freedom of course: they are

(a) a model with three nodes, with v and θ_z ($=dv/dx$) as nodal freedoms
(b) a model with two nodes, with v, θ_z, and χ_z ($=-d^2v/dx^2$) (the curvature) as nodal freedoms.

For these models the $\mathbf{N}$ matrix can again be derived indirectly as $\boldsymbol{\alpha}\mathbf{C}^{-1}$ but the individual shape functions can also be obtained directly from the Hermite polynomials. For model (b), for instance, the shape functions correspond to the Hermite polynomials of equations (6.75), shown in Fig. 6.7. With regard to this model it is clear that its use imposes excess continuity in an assemblage of elements and this will give problems in particular circumstances where curvature should not be continuous (such as at a step change in cross-section or at the point of application of a concentrated couple). In other circumstances, though, the model—like model (a)—will give highly accurate results. In particular it should be clear that the beam models based on the quintic displacement field will give the exact solution for displacement, bending moment, and shearing force for a beam of uniform section carrying a loading of uniform or linearly varying intensity along its length; thus, use of a single quintic element would give the exact solution to Example 6.3.

Example 6.4.
(a) Determine the stiffness matrix and the stress matrix for the three-node quadratic displacement) bar element shown in Fig. 6.8.
(b) Also evaluate the consistent load column matrix $\mathbf{Q}$ corresponding to a uniformly distributed axial loading (acting in the positive x direction) along the whole element length.

(a) The stiffness matrix for the bar element could be evaluated using either form of equation (6.38), as convenient. The displacement field is given in generalized displacement form by equation (6.76) whilst the shape functions are recorded in Fig. 6.8. Since the shape functions are available the stiffness matrix will be evaluated using equation (6.38a); thus, with the cross-section assumed to be uniform (and noting that the co-ordinate origin is now at the beam centre),

$$\mathbf{k} = AE \int_{-l/2}^{l/2} \mathbf{B^t B} \, dx. \tag{a}$$

The displacement field is (with $\bar{x} = x/l$)

$$u = [2\bar{x}^2 - \bar{x} \quad 1 - 4\bar{x}^2 \quad 2\bar{x}^2 + \bar{x}] \begin{Bmatrix} u_1 \\ u_2 \\ u_3 \end{Bmatrix} = \mathbf{Nd}.$$

Matrix $\mathbf{B}$, relating the axial strain $\varepsilon_x = du/dx$ to the nodal displacements, is equal to $d\mathbf{N}/dx$ (see equation (6.6)). Therefore

$$\mathbf{B} = \frac{d}{dx}[2\bar{x}^2 - \bar{x} \quad 1 - 4\bar{x}^2 \quad 2\bar{x}^2 + \bar{x}] = \frac{1}{l}[4\bar{x} - 1 \quad -8\bar{x} \quad 4\bar{x} + 1] \tag{b}$$

Substituting equation (b) into (a) gives

$$\mathbf{k} = \frac{AE}{l^2} \int_{-l/2}^{l/2} \begin{Bmatrix} 4\bar{x} - 1 \\ -8\bar{x} \\ 4\bar{x} + 1 \end{Bmatrix} [4\bar{x} - 1 \quad -8\bar{x} \quad 4\bar{x} + 1] \, dx$$

$$= \frac{AE}{l^2} \int_{-l/2}^{l/2} \begin{bmatrix} 16\bar{x}^2 - 8\bar{x} + 1 & & \text{Symmetric} \\ -32\bar{x}^2 + 8\bar{x} & 64\bar{x}^2 & \\ 16\bar{x}^2 - 1 & -32\bar{x}^2 - 8\bar{x} & 16\bar{x}^2 + 8\bar{x} + 1 \end{bmatrix} dx$$

and performing the integration gives the element stiffness matrix as

$$\mathbf{k} = \frac{AE}{3l} \begin{matrix} u_1 & u_2 & u_3 \\ \begin{bmatrix} 7 & -8 & 1 \\ -8 & 16 & -8 \\ 1 & -8 & 7 \end{bmatrix} \end{matrix}. \tag{c}$$

The stress matrix $\mathbf{S}$ is simply $\mathbf{S} = E\mathbf{B}$ with $\mathbf{B}$ given by equation (b).

(b) The consistent load column matrix for a uniformly distributed loading of p per unit length is (see equation (6.45))

$$\mathbf{Q} = p \int_{-l/2}^{l/2} \mathbf{N^t} \, dx = p \int_{-l/2}^{l/2} \begin{Bmatrix} 2\bar{x}^2 - \bar{x} \\ 1 - 4\bar{x}^2 \\ 2\bar{x}^2 + \bar{x} \end{Bmatrix} dx = \frac{pl}{6} \begin{Bmatrix} 1 \\ 4 \\ 1 \end{Bmatrix}. \tag{d}$$

Thus four-sixths of the loading is allocated to the centre node and one-sixth to each end.

Example 6.5. A uniform bar, held at one end as shown in Fig. E6.5(a), is subjected to an axial force of p/unit length where p is a constant. Determine the distribution of displacement and stress along the bar when the bar is modelled with

(i) One linear-displacement (i.e. basic) element
(ii) Two linear-displacement elements of equal length
(iii) One quadratic-displacement element.

For the basic linear-displacement element the stiffness matrix is given by

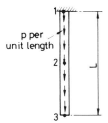

Fig. E6.5(a)

equation (6.12) and the column matrix $\mathbf{Q}$ corresponding to uniformly distributed loading has been evaluated in Example 6.1. The stress matrix $\mathbf{S} = E\mathbf{B}$ (see equations (6.64)–(6.66)) where $\mathbf{B}$ is given in equation (6.6). Thus, in summary for the linear displacement element

$$\mathbf{k} = \frac{AE}{l}\begin{bmatrix} 1 & -1 \\ -1 & 1 \end{bmatrix} \qquad \mathbf{Q} = pl\begin{Bmatrix} \frac{1}{2} \\ \frac{1}{2} \end{Bmatrix} \qquad \mathbf{S} = E[-1/l \quad 1/l]\mathbf{d}.$$

The structure stiffness equations will be of the form of equation 6.58(b), but note that $\mathbf{F}_r = \mathbf{0}$ since no applied forces actually act directly at a node.

For modelling (i) only u_3 is non-zero (there is no node at point 2) and the single stiffness equation is

$$\frac{pL}{2} = \frac{AE}{L} 1\, u_3 \qquad \text{or} \qquad u_3 = \frac{pL^2}{2AE}.$$

The uniform stress is

$$\sigma_x = E\left(\frac{1}{L}\right) u_3 = \frac{pL}{2A}.$$

For modelling (ii), u_2 and u_3 are non-zero and the equations are (now $l = L/2$)

$$\begin{Bmatrix} pL/4 + pL/4 \\ pL/4 \end{Bmatrix} = \frac{AE}{L/2}\begin{bmatrix} 1+1 & -1 \\ -1 & 1 \end{bmatrix}\begin{Bmatrix} u_2 \\ u_3 \end{Bmatrix}.$$

Thus

$$u_2 = \frac{3}{8}\frac{pL}{AE} \qquad u_3 = \frac{pL^2}{2AE}.$$

The stress (uniform for each element of course) is

$$E\left(\frac{2}{L}\right) u_2 = \frac{3}{4}\frac{pL}{A}$$

for element 1–2, and

$$E\begin{bmatrix} -\dfrac{2}{L} & \dfrac{2}{L} \end{bmatrix}\begin{Bmatrix} u_2 \\ u_3 \end{Bmatrix} = \dfrac{pL}{4A}$$

for element 2–3

For the quadratic displacement element, with three nodes, the matrices **k**, **Q**, and **S** have been determined in Example 6.4. Modelling (iii) is simply one of these elements representing the complete bar 1–2–3 and the structure stiffness equations are therefore, using equations (c) and (d) of Example 6.4,

$$p\dfrac{L}{6}\begin{Bmatrix} 4 \\ 1 \end{Bmatrix} = \dfrac{AE}{3L}\begin{bmatrix} 16 & -8 \\ -8 & 7 \end{bmatrix}\begin{Bmatrix} u_2 \\ u_3 \end{Bmatrix}.$$

Thus

$$u_2 = \dfrac{3pL^2}{8AE} \qquad u_3 = \dfrac{pL^2}{2AE}.$$

The stress at any point along the element is

$$\sigma_x = \mathbf{Sd} = E\dfrac{1}{L}\begin{bmatrix} 4\bar{x} - 1 & -8\bar{x} & 4\bar{x} + 1 \end{bmatrix}\begin{Bmatrix} 0 \\ 3/8 \\ 1/2 \end{Bmatrix}\dfrac{pL^2}{AE}.$$

This demonstrates a linear variation of stress from zero at the free end of the bar ($\bar{x} = 1/2$) to pL/A at the fixed end ($\bar{x} = -1/2$).

The results of the three finite-element analyses are shown graphically in Fig. E6.5(b) for displacement and stress. Modelling (iii), that is the single three-noded element, gives the exact solution everywhere. Modellings (i) and (ii) give the exact displacement values at the node points but are in error elsewhere. With regard to these two modellings it is seen that the finite-element stress is necessarily uniform throughout each element but that it has the exact value at the centre of the elements; also in the two-element representation the average of the stress values in the two elements meeting at node 2 is equal to the exact stress value at this point.

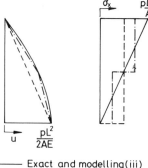

——— Exact and modelling(iii)
– – – – Modelling (i)
·—·—· Modelling (ii)

Fig. E6.5(b)

The reader will no doubt recognize that the problem considered here is exactly that considered using the Rayleigh–Ritz approach in Examples 5.4 and 5.8, and that finite–element modellings (i), (ii), and (iii) are effectively equivalent to the analyses of Examples 5.4(i), 5.4(ii) and 5.8 respectively.

6.11. Static condensation

It sometimes happens in structural analysis that it is desirable, or indeed necessary, to reduce the size of a system of stiffness equations by removing from consideration certain of the degrees of freedom. This can be achieved by a process known as static condensation (or simply condensation). The process has its uses both at the element level and at the structure level.

Static condensation at the element level is frequently employed for elements which have one or more nodes which are completely internal to the element, i.e. nodes of an element at which no connection whatsoever is made to any other elements. We have already seen examples of such elements, of the bar and beam type in Section 6.10 and other examples occur quite commonly in two- and three-dimensional analysis as will be seen in later chapters. The object of static condensation at the element level is to produce a condensed set of stiffness equations expressed in terms of those degrees of freedom at the external nodes of the element only; the procedure is not essential but does make for a more convenient analysis.

For an element having an internal node or nodes it will be possible to write the complete set of element stiffness equations in the partitioned form

$$\begin{Bmatrix} \mathbf{P}_E \\ \mathbf{P}_I \end{Bmatrix} = \begin{bmatrix} \mathbf{k}_{EE} & \mathbf{k}_{EI} \\ \mathbf{k}_{IE} & \mathbf{k}_{II} \end{bmatrix} \begin{Bmatrix} \mathbf{d}_E \\ \mathbf{d}_I \end{Bmatrix}. \tag{6.78}$$

Here the subscripts E and I are used to denote quantities related to external and internal degrees of freedom respectively. From the lower set of equation (6.78),

$$\mathbf{P}_I = \mathbf{k}_{IE}\mathbf{d}_E + \mathbf{k}_{II}\mathbf{d}_I,$$

we can obtain an expression for the internal degrees of freedom $\mathbf{d}_I$ in terms of the external degrees of freedom. Thus

$$\mathbf{d}_I = \mathbf{k}_{II}^{-1}(\mathbf{P}_I - \mathbf{k}_{IE}\mathbf{d}_E). \tag{6.79}$$

Substituting for $\mathbf{d}_I$ in the upper set of equations (6.78) gives

$$\mathbf{P}_E = \mathbf{k}_{EE}\mathbf{d}_E + \mathbf{k}_{EI}\mathbf{k}_{II}^{-1}(\mathbf{P}_I - \mathbf{k}_{IE}\mathbf{d}_E).$$

This can be written

$$\bar{\bar{\mathbf{P}}} = \bar{\bar{\mathbf{k}}}\mathbf{d}_E \tag{6.80}$$

where

$$\bar{\bar{\mathbf{P}}} = \mathbf{P}_E - \mathbf{k}_{EI}\mathbf{k}_{II}^{-1}\mathbf{P}_I \qquad (6.81)$$

and

$$\bar{\bar{\mathbf{k}}} = \mathbf{k}_{EE} - \mathbf{k}_{EI}\mathbf{k}_{II}^{-1}\mathbf{k}_{IE}. \qquad (6.82)$$

Equations (6.80)–(6.82) are the condensed set of element stiffness equations, appropriate to external degrees of freedom only. The condensed element stiffness matrix $\bar{\bar{\mathbf{k}}}$ can be used in an element assemblage in exactly the same way as any other element stiffness matrix. The column matrix of nodal forces $\bar{\bar{\mathbf{P}}}$ corresponding to the displacements $\mathbf{d}_E$ incorporates the possible presence of forces at the internal nodes. In regard to this it is clear that since the internal nodes of an element are unconnected to any other element the only forces that can be present acting on these nodes are those corresponding to externally applied loadings. Thus $\mathbf{P}_I \equiv \mathbf{F}_I + \mathbf{Q}_I$, in fact. Finally it should perhaps be emphasized that although displacements $\mathbf{d}_I$ do not appear in equations (6.80)–(6.82) they have not really been lost in the analysis procedure since they can be calculated using equation (6.79) once the displacements $\mathbf{d}_E$ are known.

Turning to the structure level, the above procedure could clearly be used on the assembled structure stiffness equations to condense these to relate effectively only to a limited number of degrees of freedom. However, the solution procedure will, in fact, require at least as many operations as would the solution of the original set of equations. Consequently this procedure is not generally recommended. There are circumstances, though, where procedures based on the concept of static condensation are necessary: these occur where the full set of structure equations is simply too large to fit into the available computer storage. Then the structure can be divided into two or more substructures and condensed equations of the form given (for an element) by equations (6.80)–(6.82) can be generated for each substructure, these equations pertaining to degrees of freedom at the boundaries of the substructures. The substructures can be effectively connected together to form the complete structure by matching nodal forces and displacements at the boundaries of adjacent substructures in accordance with conditions of equilibrium and compatibility. The procedure is commonly referred to as the method of substructures or sometimes as the method of tearing or diakoptics and is used, where storage resources are otherwise inadequate, in both MDM and FEM work; its development will not be pursued further here.

6.12. Inclusion of initial strain effects

Thus far in this chapter only loading of a mechanical nature has been considered, that is the loading has comprised external dead loading

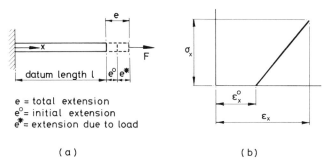

e = total extension
e⁰ = initial extension
e* = extension due to load

(a)   (b)

Fig. 6.11. Bar with initial strain: (a) the bar; (b) stress–strain relationship.

applied either directly at the nodes or along the elements. This is much the most common form of loading but, as has already been seen in Section 4.3, other effects can cause deformation and stressing of a structure. In Section 4.3 the effects of temperature change and lack of fit were incorporated into the MDM approach using a superposition procedure. Now these effects will be included in the PMPE approach in a way which will allow extension to general structures, through the use of the concept of 'initial strain'.

By way of introduction of the concept, consider the simple case of a bar held at one end as shown in Fig. 6.11(a). Such a bar was considered in Section 5.2 where the effect of an applied axial force (alone) was examined. Now imagine that prior to the application of any force the bar is initially strained with reference to its datum length l. This would arise if, say, the bar were heated from a datum temperature (at which its length is l) to give the extension e^0. No stress is associated with this change in length but the strain existing in the bar before the axial load is applied— that is the initial strain—has a value ε_x^0. When the axial load is gradually applied up to a magnitude F a further extension e^* is present and the total extension is $e^0 + e^*$. Correspondingly the total strain in the x direction is

$$\varepsilon_x = \varepsilon_x^0 + \varepsilon_x^* \tag{6.83}$$

where ε_x^* is the strain due to the applied load. The bar (total) stress σ_x is due to ε_x^* alone and is

$$\sigma_x = E\varepsilon_x^* = E(\varepsilon_x - \varepsilon_x^0). \tag{6.84}$$

The plot of total stress σ_x versus total strain ε_x is shown in Fig. 6.11(b) where linear elastic behaviour is assumed.

The strain energy per unit volume in the bar at load level F is the area under the line in Fig. 6.11(b), i.e.

$$\nu_p = \frac{\sigma_x(\varepsilon_x - \varepsilon_x^0)}{2}$$

or using equation (6.84)

$$\nu_{\rm p} = \frac{E(\varepsilon_x - \varepsilon_x^0)^2}{2} \tag{6.85}$$

It is noted that the normal strain-displacement equation holds provided that the total strain and total displacement (measured from the datum position) are used, i.e. $\varepsilon_x = {\rm d}u/{\rm d}x$ where u is the displacement at any section x.

Consider now a bar finite element of the type described in Section 6.2. We wish to augment the analysis detailed there to include the effect of an initial strain ε_x^0; this effect is accommodated in the strain energy expression, as has just been indicated. With equation (6.85) in mind the strain energy of the bar element, previously given by equation (6.7), now becomes

$$U_{\rm p} = \tfrac{1}{2} \int_{V_e} (\varepsilon_x - \varepsilon_x^0)^{\rm t} E(\varepsilon_x - \varepsilon_x^0) \, {\rm d}V_e. \tag{6.86}$$

Now $\varepsilon_x = \mathbf{Bd}$ (equation (6.6)) and so $U_{\rm p}$ becomes

$$U_{\rm p} = \tfrac{1}{2}\mathbf{d}^{\rm t}\left(\int_{V_e} \mathbf{B}^{\rm t}E\mathbf{B} \, {\rm d}V_e\right)\mathbf{d} - \mathbf{d}^{\rm t}\int_{V_e} \mathbf{B}^{\rm t}E\varepsilon_x^0 \, {\rm d}V_e + \tfrac{1}{2}\int_{V_e} \varepsilon_x^{0{\rm t}}E\varepsilon_x^0 \, {\rm d}V_e. \tag{6.87}$$

With the potential energy of the element nodal forces given by equation (6.14) the total potential energy of the element is (akin to equation (6.15))

$$\Pi_{\rm p} = \tfrac{1}{2}\mathbf{d}^{\rm t}\mathbf{kd} - \mathbf{d}^{\rm t}\mathbf{P} - \mathbf{d}^{\rm t}\mathbf{Q}^0 + \tfrac{1}{2}\int_{V_e} \varepsilon_x^{0{\rm t}}E\varepsilon_x^0 \, {\rm d}V_e. \tag{6.88}$$

Here $\mathbf{k}$ is the usual element stiffness matrix (defined by equation (6.9)) and

$$\mathbf{Q}^0 = \int_{V_e} \mathbf{B}^{\rm t}E\varepsilon_x^0 \, {\rm d}V_e \tag{6.89}$$

is the new column matrix of consistent nodal loads for the element due to the effect of initial strain. The final term on the right-hand side of equation (6.88) does not depend upon $\mathbf{d}$ and hence has no effect when the minimization procedure $\partial\Pi_{\rm p}/\partial d_i = 0$ is applied. This procedure yields

$$\mathbf{P} + \mathbf{Q}^0 = \mathbf{kd} \tag{6.90}$$

which is the set of element stiffness equations when initial strains are present. (This compares with equation (6.47) for the distributed mechanical loading case.)

For the uniform basic bar element, with $\mathbf{B}$ defined in equation (6.6), the column matrix $\mathbf{Q}^0$ (equation (6.89)) becomes

$$\mathbf{Q}^0 = AE \int_0^l \mathbf{B}^t \varepsilon_x^0 \, dx = AE \int_0^l \left\{ \begin{matrix} -1/l \\ 1/l \end{matrix} \right\} \varepsilon_x^0 \, dx. \qquad (6.91)$$

As has been mentioned the source of the initial strain may be (amongst other possibilities) a temperature change or a lack of fit. In the former case $\varepsilon_x^0 = \alpha T$, T being a temperature *increase* (assumed uniform); in the latter case $\varepsilon_x^0 = e/l$ where e is the amount by which the element is too *long*. With ε_x^0 a constant value along the element length we have

$$\mathbf{Q}^0 = AE\varepsilon_x^0 \left\{ \begin{matrix} -1 \\ 1 \end{matrix} \right\}. \qquad (6.92)$$

It should be noted that the element stress σ_x can be calculated from the nodal displacements by using the expression (from equation (6.84))

$$\sigma_x = E(\mathbf{B}\mathbf{d} - \varepsilon_x^0). \qquad (6.93)$$

Turning now to a basic beam element of the type described in Section 6.3 we have by direct analogy with the bar case that

$$\mathbf{Q}^0 = EI_z \int_0^l \mathbf{B}^t \chi_z^0 \, dx \qquad (6.94)$$

where $\mathbf{B}$ is defined in equation (6.29)). The initial pseudo-strain χ_z^0 is
  (a) $\alpha(T_1 - T_2)/d$ for the case of a thermal gradient through the beam depth
  (b) $8e/l^2$ for the case of initial out of straightness.
These cases are shown in Table 4.2 wherein T_1, T_2, and e are effectively defined. Also by direct analogy with the bar case, the moment at any point along the beam is given by

$$M_z = EI_z(\mathbf{B}\mathbf{d} - \chi_z^0). \qquad (6.95)$$

Obviously a particular application may involve the effects of both initial strain and distributed mechanical loading, the latter having been considered in Section 6.5. Where this is so the element stiffness equations, by simple combination of equations (6.47) and (6.90), are clearly

$$\mathbf{P} + \mathbf{Q} + \mathbf{Q}^0 = \mathbf{k}\mathbf{d} \qquad (6.96)$$

The above has been concerned with element relationships. Structure stiffness equations for problems involving initial strain effects are, of course, obtained via the same sort of direct assembly procedure as detailed in Section 6.6. If indeed the effects of both initial strain and distributed loading are involved the structure stiffness equations (before application of the boundary conditions) are by simple extension of equation (6.58a)

$$\mathbf{F} + \mathbf{F}_Q + \mathbf{F}_{Q^o} = \mathbf{KD} \qquad (6.97)$$

where the definition of $\mathbf{F}_{Q^o}$, the column matrix of consistent nodal loads for the structure due to initial strain effects, is obvious by analogy with that for $\mathbf{F}_Q$ given by equation (6.60).

The procedure described in this section for including the effects of temperature change and lack of fit as initial strain effects in a PMPE approach will yield exactly the same results as those obtained by the more physical approach of Section 4.3 when the problem is concerned with prismatic basic bar and beam elements having temperature distributions not varying along the element length. This is verified in Example 6.6. However, the more mathematical approach detailed here is much more general, being applicable to all types of structure and to complicated spatial distributions of initial strain.

Example 6.6. Repeat Example 4.4 using the initial strain approach of Section 6.12.

Details of the structure and of the elements used to model it are given in Example 3.2 whilst the applied temperature changes are detailed in Example 4.4.

The only element subjected to temperature change is element 2–4. This element is a combined bar–beam element and initial strain effects are present in the axial sense, owing to a rise in temperature at the centroidal (neutral) axis, and in the bending sense, owing to a temperature differential through the beam depth. In the axial sense $\varepsilon_x^0 = \alpha T$, where $T = (40 + 10)/2 = 25\,°C$, and therefore using equation (6.92)

$$\mathbf{Q}_{axial}^0 = AE\varepsilon_x^0 \begin{Bmatrix} -1 \\ 1 \end{Bmatrix} = 60 \times 10^4 \times 10^{-5} \times 25 \begin{Bmatrix} -1 \\ 1 \end{Bmatrix} = \begin{Bmatrix} -150 \\ 150 \end{Bmatrix}\,kN.$$

In the bending sense $\chi_z^0 = \alpha(T_1 - T_2)/d$ and substituting for $\mathbf{B}$ (equation (6.29)) in equation (6.94) gives

$$\mathbf{Q}_{bending}^0 = EI_z \frac{\alpha(T_1 - T_2)}{d} \int_0^l \begin{Bmatrix} 6/l^2 - 12x/l^3 \\ 4/l - 6x/l^2 \\ -6/l^2 + 12x/l^3 \\ 2/l - 6x/l^2 \end{Bmatrix} dx$$

$$= EI_z \frac{\alpha(T_1 - T_2)}{d} \begin{Bmatrix} 0 \\ 1 \\ 0 \\ -1 \end{Bmatrix} = 1.25 \times 10^4 \times 10^{-5} \frac{(40 - 10)}{0.2} \begin{Bmatrix} 0 \\ 1 \\ 0 \\ -1 \end{Bmatrix}$$

$$= \begin{Bmatrix} 0 \\ 18.75\ kN\,m \\ 0 \\ -18.75\ kN\,m \end{Bmatrix}.$$

Thus the total $\mathbf{Q}^0$ for element 2–4 in its local reference frame is

$$\mathbf{Q}^0 = \{-150 \quad 0 \quad 18.75 \quad 150 \quad 0 \quad -18.75\}.$$

This must be transformed to the global system and this is easily achieved in the standard way ($\bar{\mathbf{Q}}^0 = \boldsymbol{\tau}^t \mathbf{Q}^0$) to give

$$\bar{\mathbf{Q}}^0 = \{90 \quad 120 \quad 18.75 \quad -90 \quad -120 \quad -18.75\}.$$

It should be noted that

$$\boldsymbol{\tau} = \begin{bmatrix} \mathbf{T} & \mathbf{0} \\ \mathbf{0} & \mathbf{T} \end{bmatrix}$$

where

$$\mathbf{T} = \begin{bmatrix} -0.6 & -0.8 & 0 \\ 0.8 & -0.6 & 0 \\ 0 & 0 & 1 \end{bmatrix}$$

The structure system of equations concerns point 4 only and is of the form

$$\mathbf{F}_{Q_r^0} = \mathbf{K}_r \mathbf{D}_r$$

Here $\mathbf{F}_{Q_r^0}$ is simply the last three entries of $\bar{\mathbf{Q}}^0$ whereas the stiffness matrix $\mathbf{K}_r$ has been recorded in Example 3.2. The structure system of equations therefore become here precisely those given by equations (a) of Example 4.4. Their solution, taken from Example 4.4, is

$$\mathbf{D}_r = \{\bar{u}_4 \quad \bar{v}_4 \quad \theta_{\bar{z}4}\} = \{-4.9990 \text{ m} \quad -7.8478 \text{ m} \quad -8.2261 \text{ rad}\} \times 10^{-4}.$$

To calculate axial force and bending moment distributions for element 2–4 we require column matrix $\mathbf{d}$ for this element and this is given by

$$\mathbf{d} = \boldsymbol{\tau} \bar{\mathbf{d}}$$

where

$$\bar{\mathbf{d}} = \{0 \quad 0 \quad 0 \quad \bar{u}_4 \quad \bar{v}_4 \quad \theta_{\bar{z}4}\}.$$

Thus

$$\mathbf{d} = \{u_2 \quad v_2 \quad \theta_{z2} \quad u_4 \quad v_4 \quad \theta_{z4}\} = \{0 \quad 0 \quad 0 \quad 9.2776 \quad 0.7095 \quad -8.2261\} \times 10^{-4}.$$

Then using equation (6.93) the (uniform) axial force in element 2–4 is

$$A\sigma_x = AE\left(\begin{bmatrix} -\dfrac{1}{l} & \dfrac{1}{l} \end{bmatrix}\begin{Bmatrix} u_2 \\ u_4 \end{Bmatrix} - \alpha T\right)$$

$$= 60 \times 10^4\left(\frac{0 + 9.2776 \times 10^{-4}}{5} - 10^{-5} \times 25\right) = -38.668 \text{ kN}$$

with the negative sign denoting a compressive force. Also using equation (6.94) the bending moment in element 2–4 is

$$M_z = 1.25 \times 10^4 \left(\left[\frac{6}{25} - \frac{12x}{125} \quad \frac{4}{5} - \frac{6x}{25} \quad \frac{-6}{25} + \frac{12x}{125} \quad \frac{2}{5} - \frac{6x}{25} \right] \right.$$

$$\left. \times \left\{ \begin{array}{c} 0 \\ 0 \\ 0.7095 \\ -8.2261 \end{array} \right\} \times 10^{-4} - \frac{10^{-5}(40-10)}{0.2} \right)$$

At points 2 (where $x = 0$) and 4 ($x = 5$) this gives the values of -23.076 kN m and -10.311 kN m respectively with linear variation between these points of course. The values of the axial force and bending moment calculated here agree, as expected, with those shown in Fig. E4.4(b). (The forces in the y direction at the element ends, shown in the figure, have not been calculated directly in the present approach but can easily be determined indirectly using conditions of moment equilibrium in conjunction with the calculated moment values).

Problems

6.1. For the basic (cubic) beam element of Section 6.3 determine the column matrix of consistent nodal loads $\mathbf{Q}$ for each of the following cases.
  (a) A couple M_0 applied at $x = 3l/4$.
  (b) A symmetric lateral load varying quadratically from zero at both ends to a maximum of p at the element centre.
  (c) A uniformly distributed lateral load of intensity p per unit length acting only over the range $l/4 \leq x \leq l/2$.

6.2. A uniform beam of length L and flexural rigidity EI_z is clamped at both ends and subjected to a uniformly distributed lateral load, over its whole length, of intensity p per unit length. Model the beam with two basic beam elements and compare and contrast the solutions obtained for the bending moment distribution when using the MDM and the usual FEM approaches.

6.3. Refer to the derivation of the stiffness matrix for a uniform basic beam element in Section 6.3 and extend this to show that the stiffness matrix of a beam element of non-uniform cross-section (i.e. I_z is a function of x) is

$$
\begin{array}{cccc}
\quad v_1 & \theta_{z1} & v_2 & \theta_{z2}
\end{array}
$$

$$
\mathbf{k} = \frac{1}{l^3}
\begin{bmatrix}
9I_1 - 12I_2 + 4I_3 & & & \\
l(6I_1 - 7I_2 + 2I_3) & l^2(4I_1 - 4I_2 + I_3) & \text{Symmetric} & \\
-9I_1 + 12I_2 - 4I_3 & l(-6I_1 + 7I_2 - 2I_3) & 9I_1 - 12I_2 + 4I_3 & \\
l(3I_1 - 5I_2 + 2I_3) & l^2(2I_1 - 3I_2 + I_3) & l(-3I_1 + 5I_2 - 2I_3) & l^2(I_1 - 2I_2 + I_3)
\end{bmatrix}
$$

where

$$I_1 = \frac{4E}{l} \int_0^l I_z \, dx, \qquad I_2 = \frac{12E}{l^2} \int_0^l I_z x \, dx, \qquad I_3 = \frac{36E}{l^3} \int_0^l I_z x^2 \, dx.$$

6.4. A cantilever beam of length L has a varying cross-section with flexural rigidity varying linearly from EI_0 at the free end to $3EI_0$ at the clamped end. For

a lateral force W applied at the free end the exact solutions for deflection Δ and slope ϕ at the free end are $\Delta = 0.1373 WL^3/EI_0$, $\phi = 0.2253 WL^2/EI_0$. Compare with these results those obtained using the FEM by representing the beam with
   (i) one cubic beam element of length L
   (ii) two cubic elements each of length $L/2$.
(Use the result of Problem 6.3)

6.5. Refer to the derivation of the stiffness matrix for a uniform basic bar element in Section 6.2 and extend this to show that the stiffness matrix of a bar element of non-uniform cross-section (i.e. A is a function of x) is

$$\begin{matrix} u_1 & u_2 \end{matrix}$$
$$\mathbf{k} = \frac{E\phi}{l^2} \begin{bmatrix} 1 & -1 \\ -1 & 1 \end{bmatrix} \quad \text{where} \quad \phi = \int_0^l A \, dx.$$

6.6. A bar CD is tapered along its length L such that its cross-sectional area varies in linear fashion from $2A_0$ at end C to A_0 at end D. The end C is held whilst end D is pulled with an axial force W. In the exact solution to this problem $u_D = 0.6932 WL/A_0 E$ and the stress at distance x from C is $\sigma = WL/(A_0(2L-x))$. Obtain FEM solutions for this problem using
   (i) two uniform basic elements of length $L/2$, with the constant cross-sectional area of each element equal to the area of the true bar at the centre length of the element
   (ii) two tapered basic elements of length $L/2$ of the type described in Problem 6.5.

6.7. Starting with equation (6.76) verify the shape functions recorded in Fig. 6.8 for the three-node bar element having a quadratic displacement field.

6.8. Deduce the shape functions for the four-node cubic bar element shown in Fig. 6.9(a) and sketch these functions. Take the x co-ordinate origin to be at the centre of the element.

6.9. Determine the stiffness matrix for the prismatic bar element of Problem 6.8. Also determine column matrix $\mathbf{Q}$ corresponding to a distributed axial loading varying linearly from intensity f per unit length at node 1 to λf per unit length at node 4.

6.10. Refer to Example 6.5 but now replace the uniformly distributed axial load with a distributed loading whose intensity varies linearly from $2p$ per unit length at the held end to p per unit length at the free end. The exact solution to this problem is easily obtained and gives the result that the total extension is $2pL^2/3AE$ and that the stress at distance x from the held end is

$$\sigma = \frac{pL}{2A}\left(3 - 4\frac{x}{L} + \frac{x^2}{L^2}\right).$$

Model the bar with one four-node element of the type shown in Fig. 6.9(a) whose properties are developed in Problems 6.8 and 6.9; show that this modelling yields the exact solution. Also model the bar with three equal-length basic elements and compare the solution obtained with the exact solution.

6.11. Starting with equation (6.77) develop the shape functions for the three-node quintic beam element shown in Fig. 6.10(a). Take the co-ordinate origin to be at node 2 and the element length to be l.

6.12. Proceed from Problem 6.11 to determine the stiffness matrix of the three-node prismatic beam element. Also determine column matrix $\mathbf{Q}$ for this element corresponding to a uniformly distributed load of intensity p over the whole element.

6.13. Use a single quintic beam element of the type described in Problems 6.11 and 6.12 and illustrated in Fig. 6.10(a) to model
  (a) the cantilever beam of Example 6.3
  (b) the clamped beam of Problem 6.2.
In each case verify that the exact solution (as found using the MDM) is obtained.

6.14. In Example 6.4 matrices $\mathbf{k}$ and $\mathbf{Q}$ (for a uniformly distributed load) have been derived for the three-node bar shown in Fig. 6.8. The process of static condensation (Section 6.11) can be used to produce a system of stiffness equations relating to the external degrees of freedom u_1 and u_3 only. Show that the result is

$$\begin{Bmatrix} P_1 \\ P_3 \end{Bmatrix} = \frac{AE}{l} \begin{bmatrix} 1 & -1 \\ -1 & 1 \end{bmatrix} \begin{Bmatrix} u_1 \\ u_3 \end{Bmatrix} - P_2 \begin{Bmatrix} \frac{1}{2} \\ \frac{1}{2} \end{Bmatrix}$$

where P_2 is the force acting at node 2 corresponding to any externally applied loading. (Clearly if P_2 is zero the process has produced the basic bar element stiffness relationship.)
    Use the above result to solve Example 6.5 and verify that the same solution is obtained as was obtained using the uncondensed stiffness relationship of the three-node element.

6.15. Use the process of static condensation to produce a system of stiffness equations relating only to the external degrees of freedom v_1, θ_{z1}, v_3, and θ_{z3} of the quintic beam element shown in Fig. 6.10(a). Proceed to solve Example 6.3 using the condensed system of equations and verify that the exact solution is obtained. It should be noted that the full stiffness matrix and the column matrix $\mathbf{Q}$ for uniformly distributed load are given in the solution to Problem 6.12.

6.16. Use the process of static condensation to produce a system of stiffness equations relating only to the external degrees of freedom u_1 and u_4 of the cubic bar element shown in Fig. 6.9(a) and considered in Problems 6.8 and 6.9. (Take the element loading to be as detailed in Problem 6.9.) Proceed to resolve Problem 6.10, verifying that the same solution is obtained as when a single cubic element is used without any condensation.

6.17. Resolve Example 4.3 using the initial strain approach of Section 6.12.

6.18. Resolve Problem 4.10 using the initial strain approach.

6.19. Resolve Problem 4.13 using the initial strain approach.

6.20. Resolve Problem 4.18 using the initial strain approach.

7

INTRODUCTION TO TWO- AND THREE-DIMENSIONAL ANALYSIS

7.1. Introduction

THE ground rules for the derivation of finite-element relationships through application of the PMPE have been established. This has been done through consideration of simple one-dimensional elements but, as has been stated earlier, the extension to two- and three-dimensional elements differs from the unidirectional case only in the degree of complexity involved and not in the basic philosophy. The remainder of the text will be concerned predominantly with multidimensional elements, but before such elements can be studied in detail we need to consider the relevant concepts and equations of the solid elastic continuum and this is the purpose of the present chapter.

Consideration here is limited to linear elasticity and only a résumé of the theory is given, with the main relationships being documented. The reader who desires a fuller exposition of the elasticity theory, including rigorous derivations and proofs, is referred to specialist books, amongst which are those of Timoshenko and Goodier [1], Wang [2] and Sokolnikoff [3].

The basic equations of the three-dimensional theory expressed in relation to a Cartesian co-ordinate system are given and discussed first. This is followed by consideration of the potential energy of the three-dimensional body which leads into a description of the general finite-element formulation. The chapter continues with specialization of the elasticity equations to some important two-dimensional situations. These are the cases of plane stress, plane strain, and axisymmetric solids and the finite-element analysis of all of these cases is considered in later chapters. Finally, a further specialization of the three-dimensional elasticity equations is made to produce the equations of classical plate theory which form the basis of a range of plate-bending finite elements. Again the treatment of the theory is necessarily brief and the reader requiring fuller details is referred to the texts of Timoshenko and Woinowsky-Krieger [4] and Szilard [5].

7.2. Equations from three-dimensional elasticity theory

7.2.1. Stresses: *equilibrium equations*

Imagine that an arbitrary elastic body, or continuum, is acted upon by some system of external forces. The forces may comprise surface forces,

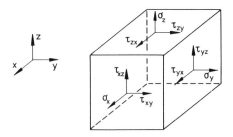

Fig. 7.1. Components of Cartesian stress.

acting over part or whole of the external surface of the body, and body forces, acting throughout the volume of the body. Owing to these forces a state of stress is set up in the body. Figure 7.1 shows the components of Cartesian stress acting on an infinitesimal volume in the interior of the continuum. The stresses σ_x, σ_y, and σ_z are direct or normal stresses and their single subscripts indicate the directions (x, y, or z) in which they act. A normal stress is taken to be positive when it produces tension. The stresses τ_{xy}, τ_{yx}, τ_{yz}, τ_{zy}, τ_{zx}, and τ_{xz} are shear stresses. It should be noted that in this double-subscript convention the first subscript letter indicates the direction of the normal to the plane on which the stress acts and the second indicates the direction in which the stress acts. The positive directions of the two shear stresses acting on a plane are in the positive directions of the corresponding two co-ordinate axes if the outward normal to the plane is in the positive sense of the third co-ordinate axis. Although six components of shear stress are indicated in Fig. 7.1 only three of these are in fact independent. By considering rotational equilibrium about the three co-ordinate axes in turn it is easy to show that $\tau_{xy} = \tau_{yx}$, $\tau_{yz} = \tau_{zy}$, and $\tau_{zx} = \tau_{xz}$. Thus the state of stress at a point is fully described by the six stress components σ_x, σ_y, σ_z, τ_{xy}, τ_{yz}, and τ_{zx}.

The equilibrium equations for all points in the interior of the elastic body can be obtained by equating to zero the sum of forces acting on an infinitesimal volume in each of the three co-ordinate directions in turn. In general the stresses will vary from point to point in the body so that the values of stress at parallel faces of the infinitesimal volume differ slightly from one another as shown in Fig. 7.2. (Values of stress also vary over a face but this produces only a higher-order effect which can be disregarded.) For clarity in Fig. 7.2 only the stress components making a contribution to the equilibrium of forces in the x direction are labelled. Also indicated in the figure is the presence of the components of a body force: these are R_x, R_y, and R_z per unit volume in the x, y, and z directions respectively. From Fig. 7.2 it is quite easy to obtain the governing equilibrium equation for the x direction. This is the first of the

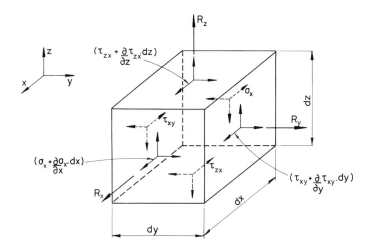

Fig. 7.2. Equilibrium of an infinitesimal volume.

three equations now given and the other two are the analagous equations for the y and z directions. The equilibrium equations are

$$\frac{\partial \sigma_x}{\partial x} + \frac{\partial \tau_{xy}}{\partial y} + \frac{\partial \tau_{zx}}{\partial z} + R_x = 0$$

$$\frac{\partial \sigma_y}{\partial y} + \frac{\partial \tau_{yz}}{\partial z} + \frac{\partial \tau_{xy}}{\partial x} + R_y = 0 \qquad (7.1)$$

$$\frac{\partial \sigma_z}{\partial z} + \frac{\partial \tau_{zx}}{\partial x} + \frac{\partial \tau_{yz}}{\partial y} + R_z = 0.$$

7.2.2. Displacements and strains: strain–displacement relationships

Under the action of applied forces and stresses the elastic body is strained so that the relative position of points in the body is altered. A particle of the body located initially at the point x, y, z undergoes a displacement having components u, v, and w in the x, y, and z directions. In general the displacement components are functions of x, y, and z, of course, and we assume that the displacements are very small.

There are six components of strain to be considered in three-dimensional analysis—three direct or normal components and three shear components—corresponding to the six components of stress. The components of strain can be expressed in terms of partial first derivatives of the displacement components. To demonstrate how this can be done using purely geometric arguments consider for the moment the deformation occurring in, say, the xy plane of an infinitesimal volume dx dy dz of the body. Figure 7.3 shows the infinitesimal volume in its original unloaded

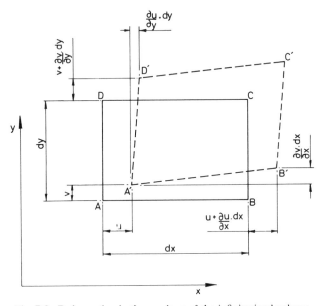

Fig. 7.3. Deformation in the xy plane of the infinitesimal volume.

configuration ABCD and in its deformed configuration A'B'C'D'. The direct strain is simply the ratio of change in length to original length. For AB, for example, the original length is dx and the final length A'B' is approximately $dx(1 + \partial u/\partial x)$ so that the component of direct strain in the x direction is $\varepsilon_x = \partial u/\partial x$. (The approximation comes in in neglecting squares of displacement derivatives in comparison with unity; this is consistent with the assumption of small deformations.) The component of engineering shearing strain in the xy plane is denoted by γ_{xy} $(=\gamma_{yx})$ and represents the change in the angle between edges AB and AD arising from the deformation process as ABCD moves to A'B'C'D'. From Fig. 7.3 it can thus be seen that $\gamma_{xy} = \partial v/\partial x + \partial u/\partial y$ (approximately). By a similar process expressions for the remaining strain components can be established and the full set of linear strain–displacement equations is

$$\varepsilon_x = \frac{\partial u}{\partial x} \qquad \varepsilon_y = \frac{\partial v}{\partial y} \qquad \varepsilon_z = \frac{\partial w}{\partial z}$$

$$\gamma_{xy} = \frac{\partial u}{\partial y} + \frac{\partial v}{\partial x} \qquad \gamma_{yz} = \frac{\partial v}{\partial z} + \frac{\partial w}{\partial y} \qquad \gamma_{zx} = \frac{\partial w}{\partial x} + \frac{\partial u}{\partial z}$$

$$(7.2)$$

Incidentally, on consideration of equations (7.2) it can be seen that the strain components will be unchanged if to any expression of the displacements u, v, w we add the terms u_r, v_r, w_r respectively where

$$u_r = a - by + cz$$
$$v_r = d + bx - ez$$
$$w_r = f - cx + ey.$$

The constants a, d, and f in these expressions represent a translatory motion of the body whilst the constants b, c, and e represent a rotary motion of the body about the co-ordinate axes. The displacements u_r, v_r, and w_r therefore represent the motion of the body as a whole and do not induce any strain in the body. These displacements are thus called the rigid-body motion displacements or the strain-free displacements.

7.2.3. Stress–strain relations (constitutive equations)

The six components of stress are related to the six components of strain through a set of constitutive equations. If column matrices of stress and strain are defined as

$$\boldsymbol{\sigma} = \{\sigma_x \quad \sigma_y \quad \sigma_z \quad \tau_{xy} \quad \tau_{yz} \quad \tau_{zx}\} \tag{7.3}$$

and

$$\boldsymbol{\varepsilon} = \{\varepsilon_x \quad \varepsilon_y \quad \varepsilon_z \quad \gamma_{xy} \quad \gamma_{yz} \quad \gamma_{zx}\}, \tag{7.4}$$

the equations have the form, in the absence of initial strains,

$$\boldsymbol{\sigma} = \mathbf{E}\boldsymbol{\varepsilon}. \tag{7.5}$$

(The effect of initial strains is discussed briefly at the end of Section 7.3.) The 6×6 matrix $\mathbf{E}$ is known as the elastic-coefficient matrix or simply the elasticity matrix, and serves the same purpose as E does in Hooke's law of one-dimensional analysis. Consequently equation (7.5) may be referred to as a generalized Hooke's law.

For the most general constitutive law there are 21 independent elastic coefficients in $\mathbf{E}$, taking account of the fact that $\mathbf{E}$ is a symmetric matrix. Thus

$$\mathbf{E} = \begin{bmatrix} e_{11} & & & & & \\ e_{21} & e_{22} & & \text{Symmetric} & & \\ e_{31} & e_{32} & e_{33} & & & \\ e_{41} & e_{42} & e_{43} & e_{44} & & \\ e_{51} & e_{52} & e_{53} & e_{54} & e_{55} & \\ e_{61} & e_{62} & e_{63} & e_{64} & e_{65} & e_{66} \end{bmatrix}. \tag{7.6}$$

The coefficients of the elasticity matrix are established by laboratory testing of the material. We shall assume that the elastic body is *homogeneous*, that is that the physical properties of the continuum are the same at all points so that the elastic coefficients are constants. Further, for most materials there exist one or more planes of symmetry in the material structure and this allows considerable simplification by reducing the number of independent coefficients. The most frequently occurring, and hence most important, type of material in use in structural engineering is the *isotropic* material in which the elastic properties are the same in all directions. This means that for isotropic materials a knowledge of only two independent elastic properties is required; these are Young's modulus E and Poisson's ratio ν.

Isotropic material behaviour is assumed throughout the remainder of this text but it should be realized that other types of material behaviour could be accommodated quite readily simply by appropriate specification of matrix **E** in equation (7.6).

It is well known that in a three-dimensional situation the application of positive, or tensile, direct stress in one direction gives rise to a positive strain, or extension, in that direction and at the same time to a negative strain, or contraction, in the other two directions at right angles to the first; the amount of the contraction is proportional to the value of Poisson's ratio. Thus if a single stress σ_x is applied to an isotropic three-dimensional body the resulting strains are

$$\varepsilon_x = \frac{\sigma_x}{E} \qquad \varepsilon_y = \frac{-\nu\sigma_x}{E} \qquad \varepsilon_z = \frac{-\nu\sigma_x}{E}.$$

If all three direct stresses are acting the direct strain–direct stress relationships become

$$\varepsilon_x = \frac{\sigma_x - \nu(\sigma_y + \sigma_z)}{E}$$

$$\varepsilon_y = \frac{\sigma_y - \nu(\sigma_z + \sigma_x)}{E}$$

$$\varepsilon_z = \frac{\sigma_z - \nu(\sigma_x + \sigma_y)}{E}.$$

These relationships are clearly independent of any shearing behaviour. The shearing strain–shearing stress relationships in fact have the simple form

$$\gamma_{xy} = \frac{1}{G}\tau_{xy} \qquad \gamma_{yz} = \frac{1}{G}\tau_{yz} \qquad \gamma_{zx} = \frac{1}{G}\tau_{zx}$$

where $G = E/2(1+\nu)$ is the shearing modulus or modulus of rigidity. From these given relationships the elasticity matrix **E** (relating stresses to strains) for a homogeneous isotropic material is found to be

$$
\mathbf{E} = \frac{E}{(1+\nu)(1-2\nu)}
\begin{bmatrix}
1-\nu & & & & & \\
\nu & 1-\nu & & & \text{Symmetric} & \\
\nu & \nu & 1-\nu & & & \\
0 & 0 & 0 & \tfrac{1}{2}(1-2\nu) & & \\
0 & 0 & 0 & 0 & \tfrac{1}{2}(1-2\nu) & \\
0 & 0 & 0 & 0 & 0 & \tfrac{1}{2}(1-2\nu)
\end{bmatrix}.
$$

$$(7.7)$$

7.2.4. Compatibility conditions

Referring to equations (7.2) it is seen that the six components of strain at a point are expressed in terms of only three displacement components. Thus the components of strain cannot be completely independent of one another and there must be some conditions imposed on the strain components in order that the six strain–displacement equations give a set of single-valued continuous solutions for the three displacement components throughout the interior of the body. The conditions are

$$\frac{\partial^2 \varepsilon_x}{\partial y^2} + \frac{\partial^2 \varepsilon_y}{\partial x^2} = \frac{\partial^2 \gamma_{xy}}{\partial x\,\partial y} \qquad 2\frac{\partial^2 \varepsilon_x}{\partial y\,\partial z} = \frac{\partial}{\partial x}\left(-\frac{\partial \gamma_{yz}}{\partial x} + \frac{\partial \gamma_{zx}}{\partial y} + \frac{\partial \gamma_{xy}}{\partial z}\right)$$

$$\frac{\partial^2 \varepsilon_y}{\partial z^2} + \frac{\partial^2 \varepsilon_z}{\partial y^2} = \frac{\partial^2 \gamma_{yz}}{\partial y\,\partial z} \qquad 2\frac{\partial^2 \varepsilon_y}{\partial x\,\partial z} = \frac{\partial}{\partial y}\left(\frac{\partial \gamma_{yz}}{\partial x} - \frac{\partial \gamma_{xz}}{\partial y} + \frac{\partial \gamma_{xy}}{\partial z}\right) \qquad (7.8)$$

$$\frac{\partial^2 \varepsilon_z}{\partial x^2} + \frac{\partial^2 \varepsilon_x}{\partial z^2} = \frac{\partial^2 \gamma_{xz}}{\partial x\,\partial z} \qquad 2\frac{\partial^2 \varepsilon_z}{\partial x\,\partial y} = \frac{\partial}{\partial z}\left(\frac{\partial \gamma_{yz}}{\partial x} + \frac{\partial \gamma_{xz}}{\partial y} - \frac{\partial \gamma_{xy}}{\partial z}\right).$$

These differential equations are called the conditions of compatibility of strains. Their only purpose is to impose restrictions on the strain components which will ensure that the displacements u, v, and w are single valued and continuous throughout the continuum. Conversely, these equations are automatically satisfied if the displacements are single valued and continuous.

7.2.5. Boundary conditions

The equations presented above have been concerned with the displacements, strains, and stresses in the interior of the elastic body. Obviously, when the boundary or surface of the body is reached certain conditions have to be satisfied. Usually, mixed boundary conditions apply in that over a part of the surface, S_1 say, surface forces or tractions are prescribed, whilst over the remainder of the surface, S_2, displacement conditions are prescribed.

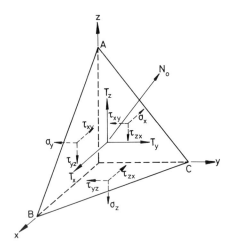

Fig. 7.4. Tractions and stresses at the boundary.

On the S_1 part of the boundary, let T_x, T_y, and T_z be the Cartesian components of prescribed traction per unit area of the surface. These tractions must be in equilibrium with the stresses in the body at the boundary. If l, m, and n are the direction cosines of the outward normal to the boundary, it can be shown by consideration of the equilibrium of the infinitesimal tetrahedron shown in Fig. 7.4, whose face ABC lies at the boundary, that the appropriate conditions are

$$T_x = l\sigma_x + m\tau_{xy} + n\tau_{zx}$$
$$T_y = l\tau_{xy} + m\sigma_y + n\tau_{yz} \qquad (7.9)$$
$$T_z = l\tau_{zx} + m\tau_{yz} + n\sigma_z.$$

The direction cosines l, m, and n are the cosines of the angles between the normal N_0 to the boundary and the x, y, and z axes respectively. Equations (7.9) are the natural or stress boundary conditions and they complete the statement of equilibrium for the body. It should be noted that these equations apply whether or not a body force is present.

On the S_2 part of the boundary the displacement conditions are simply expressed as

$$u = \bar{u} \qquad v = \bar{v} \qquad w = \bar{w} \qquad (7.10)$$

where $\bar{u}$, $\bar{v}$, and $\bar{w}$ are prescribed displacements. These equations are the geometric or displacement boundary conditions and they complete the statement of compatibility for the body.

7.2.6. Comments

In summarizing the above equations it can be said that the *exact* solution of an elasticity problem in which a body is subjected to prescribed body and surface forces requires the following.

(1) The three displacement components satisfy the displacement boundary conditions (equations (7.10)).

(2) The six strain components are linked to the three displacement components by equations (7.2) and satisfy the compatibility conditions (equations (7.8)); these latter conditions are automatically satisfied if the displacements are single valued continuous functions.

(3) The six stress components are linked to the six strain components by equations (7.5) and satisfy the equations of equilibrium, both in the interior of the body (equations (7.1)) and at the boundary of the body (equations (7.9)).

Of course, in the finite-element displacement approach we are concerned with generating *approximate* solutions through the PMPE. From the material of Chapters 5 and 6 it will be realized that only the strain–displacement and stress–strain equations are directly required for the finite-element analysis, together with appropriate assumptions for the displacement components which will meet the compatibility requirements. The equilibrium conditions in the interior and at the boundary (equations (7.1) and (7.9)) and the conditions of compatibility of strains (equations (7.8)) do not have to be considered directly.

7.3. Potential energy for the continuum: the general finite-element formulation†

In setting up expressions for the potential energy of a continuum and proceeding to the consideration of corresponding finite-element properties we can draw very heavily on the procedures used when one-dimensional situations were considered. The approach is basically identical and the difference is simply one of degree of complexity—in that now there are three components of displacement and six of stress and strain to consider—rather than of principle. Initial strain effects are ignored at the moment.

The strain energy U_p of a linear elastic solid body (by analogy with equations (5.8) or (6.7) for the one-dimensional body) is

$$U_p = \tfrac{1}{2} \int_{V_0} (\sigma_x \varepsilon_x + \sigma_y \varepsilon_y + \sigma_z \varepsilon_z + \tau_{xy} \gamma_{xy} + \tau_{yz} \gamma_{yz} + \tau_{zx} \gamma_{zx}) \, \mathrm{d}V_0$$

† See Appendix B for a brief description of the general finite-element formulation based on the use of the PVD rather than on the use of the PMPE as described here; the two approaches yield identical results.

or

$$U_p = \tfrac{1}{2} \int_{V_0} \boldsymbol{\sigma}^t \boldsymbol{\varepsilon} \, dV_0 \tag{7.11}$$

where $\boldsymbol{\sigma}$ and $\boldsymbol{\varepsilon}$ are defined in equations (7.3) and (7.4) and V_0 is the volume of the body. When the constitutive relationships (equation (7.5)) are used the strain energy can be expressed as a quadratic function of the strain components only:

$$U_p = \tfrac{1}{2} \int_{V_0} \boldsymbol{\varepsilon}^t \mathbf{E} \boldsymbol{\varepsilon} \, dV_0. \tag{7.12}$$

The potential energy of distributed body forces and surface forces for the three-dimensional geometry is (compare with equation (6.41) for the one-dimensional case)

$$V_{pd} = - \int_{V_0} \mathbf{u}^t \mathbf{R} \, dV_0 - \int_{S_1} \mathbf{u}^t \mathbf{T} \, dS. \tag{7.13}$$

Here

$$\mathbf{u} = \{u \quad v \quad w\} \tag{7.14}$$

is the column matrix of displacement components,

$$\mathbf{R} = \{R_x \quad R_y \quad R_z\} \tag{7.15}$$

is the column matrix of the components of body force, and

$$\mathbf{T} = \{T_x \quad T_y \quad T_z\} \tag{7.16}$$

is the column matrix of the components of surface force (or tractions). The body force and surface force components are prescribed externally applied forces. It should be noted particularly that the integral involving surface forces in equation (7.13) is evaluated only over that part of the surface S_1 where the external tractions are prescribed.

For the three-dimensional continuum subjected to distributed body and surface forces the total potential energy is the familiar expression

$$\Pi_p = U_p + V_{pd}. \tag{7.17}$$

It is noted in passing that if we were to make a formal application of the procedures of the calculus of variations to the potential energy functional defined by equations (7.12), (7.13), and (7.17) we would generate the conditions of equilibrium in the interior (equations (7.1)) and at the boundary (equations (7.9)) of the body. This is not our purpose here, of course, for we are concerned with the direct generation of approximate solutions through procedures of the Rayleigh–Ritz type. It is possible to generate such direct solutions to elasticity problems for simple geometries and loadings by the classical Rayleigh–Ritz approach (wherein continuous displacement fields are assumed for the whole body) and examples of this are described in many text books, such as those by Timoshenko and Goodier [1], Sokolnikoff [3], and Richards [6]. However, we shall not concern ourselves here with the classical Rayleigh–Ritz approach since the piecewise Rayleigh–Ritz approach, which constitutes the finite-element method, is much more versatile in two- and three-dimensional situations.

When finite-sized regions, or finite elements, of the continuum are considered we require to evaluate U_p and V_{pd} over each and every element, based—as has been seen for the one-dimensional case—on the assumption of an appropriate local displacement field. (The actual shape of the individual region or element need not concern us for the present.) The displacement field for the element will be written either directly or indirectly in the now familiar form

$$\mathbf{u} = \mathbf{Nd} \tag{7.18}$$

where $\mathbf{d}$ is the usual column matrix of nodal degrees of freedom and where assumptions have now to be made for the variation of each of u, v, and w throughout the element. Using equations (7.2) the six strain components can be expressed in terms of the nodal freedoms as

$$\boldsymbol{\varepsilon} = \mathbf{Bd} \tag{7.19}$$

where $\mathbf{B}$ is obtained from $\mathbf{N}$ by appropriate differentiation. Substituting this last equation into an expression of the form of equation (7.12), but restricting the integration to the element volume V_e, gives the strain energy of an element e as

$$U_p^e = \tfrac{1}{2}\mathbf{d}^t\mathbf{kd}. \tag{7.20}$$

This is the standard quadratic form of element strain energy and $\mathbf{k}$ is the element stiffness matrix defined by

$$\mathbf{k} = \int_{V_e} \mathbf{B}^t\mathbf{EB}\, dV_e. \tag{7.21}$$

Now, substituting equation (7.18) into an expression of the form of equation (7.13) but again restricting the integration to the element gives the potential energy of distributed forces as

$$V_{pd}^e = -\int_{V_e} \mathbf{u}^t\mathbf{R}\, dV_e - \int_{S_e} \mathbf{u}^t\mathbf{T}\, dS$$

or

$$V_{pd}^e = -\mathbf{d}^t\mathbf{Q} \tag{7.22}$$

where

$$\mathbf{Q} = \int_{V_e} \mathbf{N^t R} \, dV_e + \int_{S_e} \mathbf{N^t T} \, dS \qquad (7.23)$$

is the column matrix of consistent loads corresponding to the given distributed body forces and surface tractions. The integral on the right-hand side of equation (7.23) containing the surface tractions is evaluated only over that part S_e of the surface of the element which is an exterior surface for the body as a whole and which has tractions applied to it.

Equations (7.21) and (7.23) give expressions for the stiffness matrix and consistent load column matrix of a general finite element; they compare directly with equations (6.37a) and (6.45) for the one-dimensional element. Since $\mathbf{R}$ and $\mathbf{T}$ are known, or prescribed, quantities it is clear that $\mathbf{k}$ and $\mathbf{Q}$ can be generated for any element on specification of the three matrices $\mathbf{E}$, $\mathbf{B}$, and $\mathbf{N}$. The first of these reflects the constitutive relationships and these are taken to be known for any particular class of problem, whilst the second is obtained from the third through use of the strain–displacement relationships and, again, these latter relationships are known for a particular class of problem. Consequently, within a particular class of structural problem, it is only matrix $\mathbf{N}$—the shape function matrix—which is effectively open to choice and it is on this matrix, of course, that the efficiency or otherwise of a particular element type depends.

Just as in Chapter 6 the choice of displacement field could, if convenient, be initially expressed in terms of generalized displacements, in the form of equation (6.1) say, and then $\mathbf{N}$ and $\mathbf{B}$ could be written as

$$\mathbf{N} = \boldsymbol{\alpha} \mathbf{C}^{-1} \qquad \text{and} \qquad \mathbf{B} = \boldsymbol{\beta} \mathbf{C}^{-1}$$

where $\boldsymbol{\alpha}$, $\boldsymbol{\beta}$, and $\mathbf{C}$ have the same meaning as in Chapter 6. This allows the re-expression of $\mathbf{k}$ and $\mathbf{Q}$ for the general element (from equations (7.21) and (7.23)) in the form

$$\mathbf{k} = (\mathbf{C}^{-1})^t \left(\int_{V_e} \boldsymbol{\beta}^t \mathbf{E} \boldsymbol{\beta} \, dV_e \right) \mathbf{C}^{-1} \qquad (7.24)$$

and

$$\mathbf{Q} = (\mathbf{C}^{-1})^t \int_{V_e} \boldsymbol{\alpha}^t \mathbf{R} \, dV_e + (\mathbf{C}^{-1})^t \int_{S_e} \boldsymbol{\alpha}^t \mathbf{T} \, dS. \qquad (7.25)$$

(Compare with equations (6.37b) and (6.48) for the one-dimensional element.)

With the element stiffness and consistent-load matrices established, according to equations (7.21) and (7.23) or (7.24) and (7.25), the assembly process within the PMPE approach proceeds precisely as already described in Section 6.6. The structure stiffness and consistent load matrices, $\mathbf{K}$ and $\mathbf{F_O}$ respectively, are assembled from the element stiffness and consistent-load matrices by direct superposition (equations (6.59) and (6.60)), and any externally applied concentrated forces acting at the structure nodes themselves are accommodated in column matrix $\mathbf{F}$; the structure stiffness equations are then represented by equation (6.58) (by (6.58a) before the geometric boundary conditions are applied and by (6.58b) after) and can be solved to yield the structure nodal displacements.

When the nodal displacements have been determined the element stresses can be calculated by a procedure similar to that described in Section 6.7 for the one-dimensional element. At any point in an element the stresses are related to the strains by equation (7.5) and the strains are expressed in terms of the element nodal displacements by equation (7.19) in which the individual terms of $\mathbf{B}$ are usually functions of the co-ordinate position of the point. The stresses at a point are therefore

$$\boldsymbol{\sigma} = \mathbf{E}\boldsymbol{\varepsilon} = \mathbf{EBd} = \mathbf{Sd} \tag{7.26}$$

where $\mathbf{S}$ is the stress matrix (see equations (6.64) and (6.66) for the one-dimensional case) and $\mathbf{B}$ could be written as $\boldsymbol{\beta}\mathbf{C}^{-1}$ if convenient.

In the above development the effects of initial strains have been ignored both in the interests of simplicity and since such effects are not present in the great majority of applications. However, these effects can be included without undue complication by direct extension of the procedure described for the one-dimensional situation in Section 6.12, and this will now be briefly described.

Where initial strains are present in the three-dimensional situation the constitutive equations become

$$\boldsymbol{\sigma} = \mathbf{E}(\boldsymbol{\varepsilon} - \boldsymbol{\varepsilon}^0) \tag{7.27}$$

where $\boldsymbol{\sigma}$, $\boldsymbol{\varepsilon}$ (the *total* strains), and $\mathbf{E}$ are as defined in Section 7.2.3, and

$$\boldsymbol{\varepsilon}^0 = \{\varepsilon_x^0 \quad \varepsilon_y^0 \quad \varepsilon_z^0 \quad \gamma_{xy}^0 \quad \gamma_{yz}^0 \quad \gamma_{zx}^0\} \tag{7.28}$$

is the column matrix of initial strains. Much the most common type of initial strain is thermal strain due to temperature change. Such temperature change does not affect the shear strain components in an isotropic material so that the column matrix of initial strains becomes

$$\boldsymbol{\varepsilon}^0 = \alpha T\{1 \quad 1 \quad 1 \quad 0 \quad 0 \quad 0\}. \tag{7.29}$$

Inclusion of initial strain effects in the three-dimensional case will result in the same type of element stiffness equations as in the one-dimensional case of Section 6.12 and it is superfluous to elaborate further. Thus the element equations are

$$\mathbf{P} + \mathbf{Q} + \mathbf{Q}^0 = \mathbf{kd} \tag{6.96}$$

where the presence of distributed mechanical loading is assumed (hence the appearance of $\mathbf{Q}$) in addition to that of initial strains. The definition of column matrix $\mathbf{Q}^0$ which accounts for initial strains is, by analogy with equation (6.89),

$$\mathbf{Q}^0 = \int_{V_e} \mathbf{B}^t \mathbf{E} \boldsymbol{\varepsilon}^0 \, dV_e. \tag{7.30}$$

Stresses within the element can be obtained from the nodal displacements by the equations (see equation (6.93) for the one-dimensional case)

$$\boldsymbol{\sigma} = \mathbf{E}(\mathbf{Bd} - \boldsymbol{\varepsilon}^0). \tag{7.31}$$

Once the initial strains $\boldsymbol{\varepsilon}^0$ are prescribed it is clearly a quite standardized procedure to construct $\mathbf{Q}^0$ for three-dimensional analysis or for any specializations thereof which are discussed in what follows in Sections 7.4–7.6. Having said this, the effect of initial strains will not be considered further during the remainder of this text.

7.4. Plane stress and plane strain

The finite-element analysis of three-dimensional continua is computationally expensive and where it can be avoided by introducing specializations to reduce the problem to two dimensions (or even one) it is important to do so. Two well-known specializations to two dimensions are the cases of plane stress and plane strain and these will now be considered. (Another two-dimensional specialization is considered in Section 7.5: this is the axisymmetric solid under axisymmetric loading.) Although the plane stress and plane strain problems are in a physical sense quite distinct and different, their analysis is fundamentally the same, with the only significant difference between the two types of problem occurring in the statements of the constitutive relationships.

In *plane stress* problems the dimension of the elastic body in one co-ordinate direction—the z direction—is very small compared with its dimensions in the directions of the other two co-ordinates, x and y; Fig. 7.5 shows the thin plane stress body lying in the xy plane. It is assumed that no loading whatsoever is applied in the z direction and that there are no surface forces acting on the top and bottom faces (i.e. on the exterior surfaces which lie parallel to the xy plane). Consequently the only external forces considered are surface forces T_x and T_y acting at the curved boundary and body forces R_x and R_y acting throughout the volume. All these forces act in the xy plane, are uniform through the

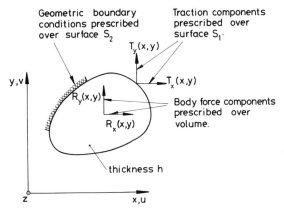

Fig. 7.5. Plane stress body.

thickness, and are thus functions of x and y only. Because of these assumptions it follows that σ_z, τ_{zx}, and τ_{yz} are zero in plane stress analysis and that the displacement components u and v, the strain components ε_x, ε_y, and γ_{xy}, and the stress components σ_x, σ_y, and τ_{xy} are functions of x and y only. For the plane stress problem, then, the pertinent column matrices of stress and strain components are obtained from equations (7.3) and (7.4) by appropriate reduction as

$$\boldsymbol{\sigma} = \{\sigma_x \quad \sigma_y \quad \tau_{xy}\} \tag{7.32}$$

$$\boldsymbol{\varepsilon} = \{\varepsilon_x \quad \varepsilon_y \quad \gamma_{xy}\}. \tag{7.33}$$

The pertinent strain–displacement equations are, from equations (7.2),

$$\varepsilon_x = \frac{\partial u}{\partial x} \qquad \varepsilon_y = \frac{\partial v}{\partial y} \qquad \gamma_{xy} = \frac{\partial u}{\partial y} + \frac{\partial v}{\partial x}. \tag{7.34}$$

The stress–strain relations for an isotropic material are of the form of equation (7.5) with the elasticity matrix for the plane stress case defined as

$$\mathbf{E} = \frac{E}{1-\nu^2} \begin{bmatrix} 1 & \nu & 0 \\ \nu & 1 & 0 \\ 0 & 0 & \dfrac{1-\nu}{2} \end{bmatrix}. \tag{7.35}$$

Matrix $\mathbf{E}$ given here for plane stress is obtained from $\mathbf{E}$ for the three-dimensional case (equation (7.7)) by using the conditions that $\sigma_z = \tau_{zx} = \tau_{yz} = 0$. (It should be noted that $\varepsilon_z = -\nu(\varepsilon_x + \varepsilon_y)/(1-\nu)$ is not zero but that it makes no contribution to the plane-stress strain energy since it is associated with the zero σ_z.)

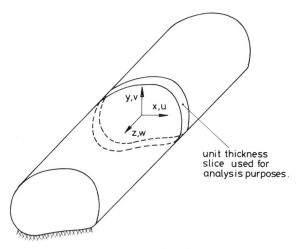

Fig. 7.6. Plane strain body.

In *plane strain* problems the dimension of the elastic body in one co-ordinate direction—the z direction—is large compared with its dimensions in the directions of the other two co-ordinates. The plane strain body, shown in Fig. 7.6, is a long cylindrical or prismatic body loaded by surface forces T_x and T_y and body forces R_x and R_y acting perpendicular to its axis; neither the geometry nor the loading varies along the length and it may therefore be assumed that all cross sections are in the same condition. Also the ends are assumed to be restrained by fixed rigid planes so that all displacement w in the axial direction is prevented. It follows from these assumptions that the strain components ε_z, γ_{yz}, and γ_{zx} are zero, and that the displacements u and v, the strain components ε_x, ε_y, and γ_{xy}, and the stress components σ_x, σ_y, and τ_{xy} are again the pertinent variables of the problem and are all functions of x and y only. Thus, equations (7.32)–(7.34) apply once more, and only in the stress–strain relations does there arise a difference from the plane stress case. For isotropic material the elasticity matrix for plane strain, linking the stresses $\boldsymbol{\sigma}$ and strains $\boldsymbol{\varepsilon}$ defined in equations (7.32) and (7.33), is

$$
\mathbf{E} = \frac{E}{(1+\nu)(1-2\nu)}
\begin{bmatrix}
1-\nu & \nu & 0 \\
\nu & 1-\nu & 0 \\
0 & 0 & \dfrac{1-2\nu}{2}
\end{bmatrix}
$$

$$
= \frac{E^*}{1-\nu^{*2}}
\begin{bmatrix}
1 & \nu^* & 0 \\
\nu^* & 1 & 0 \\
0 & 0 & \dfrac{1-\nu^*}{2}
\end{bmatrix}
\tag{7.36}
$$

where $E^* = E/(1-\nu^2)$ and $\nu^* = \nu/(1-\nu)$. This is obtained directly from equation (7.7) by setting $\varepsilon_z = \gamma_{yz} = \gamma_{zx} = 0$. (It should be noted that in the plane strain problem $\sigma_z = \nu(\sigma_x + \sigma_y)$ is not zero but that again, as in the plane stress case, the product $\sigma_z \varepsilon_z$ appearing in the strain energy expression is zero.) In plane strain analysis it is usual to consider a slice of body of unit thickness.

7.5. Bodies of revolution with axisymmetric loading

Bodies of revolution which are subjected *only to axisymmetric loading* occur quite frequently and constitute another important category of structure which is essentially two dimensional in nature; such structures are termed axisymmetric continua. A typical body of revolution is shown, with part of the body cut away for clarity, in Fig. 7.7(a). The y axis is the longitudinal axis about which the geometry and applied loading is symmetric, whilst r is directed radially outwards and θ is the polar angle; r, θ, and y are a system of cylindrical co-ordinates which are clearly much more convenient for the analysis of axisymmetric continua than would be the system of Cartesian co-ordinates considered hitherto. The displacement components in the longitudinal and radial directions are v and u respectively; these displacements do not vary with θ and are the only two

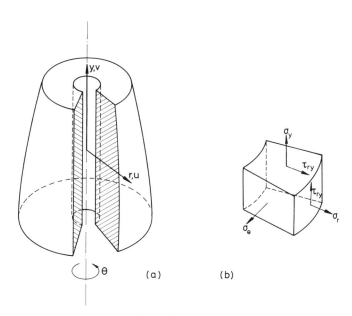

Fig. 7.7. Axisymmetric body.

relevant displacements since the prescription of symmetry means that the tangential component of displacement is zero everywhere.

The six components of stress for an arbitrary three-dimensional body expressed in terms of a cylindrical co-ordinate system comprise σ_r, σ_y, σ_θ, τ_{ry}, $\tau_{y\theta}$, and $\tau_{\theta r}$. For the axisymmetric continua, though, the two shearing stresses $\tau_{y\theta}$ and $\tau_{\theta r}$ are zero and so the number of stress components reduces to four. The four non-zero stress components are shown acting on an infinitesimal volume of the axisymmetric continua in Fig. 7.7(b) and the column matrix of stresses is defined as

$$\boldsymbol{\sigma} = \{\sigma_r \quad \sigma_y \quad \sigma_\theta \quad \tau_{ry}\}. \tag{7.37}$$

The corresponding column matrix of strains is

$$\boldsymbol{\varepsilon} = \{\varepsilon_r \quad \varepsilon_y \quad \varepsilon_\theta \quad \gamma_{ry}\}. \tag{7.38}$$

The strain–displacement relationships, obtainable from the textbooks on elasticity theory mentioned in Section 7.1, are

$$\varepsilon_r = \frac{\partial u}{\partial r} \qquad \varepsilon_y = \frac{\partial v}{\partial y} \qquad \varepsilon_\theta = \frac{u}{r} \qquad \gamma_{ry} = \frac{\partial u}{\partial y} + \frac{\partial v}{\partial r}. \tag{7.39}$$

The stresses (equation (7.37)) and strains (equation (7.38)) are related by an elasticity matrix in the usual manner. For an isotropic material the elasticity matrix is the following reduced form of that given by equation (7.7):

$$\mathbf{E} = \frac{E}{(1+\nu)(1-2\nu)} \begin{bmatrix} 1-\nu & & \text{Symmetric} & \\ \nu & 1-\nu & & \\ \nu & \nu & 1-\nu & \\ 0 & 0 & 0 & \frac{1}{2}(1-2\nu) \end{bmatrix}. \tag{7.40}$$

The stresses and strains, like the u and v displacement components, are obviously functions of r and y only.

Despite the presence of four (rather than three) stresses and strains the analysis of axisymmetric continua has much in common with plane stress and plain strain analysis. This will be made more evident in the next two chapters.

7.6. Plate bending

A plate is a structural body bounded by two parallel planes—the faces of the plate—and by a cylindrical boundary which is perpendicular to the faces, i.e. the edge of the plate. The distance between the faces of the plate is the plate thickness h and the surface lying equidistant from the two plate faces is known as the middle surface or mid-surface. Figure 7.8

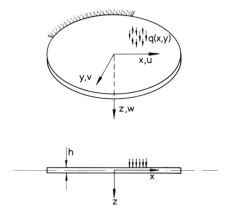

Fig. 7.8. The plate bending problem.

shows a plate of arbitrary shape whose middle surface is the xy plane. (It should be noted that in Fig. 7.8 the x, y, and z axes are arranged so that z acts downwards as is traditional in plate analysis; since the xyz co-ordinates still form a right-handed orthogonal system they, of course, do not differ in any meaningful way from the co-ordinate system of three-dimensional elasticity shown in Fig. 7.1.) The geometry of the plate of Fig. 7.8 is the same as that of the plane stress body considered previously (see Fig. 7.5); it is in the nature of the loading that the two types of problem differ, of course. In the plate bending problem the applied loads act normal to the plate middle surface, i.e. in the z direction. These applied loads result in bending of the plate with displacement of the middle surface of the plate in the z direction; this displacement is usually called the deflection of the plate and is denoted by w. The plate bending problem, when formulated through the classical plate theory, is the equivalent in two dimensions of the Bernoulli–Euler beam problem in one dimension.

The equations of plate theory can be obtained from those of three-dimensional elasticity—presented in Section 7.2—by introducing certain assumptions. These assumptions simplify the problem but at the same time introduce approximations, and indeed contradictions, so that the plate theory is not exact in the sense that elasticity theory is. However, there is of course a wide range of problems for which plate theory provides solutions of high accuracy.

The assumptions of plate theory include the following.
   (1) The thickness of the plate is small compared with its lateral dimensions.
   (2) The deflection of the plate is small compared with its thickness and

the slopes of the deflected middle plane are small compared with unity.

(3) The deformation of the plate is such that straight lines which are initially normal to the middle surface remain straight and normal to the middle surface in the loaded configuration. This is the Kirchhoff hypothesis.

(4) There is no straining of the middle surface of the plate.

The plate theory being considered here, to which the above assumptions apply, is known as the Kirchhoff–Love plate theory or the classical plate theory. It is the generally accepted thin-plate theory whose main virtue, as will be seen, is that it allows plate deformation and stresses to be expressed in terms of the deflection w of the middle surface alone. It should be noted, though, that other 'improved' plate theories exist which do not require all the assumptions of classical plate theory (see Section 11.9).

The development of those portions of classical plate theory that are relevant to this text will now be briefly described and in doing so it will be further assumed that the plate material is homogeneous and isotropic.

We begin development of the classical plate theory by noting that assumption (1) allows us to assume that the vertical movement of all points along any normal to the middle surface is the same, that is

$$w(x, y, z) = w(x, y) \qquad (7.41)$$

where $w(x, y)$ is the deflection of the middle surface at co-ordinate position x, y. Assumption (2) allows us to base the generation of the plate equations on the undeformed geometry rather than on the deformed geometry as would be strictly required. Because of this, when considering assumption (3) we can, for instance, ignore the difference between the co-ordinate position z of a point on the typical plate normal in the undeformed and deformed geometries. Figure 7.9 illustrates what as-sumptions (3) and (4) mean as far as plate deformation taking place in a

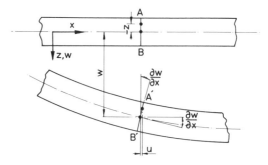

Fig. 7.9. Deformation of the classical plate.

plane lying parallel to the xz plane is concerned; a similar figure could of course be drawn to represent deformation in a plane parallel to the yz plane. The figure shows that a typical normal to the plate middle surface, designated AB, translates in the vertical direction by the amount w and also rotates as a result of bending by an amount equal to the slope $\partial w/\partial x$ of the deformed middle surface. The displacement u in the x direction of a point lying on the normal to the plate can now be written

$$u(x, y, z) = -z \frac{\partial w(x, y)}{\partial x}. \tag{7.42}$$

Clearly the displacement v in the y direction of such a point will similarly be

$$v(x, y, z) = -z \frac{\partial w(x, y)}{\partial y}. \tag{7.43}$$

Equations (7.41)–(7.43) define the three displacement components at any point in the plate in terms of the lateral deflection w of the middle surface.

Having defined the displacement components, the strains can be determined using the three-dimensional equations (7.2). Thus substituting equations (7.41)–(7.43) into equations (7.2) gives

$$\varepsilon_x = -z \frac{\partial^2 w}{\partial x^2}$$

$$\varepsilon_y = -z \frac{\partial^2 w}{\partial y^2} \tag{7.44}$$

$$\gamma_{xy} = -2z \frac{\partial^2 w}{\partial x \, \partial y} = \gamma_{yx}$$

and the other three strains, namely ε_z, γ_{yz}, and γ_{zx}, are zero.

The three non-zero types of strain defined by equations (7.44) are those that occur in plane stress analysis and it is assumed therefore that the corresponding stresses σ_x, σ_y, and τ_{xy} can be obtained from these strains via the elasticity matrix $\mathbf{E}$ defined by equation (7.35). This infers that thin layers of the plate (of thickness dz) are in a state of plane stress. The stress components are then

$$\sigma_x = \frac{-Ez}{1 - \nu^2} \left(\frac{\partial^2 w}{\partial x^2} + \nu \frac{\partial^2 w}{\partial y^2} \right)$$

$$\sigma_y = \frac{-Ez}{1 - \nu^2} \left(\frac{\partial^2 w}{\partial y^2} + \nu \frac{\partial^2 w}{\partial x^2} \right) \tag{7.45}$$

$$\tau_{xy} = \frac{-Ez}{1 + \nu} \left(\frac{\partial^2 w}{\partial x \, \partial y} \right) = \tau_{yx}.$$

It should be recognized that in fact this introduces a contradiction since although ε_z is zero, owing to the assumption of equation (7.41), the corresponding stress σ_z is non-zero† and hence the assumption of a plane stress situation occurring in the layers of thickness dz is not strictly correct. Another contradiction is also present in that γ_{yz} and γ_{zx} have been stated to be zero everywhere and consequently τ_{yz} ($=G\gamma_{yz}$) and τ_{zx} ($=G\gamma_{zx}$) will be zero, i.e. the transverse shear strains and stresses are theoretically zero. This follows directly from the assumption of the Kirchhoff hypothesis and clearly cannot be true since transverse shear stresses acting in the yz and zx planes when integrated over the plate thickness become the shearing forces which are needed to provide vertical support for the plate. We can interpret the presence of non-zero stresses τ_{yz} and τ_{zx} whilst the corresponding strains γ_{yz} and γ_{zx} are zero if we imagine that there is absolutely rigid constraint against transverse shearing, i.e. if we imagine G to have infinite value. The contradictions in classical plate theory are the price paid to achieve a simplified solution but fortunately their effect on the accuracy of the solution is small when the small deflection of thin isotropic plates is considered.

As is the case in the bending of slender beams it is traditional and convenient in plate bending analysis to work in terms of bending (and twisting) moments and transverse shearing forces rather than in terms of stresses; further it is usual to express moments and forces as distributions per unit length. The quantities that contribute to the strain energy of the deformed plate are the bending and twisting moments. To derive expressions for these quantities consider Fig. 7.10 which shows the stresses acting on a plate layer. For convenience the lateral dimensions of the plate layer are taken to be unity. The definitions of the bending moments per unit length are

$$M_x = \int_{-h/2}^{h/2} \sigma_x z \, dz \qquad M_y = \int_{-h/2}^{h/2} \sigma_y z \, dz \qquad (7.46\text{a,b})$$

and additionally the presence of the τ_{xy} shear stress gives rise to a twisting moment per unit length defined as

$$M_{xy} = M_{yx} = \int_{-h/2}^{h/2} \tau_{xy} z \, dz. \qquad (7.46\text{c})$$

The transverse shearing forces do not contribute to the strain energy in classical plate theory since the corresponding shearing strains are zero.

---

† Using the stress–strain relations for three-dimensional elasticity (equations (7.3)–(7.5) and (7.7)) it is easy to see that

$$\sigma_z = \frac{E\nu}{(1+\nu)(1-2\nu)} (\varepsilon_x + \varepsilon_y)$$

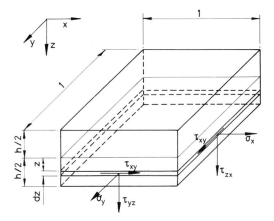

Fig. 7.10. Stresses acting on a plate layer.

Nevertheless they do exist, of course, and their formal definitions are (again expressed per unit length)

$$Q_x = \int_{-h/2}^{h/2} \tau_{zx}\, dz, \qquad Q_y = \int_{-h/2}^{h/2} \tau_{yz}\, dz \qquad (7.47)$$

although they cannot be evaluated in this form since the distributions of τ_{zx} and τ_{yz} are not established. The quantities defined in equations (7.46) and (7.47) are shown in Fig. 7.11 which should make their meaning clear. It should be noted that as is traditional in plate theory, M_x is the moment per unit length acting *along* the x direction not *about* the x axis, and so on.

Q_x and Q_y can in fact be related to the moments using equilibrium conditions by the relations

$$Q_x = \frac{\partial M_x}{\partial x} + \frac{\partial M_{xy}}{\partial y} \qquad Q_y = \frac{\partial M_{xy}}{\partial x} + \frac{\partial M_y}{\partial y} \qquad (7.48)$$

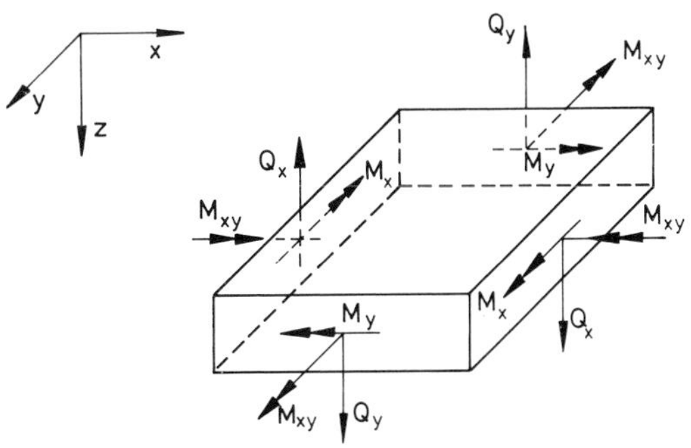

Fig. 7.11. Plate moments and shearing forces.

and hence can be evaluated once definitions of M_x, M_y, and M_{xy} are established (as in equations (7.49) which follow).

To evaluate the bending and twisting moments in terms of second derivatives of w we substitute equations (7.45) into equations (7.46). The result is

$$M_x = -D\left(\frac{\partial^2 w}{\partial x^2} + \nu \frac{\partial^2 w}{\partial y^2}\right)$$

$$M_y = -D\left(\frac{\partial^2 w}{\partial y^2} + \nu \frac{\partial^2 w}{\partial x^2}\right) \tag{7.49}$$

$$M_{xy} = -D(1-\nu)\frac{\partial^2 w}{\partial x \, \partial y}$$

where

$$D = \frac{E}{(1-\nu^2)} \int_{-h/2}^{h/2} z^2 \, dz = \frac{Eh^3}{12(1-\nu^2)}. \tag{7.50}$$

The quantity D is the *flexural rigidity* of the plate (the equivalent of EI in beam theory).

The second derivatives of w that occur here are related to the direct curvatures χ_x and χ_y in the x and y directions and to the twisting curvature χ_{xy} of the deformed middle surface. These quantities are defined as

$$\chi_x = -\frac{\partial^2 w}{\partial x^2} \qquad \chi_y = -\frac{\partial^2 w}{\partial y^2} \qquad \chi_{xy} = -\frac{\partial^2 w}{\partial x \, \partial y} \tag{7.51}$$

We define now a column matrix of moments

$$\boldsymbol{\sigma}_p = \{M_x \quad M_y \quad M_{xy}\} \tag{7.52}$$

and a column matrix of curvatures

$$\boldsymbol{\varepsilon}_p = \{\chi_x \quad \chi_y \quad 2\chi_{xy}\} = -\left\{\frac{\partial^2 w}{\partial x^2} \quad \frac{\partial^2 w}{\partial y^2} \quad 2\frac{\partial^2 w}{\partial x \, \partial y}\right\}. \tag{7.53}$$

Here the symbolism $\boldsymbol{\sigma}_p$ and $\boldsymbol{\varepsilon}_p$ is used since the moments and curvatures whilst being directly related to the true stresses and strains respectively are of course pseudo-stresses and pseudo-strains. It should be noted that the factor 2 associated with χ_{xy} in equation (7.53) is introduced in line with the factor 2 appearing in the expression for γ_{xy} in equation (7.44).

Using equations (7.49)–(7.51) the moment–curvature relationships become

$$\begin{Bmatrix} M_x \\ M_y \\ M_{xy} \end{Bmatrix} = D \begin{bmatrix} 1 & \nu & 0 \\ \nu & 1 & 0 \\ 0 & 0 & \frac{1}{2}(1-\nu) \end{bmatrix} \begin{Bmatrix} \chi_x \\ \chi_y \\ 2\chi_{xy} \end{Bmatrix}$$

or

$$\boldsymbol{\sigma}_p = \mathbf{E}_p \boldsymbol{\varepsilon}_p \qquad (7.54)$$

which is of the form of the general constitutive equation (7.5).

The relationship between pseudo-stresses (moments), pseudo-strains (curvatures),and the plate deflection w are now established. These relationships are necessary ingredients in the construction of stiffness matrices for plate bending elements. We note that the strain energy of an elastic body in which only the direct stresses and strains in the x and y directions and the shearing stress and strain in the xy plane are of importance is, from equations (7.11) and (7.12) of Section 7.3,

$$U_p = \frac{1}{2} \iiint_{V_0} [\sigma_x \quad \sigma_y \quad \tau_{xy}] \begin{Bmatrix} \varepsilon_x \\ \varepsilon_y \\ \gamma_{xy} \end{Bmatrix} dx \, dy \, dz$$

$$= \frac{1}{2} \iiint_{V_0} \boldsymbol{\sigma}^t \boldsymbol{\varepsilon} \, dx \, dy \, dz = \frac{1}{2} \iiint_{V_0} \boldsymbol{\varepsilon}^t \mathbf{E} \boldsymbol{\varepsilon} \, dx \, dy \, dz \qquad (7.55)$$

where V_0, as before, is the volume of the body, $\mathbf{E}$ is defined by equation (7.35) and the (true) strains are given by equation (7.44). For the plate bending situation, when using pseudo-stresses and pseudo-strains, this becomes

$$U_p = \frac{1}{2} \iint_{A_0} \boldsymbol{\varepsilon}_p^t \mathbf{E}_p \boldsymbol{\varepsilon}_p \, dx \, dy \qquad (7.56)$$

where A_0 is the area of the plate middle surface, $\mathbf{E}_p$ is defined by equation (7.54), and the pseudo-strains are given by equation (7.53). The reader can easily check that the two expressions for U_p (equations (7.55) and (7.56)) are identical. From the strain energy expression (equation (7.56)) it is a simple matter to set up an expression for the stiffness matrix of any plate bending element which will be a modified form of equation (7.21) or equation (7.24). One particular point that should be noted is that the strains (pseudo or true) in classical plate theory are expressed in terms of second derivatives of the deflection w, just as was the case with the beam analysis considered earlier. This has important repercussions from the point of view of the compatibility conditions as will be discussed in Chapter 11.

In developing plate bending element properties later we shall also be concerned with evaluating consistent load column matrices corresponding to applied loadings. The applied loadings will act in the z direction and may be of the body force or surface force type. However, in classical plate theory the loads, such as $q(x, y)$ shown in Fig. 7.8, are assumed to act on the middle surface of the plate so that both types of force are in fact accommodated in the same fashion. If $q(x, y)$ is the force per unit area of middle surface the corresponding potential energy V_{pd} (given for the continuum by equation (7.13)) is

$$V_{pd} = -\iint_{A_0} q(x, y)w(x, y)\, dx\, dy. \tag{7.57}$$

A further item that has to be considered in finite-element analysis of plates is the specification of geometric boundary conditions. (The natural or stress boundary conditions need not be considered directly.) For the three-dimensional continuum such prescription on the S_2 part of the boundary has the simple form of equation (7.10). For the plate bending problem the geometric boundary conditions at standard types of supported edge involve specification of first derivatives of w as well as of w itself. Two types of supported edge will be identified here and it is assumed that the edge in question lies parallel to the x axis.

(i) For a *simply supported* edge (i.e. where no deflection is permitted but rotation can take place about the edge)

$$w = 0 \qquad \frac{\partial w}{\partial x} = 0. \tag{7.58}$$

(ii) For a *clamped* edge (i.e. where no deflection or rotation is allowed)

$$w = 0 \qquad \frac{\partial w}{\partial x} = 0 \qquad \frac{\partial w}{\partial y} = 0. \tag{7.59}$$

Finally, although it is not directly required in a finite-element analysis, it is of interest to record that the governing differential equation (in terms of the deflection) of the classical thin plate under lateral load is

$$\frac{\partial^2 w}{\partial x^4} + 2\frac{\partial^4 w}{\partial x^2\, \partial y^2} + \frac{\partial^4 w}{\partial y^4} = \frac{q(x, y)}{D}. \tag{7.60}$$

It should be noted that this equation is a *fourth*-order partial differential equation whereas the governing differential equations of the solid continuum (and of the special cases covered in Sections 7.4 and 7.5) are of only *second*-order when written in terms of displacements.

References

1 TIMOSHENKO, S. and GOODIER, J. *Theory of elasticity* (3rd ed). McGraw-Hill, New York (1970).
2 WANG, C. T. *Applied elasticity*. McGraw-Hill, New York (1953).
3 SOKOLNIKOFF, I. S. *Mathematical theory of elasticity* (2nd ed). McGraw-Hill, New York (1956).

4   TIMOSHENKO, S. and WOINOWSKY-KRIEGER, S. *Theory of plates and shells* (2nd ed). McGraw-Hill, New York (1969).
5   SZILARD, R. *Theory and analysis of plates.* Prentice-Hall, Englewood Cliffs, NJ (1974).
6   RICHARDS, T. H. *Energy methods in stress analysis.* Ellis Horwood, Chichester (1977).

8

SOME FINITE ELEMENTS FOR
PLANAR ANALYSIS

8.1. Introduction

In the analysis of bars, beams, frames, etc., the one-dimensional elements making up the complete structure usually correspond to discrete well-defined physical parts of the structure; it is then quite obvious what elements are required and how they should be arranged to achieve a solution. Such is not the case when we come to consider the finite-element analysis of two- and three-dimensional continua. Here there is generally no unique way of modelling or idealizing the structure with finite elements because usually such elements are simply mathematical regions of the continuum rather than discrete physical parts. In almost all circumstances the derived element relationships, and the structure relationships assembled from them, are approximate. How accurate the solution of the structural problem will be will depend on the accuracy of the individual element relationships and on the number, type, and arrangement of the elements from which the structure is assembled. There is considerable choice available to the analyst in the selection of the basic shape of the element, the displacement field assumed for the element, and the arrangement of the elements.

In two-dimensional analysis one shape that could be employed is the plane straight-sided triangle. Fig. 8.1(a) shows an arbitrary planar problem, and two posible finite-element idealizations of the problem using triangular elements are shown in Figs. 8.1(b) and 8.1(c). The choice of the element mesh is not clear cut and the idealizations shown are only two of an infinite number that could be employed. Furthermore, there is variety in the type of individual triangular element that may be imagined to be illustrated in Fig. 8.1. The basic triangular element has nodal points at its vertices only, but a range of refined elements exists having additional nodes and consequently a more sophisticated representation of displacement and stress (akin to the one-dimensional elements of Section 6.10). Following a discussion of the requirements of element displacement fields in Section 8.2, triangular planar elements are discussed in detail in Section 8.3.

Element shapes other than the straight-sided triangle can be used to solve the plane problem. The rectangle, considered in Section 8.4, is an obvious possibility but if used exclusively it is clearly limited in its

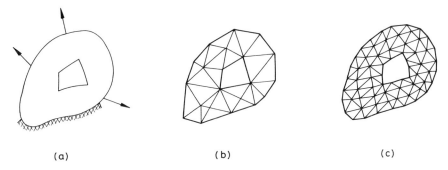

(a)   (b)   (c)

Fig. 8.1. Plane stress structure modelled with triangular finite elements: (a) the structure; (b) coarse mesh; (c) fine mesh.

capacity to model accurately other than very simple structural geometry. However, some types of rectangle have better accuracy than the corresponding triangles, and in any case there is no reason why rectangles cannot be used in association with triangles. By this is meant that the interior of a plane body could be modelled with rectangular elements, with triangular elements used just where necessary to provide a good match to the structure shape at its boundary. This mixing of elements is perfectly permissible so long as the elements can be properly 'connected' at their nodes, that is so long as the common edges of adjacent elements have the same number and location of nodes and the same number of degrees of freedom at each node.

The triangles and rectangles considered in this chapter have the disadvantage of straight sides, which makes it impossible to model curved boundaries precisely, although the triangle does allow a good approximation to such boundaries if a sufficient number of elements are used. It is noted that other more sophisticated elements exist which have curved sides and whose shapes are best described as 'curvilinear quadrilaterals' and 'curvilinear triangles'; these elements will be described in Chapter 10. In connection with this it is noted also that whilst in this chapter the properties of elements are derived on the basis of the use of a conventional Cartesian co-ordinate system, there exist other co-ordinate systems which are often more convenient to use; again Chapter 10 will provide discussion of these systems.

On the general point of the selection of an element mesh we would expect, in a properly formulated analysis, to increase the solution accuracy by using more elements, of a particular type, but this would be achieved only by increasing the effort and cost of obtaining the solution, since the number of structure freedoms is increased. In choosing an element mesh these two contradictory factors clearly have to be borne in mind and the analyst has to use his judgement, intuition, and experience to produce a mesh which will give a result of the desired accuracy for a given computational effort. Obviously the geometry of the actual structure needs to be matched as closely as possible by the element assemblage, and the expected response of the structure to the applied loading must also be considered in deciding the detailed arrangement of elements. This latter point is important, since there will perhaps be regions of the structure where displacements, strains, and stresses vary in a complicated fashion and other regions where the variation is of a simple nature; clearly, relatively more elements should be used in regions of complicated behaviour.

8.2. Requirements for element displacement fields

The necessary requirements for element displacement fields have been discussed in general in Section 6.8 and four specific criteria have been noted. In a Cartesian formulation of the planar problem the first two criteria—relating to rigid body motion and constant strain states—are satisfied if the displacement field *at least* includes the terms

$$u = u(x, y) = A_1 + A_2 x + A_3 y$$
$$v = v(x, y) = A_4 + A_5 x + A_6 y. \tag{8.1}$$

This is easily seen to be so, for with the strain–displacement relationships given by equation (7.34) we then have

$$\{\varepsilon_x \quad \varepsilon_y \quad \gamma_{xy}\} = \{A_2 \quad A_6 \quad (A_3 + A_5)\}$$

so that all strain components are precisely uniform under all conditions. Further, if nodal displacements are such that $A_2 = A_6 = 0$ and $A_3 + A_5 = 0$, then all strain components will have zero value corresponding to the general planar rigid body motion (see end of Section 7.2.2):

$$u_r = A_1 - A_3 y$$
$$v_r = A_4 + A_3 x.$$

This general rigid-body motion comprises a translational component in the x direction ($u = A_1$), a translational component in the y direction, ($v = A_4$), and a rotational component in the xy plane ($u = -A_3 y$, $v = A_3 x$). These components are illustrated in Fig. 8.2, with the displacements greatly enlarged for clarity. (It should be noted that in finite-element analysis of planar structures the element assemblage has to be constrained by the application of appropriate boundary conditions such that none of the components of the rigid-body motion can occur in the body as a whole.)

The third criterion, concerned with displacement compatibility within

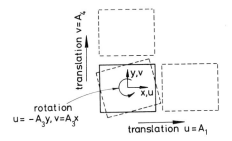

Fig. 8.2. Components of the planar rigid body motion.

the element, poses no difficulty as long as u and v are single valued inside the element. The fourth criterion requires the satisfaction of compatibility at interelement boundaries. It will be recalled from Section 6.8 that the elasticity problem is one in which only C_0 continuity of the overall displacement field is required, that is only continuity of the displacements themselves is necessary corresponding to the fact that only first derivatives of displacement occur in the strain-displacement equations.

It is emphasized that the C_0 continuity condition has to be satisfied all along the common edges of adjacent elements and not just at the common node points, but it will be demonstrated in what follows that this can be achieved quite readily.

A further desirable, but not strictly necessary, requirement of the displacement field for two- and three-dimensional finite elements is *invariance*. This means that the displacement field should be such that the element has no preferential directions, i.e. that its response to any given loading which has a fixed alignment to it should not vary with the orientation of the element and its loading in the global co-ordinate system. This is achieved if the components of the displacement field are assumed to be complete polynomials; for instance the expressions for u and v given above in equation (8.1) are complete linear polynomials in two dimensions and thus the displacement field is invariant. Later in this chapter the use of higher-order complete Cartesian polynomials in the formulation of the properties of planar elements of triangular form will be discussed. Invariance can also be achieved using incomplete polynomials to represent the displacement components if certain conditions are met. These conditions are that if the displacement field is expressed in terms of a local co-ordinate system then it must have a representation of terms that is 'symmetric' or 'balanced' with respect to the co-ordinates. In two dimensions this means that if x and y are assumed to be local co-ordinates then the presence of an $x^m y^n$ term in the displacement field, where m and n are given integers, must be balanced by the corresponding presence of an $x^n y^m$ term. Such incomplete but balanced displacement fields have been widely used in deriving planar elements of rectangular form, as will be described later in this chapter.

8.3. Triangular elements for plane stress or strain

8.3.1. The constant strain triangle: displacement field, stiffness matrix, and stress matrix

The simplest possible triangular element for planar analysis is shown in Fig. 8.3. It has nodes at the vertices of the triangle only, with the nodes numbered in increasing order anticlockwise and with the u and v components of displacement used as nodal degrees of freedom; thus the element has a total of six degrees of freedom with the column matrix of nodal

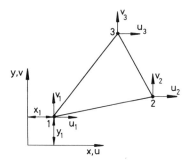

Fig. 8.3. The constant strain triangular element.

displacements for the element defined as

$$\mathbf{d} = \{u_1 \quad v_1 \quad u_2 \quad v_2 \quad u_3 \quad v_3\}. \tag{8.2}$$

It is noted that the orientation of the element with respect to the xy co-ordinate system is completely arbitrary and that u and v are displacement components in the x and y directions. Thus the desired element properties will be expressed directly in the xy global system.† This element was the first successful two-dimensional finite element and was introduced by Turner, Clough, Martin, and Topp [1] (see also [2]).

The satisfaction of rigid-body motion and constant strain states has been seen to require the assumption of a displacement field in which both u and v are written as complete linear polynomials in x and y, as in

---

† In Chapters 3 and 4 global co-ordinates and displacements were indicated by a bar since it was generally necessary to distinguish clearly between local and global systems. This will not often be the case in the remainder of the text and consequently, as long as no possibility of misinterpretation arises, the unbarred coordinates x, y, and z and displacements u, v, and w will be used as global quantities.

equation (8.1). Since this displacement field is expressed in terms of six generalized displacements it can be used directly in the formation of properties for the element of Fig. 8.3.

The displacement field in matrix form is therefore

$$\left\{\begin{matrix} u \\ v \end{matrix}\right\} = \begin{bmatrix} 1 & x & y & 0 & 0 & 0 \\ 0 & 0 & 0 & 1 & x & y \end{bmatrix} \left\{\begin{matrix} A_1 \\ A_2 \\ \vdots \\ A_6 \end{matrix}\right\} \qquad \text{or} \qquad \mathbf{u} = \boldsymbol{\alpha}\mathbf{A}. \tag{8.3}$$

The generalized displacements $\mathbf{A}$ can be expressed in terms of the nodal displacements $\mathbf{d}$ by using the 'boundary conditions' that at each node i $(i = 1, 2, 3)$,

$$u = u_i \qquad \text{and} \qquad v = v_i \qquad \text{at} \qquad x = x_i, y = y_i.$$

This can be expressed as

$$\left\{\begin{matrix} u_1 \\ v_1 \\ u_2 \\ v_2 \\ u_3 \\ v_3 \end{matrix}\right\} = \begin{bmatrix} 1 & x_1 & y_1 & 0 & 0 & 0 \\ 0 & 0 & 0 & 1 & x_1 & y_1 \\ 1 & x_2 & y_2 & 0 & 0 & 0 \\ 0 & 0 & 0 & 1 & x_2 & y_2 \\ 1 & x_3 & y_3 & 0 & 0 & 0 \\ 0 & 0 & 0 & 1 & x_3 & y_3 \end{bmatrix} \left\{\begin{matrix} A_1 \\ A_2 \\ A_3 \\ A_4 \\ A_5 \\ A_6 \end{matrix}\right\} \qquad \text{or} \qquad \mathbf{d} = \mathbf{C}\mathbf{A}. \tag{8.4}$$

The inverse relationship to that of equation (8.4) can be obtained algebraically. Noting that u_1, u_2, and u_3 depend only upon A_1, A_2, and A_3 and v_1, v_2, and v_3 depend only upon A_4, A_5, and A_6, we can separate out two independent sets of three equations, each of which can be solved separately. This requires finding the inverse of the matrix

$$\begin{bmatrix} 1 & x_1 & y_1 \\ 1 & x_2 & y_2 \\ 1 & x_3 & y_3 \end{bmatrix}$$

which can be achieved using the formal definition of a matrix inverse given in Appendix 1. The inverse is found to be

$$\frac{1}{2\Delta}\begin{bmatrix} \alpha_1 & \alpha_2 & \alpha_3 \\ \beta_1 & \beta_2 & \beta_3 \\ \gamma_1 & \gamma_2 & \gamma_3 \end{bmatrix}. \tag{8.5}$$

Here

$$\alpha_1 = \begin{vmatrix} x_2 & y_2 \\ x_3 & y_3 \end{vmatrix} = x_2 y_3 - x_3 y_2$$

$$\beta_1 = -\begin{vmatrix} 1 & y_2 \\ 1 & y_3 \end{vmatrix} = y_2 - y_3 \qquad (8.6)$$

$$\gamma_1 = \begin{vmatrix} 1 & x_2 \\ 1 & x_3 \end{vmatrix} = x_3 - x_2$$

with the other coefficients $\alpha_2 \ldots \gamma_3$ defined by cyclic interchange of the subscripts in the order 1, 2, 3 (i.e. $\alpha_2 = x_3 y_1 - x_1 y_3$ etc.) Also

$$2\Delta = \begin{vmatrix} 1 & x_1 & y_1 \\ 1 & x_2 & y_2 \\ 1 & x_3 & y_3 \end{vmatrix} = \alpha_1 + \alpha_2 + \alpha_3 \qquad (8.7)$$

where the physical meaning of Δ is that it is the area of the triangle 1–2–3. Thus, finally, the complete inverse relationship is

$$\mathbf{A} = \mathbf{C}^{-1} \mathbf{d}$$

where

$$\mathbf{C}^{-1} = \frac{1}{2\Delta} \begin{bmatrix} \alpha_1 & 0 & \alpha_2 & 0 & \alpha_3 & 0 \\ \beta_1 & 0 & \beta_2 & 0 & \beta_3 & 0 \\ \gamma_1 & 0 & \gamma_2 & 0 & \gamma_3 & 0 \\ 0 & \alpha_1 & 0 & \alpha_2 & 0 & \alpha_3 \\ 0 & \beta_1 & 0 & \beta_2 & 0 & \beta_3 \\ 0 & \gamma_1 & 0 & \gamma_2 & 0 & \gamma_3 \end{bmatrix} \qquad (8.8)$$

From equations (8.3) and (8.8) we have the standard form

$$\mathbf{u} = \boldsymbol{\alpha} \mathbf{C}^{-1} \mathbf{d} = \mathbf{N} \mathbf{d} \qquad (8.9)$$

where now, using the definitions of $\boldsymbol{\alpha}$ and $\mathbf{C}^{-1}$ given above,

$$\mathbf{N} = \begin{bmatrix} N_1 & 0 & N_2 & 0 & N_3 & 0 \\ 0 & N_1 & 0 & N_2 & 0 & N_3 \end{bmatrix} \qquad (8.10)$$

is the matrix of shape functions, and

$$N_i = N_i(x, y) = \frac{1}{2\Delta}(\alpha_i + x\beta_i + y\gamma_i) \qquad i = 1, 2, 3. \qquad (8.11)$$

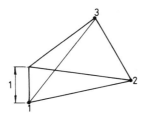

Fig. 8.4. Shape function N_1 for the CST element.

This completes the definition of the element displacement field. Figure 8.4 illustrates the form of a typical shape function (N_1 in fact) corresponding to the definition of equation (8.11). (Compare with Fig. 6.4 for the basic bar element.)

The strain–displacement equations for the plane stress or plane strain problem are given by equation (7.34) and are now used to evaluate matrix $\mathbf{B}$. With $\boldsymbol{\varepsilon}$ defined by equation (7.33) we have that

$$\boldsymbol{\varepsilon} = \begin{bmatrix} \partial/\partial x & 0 \\ 0 & \partial/\partial y \\ \partial/\partial y & \partial/\partial x \end{bmatrix} \begin{Bmatrix} u \\ v \end{Bmatrix} =$$

$$\begin{bmatrix} \partial/\partial x & 0 \\ 0 & \partial/\partial y \\ \partial/\partial y & \partial/\partial x \end{bmatrix} \begin{bmatrix} N_1 & 0 & N_2 & 0 & N_3 & 0 \\ 0 & N_1 & 0 & N_2 & 0 & N_3 \end{bmatrix} \mathbf{d} = \mathbf{Bd} \quad (8.12)$$

where equations (8.9) and (8.10) have been used. Then, with N_i defined by equation (8.11), matrix $\mathbf{B}$ becomes

$$\mathbf{B} = \frac{1}{2\Delta} \begin{bmatrix} \beta_1 & 0 & \beta_2 & 0 & \beta_3 & 0 \\ 0 & \gamma_1 & 0 & \gamma_2 & 0 & \gamma_3 \\ \gamma_1 & \beta_1 & \gamma_2 & \beta_2 & \gamma_3 & \beta_3 \end{bmatrix}. \quad (8.13)$$

The individual terms of $\mathbf{B}$ are seen to be all constants for this element, reflecting the fact that all three strain components have a constant value over the whole element.

The elasticity matrix $\mathbf{E}$, linking stresses to strains, is given by equation (7.35) for plane stress or equation (7.36) for plane strain, assuming isotropic material.

With matrices $\mathbf{B}$ and $\mathbf{E}$ established the stiffness matrix for the constant-strain triangle (CST) can now be determined using equation (7.21). Thus

$$\mathbf{k} = \int_{V_e} \mathbf{B}^t \mathbf{E} \mathbf{B} \, dV_e = \iint_\Delta h \mathbf{B}^t \mathbf{E} \mathbf{B} \, dx \, dy \quad (8.14)$$

where Δ, as stated before, is the area of the triangle and h is the element thickness. Assuming the thickness is constant and performing the matrix multiplication gives, for the plane stress problem (for plane strain E^* and ν^* defined following equation (7.36) replace E and ν)),

$$
\mathbf{k} = \begin{bmatrix} \mathbf{k}_{11} & \mathbf{k}_{12} & \mathbf{k}_{13} \\ \mathbf{k}_{21} & \mathbf{k}_{22} & \mathbf{k}_{23} \\ \mathbf{k}_{31} & \mathbf{k}_{32} & \mathbf{k}_{33} \end{bmatrix} \tag{8.15}
$$

where the $\mathbf{k}_{mn}(m, n = 1, 2, 3)$ are 2×2 submatrices given by

$$
\mathbf{k}_{mn} = \mathbf{k}_{nm}^t = \frac{Eh}{4\Delta(1-\nu^2)} \left[\begin{array}{c:c} \beta_m\beta_n + \frac{1}{2}(1-\nu)\gamma_m\gamma_n & \nu\beta_m\gamma_n + \frac{1}{2}(1-\nu)\gamma_m\beta_n \\ \hdashline \nu\gamma_m\beta_n + \frac{1}{2}(1-\nu)\beta_m\gamma_n & \gamma_m\gamma_n + \frac{1}{2}(1-\nu)\beta_m\beta_n \end{array} \right] \tag{8.16}
$$

Equations (8.15) and (8.16) explicitly define the stiffness matrix for the constant-strain triangular element of uniform thickness.

It should be noted that if the element thickness does vary then $h = h(x, y)$ can be included under the integral sign on the right-hand side of equation (8.14). For linearly varying thickness we could express the thickness in the same manner as a displacement component, that is as

$$h(x, y) = N_1 h_1 + N_2 h_2 + N_3 h_3$$

where $h_i(i = 1, 2, 3)$ is the thickness at node i and N_i is the shape function of equation (8.11). However, it would be consistent with the simplicity of the constant-strain element if a thickness equal to the average of the thicknesses at the nodal points were used in generating $\mathbf{k}$.

The stress matrix $\mathbf{S}$ for the constant-strain triangle is easily obtained using the general expression (equation (7.26)). Thus

$$
\mathbf{S} = \mathbf{EB} = \frac{E}{2\Delta(1-\nu^2)} \times
$$

$$
\begin{bmatrix} \beta_1 & \nu\gamma_1 & \beta_2 & \nu\gamma_2 & \beta_3 & \nu\gamma_3 \\ \nu\beta_1 & \gamma_1 & \nu\beta_2 & \gamma_2 & \nu\beta_3 & \gamma_3 \\ \frac{1}{2}(1-\nu)\gamma_1 & \frac{1}{2}(1-\nu)\beta_1 & \frac{1}{2}(1-\nu)\gamma_2 & \frac{1}{2}(1-\nu)\beta_2 & \frac{1}{2}(1-\nu)\gamma_3 & \frac{1}{2}(1-\nu)\beta_3 \end{bmatrix} \tag{8.17}
$$

and $\mathbf{S}$ relates the stresses $\boldsymbol{\sigma} = \{\sigma_x \; \sigma_y \; \tau_{xy}\}$ to the nodal displacements $\mathbf{d}$ of course. Each stress component has uniform value over the triangle and this causes some difficulty in interpreting the stresses in a mesh of constant-strain elements. This will be referred to again later.

Before considering the representation of distributed loads in a consistent load column matrix we need to look at the question of interelement compatibility for an assemblage of constant-strain triangles. To do this, consider the situation shown in Fig. 8.5 wherein two elements, (1) and (2), have a common edge joining structure nodes I and J. To satisfy the compatibility requirement we have to be sure that when the elements are connected by matching their values of u and v at points I and J in the

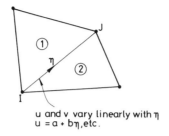

u and v vary linearly with η
u = a + bη, etc.

Fig. 8.5. Compatibility of CST elements.

assembly process, then their values of u and v are also matched at all points along the edge IJ. For this to be possible, obviously the displacement at any point along an edge must depend *only* upon values of displacement at nodal points located on that edge. Now, for the constant-strain triangle the assumed distribution of each displacement component u and v over the triangle is a complete linear polynomial in x and y (see equation (8.3)). It follows that the variation of a displacement component along any straight line will be linear (i.e. a polynomial of order unity), as indicated in Fig. 8.5. Since a linear variation is uniquely defined by the specification of two values, the matching of values of u and v at the two common vertex nodes of adjacent elements does ensure that u and v are also matched at all other points on the common edge. Thus the compatibility requirement is met and the element is conforming. Looking ahead to consideration of other elements this argument can readily be extended to other C_0-continuity elements having node points placed along their edges and not just at the ends. If in a particular element the variation of displacement components along an edge is a polynomial of order $p-1$ and p nodes are present on the edge then the element is conforming.

8.3.2. The constant-strain triangle: consistent load column matrix

The general expression for the consistent load column matrix corresponding to given applied body forces and surface tractions is given by equation (7.23). If a uniform body force is present which has components R_x and R_y in the x and y directions respectively, the appropriate consistent load matrix for the basic triangular element of uniform thickness is (using equations (7.23) and (8.10))

$$\mathbf{Q} = \int_{V_e} \mathbf{N}^t \mathbf{R}\, dV_e = \int_{V_e} \begin{bmatrix} N_1 & 0 \\ 0 & N_1 \\ N_2 & 0 \\ 0 & N_2 \\ N_3 & 0 \\ 0 & N_3 \end{bmatrix} \begin{Bmatrix} R_x \\ R_y \end{Bmatrix} dV_e = h \iint_{\Delta} \begin{Bmatrix} N_1 R_x \\ N_1 R_y \\ N_2 R_x \\ N_2 R_y \\ N_3 R_x \\ N_3 R_y \end{Bmatrix} dx\, dy. \quad (8.18)$$

Now the N_i appearing within the integral are defined by equation (8.11) and are linear functions of x and y (whereas only constants were integrated earlier in evaluating $\mathbf{k}$ (equation (8.14)).). Simply to ease the integration it is convenient to assume now that the co-ordinate origin coincides with the centroid of the element; this will not affect the result since no *rotation* of axes is involved. In such a case we have that

$$\iint_\Delta x \, dx \, dy = \iint_\Delta y \, dx \, dy = 0 \tag{8.19}$$

and

$$\alpha_1 = \alpha_2 = \alpha_3 = 2\Delta/3. \tag{8.20}$$

Then, using equations (8.11), (8.19), and (8.20) in equation (8.18) gives

$$\mathbf{Q} = \frac{\Delta h}{3} \{ R_x \quad R_y \quad R_x \quad R_y \quad R_x \quad R_y \} \tag{8.21}$$

as the consistent load matrix corresponding to the uniform body force. We see that this result indicates that the total body force acting in the x direction is allocated to the three nodes as equal forces in the x direction of one-third of the magnitude of the total force; similar remarks apply to the y direction body force. This result is what would be expected by direct physical reasoning.

Consider now the form of the column matrix of consistent loads corresponding to a given surface traction. The formal expression for this is, from equation (7.23),

$$\mathbf{Q} = \int_{S_e} \mathbf{N}^t \mathbf{T} \, dS. \tag{8.22}$$

It has been emphasized that *this integral is evaluated only over that part of the surface S_e of an element which coincides with a part of the exterior surface of the body as a whole to which tractions are applied.* It is recalled that the expression arises from the potential energy of the tractions at the exterior surface, this energy being expressed as the negative of the integral over the surface S_e of products of components of displacement and traction (i.e. the potential energy is $-\int_{S_e} \mathbf{u}^t \mathbf{T} \, dS$ as given in equation (7.22)). Thus, in evaluating the non-zero consistent loads corresponding to applied traction we need only be concerned with the form of the displacements pertaining to the surface S_e at which tractions are applied and need not be concerned about the way in which displacements vary over the element as a whole. Bearing this in mind will often lead to a rather simpler calculation of the consistent loads than would be the case if the formal definition (equation (8.22)) were applied. A specific example will best illustrate what is meant by this.

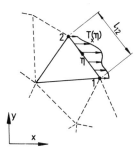

Fig. 8.6. Edge traction on the CST element.

Figure 8.6 shows a basic triangular plane stress element which is arbitrarily oriented to the global axes and whose edge 1–2 lies at the boundary of the structure. An applied traction acts at the edge 1–2. This traction could be of a very general nature but we shall consider first the specific case of a variable traction distributed over the edge 1–2 and acting in the x direction; this traction $T_x(\eta)$ is a specified function of the co-ordinate η which is measured along the boundary edge 1–2 of length l_{12}. Since T_x is the only non-zero traction, the only displacement component we need consider is the u component. For the basic triangular element it has been seen that the displacement components vary linearly along any straight line. Therefore, along edge 1–2 the u component will have the form

$$u^{12} = A + B\eta$$

where A and B are constant coefficients. Since $u = u_1$ at $\eta = 0$ and $u = u_2$ at $\eta = l_{12}$ this can be expressed in the shape function form

$$u^{12} = n_1(\eta)u_1 + n_2(\eta)u_2 = \mathbf{n}_u^{12}\mathbf{d}_u^{12}. \tag{8.23}$$

Here

$$\mathbf{n}_u^{12} = [n_1(\eta)\ n_2(\eta)] \qquad \mathbf{d}_u^{12} = \{u_1\ u_2\}$$

and

$$n_1(\eta) = 1 - \frac{\eta}{l_{12}} \qquad n_2(\eta) = \frac{\eta}{l_{12}}.$$

Regarding the subscripts and superscripts used here, it should be noted that $\mathbf{d}_u^{12}$ denotes the column matrix of the u components of displacement at all nodes along edge 1–2 and $\mathbf{n}_u^{12}$ is the matrix of the corresponding shape functions.

The general form of expression for the consistent loading matrix corresponding to applied tractions, given in equation (8.22), can now be replaced by the expression

$$\mathbf{Q}_u^{12} = \int_{\eta=0}^{l_{12}} (\mathbf{n}_u^{12})^t T_x (h \ d\eta) \tag{8.24}$$

for the particular circumstance shown in Fig. 8.6. Here $\mathbf{Q}_u^{12}$ is the column matrix of those consistent loads which act in the u direction at all the nodes along edge 1–2 only. Performing the operation indicated in equation (8.24) gives

$$\mathbf{Q}_u^{12} = \int_0^{l_{12}} \begin{Bmatrix} 1 - \dfrac{\eta}{l_{12}} \\ \dfrac{\eta}{l_{12}} \end{Bmatrix} T_x h \ d\eta. \tag{8.25}$$

$T_x(\eta)$ is a single (scalar) quantity and its distribution along 1–2 is assumed to be known. For the case of a linear variation of $T_x(\eta)$ from the value t_{x1} at node 1 to t_{x2} at node 2 we have that $T_x(\eta) = (1 - \eta/l_{12})t_{x1} + (\eta/l_{12})t_{x2}$. Substituting this expression into equation (8.25) gives

$$\mathbf{Q}_u^{12} = \frac{hl_{12}}{6} \begin{Bmatrix} 2t_{x1} + t_{x2} \\ t_{x1} + 2t_{x2} \end{Bmatrix}. \tag{8.26}$$

Clearly these two loads will be located in the element consistent load column matrix $\mathbf{Q}$ in locations 1 and 3 corresponding to the positions of u_1 and u_2 in column matrix $\mathbf{d}$.

The following should be noted.

(i) If $t_{x1} = t_{x2} = t^*$ the applied traction is uniform between nodes 1 and 2 and

$$\mathbf{Q}_u^{12} = hl_{12} \begin{Bmatrix} 1/2 \\ 1/2 \end{Bmatrix} t^*, \tag{8.27}$$

that is half the total applied load is concentrated at each of nodes 1 and 2.

(ii) If $t_{x1} = 0$ the applied traction has a triangular distribution varying linearly from zero at node 1 to a maximum of t_{x2} at node 2. Then

$$\mathbf{Q}_u^{12} = \frac{hl_{12}}{6} \begin{Bmatrix} 1 \\ 2 \end{Bmatrix} t_{x2}, \tag{8.28}$$

that is the total applied load is split in the ratio 2 to 1 between nodes 2 and 1.

Both these results would probably have been anticipated by a direct lumping approach but such would not be the case for more complicated displacement distributions where the procedures described here are of greater importance.

So far only the traction T_x has been considered but, in general, a traction at an edge such as 1–2 will have components in both the x and y directions. Where a T_y component is present this will give rise to consistent nodal loads in the y (or v) direction defined, in similar fashion to equation (8.24), as

$$\mathbf{Q}_v^{12} = \int_0^{l_{12}} (\mathbf{n}_v^{12})^t T_y (h \ \mathrm{d}\eta). \tag{8.29}$$

For the basic plane stress element the u and v components vary in the same manner (i.e. linearly) along the edge 1–2 and therefore $\mathbf{n}_v^{12}$ is identical to $\mathbf{n}_u^{12}$.

Although the constant-strain triangle has been considered here it should be realized that as far as the calculation of the consistent load column matrix for a surface traction of the type shown in Fig. 8.6 is concerned, it is immaterial what the shape of the element is. All that concerns us is the way in which displacements (and the applied tractions of course) vary along the edge 1–2; this edge could be the edge of a triangle, as shown in the figure, but could equally be the edge of a rectangle or quadrilateral. The same result would be obtained for any shape of element so long as the manner of variation of the displacements along the particular edge in question is the same.

If body and surface forces act simultaneously on an element then obviously the individual contributions of both types of force have to be added into the column matrix $\mathbf{Q}$.

Example 8.1. The simple plane stress structure shown in Fig. E8.1 is of uniform thickness and is to be modelled using two basic triangular elements as shown in the figure. Neglecting any consideration of boundary conditions set up a stiffness matrix for the structure.

At the outset it should be noted that the chosen element mesh shown in Fig. E8.1 is not the only two-element mesh that could be chosen since clearly a mesh comprising two elements having a common boundary along diagonal 2–4 is equally possible; in most circumstances the answers obtained using the two different element assemblies would not be exactly the same.

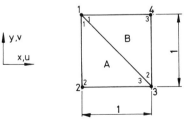

Fig. E8.1

To set up the structure stiffness matrix requires evaluation of the stiffness matrix for each element according to equations (8.15) and (8.16), and then application of the direct stiffness procedure exactly as has been done earlier for skeletal structures. The structure stiffness matrix will correspond to the column matrix of displacements $\mathbf{D}$ defined as

$$\mathbf{D} = \{u_1 \quad v_1 \quad u_2 \quad v_2 \quad u_3 \quad v_3 \quad u_4 \quad v_4\}.$$

In evaluating the stiffness matrix of each element the nodes 1, 2, 3 of the typical element of Fig. 8.3 can be related to the nodes of the present structure in arbitrary fashion as long as it is remembered that the element numbering system proceeds in anticlockwise fashion around the element. Here the connection between the individual element nodes and the structure nodes is chosen to be as follows:

Element	1	2	3	← element nodes
A	1	2	3	← structure nodes.
B	1	3	4	

The numbers shown within the elements in Fig. E8.1 are the node numbers in the element systems.

For element A (see equation (8.6)),

$$\beta_1 = 0, \quad \beta_2 = -1, \quad \beta_3 = 1, \quad \gamma_1 = 1, \quad \gamma_2 = -1, \quad \gamma_3 = 0.$$

Also $\Delta = $ area of element $A = \frac{1}{2}$.
The stiffness matrix for element A is then

$$\mathbf{k}^A = \frac{Eh}{2(1-\nu^2)} \begin{bmatrix} \frac{1}{2}(1-\nu) & & & & & \\ 0 & 1 & & \text{Symmetric} & & \\ -\frac{1}{2}(1-\nu) & -\nu & 1+\frac{1}{2}(1-\nu) & & & \\ -\frac{1}{2}(1-\nu) & -1 & \nu+\frac{1}{2}(1-\nu) & 1+\frac{1}{2}(1-\nu) & & \\ 0 & \nu & -1 & -\nu & 1 & \\ \frac{1}{2}(1-\nu) & 0 & -\frac{1}{2}(1-\nu) & -\frac{1}{2}(1-\nu) & 0 & \frac{1}{2}(1-\nu) \end{bmatrix} \begin{matrix} u_1 \\ v_1 \\ u_2 \\ v_2 \\ u_3 \\ v_3 \end{matrix}$$

Here, as in earlier chapters, the column headings u_1 etc. are written to indicate the *structure* degree of freedom associated with each column and corresponding row.

For element B

$$\beta_1 = -1, \quad \beta_2 = 0, \quad \beta_3 = 1, \quad \gamma_1 = 0, \quad \gamma_2 = -1, \quad \gamma_3 = 1$$

and

$$\Delta = \frac{1}{2} \quad \text{again.}$$

The stiffness matrix for element B is

$$
\mathbf{k}^B = \frac{Eh}{2(1-\nu^2)}
\begin{bmatrix}
1 & & & & \text{Symmetric} & \\
0 & \frac{1}{2}(1-\nu) & & & & \\
0 & \frac{1}{2}(1-\nu) & \frac{1}{2}(1-\nu) & & & \\
\nu & 0 & 0 & 1 & & \\
-1 & -\frac{1}{2}(1-\nu) & -\frac{1}{2}(1-\nu) & -\nu & 1+\frac{1}{2}(1-\nu) & \\
-\nu & -\frac{1}{2}(1-\nu) & -\frac{1}{2}(1-\nu) & -1 & \nu+\frac{1}{2}(1-\nu) & 1+\frac{1}{2}(1-\nu)
\end{bmatrix}
$$

(columns: u_1, v_1, u_3, v_3, u_4, v_4)

For the complete structure the direct stiffness procedure gives the assembled stiffness matrix as

$$
\mathbf{K} = \frac{Eh}{2(1-\nu^2)} \times
$$

(columns: u_1, v_1, u_2, v_2, u_3, v_3, u_4, v_4)

$$
\begin{bmatrix}
\frac{3-\nu}{2} & & & & & & & \\
0 & \frac{3-\nu}{2} & & & & \text{Symmetric} & & \\
-\frac{1}{2}(1-\nu) & -\nu & \frac{3-\nu}{2} & & & & & \\
-\frac{1}{2}(1-\nu) & -1 & \frac{1+\nu}{2} & \frac{3-\nu}{2} & & & & \\
0 & \frac{1+\nu}{2} & -1 & -\nu & \frac{3-\nu}{2} & & & \\
\frac{1+\nu}{2} & 0 & -\frac{1}{2}(1-\nu) & -\frac{1}{2}(1-\nu) & 0 & \frac{3-\nu}{2} & & \\
-1 & -\frac{1}{2}(1-\nu) & 0 & 0 & -\frac{1}{2}(1-\nu) & -\nu & \frac{3-\nu}{2} & \\
-\nu & -\frac{1}{2}(1-\nu) & 0 & 0 & -\frac{1}{2}(1-\nu) & -1 & \frac{1+\nu}{2} & \frac{3-\nu}{2}
\end{bmatrix}
$$

Note that for each of $\mathbf{k}^A$, $\mathbf{k}^B$, and $\mathbf{K}$ the terms appearing in any individual column should be such as to reflect the overall equilibrium of the element or structure (see Section 2.6) and this provides a partial check on manual calculations. Overall equilibrium in the x and y direction respectively requires that for any column of the stiffness matrix the sum of terms in the odd-numbered row locations is zero and the sum of terms in the even-numbered row locations is zero.

Example 8.2. Use the result of Example 8.1 to obtain FEM solutions to the two problems shown in Fig. E8.2(a). This figure illustrates a doubly symmetric situation in which the symmetry allows consideration of just a quarter of the plate

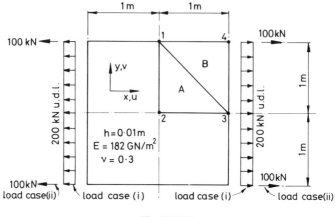

Fig. E8.2(a)

structure, which is shown modelled using the same element mesh as shown in Fig. E8.1. Two forms of applied loading are to be separately considered: in load case (i) uniform tractions act in the x direction at the structure edges lying parallel to the y axis whereas in load case (ii) these tractions are replaced with concentrated forces acting at the plate corners.

For both loading cases the structure stiffness matrix will be identical. Before application of the pertinent boundary conditions this stiffness matrix is $\mathbf{K}$, given above. The nodal boundary conditions which reflect the symmetry of the problems are that $u_1 = u_2 = v_2 = v_3 = 0$. Since displacements vary linearly on the edge of the constant-strain triangle, application of these nodal conditions in fact means that u is zero all along edge 1–2 and v is zero all along edge 2–3, as required. The problems have the four degrees of freedom defined by

$$\mathbf{D}_r = \{v_1 \quad u_3 \quad u_4 \quad v_4\}$$

The reduced structure stiffness matrix $\mathbf{K}_r$ is obtained from $\mathbf{K}$ in the usual manner by eliminating rows and columns corresponding to the zero displacements. For the given values of E, ν, and h the result is (in kilonewton and metre units)

$$\mathbf{K}_r = 10^6 \begin{bmatrix} v_1 & u_3 & u_4 & v_4 \\ 1.35 & & & \\ 0.65 & 1.35 & \text{Symmetric} & \\ -0.35 & -0.35 & 1.35 & \\ -0.35 & -0.35 & 0.65 & 1.35 \end{bmatrix}.$$

Load case (i). The loading applied to the structure is distributed and it is therefore necessary to set up the column matrix $\mathbf{F}_Q$ of consistent nodal loads for the structure from the contributions $\mathbf{Q}_e$ of the individual elements. It is obviously only element B that makes a contribution to $\mathbf{F}_Q$ and the non-zero consistent nodal loads for this element are simply half of the total traction applied to edge 3–4 (see

Section 8.3.2). Consequently

$$\begin{array}{cccccc} u_1 & v_1 & u_3 & v_3 & u_4 & v_4 \end{array}$$
$$\mathbf{Q}^B = \{0 \quad 0 \quad 50 \quad 0 \quad 50 \quad 0\}.$$

The reduced structure consistent load column matrix is then

$$\begin{array}{cccc} v_1 & u_3 & u_4 & v_4 \end{array}$$
$$\mathbf{F}_{Qr} = \{0 \quad 50 \quad 50 \quad 0\}.$$

Applying equation (6.58b), with $\mathbf{F}_r = \mathbf{0}$ since no external forces are applied directly to the nodes, gives

$$\begin{Bmatrix} 0 \\ 50 \\ 50 \\ 0 \end{Bmatrix} = 10^6 \begin{bmatrix} 1.35 & & & \\ 0.65 & 1.35 & \text{Symmetric} & \\ -0.35 & -0.35 & 1.35 & \\ -0.35 & -0.35 & 0.65 & 1.35 \end{bmatrix} \begin{Bmatrix} v_1 \\ u_3 \\ u_4 \\ v_4 \end{Bmatrix}.$$

The solution of these stiffness equations is

$$\{v_1 \quad u_3 \quad u_4 \quad v_4\} = \frac{1}{18200}\{-0.3 \quad 1 \quad 1 \quad -0.3\}\,\text{m}$$

Having found the structure displacements the calculation of the stresses in the individual elements proceeds using the stress matrix $\mathbf{S}$ defined by equation (8.17). For element A we have that

$$\mathbf{d}^A = \{u_1 \ v_1 \ u_2 \ v_2 \ u_3 \ v_3\} = \frac{1}{18200}\{0 \ -0.3 \ 0 \ 0 \ 1 \ 0\}$$

and the parameters β_1, γ_1 etc. have been recorded earlier. Thus

$$\begin{Bmatrix} \sigma_x \\ \sigma_y \\ \tau_{xy} \end{Bmatrix} = \frac{182 \times 10^6}{0.91} \begin{bmatrix} 0 & 0.3 & -1 & -0.3 & 1 & 0 \\ 0 & 1 & -0.3 & -1 & 0.3 & 0 \\ 0.35 & 0 & -0.35 & -0.35 & 0 & 0.35 \end{bmatrix} \times$$

$$\begin{Bmatrix} 0 \\ -0.3 \\ 0 \\ 0 \\ 1 \\ 0 \end{Bmatrix} \frac{1}{18200} = \begin{Bmatrix} 10^4 \\ 0 \\ 0 \end{Bmatrix} \text{kN m}^{-2}$$

Similarly for element B

$$\mathbf{d}^B = \{u_1 \ v_1 \ u_3 \ v_3 \ u_4 \ v_4\} = \frac{1}{18200}\{0 \ -0.3 \ 1 \ 0 \ 1 \ -0.3\}$$

and the stresses become

$$\{\sigma_x \quad \sigma_y \quad \tau_{xy}\}^B = \{10^4 \quad 0 \quad 0\}\,\text{kN m}^{-2}.$$

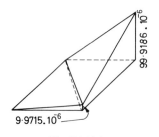

Fig. E8.2(b)

Thus it is seen that the finite-element solution for load case (i) predicts a uniform stretching of the whole plate in the x direction of total magnitude $2u_3 = 2u_4 = 2/18200$ m, a uniform contraction in the y direction of total magnitude $2v_1 = 2v_4 = 0.6/18200$ m, and a stress field in which $\sigma_x = 10^4$ kN m^{-2} everywhere, with the other stress components having zero value. The exact solution to this problem corresponding to a uniform traction in the x direction of magnitude σ is readily obtained from elasticity theory as $u = \sigma x/E$, $v = -\nu\sigma y/E$, $\sigma_x = \sigma$, and $\sigma_y = \tau_{xy} = 0$, and so it is concluded that the finite-element solution is the exact solution for the given simple loading state. This was to be expected, of course, since the finite element used does precisely represent states of constant stress and strain.

Load case (*ii*). Here the applied load acts directly at structure node 4 so that

$$\mathbf{F}_r = \{0 \quad 0 \quad 100 \quad 0\}$$

and $\mathbf{F}_{Or}$ is zero, of course. The stiffness equations for this load case are as for load case (i) but with this $\mathbf{F}_r$ replacing the $\mathbf{F}_{Or}$ of case (i). The solution of the stiffness equations is then

$$\{v_1 \quad u_3 \quad u_4 \quad v_4\} = 10^{-6}\{9.9715 \quad 9.9715 \quad 99.9186 \quad -42.9385\} \text{ m}.$$

In Fig. E8.2(b) the way in which the u component of displacement varies over the analysis quadrant is illustrated. It should be noted that the displacement is single valued everywhere but that the rate of change of u alters at the interelement boundary of course.

The element stresses can be calculated as for case (i) and the results are

$$\{\sigma_x \quad \sigma_y \quad \tau_{xy}\}^A = \{2592.6 \quad 2592.6 \quad 0\} \text{ kN m}^{-2}$$
$$\{\sigma_x \quad \sigma_y \quad \tau_{xy}\}^B = \{17407.4 \quad -2592.6 \quad 2592.6\} \text{ kN m}^{-2}.$$

The major stress component is σ_x. The variation of this stress component over the quarter plate is shown diagrammatically in Fig. E8.2(c) where it is seen that

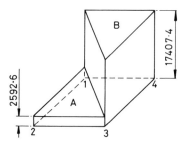

Fig. E8.2(c)

the rather extreme corner-load case results in a finite-element stress distribution which exhibits a gross discontinuity at the interelement boundary. This is hardly surprising when using only two basic elements in a problem where the real stress distribution is quite complex. Obviously the solution obtained here is not very precise and considerably more elements would be required to obtain an accurate solution wherein the finite-element stress discontinuities at interelement boundaries would be small. This will not be pursued here but it is noted that some convergence studies using the basic and other planar elements are described in Section 8.3.6.

Although the present two-element solution is inaccurate in any detailed sense it is worth noting the following points related to stress prediction in an overall sense.

(a) The average finite-element stress σ_x in the analysis region is the average of 2592.6 and 17407.4, that is it is exactly $10\,000\ \text{kN m}^{-2}$. This is what the average stress should be in the real problem of course.

(b) The average finite-element stress σ_y in the analysis region is zero, again coinciding with the real average stress value.

8.3.3. The linear strain triangle

The first of the refined triangular elements considered here (and the first historically speaking, presented by de Veubeke [3] and Argyris [4, 5]) is based on the assumption of complete second-order polynomial fields, that is quadratic fields, for the displacement components. The displacement field is thus

$$\begin{Bmatrix} u \\ v \end{Bmatrix} = \begin{bmatrix} 1 & x & y & x^2 & xy & y^2 & 0 & 0 & 0 & 0 & 0 & 0 \\ 0 & 0 & 0 & 0 & 0 & 0 & 1 & x & y & x^2 & xy & y^2 \end{bmatrix} \begin{Bmatrix} A_1 \\ A_2 \\ \vdots \\ A_{12} \end{Bmatrix}$$

$$\text{or} \quad \mathbf{u} = \boldsymbol{\alpha}\mathbf{A} \tag{8.30}$$

There are six generalized displacements associated with each displacement component and thus six nodal values pertaining to both u and v need to be specified. A suitable arrangement of nodal freedoms for the triangle is shown in Fig. 8.7; nodes are introduced along the edges, conveniently at the mid-points, and values of u and v at each of the six nodes are the element degrees of freedom. The presence of three nodes

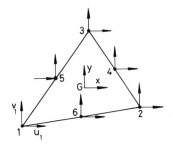

Fig. 8.7. The linear strain triangular element.

along any edge, where the components of displacement vary quadratically, ensures that the element is a conforming one. Also, the assumed
displacement field obviously includes the representation of rigid-body
motion and constant strain states. Thus all required conditions are met by
the displacement field.

It will be observed in Fig. 8.7 that although the co-ordinate axes x, y
are arbitrarily oriented to the triangle, the co-ordinate origin is located at
a specific point G. This point is the centroid of the triangle and it is
chosen as the location of the origin simply to facilitate the integration
over the area of the triangle of those simple functions of x and y that
arise in the development of element properties such as stiffness etc. (This
was done for this same reason in Section 8.3.2 when calculating the
consistent load column matrix for a body force. The choice of origin has
no significance other than that of convenience since the stiffness etc. is
independent of the *position* of the coordinate origin.) Before proceeding,
it is as well to record all the formulae which are required to evaluate the
necessary integrations—with the coordinate origin at the centroid—which
arise in the development of the properties of the linear strain triangle
(LST). These are, with Δ again denoting the area of the triangle,

$$\iint_\Delta dx\,dy = \Delta \qquad\qquad \iint_\Delta x\,dx\,dy = \iint_\Delta y\,dx\,dy = 0$$

$$\iint_\Delta x^2\,dx\,dy = \frac{\Delta}{12}(x_1^2 + x_2^2 + x_3^2)$$

$$\iint_\Delta y^2\,dx\,dy = \frac{\Delta}{12}(y_1^2 + y_2^2 + y_3^2)$$

$$\iint_\Delta xy\,dx\,dy = \frac{\Delta}{12}(x_1y_1 + x_2y_2 + x_3y_3) \qquad\qquad (8.31)$$

where x_1, y_1 are the co-ordinate positions of node 1 etc.

A further advantageous consequence of locating the co-ordinate
origin at the centroid is that with the co-ordinates of a corner node i
($i = 1, 2$, or 3) being x_i, y_i the co-ordinates of a corresponding mid-side
node $(i + 3)$ are

$$x_{i+3} = -\tfrac{1}{2}x_i \qquad\qquad y_{i+3} = -\tfrac{1}{2}y_i. \qquad\qquad (8.32)$$

The column matrix of nodal displacements is defined as

$$\mathbf{d} = \{u_1 \quad v_1 \quad u_2 \quad v_2 \ldots u_6 \quad v_6\}. \tag{8.33}$$

In the usual way the nodal displacements can be related to the generalized displacements by using the conditions at the nodes. This gives

$$\begin{Bmatrix} u_1 \\ v_1 \\ \vdots \\ u_4 \\ v_4 \\ \vdots \\ v_6 \end{Bmatrix} = \begin{bmatrix} 1 & x_1 & y_1 & x_1^2 & x_1y_1 & y_1^2 & 0 & 0 & 0 & 0 & 0 & 0 \\ 0 & 0 & 0 & 0 & 0 & 0 & 1 & x_1 & y_1 & x_1^2 & x_1y_1 & y_1^2 \\ \vdots & \vdots & \vdots & \vdots & \vdots & \vdots & \vdots & \vdots & \vdots & \vdots & \vdots & \vdots \\ 1 & -\frac{1}{2}x_1 & -\frac{1}{2}y_1 & \frac{1}{4}x_1^2 & \frac{1}{4}x_1y_1 & \frac{1}{4}y_1^2 & 0 & 0 & 0 & 0 & 0 & 0 \\ 0 & 0 & 0 & 0 & 0 & 0 & 1 & -\frac{1}{2}x_1 & -\frac{1}{2}y_1 & \frac{1}{4}x_1^2 & \frac{1}{4}x_1y_1 & \frac{1}{4}y_1^2 \\ \vdots & \vdots & \vdots & \vdots & \vdots & \vdots & \vdots & \vdots & \vdots & \vdots & \vdots & \vdots \\ 0 & 0 & 0 & 0 & 0 & 0 & 1 & -\frac{1}{2}x_3 & -\frac{1}{2}y_3 & \frac{1}{4}x_3^2 & \frac{1}{4}x_3y_3 & \frac{1}{4}y_3^2 \end{bmatrix} \begin{Bmatrix} A_1 \\ A_2 \\ \vdots \\ \vdots \\ \vdots \\ A_{12} \end{Bmatrix}$$

$$\tag{8.34}$$

or

$$\mathbf{d} = \mathbf{CA}.$$

The matrix $\mathbf{C}$ can be inverted, though this inverse will not be given here, and the standard form $\mathbf{u} = \alpha\mathbf{C}^{-1}\mathbf{d}$ or $\mathbf{u} = \mathbf{Nd}$ can be used to complete the definition of the displacement field. The form of the shape functions is illustrated in Fig. 8.8 for a typical corner node (node 1) and a typical mid-side node (node 6); see Fig. 6.10 for comparison with the shape functions of the three-node bar element.

The elasticity matrix $\mathbf{E}$ is, of course, again given by equation (7.35) for plane stress or equation (7.36) for plane strain.

The 12×12 stiffness matrix can be generated using either of the standard expressions (equation (7.21) or equation (7.24)). If we choose to use the latter form the matrix $\boldsymbol{\beta}$, which occurs in the relationship $\mathbf{E} = \boldsymbol{\beta}\mathbf{A}$, is

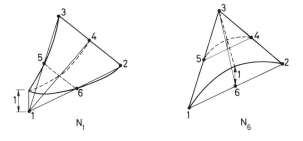

Fig. 8.8. Shape functions for the LST element.

obtained by appropriate differentiation of terms in $\boldsymbol{\alpha}$; that is

$$\boldsymbol{\beta} = \begin{bmatrix} \partial/\partial x & 0 \\ 0 & \partial/\partial y \\ \partial/\partial y & \partial/\partial x \end{bmatrix} \begin{bmatrix} 1 & x & y & x^2 & xy & y^2 & 0 & 0 & 0 & 0 & 0 & 0 \\ 0 & 0 & 0 & 0 & 0 & 0 & 1 & x & y & x^2 & xy & y^2 \end{bmatrix}$$

or

$$\boldsymbol{\beta} = \begin{bmatrix} 0 & 1 & 0 & 2x & y & 0 & 0 & 0 & 0 & 0 & 0 & 0 \\ 0 & 0 & 0 & 0 & 0 & 0 & 0 & 0 & 1 & 0 & x & 2y \\ 0 & 0 & 1 & 0 & x & 2y & 0 & 1 & 0 & 2x & y & 0 \end{bmatrix}. \tag{8.35}$$

The individual terms of $\boldsymbol{\beta}$ (and hence of $\mathbf{B} = \boldsymbol{\beta}\mathbf{C}^{-1}$) are either constants or are proportional to x or to y. This reflects the linear nature of the strain (and hence stress) variation over the element. The required integration involved in generating $\mathbf{k}$ using equation (7.24) can be accommodated quite easily with the use of equations (8.31).

The element stress matrix $\mathbf{S}$ can easily be determined as the triple product $\mathbf{E}\boldsymbol{\beta}\mathbf{C}^{-1}$ for a general point x, y. Numerical specification of the co-ordinate position of a point will allow the calculation of σ_x, σ_y, and τ_{xy} at that point; usually the stresses are calculated at the vertices of the element (and are known to vary linearly between the vertices).

Consistent load column matrices corresponding to body and surface forces can be determined using the procedures outlined in Section 8.3.2. These procedures are more important for refined elements than for more basic elements since it becomes increasingly difficult to allocate loading to the nodes by intuitive reasoning as the number of nodes increases. Some problems for solution involving the consistent representation of applied loadings are given at the end of the chapter.

8.3.4. Other triangular elements

The two types of planar triangular element considered so far are simply the first two of an unlimited series, or family, of elements that could be developed for the C_0-continuity problem. The basic element, with constant strain properties, is derived from a displacement field in which each of the two components is a complete first-order (or linear) polynomial in the Cartesian co-ordinates x and y. The next element, with linear strain properties, is based on the assumption of complete second-order (or quadratic) polynomials for u and v. Other refined triangular elements can be based on expressing u and v as complete Cartesian polynomials of progressively higher order, as will now be discussed. Further details of these elements are given in references 6–10.

In two dimensions the various levels of terms corresponding to complete polynomials are conveniently represented by what is known as the

Pascal triangle, in the form

$$
\begin{array}{c}
1 \\
x \qquad y \\
x^2 \qquad xy \qquad y^2 \\
x^3 \qquad x^2y \qquad xy^2 \qquad y^3 \\
x^4 \qquad x^3y \qquad x^2y^2 \qquad xy^3 \qquad y^4 \\
x^5 \qquad x^4y \qquad x^3y^2 \qquad x^2y^3 \qquad xy^4 \qquad y^5
\end{array}
$$

complete linear (3 terms)

complete quadratic (6 terms)

complete cubic (10 terms)

complete quartic (15 tems)

complete quintic (21 terms).

etc.

$$(8.36)$$

This scheme readily shows how many terms are present in a polynomial of particular order, and what these terms are.

We have seen the configuration of the nodes of a triangular element corresponding to the complete linear and complete quadratic representation of the u and v displacement components in Figs. 8.3 and 8.7. By direct extension we can envisage a family of elements based on the use of complete polynomial displacement fields with nodes located in positions which effectively correspond to the positions of individual terms in the Pascal triangle. The first five members of such a family of elements (including the two already considered in detail) are illustrated in Fig. 8.9. The degrees of freedom at each and every node are the values of u and v, and so the total number of degrees of freedom for the various elements, in order, is 6, 12, 20, 30, 42, and so on.

Each member of the family of elements is perfectly valid in that all necessary and desirable conditions for convergence are met; rigid-body

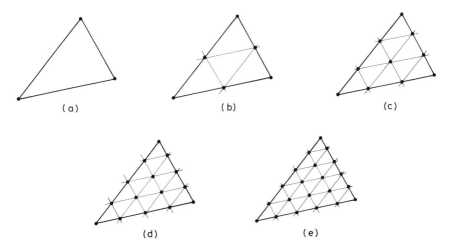

Fig. 8.9. First five members of a family of planar triangular elements.

314   ANALYSIS OF STRUCTURES

motion and constant strain states are properly represented, the elements
are conforming, and the displacement fields are geometrically invariant.
In all cases, if the displacement field is initially expressed in terms of
generalized displacements the inverse matrix $\mathbf{C}^{-1}$ can be determined.

It can be seen from Fig. 8.9 that with the exception of the two
lowest-order members—the CST and LST triangles—the elements of the
family have one or more interior nodes. The degrees of freedom associated with these interior nodes can be removed at the element level
(i.e. before assembling the structure equations) by the process of static
condensation described in Section 6.11. This effectively reduces the
number of degrees of freedom of the element (from 20 to 18, 30 to 24,
and 42 to 30 for the elements shown in Figs. 8.9(c), 8.9(d) and 8.9(e)
respectively) at the expense of some increased work and complexity in
formulating the element properties.

It should be noted that there are alternative arrangements of nodal
freedoms, other than the use of values of just u and v, that could be used
in developing triangular elements for planar analysis. For example, as an
alternative to the element illustrated in Fig. 8.9(c), which corresponds to
the cubic variation of displacement or quadratic variation of strains
(hence referred to as the QST element) the arrangement of nodal
variables shown in Fig. 8.10 could be adopted. Here no nodes are present
on the element sides between the vertices but, of course, there is an
increased number of freedoms present at the vertice nodes. These freedoms can be either

$$u, \quad v, \quad \partial u/\partial x, \quad \partial u/\partial y, \quad \partial v/\partial x, \quad \partial v/\partial y$$

as shown in the figure, or can be

$$u, \quad v, \quad \varepsilon_x, \quad \varepsilon_y, \quad \varepsilon_{xy}, \quad \omega_{xy}$$

where $\omega_{xy} = \frac{1}{2}(\partial u/\partial y - \partial v/\partial x)$. The element is still conforming with the use
of either of these sets of nodal freedoms since for each component of
displacement there are effectively four quantities available at the nodes

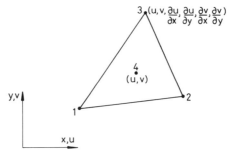

Fig. 8.10. Alternative form of the QST element.

which pertain to each element edge, along which the displacement varies as a cubic polynomial.

In some applications this modified type of formulation of the cubic displacement element has some considerable advantages in that the concentration of element degrees of freedom at the vertices often substantially reduces both the total number of structure degrees of freedom and the bandwith (see Example 8.3 below). However, it is clear that an assemblage of such elements will be connected with excess continuity (see Section 6.10) and this would lead to some difficulty where step changes of load or material property occur. Alternative arrangements of the nodal freedoms for the quartic and quintic displacement models are also possible but these will not be detailed here.

8.3.5. Comments on the triangular elements

Given the range of straight-sided triangular elements that exist (or could be derived) the question which naturally arises is: Which element is the best one to use? Unfortunately there is no clear-cut answer that can be given to this, since relative performance of the various element types is very much dependent on the type of problem being studied. Choice of a particular type of element for use in the solution of a particular problem would be influenced by the following considerations.

In most circumstances the more refined or sophisticated the element is, i.e. the higher the order of the displacement field, the less will be both the number of elements (of course) and (more pertinently) the total number of structure degrees of freedom required to achieve a specific level of accuracy of solution. This points to the use of rather refined elements but against this viewpoint three factors must be taken into account. Firstly, with refined elements there is more computational effort required in formulating the properties of individual elements, as compared with simple elements: this may well nullify some of the saving occurring from the reduction in the number of overall degrees of freedom. Secondly, for the same overall number of degrees of freedom the bandwidth of the SSM may be considerably greater when using refined elements than when using simple elements. (This is illustrated in Example 8.3 below). Thirdly, if the shape of the plane stress or strain structure is rather complicated, then when using straight-sided elements it will be easier to idealize the structure accurately using a relatively large number of simple elements than it will be using a relatively small number of refined elements. A reasonable compromise in *most* applications would be to use the elements based on either quadratic displacements (linear strains) or cubic displacements (quadratic strains). Normally, these elements would be significantly more efficient than would be the basic (constant strain) element, whilst there would be little, if any, gain in efficiency in using elements with cubic or higher-order strain variation. (Two particular comparison studies involving CST and LST elements are detailed in Section 8.3.6).

In the above we have been concerned with developing triangular elements for planar analysis based on the assumption of a displacement field whose components are expressed as functions of the *Cartesian* co-ordinates x and y. Although this is perfectly satisfactory in the sense that valid elements, complying with all necessary and desirable criteria, can be generated in this way, it is appropriate to point out that another type of co-ordinate system has frequently been used in the derivation of triangular element properties. The alternative co-ordinates were specifically developed for this purpose and are known as 'triangular' or 'area' co-ordinates. They provide an elegant alternative to the Cartesian system and allow the direct specification of element shape functions. Area co-ordinates are a particular form of 'natural' co-ordinates; these co-ordinate systems will be discussed in Chapter 10.

Example 8.3. Compare the semi-bandwidth and sizes of the SSMs of the regular meshes of different types of planar triangle shown in Fig. E8.3.

The object here is to provide amplification of remarks made earlier concerning the form of the SSM when using triangular elements based on different orders of interpolation.

In Figs. E8.3(a), E8.3(b), and E8.3(c) are shown regular arrays of CST, LST, and QST elements in which the type, location, and number ($49 \times 2 = 98$) of overall structure degrees of freedom is the same in each instance. (For the QST element the degree of freedom at the internal node is assumed not to have been removed by static condensation.) Because of the nature of the three types of element a force at a typical vertex node in the interior is linked to the displacements at differing numbers of nodes in each case. In Figs. E8.3(a), E8.3(b), and E8.3(c) the nodes indicated by full circles are those which are connected by stiffness relationships to a typical interior vertex node (node 33 for (a) and (b), node 25 for (c)). The effect of the different patterns of connection on the size of the semi-bandwidth is considerable. For the CST, LST, and QST elements this size is 16, 30, and 44 respectively. (For the CST element of Fig. E8.3(a), for instance, the semi-bandwidth is calculated as $2(40-33)+2$ in the manner of Section 3.8.) Thus

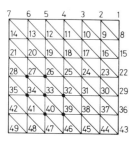

Fig. E8.3(a)

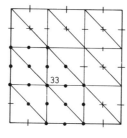

Fig. E8.3(b)

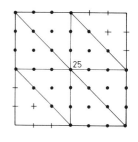

Fig. E8.3(c)

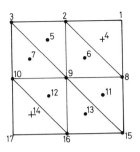

Fig. E8.3(d)

it can be seen that for the same size problem the use of higher-order elements gives increased bandwidth and hence requires more computational effort to achieve a solution than does the use of lower-order elements.

Fig. E8.3(d) shows an array of the alternative (four-noded) form of the QST element in which derivative degrees of freedom are used so that six quantities are present at vertex nodes (with two degrees of freedom still present at the interior node of each element). The total number of degrees of freedom with this arrangement is $9 \times 6 + 8 \times 2 = 70$, assuming that all freedoms are connected at the vertex nodes. This is a considerable reduction compared with the 98 degrees of freedom associated with the ten-noded QST version shown in Fig. E8.3(c). Furthermore the semi-bandwidth is now 32 compared with 44 when using the ten-noded version.

8.3.6. Numerical studies using the constant strain triangle and linear strain triangle elements

In this section the solution of two fairly typical test problems is considered using in turn CST and LST plane triangular elements. The first problem concerns a deep beam and is analysed as a plane stress problem using regular element meshes. (The beam is too deep to allow accurate predictions using an analysis based on Bernoulli–Euler theory.) The second problem concerns a rectangular plate containing a small circular hole at the plate centre and loaded with uniform traction at the edges. Steep stress gradients occur local to the hole and necessitate the use of irregular, or graded, element meshes which are refined in the region of the hole. The two problems have known solutions based on elasticity theory [11] and were chosen with this in mind.

In both problems the applied loading is uniform edge traction; this is applied as a column matrix of consistent nodal loads of course.

The presentation of finite-element stress results, in the form of distributions along some line or other, presents some difficulty owing to the fact that any stress component is not single valued (apart from exceptional cases) at a node common to two or more elements. This is most markedly so for the CST element for which the distribution of stresses strictly presents a histogram-type appearance, often with large step changes at interelement boundaries when few elements are used (see Example 8.2 for instance). Some 'smoothing-out' process is desirable in interpreting the finite-element stresses for graphical presentation and a number of schemes—some simple and some complex—have been suggested for this. One simple scheme which is often suitable for elements in which the components of stress are restricted to uniform values (as with the CST and with the basic bar element) is to plot values at the centroids of elements. Another simple scheme is to represent stress component values at nodes where elements join as the average of the stress components in all such joining elements. This scheme, like all the others, is not perfect but it is simple to implement and is consequently used here. The errors involved in any valid stress interpretation scheme will, of course, reduce with the use of an increasingly fine element mesh.

The two problems presented here will be reconsidered using other element types in Section 10.4.5.

Uniformly loaded deep beam. The deep beam is shown in Fig. 8.11(a).

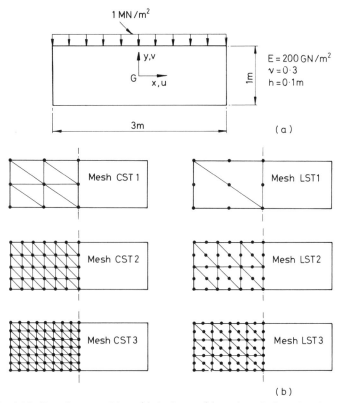

Fig. 8.11. Deep beam problem: (a) the beam; (b) meshes of triangular elements.

It carries a load uniformly distributed along its top edge and is supported at its ends by vertical shearing forces whose presence is deemed in the finite-element analysis to be equivalent to setting to zero the vertical displacement at all nodes lying at an end. The regular meshes of finite elements used in generating numerical results are shown in Fig. 8.11(b); each mesh of CST elements is matched by a mesh of LST elements containing the same number of nodal points (and hence the same number of degrees of freedom) but of course only a quarter of the number of elements. The problem is obviously symmetric about the vertical centre line of the beam and hence only half of the beam is considered, with the appropriate symmetry condition applied by setting the horizontal displacement component to zero at all nodes lying on the vertical centre line.

A quality which can be used as a measure of the relative accuracy of the finite-element results is the vertical deflection at the beam centre (point G). Figure 8.12(a) shows the results together with the elasticity theory solution and the Bernoulli–Euler beam theory solution (which is in considerable error). It can be seen that on the basis of the number of nodes (or degrees of freedom) used the performance of the LST element is significantly superior to that of the CST element. It should be noted that the number of nodes referred to in Fig 8.12(a) is that pertaining to the whole beam even though only a symmetric half of the beam was used in the analyses. This is done so as to correspond to further results presented in Section 10.4.5 for this problem.

The major stress component is the longitudinal stress at the upper and lower edges of the beam at the vertical centre line. Fig. 8.12(b) shows how the calculated finite-element results for longitudinal stress σ_x compare with the elasticity theory solution at the vertical centre line.

Rectangular plate with central hole. Figure 8.13(a) shows a rectangular plate having a small circular hole at its centre with the plate being subjected to uniform edge tractions in one or two directions (f_1 is non-zero but f_2 may be zero or non-zero). The problem is doubly symmetric and hence only a quarter of the plate need be considered if appropriate displacement conditions are applied to nodes lying on the horizontal and vertical centre lines of the plate (i.e. $v = 0$ along the horizontal centre line, $u = 0$ along the vertical centre line). The element meshes used in the analysis quadrant (the top right-hand quadrant) are shown in Fig. 8.13(b); note that the meshes shown there apply for both CST and LST elements, with nodes at the element vertices only in the former case and at the vertices and mid-side points in the latter case.

The meshes used in this application are irregular with considerable refinement employed local to the hole. Such refinement is required in part to represent the curved edge of the hole adequately but also to accommodate the steep stress gradients that occur near the hole. In deciding upon

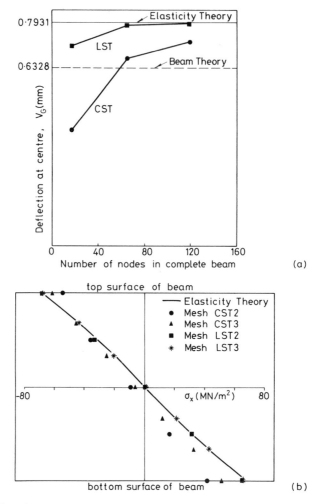

Fig. 8.12. Deep beam problem: results using triangular elements: (a) central deflection; (b) σ_x at centre line.

an element mesh it should be realized that accuracy is related to the shape of the individual elements; the elements should be kept as compact as possible, i.e. long thin triangles should be avoided.

The largest stress component is the circumferential component at the point A, that is σ_x^A. Calculated values of this quantity are given in the table of Fig. 8.13(c) where the exact solution of elasticity theory is also indicated. (In fact this exact solution strictly pertains to an infinitely large plate but the error involved in applying it to the plate considered here is very slight.) It can be seen that results for the $f_1 = f_2$ case are quite

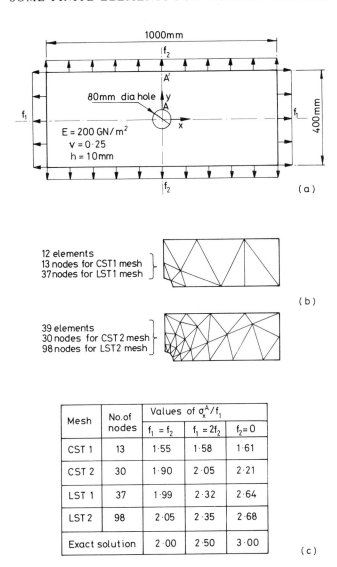

Fig. 8.13. Hole-in-plate problem: (a) the plate; (b) meshes of triangular elements; (c) results for stress concentration factor σ_x^A/f_1.

accurate, except for the CST1 mesh, but that more elements, especially of the CST type, would need to be used in the other cases if high accuracy were required. Graphical representation of the distribution of the σ_x stress component along the line AA′ is given in Fig. 8.14 for the case in which f_2 is zero. This shows the steep stress gradient which occurs local to the hole.

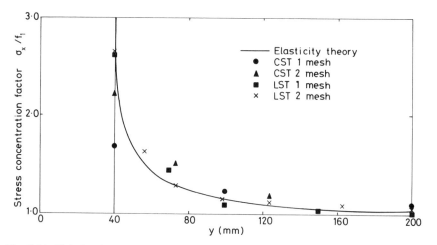

Fig. 8.14. Hole-in-plate problem: variation of σ_x along AA' when using triangular elements.

8.4. Rectangular elements

8.4.1. The basic rectangular element

The simplest possible rectangular element for planar analysis is shown in Fig. 8.15; this element was introduced by Argyris and Kelsey [12]. It has nodes at the four corner points with u and v as the nodal degrees of freedom and hence has eight degrees of freedom. The Cartesian co-ordinate axes x and y are particularly oriented so that they run parallel to the element edges and thus these axes are local ones. (Where the x, y axes do not lie in the same directions as do the global axes there will clearly need to be a transformation of element properties from the local to the global axes; this is achieved in the same way as for the bar element and is illustrated in Example 8.4 below.) The origin of the x, y axes is chosen for convenience to be at the centre of the rectangle but could equally well be located at some other point without affecting the element properties in any way.

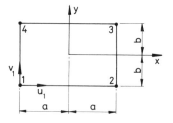

Fig. 8.15. The basic rectangular planar element.

We know that to satisfy the rigid-body motion and constant strain criteria the assumed element displacement field must include the terms given in equation (8.1), and also that to satisfy the compatibility condition—that is to provide a conforming element—each displacement component must vary linearly along an edge since only two nodal values are available per edge. Thus it is easy to see that the appropriate displacement field for the basic rectangular element is

$$\begin{Bmatrix} u \\ v \end{Bmatrix} = \begin{bmatrix} 1 & x & y & xy & 0 & 0 & 0 & 0 \\ 0 & 0 & 0 & 0 & 1 & x & y & xy \end{bmatrix} \begin{Bmatrix} A_1 \\ A_2 \\ \vdots \\ A_8 \end{Bmatrix} \quad \text{or} \quad \mathbf{u} = \boldsymbol{\alpha} \mathbf{A}. \qquad (8.37)$$

The column matrix of nodal displacements is

$$\mathbf{d} = \{ u_1 \quad v_1 \quad u_2 \quad v_2 \quad u_3 \quad v_3 \quad u_4 \quad v_4 \} \qquad (8.38)$$

and it is a straightforward matter to relate $\mathbf{d}$ to $\mathbf{A}$ using the conditions at the nodes that, for $i = 1, 2, 3$, and 4,

$$u = u_i \quad \text{and} \quad v = v_i \quad \text{at} \quad x = x_i, \quad y = y_i.$$

This standard procedure establishes matrix $\mathbf{C}$ in the relationship $\mathbf{d} = \mathbf{C}\mathbf{A}$ and the reader can check that in the subsequent solution for the generalized displacements, given by the form $\mathbf{A} = \mathbf{C}^{-1}\mathbf{d}$, the matrix $\mathbf{C}^{-1}$ is defined as

$$\mathbf{C}^{-1} = \frac{1}{4ab} \begin{bmatrix} ab & 0 & ab & 0 & ab & 0 & ab & 0 \\ -b & 0 & b & 0 & b & 0 & -b & 0 \\ -a & 0 & -a & 0 & a & 0 & a & 0 \\ 1 & 0 & -1 & 0 & 1 & 0 & -1 & 0 \\ 0 & ab & 0 & ab & 0 & ab & 0 & ab \\ 0 & -b & 0 & b & 0 & b & 0 & -b \\ 0 & -a & 0 & -a & 0 & a & 0 & a \\ 0 & 1 & 0 & -1 & 0 & 1 & 0 & -1 \end{bmatrix} \qquad (8.39)$$

The definition of the displacement field is completed by calculating the matrix of shape functions as

$$\mathbf{N} = \boldsymbol{\alpha}\mathbf{C}^{-1} = \begin{bmatrix} N_1 & 0 & N_2 & 0 & N_3 & 0 & N_4 & 0 \\ 0 & N_1 & 0 & N_2 & 0 & N_3 & 0 & N_4 \end{bmatrix} \qquad (8.40)$$

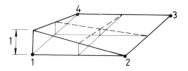

Fig. 8.16. Shape function N_1 for the basic rectangular element.

where

$$N_1 = \frac{1}{4}\left(1 - \frac{x}{a}\right)\left(1 - \frac{y}{b}\right) \qquad N_2 = \frac{1}{4}\left(1 + \frac{x}{a}\right)\left(1 - \frac{y}{b}\right)$$

$$N_3 = \frac{1}{4}\left(1 + \frac{x}{a}\right)\left(1 + \frac{y}{b}\right) \qquad N_4 = \frac{1}{4}\left(1 - \frac{x}{a}\right)\left(1 + \frac{y}{b}\right) \tag{8.41}$$

The shape function N_1 is shown, as typical of the rest, in Fig. 8.16.

With **E**, the elasticity matrix, again given by equation (7.35) for plane stress or by equation (7.36) for plain strain, we need only to establish matrix **B** for use in equation (7.21) or matrix $\boldsymbol{\beta}$ for use in equation (7.24), in order to have all the 'building blocks' needed for the calculation of the stiffness matrix. If the stiffness matrix is calculated explicitly by hand then equation (7.24) probably offers the easier formulation. This being so the matrix $\boldsymbol{\beta}$ is

$$\boldsymbol{\beta} = \begin{bmatrix} \partial/\partial x & 0 \\ 0 & \partial/\partial y \\ \partial/\partial y & \partial/\partial x \end{bmatrix} \begin{bmatrix} 1 & x & y & xy & 0 & 0 & 0 & 0 \\ 0 & 0 & 0 & 0 & 1 & x & y & xy \end{bmatrix}$$

or

$$\boldsymbol{\beta} = \begin{bmatrix} 0 & 1 & 0 & y & 0 & 0 & 0 & 0 \\ 0 & 0 & 0 & 0 & 0 & 0 & 1 & x \\ 0 & 0 & 1 & x & 0 & 1 & 0 & y \end{bmatrix}. \tag{8.42}$$

The element stiffness matrix can now be calculated using equation (7.24). If the thickness h is constant, then

$$\mathbf{k} = h(\mathbf{C}^{-1})^t \left(\int_{y=-b}^{b} \int_{x=-a}^{a} \boldsymbol{\beta}^t \mathbf{E} \boldsymbol{\beta} \, dx \, dy \right) \mathbf{C}^{-1}. \tag{8.43}$$

The result, for the plane stress case, is

$$\mathbf{k} = \frac{Eh}{12(1-v^2)}
\begin{array}{cccccccc}
u_1 & v_1 & u_2 & v_2 & u_3 & v_3 & u_4 & v_4
\end{array}$$

$$\mathbf{k} = \frac{Eh}{12(1-v^2)}
\begin{bmatrix}
4r^{-1}+4\rho r & 4r+4\rho r^{-1} & 4r^{-1}+4\rho r & 4r+4\rho r^{-1} & & & & \\
\mu & \lambda & -\mu & \lambda & 4r^{-1}+4\rho r & & & \\
-4r^{-1}+2\rho r & 2r-4\rho r^{-1} & 2r^{-1}-4\rho r & -4r+2\rho r^{-1} & \mu & 4r+4\rho r^{-1} & & \\
-\lambda & -\mu & -\lambda & \mu & -4r^{-1}+2\rho r & \lambda & 4r^{-1}+4\rho r & \\
-2r^{-1}-2\rho r & -2r-2\rho r^{-1} & -2r^{-1}-2\rho r & -2r-2\rho r^{-1} & -\lambda & 2r-4\rho r^{-1} & -\mu & 4r+4\rho r^{-1}
\end{bmatrix}$$

Symmetric

(8.44)

where $r = a/b$, $\rho = (1-\nu)/2$, $\mu = 3(1+\nu)/2$, and $\lambda = 3(1-3\nu)/2$. (For plane strain E^* and ν^*, defined following equation (7.36), replace E and ν respectively).

The stress matrix (relating $\boldsymbol{\sigma}$, defined by equation (7.32), to $\mathbf{d}$) for the basic conforming plane-stress rectangle is

$$\mathbf{S} = \mathbf{E}\boldsymbol{\beta}\mathbf{C}^{-1} = \frac{E}{4ab(1-\nu^2)} \begin{bmatrix} y-b & \nu(x-a) & b-y & -\nu(a+x) \\ \nu(y-b) & x-a & \nu(b-y) & -a-x \\ \rho(x-a) & \rho(y-b) & -\rho(a+x) & \rho(b-y) \end{bmatrix}$$

$$\left. \begin{array}{cccc} b+y & \nu(a+x) & -b-y & \nu(a-x) \\ \nu(b+y) & a+x & -\nu(b+y) & a-x \\ \rho(a+x) & \rho(b+y) & \rho(a-x) & -\rho(b+y) \end{array} \right] \quad (8.45)$$

Some further points should be noted regarding the assumed displacement field of the rectangular element. From equation (8.37) it can be seen that, in contrast to the situation described above for triangular elements, each displacement component is represented by an *incomplete* second-order polynomial in x and y (since the xy term is present but the x^2 and y^2 terms are absent). It follows that the direct strains ε_x and ε_y are represented as incomplete first-order polynomials, as is indicated by the form of $\boldsymbol{\beta}$ in equation (8.42); thus ε_x varies linearly with y but not with x, whilst ε_y varies linearly with x but not with y. All three in-plane stress components vary linearly with x and y throughout the element, as can be seen from the form of the stress matrix in equation (8.45). Because the strains and stresses are not restricted to a uniform value over the complete elements, the basic rectangular element is generally more efficient than the basic triangular element although it is less versatile. The form of the shape functions of equation (8.41) gives an idea of how higher-order planar rectangular elements might be generated and this will be investigated in Section 8.4.2.

Both the basic planar rectangle with eight degrees of freedom described here and the basic planar triangle with six degrees of freedom of Section 8.3.1 are based on displacement fields that give linear variation of each of u and v at element edges. It follows that these two types of element can be connected one with another without any loss of compatibility and can be used together as convenient to model a planar structure. Further, the basic bar element also has linear variation of displacements along its length and thus can be connected to either planar element at an edge to simulate a reinforcing member, say. The whole-structure displacement field will be kinematically admissible in the piecewise Rayleigh–Ritz sense just as it would be if only one type of element were used in the structure assembly. The direct stiffness method of assembly of the SSM proceeds in the usual manner; Example 8.5 is included to emphasize this point.

Example 8.4. The basic rectangular element described in Section 8.4.1 and illustrated in Fig. 8.15 is aligned such that its local axes lies at an angle α to a set of global Cartesian axes. Set up an expression for the stiffness matrix of the element in the global system.

The element is shown oriented to the global axes in Fig. E8.4. In the manner of Chapter 3 it is necessary to distinguish between local and global co-ordinates and the convention re-adopted here is that global co-ordinates, force components, and displacement components are identified by a superior bar.

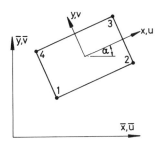

Fig. E8.4

The stiffness matrix $\mathbf{k}$ of the element in its local frame of reference is given explicitly in equation (8.44) and relates nodal forces $\mathbf{P}$ and nodal displacements $\mathbf{d}$. The local element stiffness equations can be expressed in partitioned form as

$$\begin{Bmatrix} \mathbf{P}_1 \\ \mathbf{P}_2 \\ \mathbf{P}_3 \\ \mathbf{P}_4 \end{Bmatrix} = \begin{bmatrix} \mathbf{k}_{11} & \mathbf{k}_{12} & \mathbf{k}_{13} & \mathbf{k}_{14} \\ \mathbf{k}_{21} & \mathbf{k}_{22} & \mathbf{k}_{23} & \mathbf{k}_{24} \\ \mathbf{k}_{31} & \mathbf{k}_{32} & \mathbf{k}_{33} & \mathbf{k}_{34} \\ \mathbf{k}_{41} & \mathbf{k}_{42} & \mathbf{k}_{43} & \mathbf{k}_{44} \end{bmatrix} \begin{Bmatrix} \mathbf{d}_1 \\ \mathbf{d}_2 \\ \mathbf{d}_3 \\ \mathbf{d}_4 \end{Bmatrix}. \tag{a}$$

Here

$$\mathbf{d}_i = \{u_i \quad v_i\} \qquad \mathbf{P}_i = \{U_i \quad V_i\}$$

and $\mathbf{k}_{ij}$ etc. (i and j run from 1 to 4 of course) are the 2×2 stiffness submatrices whose explicit form is obvious on comparing with equation (8.44): for instance

$$\mathbf{k}_{11} = \frac{Eh}{12(1-\nu^2)} \begin{bmatrix} 4r^{-1}+4\rho r & 3(1+\nu)/2 \\ 3(1+\nu)/2 & 4r+4\rho r^{-1} \end{bmatrix}.$$

In the global reference frame the required nodal displacements and forces will be

$$\bar{\mathbf{d}} = \{\bar{\mathbf{d}}_1 \quad \bar{\mathbf{d}}_2 \quad \bar{\mathbf{d}}_3 \quad \bar{\mathbf{d}}_4\} \qquad \bar{\mathbf{P}} = \{\bar{\mathbf{P}}_1 \quad \bar{\mathbf{P}}_2 \quad \bar{\mathbf{P}}_3 \quad \bar{\mathbf{P}}_4\}$$

where

$$\bar{\mathbf{d}}_i = \{\bar{u}_i \quad \bar{v}_i\} \qquad \bar{\mathbf{P}}_i = \{\bar{U}_i \quad \bar{V}_i\}.$$

The transformation from components of displacement and force at a node i in the local system to those in the global system is of identical form with that derived in Section 3.3 for planar pin-jointed frames. Thus only the result need be quoted here and the reader can refer to Section 3.3 for details. We have that

$$\mathbf{d}_i = \mathbf{T}\bar{\mathbf{d}}_i \qquad \mathbf{P}_i = \mathbf{T}\bar{\mathbf{P}}_i$$

where

$$\mathbf{T} = \begin{bmatrix} \cos\alpha & \sin\alpha \\ -\sin\alpha & \cos\alpha \end{bmatrix} = \begin{bmatrix} e & f \\ -f & e \end{bmatrix}.$$

It follows, by the reasoning employed in Section 3.3, that the stiffness relationships in the global system become

$$\bar{\mathbf{P}} = \bar{\mathbf{k}}\bar{\mathbf{d}}$$

where $\bar{\mathbf{k}}$ is the transformed stiffness matrix defined as

$$\bar{\mathbf{k}} = \boldsymbol{\tau}^t\mathbf{k}\boldsymbol{\tau}$$

and where

$$\boldsymbol{\tau} = \begin{bmatrix} \mathbf{T} & \mathbf{0} & \mathbf{0} & \mathbf{0} \\ \mathbf{0} & \mathbf{T} & \mathbf{0} & \mathbf{0} \\ \mathbf{0} & \mathbf{0} & \mathbf{T} & \mathbf{0} \\ \mathbf{0} & \mathbf{0} & \mathbf{0} & \mathbf{T} \end{bmatrix}$$

Alternatively, if $\bar{\mathbf{k}}$ is regarded as being partitioned in a similar way to that shown in equation (a) for $\mathbf{k}$, then the typical 2×2 submatrix $\bar{\mathbf{k}}_{ij}$ is defined as

$$\bar{\mathbf{k}}_{ij} = \mathbf{T}^t\mathbf{k}_{ij}\mathbf{T}$$

and this provides an easy way of explicitly determining $\bar{\mathbf{k}}$. For example

$$\bar{\mathbf{k}}_{11} = \frac{Eh}{12(1-\nu^2)} \left[\begin{array}{c|c} \begin{array}{c} 4e^2(r^{-1}+\rho r) \\ +4f^2(r+\rho r^{-1}) \\ +6ef(\rho-1) \end{array} & \begin{array}{c} 4ef(1-\rho)(r^{-1}-r) \\ +3(1-\rho)(e^2-f^2) \end{array} \\ \hline \begin{array}{c} 4ef(1-\rho)(r^{-1}-r) \\ +3(1-\rho)(e^2-f^2) \end{array} & \begin{array}{c} 4e^2(r+\rho r^{-1}) \\ +4f^2(r^{-1}+\rho r) \\ +6ef(1-\rho) \end{array} \end{array} \right]$$

Example 8.5. Figure E8.5 represents a planar sheet of material reinforced at its upper edge with a concentrated line member. This structure is to be modelled

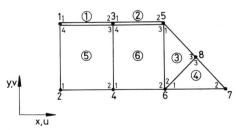

Fig. E8.5

with two basic rectangular elements (numbered ⑤ and ⑥ in the figure), two basic triangular elements (③ and ④) and two basic bar elements (① and ②). Given that the structure nodes are to be numbered as shown in the figure, assemble the SSM **K** using the direct stiffness procedure and working symbolically in terms of stiffness submatrices.

The stiffness matrices of each of the three different types of element can be represented in partitioned form in terms of 2×2 submatrices $\mathbf{k}_{ij}$ where i and j run from 1 to 2 in the case of the bar, 1 to 3 for the triangle, and 1 to 4 for the rectangle (see equation (a) of Example 8.4 for example). Each type of element meeting at a structure node has the same nodal degrees of freedom, u and v of course, so that the nodal connection process is straightforward, and as mentioned earlier the element displacement fields are such that compatibility is then maintained at all interelement boundaries. It should be noted that in this example the orientation of the rectangular and bar elements is such that their local axes coincide with the global axes indicated in Fig. E8.5; if this were not so then it would simply mean that submatrices $\bar{\mathbf{k}}_{ij}$ would be used in place of $\mathbf{k}_{ij}$.

Each element has its own local node-numbering scheme which is indicated in the figure, and to set up **K** we need to relate these element node-numbering schemes to the structure node-numbering scheme. The relationship is given by the following table.

Element	1	2	3	4	← Element nodes
①	1	3	—	—	
②	3	5	—	—	
③	5	6	8	—	← Structure nodes
④	6	7	8	—	
⑤	2	4	3	1	
⑥	4	6	5	3	

The direct stiffness procedure can now be used to assemble the SSM in exactly the same fashion as used earlier in dealing with skeletal structures except of course that here most of the elements have more than two nodes. The result is given in Table E8.5 (locations where only null submatrices occur are left blank and $\mathbf{k}_{41}^5$ for instance is submatrix $\mathbf{k}_{41}$ for element 5.)

TABLE E8.5

	1	2	3	4	5	6	7	8
	$\mathbf{k}_{11}^1 + \mathbf{k}_{44}^5$	$\mathbf{k}_{41}^5$	$\mathbf{k}_{12}^1 + \mathbf{k}_{43}^5$	$\mathbf{k}_{42}^5$				
	$\mathbf{k}_{14}^5$	$\mathbf{k}_{11}^5$	$\mathbf{k}_{13}^5$	$\mathbf{k}_{12}^5$				
	$\mathbf{k}_{21}^1 + \mathbf{k}_{34}^5$	$\mathbf{k}_{31}^5$	$\mathbf{k}_{22}^1 + \mathbf{k}_{33}^5 + \mathbf{k}_{11}^2 + \mathbf{k}_{44}^6$	$\mathbf{k}_{32}^5 + \mathbf{k}_{41}^6$	$\mathbf{k}_{12}^2 + \mathbf{k}_{43}^6$	$\mathbf{k}_{42}^6$		
$\mathbf{K} =$	$\mathbf{k}_{24}^5$	$\mathbf{k}_{21}^5$	$\mathbf{k}_{23}^5 + \mathbf{k}_{14}^6$	$\mathbf{k}_{22}^5 + \mathbf{k}_{11}^6$	$\mathbf{k}_{13}^6$	$\mathbf{k}_{12}^6$		
			$\mathbf{k}_{13}^2 + \mathbf{k}_{34}^6$	$\mathbf{k}_{31}^6$	$\mathbf{k}_{14}^2 + \mathbf{k}_{33}^6 + \mathbf{k}_{11}^3$	$\mathbf{k}_{32}^6 + \mathbf{k}_{12}^3$		$\mathbf{k}_{13}^3$
			$\mathbf{k}_{24}^6$	$\mathbf{k}_{21}^6$	$\mathbf{k}_{23}^6 + \mathbf{k}_{21}^3$	$\mathbf{k}_{22}^6 + \mathbf{k}_{22}^3 + \mathbf{k}_{11}^4$	$\mathbf{k}_{12}^4$	$\mathbf{k}_{23}^3 + \mathbf{k}_{13}^4$
						$\mathbf{k}_{21}^4$	$\mathbf{k}_{22}^4$	$\mathbf{k}_{23}^4$
					$\mathbf{k}_{31}^3$	$\mathbf{k}_{32}^3 + \mathbf{k}_{31}^4$	$\mathbf{k}_{32}^4$	$\mathbf{k}_{33}^3 + \mathbf{k}_{33}^4$

8.4.2. Higher-order rectangular elements: Lagrange element family

The shape functions of the basic rectangular element are given by equation (8.41) in the form of products of a function of x and a function of y. These latter unidirectional functions can be identified as one-dimensional Lagrange interpolation polynomials of the first order (i.e. linear polynomials). That this is so can be seen on referring back to Section 6.9 and using equation (6.72) with $x_1 = -a$, $x_2 = +a$ (i.e. the co-ordinate origin of the one-dimensional region of length $2a$ is taken to be at the centre). Thus

$$l_1(x) = \frac{1}{2}\left(1 - \frac{x}{a}\right) \tag{8.46a}$$

$$l_2(x) = \frac{1}{2}\left(1 + \frac{x}{a}\right) \tag{8.46b}$$

and similarly, by substituting y for x and b for a,

$$l_1(y) = \frac{1}{2}\left(1 - \frac{y}{b}\right) \tag{8.46c}$$

$$l_2(y) = \frac{1}{2}\left(1 + \frac{y}{b}\right). \tag{8.46d}$$

The expression for u for the basic rectangle can therefore be written

$$u = l_1(x)l_1(y)u_1 + l_2(x)l_1(y)u_2 + l_2(x)l_2(y)u_3 + l_1(x)l_2(y)u_4 \tag{8.47}$$

with a similar expression applying for v. For obvious reasons, if l_1 and l_2 are defined by equations (8.46a)–(8.46d) then the interpolation scheme represented in equation (8.47) is referred to as bilinear interpolation.

Clearly there is scope to extend the ideas represented by equation (8.47) to generate higher-order rectangles directly using appropriate products of one-dimensional interpolation polynomials as the shape functions in the expressions for the u and v displacement components. For instance, we can produce an element based on the use of biquadratic interpolation of the displacement components, i.e. on the direct use of shape functions which are products of one-dimensional Lagrange interpolation polynomials of the second order. These polynomials are again obtained from equation (6.72) (with $x_1 = -a$, $x_2 = 0$ and $x_3 = +a$ for the polynomial in the x direction) and are

$$l_1(x) = \frac{1}{2}\left(\frac{x^2}{a^2} - \frac{x}{a}\right) \qquad l_2(x) = 1 - \frac{x^2}{a^2} \qquad l_3(x) = \frac{1}{2}\left(\frac{x^2}{a^2} + \frac{x}{a}\right) \tag{8.48}$$

with similar expressions applying for $l_1(y)$, $l_2(y)$ and $l_3(y)$ when y and b replace x and a respectively. The expression for u (with a similar

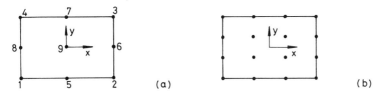

Fig. 8.17. Two Lagrangian rectangular elements: (a) second order; (b) third order.

expression applying for v) is

$$u = l_1(x)l_1(y)u_1 + l_3(x)l_1(y)u_2 + l_3(x)l_3(y)u_3 + l_1(x)l_3(y)u_4$$
$$+ l_2(x)l_1(y)u_5 + l_3(x)l_2(y)u_6 + l_2(x)l_3(y)u_7 \qquad (8.49)$$
$$+ l_1(x)l_2(y)u_8 + l_2(x)l_2(y)u_9$$

and the arrangement of nodes for this element is shown in Fig. 8.17(a). A pictorial representation of the shape function for a corner node, a mid-side node and an interior node is given in Fig. 8.18. Since we are dealing here with a complete biquadratic interpolation we would expect to obtain equation (8.49) indirectly by starting from

$$u = (A_0 + A_1 x + A_2 x^2)(B_0 + B_1 y + B_2 y^2) \qquad (8.50)$$

and evaluating the coefficients by applying the 'boundary' conditions at the nine nodal points; this is indeed the case. The representation of the coefficients of the biquadratic interpolation is shown on the Pascal triangle in Fig. 8.19 as is the bilinear interpolation. (The nine-noded element was first described in reference 5).

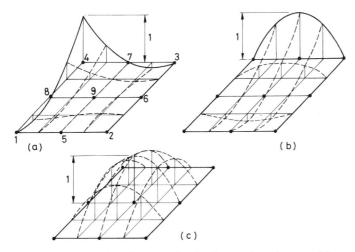

Fig. 8.18. Shape functions for the second-order Lagrangian element: (a) corner-node function N_4; (b) side-node function N_7; (c) centre-node function N_9.

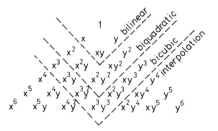

Fig. 8.19. Representation of the displacement fields of Lagrangian rectangular elements on the Pascal triangle.

At the next level of interpolation we come to bicubic interpolation as the basis for a possible rectangular element. This is also represented symbolically on the Pascal triangle in Fig. 8.19 and the arrangement of nodes is shown in Fig. 8.17(b). Even higher levels of interpolation could also be used to generate elements of even greater refinement in a straightforward fashion. For all the elements in this category it is clear from the Pascal triangle that the assumed displacement fields are incomplete two-dimensional polynomials. In fact if the fields are constructed as products of one-dimensional Lagrange interpolation polynomials of order m then these fields are seen to be complete in two dimensions to order m and incomplete to order $2m$.

The above remarks have been concerned with rectangular elements whose displacements are interpolated in exactly the same way in the x and y directions. This need not be so, however, since it is a simple matter to generate Lagrangian rectangular elements along the lines indicated above but using a different order of interpolation in the x direction to that used in the y direction. This will result in a series of elements which have different numbers of nodal points in the x direction to those in the y direction. A few such elements are illustrated in Fig. 8.20.

The Lagrangian rectangular elements have the advantage of comparative simplicity of formulation in that the shape functions of the displacement fields are readily developed in direct and logical fashion from one-dimensional interpolation polynomials. Unfortunately they also have some practical disadvantages. The overall rate of convergence of finite

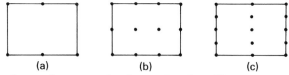

Fig. 8.20. Some Lagrangian rectangular elements based on different orders of interpolation in the x and y directions.

SOME FINITE ELEMENTS FOR PLANAR ANALYSIS   333

elements is largely dependent on the highest order of *complete* polynomial in the displacement field and the presence of only some terms belonging to higher-order polynomials is computationally inefficient. Further, the presence of the considerable number of internal nodes in all but the simplest Lagrangian rectangles is a disadvantage since, although these can be eliminated by static condensation, this does require extra operations. These disadvantages led to the search for an alternative family of rectangular planar elements and to the development of the 'serendipity' family of elements. The most useful members of this element family do not have any internal nodes and possess better curve-fitting properties than do the members of the Lagrangian family. The development of the shape functions for the serendipity elements is much less direct than is the case for the Lagrangian elements and for this reason consideration of the serendipity elements is deferred to Chapter 10.

It should be noted that it is quite possible to generate another family of rectangular elements for planar analysis along the lines described above for the Lagrangian elements but now using products of one-dimensional Hermitian interpolation polynomials (see Section 6.9). This would produce elements using displacement derivatives as nodal freedoms (analogous to the triangular element of Fig. 8.10 for instance) which would be advantageous in some circumstances (smaller bandwidth etc.) but disadvantageous in others (over-constraint etc.). Such elements will not be considered further here but will occur later in the text when plate bending is considered.

References

1 TURNER, M. J., CLOUGH, R. W., MARTIN, H. C., and TOPP, L. J. Stiffness and deflection analysis of complex structures. *J. Aeronaut. Sci.* **23,** 805–23 (1956).
2 CLOUGH, R. W. The finite element method in plane stress analysis. *Proc. 2nd Am. Soc. Civ. Eng.* Conf. on Electronic Computation, Pittsburg, Pa. (1960).
3 FRAEIJS DE VEUBEKE, B. Displacement and equilibrium models in the finite element method. In Zienkiewicz, O. C. and Holister, G. S. (eds.), *Stress analysis,* Chapter 9. Wiley, New York (1965).
4 ARGYRIS, J. H. Triangular elements with linearly varying strain for the matrix displacement method. *J. R. Aeronaut. Soc.* **69,** 711–13 (1965).
5 ARGYRIS, J. H. Continua and discontinua. *Conf. on Matrix Methods in Structural Mechanics, Air Force Institute of Technology, Wright–Patterson Air Force Base,* Dayton, Ohio (1965).
6 FELIPPA, C. A. Refined finite element analysis of linear and nonlinear two-dimensional structures, *Rep. 66–22,* University of California, Berkeley (1966).
7 TOCHER, J. L. and HARTZ, B. J. Higher-order finite element for plane stress. *Proc. Am. Soc. Civ. Eng., J. Eng. Mech. Div.* **93** (EM4), 149–74 (1967).
8 HOLAND, I. and BERGAN, P. G. *Proc. Am. Soc. Civ. Eng., J. Eng. Mech. Div.,* **94** (EM2), 698–702 (1968). (Comment on ref. 7.)

9   HANSTEEN, O. E. Analysis of stress distribution in shear walls by the finite element displacement method. *Int. Kongr. über Anwendungen der Mathematik in den Ingeneirwissenschaften, Weimar* (1967).

10 HOLAND, I. (1969). The finite element method in plane stress analysis, In *The finite element method in stress analysis* (eds. I. Holand and K. Bell), Chapter 9. Tapir Press, Trondheim (1969).

11 TIMOSHENKO, S. and GOODIER, J. N. *Theory of elasticity.* McGraw-Hill, New York (1951).

12 ARGYRIS, J. H. and KELSEY, S. *Energy theorems and structural analysis.* Butterworths, London (1960). (Previously published in *Aircr. Eng.*, **26** (1954); **27** (1955).)

Problems

8.1. Resolve Example 8.2 with the region 1–2–3–4 in Fig. E8.2(a) divided into two CST elements by diagonal line 2–4 rather than by diagonal line 1–3.

8.2. Resolve Example 8.2 using now a single basic (four-node) rectangular element to model the region 1–2–3–4 shown in Fig. E8.2(a). Demonstrate that the exact solution is obtained for load case (i). Compare the solution for load case (ii) with that obtained using CST elements in Example 8.2.

8.3. Consider Example 8.2 again and now imagine that reinforcing bars are rigidly connected to the plate along the top and bottom edges. These bars each have a cross-sectional area of 10 cm^2 and are made of the same material as the plate. For analysis purposes consider that region 1–2–3–4 of Fig. E8.2(a) is modelled with two CST elements as shown in the figure, together with a basic bar element 1–4. Assume load case (ii) of Example 8.2 and solve for the nodal displacements and the stress distributions.

8.4. The plane sheet of material shown in Fig. P8.4 is of uniform thickness 1 cm, with $E = 200 \text{ GN m}^{-2}$, $\nu = 0.3$. No movement takes place at edges 6–7 and 5–7 under the given loading. The sheet is to be modelled with six CST elements as shown. Set up the set of eight stiffness equations for this modelling.

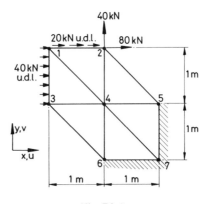

Fig. P8.4

8.5. Figure P8.5 shows a plane sheet of material of uniform thickness 1 cm which is subjected in turn to the indicated two types of loading distributed along edge MQS. The boundary conditions are that $u = 0$ on edge LPR and that $v = 0$ on edges LM and RS. A symmetric half of the structure is modelled with two CST elements, as shown. Determine the nodal displacements and the distribution of stresses for each of the loading cases. Take $E = 200$ GN m^{-2} and $\nu = 0.3$.

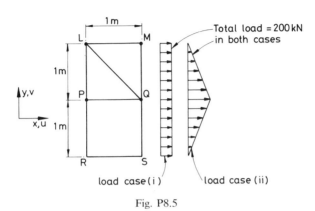

Fig. P8.5

8.6. Repeat Problem 8.5 using one basic rectangular element in place of the two CST elements.

8.7. Consider again the structure described in Problem 8.5 with loading case (ii) applying. Now model the half-structure LMPQ with, in turn, (a) four CST elements as in Fig. P8.7(a) and (b) two basic rectangular elements as in Fig. P8.7(b). Obtain solutions for nodal displacements and distributions of stresses.

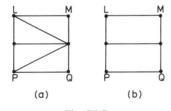

(a)   (b)

Fig. P8.7

8.8. In Sections 8.3.1 and 8.3.2 the properties of the linear-displacement CST element have been developed for arbitrary orientation of the triangle, i.e. have been developed directly in a global context. As an alternative to this, consider now the element shown in Fig. P8.8(a) and, using linear displacement fields, derive the 6×6 stiffness matrix $\mathbf{k}$ referred to the specially orientated (i.e. local) co-ordinate system xy shown in the figure.

Proceed to obtain the stiffness matrix for the element in the global $\bar{x}\bar{y}$ system (see Fig. P8.8(b)) by a transformation procedure in the manner of Example 8.4. Verify that this stiffness matrix is the same as that derived in Section 8.3.1. (See equations (8.15) and (8.16).)

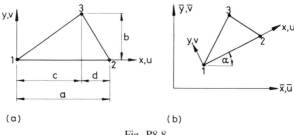

(a)   (b)

Fig. P8.8

8.9. In Fig. P8.9 the line PQR represents an edge of a plane stress element which is loaded with a distributed normal loading $f(x)$. The displacement components vary quadratically along the edge; in particular $v = A + Bx + Cx^2$. If the loading also varies quadratically as $f(x) = D + Ex + Fx^2$ show that the nodal loads which replace the distributed applied loading in consistent fashion are

$$\begin{Bmatrix} V_P \\ V_Q \\ V_R \end{Bmatrix} = \frac{l}{30} \begin{bmatrix} 4 & 2 & -1 \\ 2 & 16 & 2 \\ -1 & 2 & 4 \end{bmatrix} \begin{Bmatrix} f_P \\ f_Q \\ f_R \end{Bmatrix}$$

where f_P, f_Q, and f_R are the values of $f(x)$ at nodes P, Q, and R respectively.

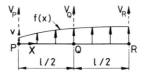

Fig. P8.9

8.10. A column has a rectangular cross-section 0.3 m by 0.2 m and at a particular section AA′ is subjected to a vertical compressive force of $P = 540$ kN, a bending moment of $M = 13.5$ kN m and a horizontal shearing force of $F = 81$ kN, as shown in Fig. P8.10. These forces and the moment are assumed to be distributed across the section in accordance with the usual engineering bending theory assumptions; thus the vertical stress due to P is uniformly distributed, that due to M varies linearly from A to A′ through zero at the centre, and the horizontal stress due to F varies quadratically along AA′, having zero value at A and A′ and maximum value at the centre. The region below AA′ is modelled with a mesh of LST elements, with three identical elements across the width of section AA′ as indicated in Fig. P8.10. Determine the loads which must be applied at the seven nodes of section AA′ to represent the given loading in consistent fashion.

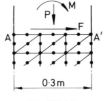

0·3m

Fig. P8.10

8.11. Starting with equation (8.50) show that the displacement field for the rectangular element illustrated in Fig. 8.17(a) is given (for u) in shape function form by equations (8.48) and (8.49).

8.12. For the rectangular element shown in Fig. 8.17(b) the displacement field is based on bicubic Lagrangian interpolation. Write down the shape functions corresponding to, for example, nodes located at

   (i) $x/a = -1,$ $y/b = +1$
   (ii) $x/a = +1/3,$ $y/b = -1$
   (iii) $x/a = +1/3,$ $y/b = -1/3$

Sketch these shape functions. (The element side lengths are $2a$ and $2b$ as in Fig. 8.15.)

8.13. For the rectangular element shown in Fig. 8.20(a) the displacement field is based on unequal interpolation in the two co-ordinate directions (see Fig. 8.15 for details of element geometry and origin of co-ordinates). Write down the shape functions for a corner node and a mid-side node and sketch these functions.

8.14. Imagine that the uniform thickness nine-noded rectangular element shown in Fig. 8.17(a) is subjected to a body force whose intensity varies linearly with x, i.e. $R_x = A + Bx$. Determine column matrix $\mathbf{Q}$ for this loading.

8.15. Figure P8.15 shows part of the boundary of a plane stress structure at which AB represents the edge of a single element over which the displacement components have a cubic variation and four equi-spaced nodes are present. At this edge the applied loading comprises a uniformly distributed shear loading of intensity p per unit length and a linearly distributed normal loading varying in intensity from p per unit length at A to $2p$ per unit length at B. Determine the nodal loads which will replace the applied loading in a consistent manner.

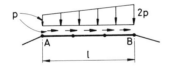

Fig. P8.15

9

SOME FINITE ELEMENTS FOR THREE-DIMENSIONAL AND AXISYMMETRIC ANALYSIS

9.1. Introduction

THE finite-element analysis of general three-dimensional bodies and of axisymmetric solids follows on naturally from that described in Chapter 8 for two-dimensional planar bodies. In the present chapter the general three-dimensional problem is considered first, followed by consideration of axisymmetric bodies in which reduction of the solid body problem can again be made to two dimensions.

In extending to three-dimensional analysis the fundamental ideas governing the development of element properties are unchanged and some families of elements for three-dimensional analysis are logical extensions of the families of elements for two-dimensional analysis which have already been described. The triangle and rectangle of two-dimensional analysis become the tetrahedron and rectangular hexahedron respectively of three-dimensional analysis. As with the previous two-dimensional planar analysis the elements described here for three-dimensional applications are based on the use of the conventional Cartesian co-ordinate system and have straight edges. Other three-dimensional element formulations are discussed in the next chapter.

With regard to three-dimensional finite-element analysis it is perhaps appropriate at the very beginning to emphasize the effect that increasing the number of dimensions has on the size of a problem (i.e. on the total number of degrees of freedom of the problem). For example, imagine that in a one-dimensional analysis it is necessary to use n nodes (at which u is the degree of freedom) to achieve a satisfactory level of accuracy of solution, i.e. n unknowns are required. Then, for a similar level of accuracy in two-dimensional analysis we would need to use around n^2 nodes with u and v as degrees of freedom at each node, giving a total of about $2n^2$ unknowns. Proceeding to three dimensions we would use about n^3 nodes with u, v, and w as degrees of freedom for the same sort of accuracy, giving a total of about $3n^3$ unknowns. If $n = 10$, say, the necessary degrees of freedom increase from 10 to 200 to 3000 as we proceed from one to two to three dimensions. It is clear that the finite-element analysis of practical three-dimensional solids will often result in the generation of very large numbers of equations.

For axisymmetric analysis the link with the planar analysis of Chapter 8 is very close despite the apparently great difference in structural form. The planar elements of Chapter 8 can be modified easily for use in axisymmetric analysis, as will be seen in the latter part of this chapter.

9.2. Requirements for element displacement fields in three-dimensional analysis

In a Cartesian formulation of the three-dimensional problem the representation of rigid-body motion and constant strain states—the first two criteria of Section 6.8—requires that the displacement field includes the terms

$$u = u(x, y, z) = A_1 + A_2x + A_3y + A_4z$$
$$v = v(x, y, z) = A_5 + A_6x + A_7y + A_8z \tag{9.1}$$
$$w = w(x, y, z) = A_9 + A_{10}x + A_{11}y + A_{12}z.$$

It is easily seen that the constant strain states are present in this field given that the relevant strain–displacement equations are equations (7.2). The general rigid-body motion occurs if the nodal displacements are specialized such that $A_2 = A_7 = A_{12} = 0$ and $A_8 + A_{11} = A_4 + A_{10} = A_3 + A_6 = 0$, and then all strains are zero of course. Thus the general rigid-body motion is (see Section 7.2.2).

$$u_r = A_1 - A_3y + A_4z$$
$$v_r = A_5 + A_3x - A_8z \tag{9.2}$$
$$w_r = A_9 - A_4x + A_8y.$$

This motion comprises three translational components in the x, y, and z directions and three rotation components about the x, y, and z axes. (Whilst it is perfectly permissible for an individual element to move as a rigid body, it should be noted that a complete three-dimensional body must be restrained against all of the six components of rigid-body motion by proper application of boundary conditions.)

Considering displacement compatibility we again have a situation in which only C_0 continuity is required. The third criterion of Section 6.8 poses no difficulty as long as u, v, and w are single valued within the three-dimensional element. The fourth criterion demands continuity of u, v, and w at interelement boundaries and here this means continuity over all common *surfaces* of adjacent elements and not just at node points or at edges.

Again, as with the two-dimensional analysis, it will be advantageous if the assumed displacement fields are geometrically invariant.

9.3. Tetrahedral elements

9.3.1. The basic (constant strain) tetrahedron

The basic tetrahedral element is illustrated in Fig. 9.1. This is the three-dimensional equivalent of the basic planar triangle of Section 8.3.1. The basic tetrahedron has nodes at its vertices only with the u, v, and w components of displacement (in the x, y, and z directions) as the nodal degrees of freedom. Thus 12 degrees of freedom are present and the column matrix of nodal displacements is

$$\mathbf{d} = \{u_1 \quad v_1 \quad w_1 \quad u_2 \quad v_2 \quad w_2 \quad u_3 \quad v_3 \quad w_3 \quad u_4 \quad v_4 \quad w_4\}. \tag{9.3}$$

The convention for numbering the nodes of a tetrahedral element is that the second, third, and fourth nodes are numbered in anticlockwise fashion when viewed from the first node.

The element is arbitrarily oriented with respect to the co-ordinate directions and thus x, y, and z are global co-ordinates. Further, as far as the generation of element stiffness is concerned the position of the origin of the co-ordinates is unimportant. (When considering other element properties, for example the generation of a consistent load column matrix corresponding to a body force, it is convenient for the purpose of performing the necessary integrations to take the origin at the centroid of the tetrahedron.) The basic tetrahedron appears to have been introduced in the open literature by Gallagher, Padlog, and Bijlaard [1], with contributions at about the same time by Melosh [2] and Argyris [3].

The number of degrees of freedom of the basic tetrahedral element

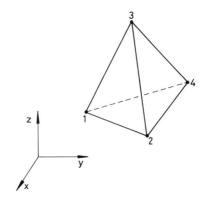

Fig. 9.1. Constant strain tetrahedral element.

(12) matches that of the number of coefficients in the minimum displacement field given by equation (9.1) which allows proper representation of rigid-body motion and constant strain states. Thus this minimum field can be used directly in formulating the properties of the basic tetrahedron. In matrix form the assumed displacement field is (compare with equation (8.3) for the basic planar triangle)

$$
\begin{Bmatrix} u \\ v \\ w \end{Bmatrix} = \begin{bmatrix} 1 & x & y & z & 0 & 0 & 0 & 0 & 0 & 0 & 0 & 0 \\ 0 & 0 & 0 & 0 & 1 & x & y & z & 0 & 0 & 0 & 0 \\ 0 & 0 & 0 & 0 & 0 & 0 & 0 & 0 & 1 & x & y & z \end{bmatrix} \begin{Bmatrix} A_1 \\ A_2 \\ \vdots \\ A_{12} \end{Bmatrix}
$$

$$\text{or} \qquad \mathbf{u} = \boldsymbol{\alpha}\mathbf{A}. \qquad (9.4)$$

Following the standard procedure we now use the 'boundary conditions' that, for $i = 1$–4,

$$u = u_i, \quad v = v_i, \quad w = w_i \text{ at each node } i \text{ where } x = x_i, \ y = y_i, \ z = z_i$$

This leads to the equations (compare with equations (8.4) for the basic triangle)

$$
\begin{Bmatrix} u_1 \\ v_1 \\ w_1 \\ u_2 \\ \vdots \\ w_4 \end{Bmatrix} = \begin{bmatrix} 1 & x_1 & y_1 & z_1 & 0 & 0 & 0 & 0 & 0 & 0 & 0 & 0 \\ 0 & 0 & 0 & 0 & 1 & x_1 & y_1 & z_1 & 0 & 0 & 0 & 0 \\ 0 & 0 & 0 & 0 & 0 & 0 & 0 & 0 & 1 & x_1 & y_1 & z_1 \\ 1 & x_2 & y_2 & z_2 & 0 & 0 & 0 & 0 & 0 & 0 & 0 & 0 \\ \vdots & & & & & & & & & & & \vdots \\ 0 & 0 & 0 & 0 & 0 & 0 & 0 & 0 & 1 & x_4 & y_4 & z_4 \end{bmatrix} \begin{Bmatrix} A_1 \\ A_2 \\ A_3 \\ \vdots \\ A_{12} \end{Bmatrix}
$$

$$\text{or} \qquad \mathbf{d} = \mathbf{C}\mathbf{A}. \qquad (9.5)$$

It is not essential to determine the inverse of $\mathbf{C}$ algebraically since $\mathbf{C}$ could be inverted numerically within a computer for any given set of values $x_1 \ldots z_4$ which define an element geometry. However, the inverse can be found algebraically in similar, but extended, fashion to that described earlier for the basic planar triangle (following equation (8.4)) and so will be given here. The set of 12 equations falls naturally into three blocks of four equations (concerning the u, v, and w displacements respectively) which can be solved separately by inverting the 4×4 matrix which is common to each block; finally a simple rearrangement then gives

$\mathbf{C}^{-1}$ which is defined as

$$
\mathbf{C}^{-1} = \frac{1}{6V_e}
\begin{bmatrix}
\alpha_1 & 0 & 0 & \alpha_2 & 0 & 0 & \alpha_3 & 0 & 0 & \alpha_4 & 0 & 0 \\
\beta_1 & 0 & 0 & \beta_2 & 0 & 0 & \beta_3 & 0 & 0 & \beta_4 & 0 & 0 \\
\gamma_1 & 0 & 0 & \gamma_2 & 0 & 0 & \gamma_3 & 0 & 0 & \gamma_4 & 0 & 0 \\
\delta_1 & 0 & 0 & \delta_2 & 0 & 0 & \delta_3 & 0 & 0 & \delta_4 & 0 & 0 \\
0 & \alpha_1 & 0 & 0 & \alpha_2 & 0 & 0 & \alpha_3 & 0 & 0 & \alpha_4 & 0 \\
0 & \beta_1 & 0 & 0 & \beta_2 & 0 & 0 & \beta_3 & 0 & 0 & \beta_4 & 0 \\
0 & \gamma_1 & 0 & 0 & \gamma_2 & 0 & 0 & \gamma_3 & 0 & 0 & \gamma_4 & 0 \\
0 & \delta_1 & 0 & 0 & \delta_2 & 0 & 0 & \delta_3 & 0 & 0 & \delta_4 & 0 \\
0 & 0 & \alpha_1 & 0 & 0 & \alpha_2 & 0 & 0 & \alpha_3 & 0 & 0 & \alpha_4 \\
0 & 0 & \beta_1 & 0 & 0 & \beta_2 & 0 & 0 & \beta_3 & 0 & 0 & \beta_4 \\
0 & 0 & \gamma_1 & 0 & 0 & \gamma_2 & 0 & 0 & \gamma_3 & 0 & 0 & \gamma_4 \\
0 & 0 & \delta_1 & 0 & 0 & \delta_2 & 0 & 0 & \delta_3 & 0 & 0 & \delta_4
\end{bmatrix}
\tag{9.6}
$$

(compare with equation (8.8) for the basic triangle). Here

$$
\alpha_1 = \begin{vmatrix} x_2 & y_2 & z_2 \\ x_3 & y_3 & z_3 \\ x_4 & y_4 & z_4 \end{vmatrix}
\qquad
\beta_1 = - \begin{vmatrix} 1 & y_2 & z_2 \\ 1 & y_3 & z_3 \\ 1 & y_4 & z_4 \end{vmatrix}
$$

$$
\gamma_1 = \begin{vmatrix} 1 & x_2 & z_2 \\ 1 & x_3 & z_3 \\ 1 & x_4 & z_4 \end{vmatrix}
\qquad
\delta_1 = - \begin{vmatrix} 1 & x_2 & y_2 \\ 1 & x_3 & y_3 \\ 1 & x_4 & y_4 \end{vmatrix}
\tag{9.7}
$$

with the other coefficients $\alpha_2 \dots \delta_4$ defined by cyclic interchange of the subscripts in the order 1, 2, 3, 4 (compare with equations (8.6)). Also

$$
6V_e = \begin{vmatrix} 1 & x_1 & y_1 & z_1 \\ 1 & x_2 & y_2 & z_2 \\ 1 & x_3 & y_3 & z_3 \\ 1 & x_4 & y_4 & z_4 \end{vmatrix}
\tag{9.8}
$$

where the physical meaning of V_e is that it is the volume of the tetrahedron 1–2–3–4 (compare with equation (8.7)).

With $\boldsymbol{\alpha}$ (equation (9.4)) and $\mathbf{C}^{-1}$ (equation (9.6)) defined the shape function matrix can be established as

$$
\mathbf{N} = \boldsymbol{\alpha}\mathbf{C}^{-1} =
$$

$$
\begin{bmatrix}
N_1 & 0 & 0 & N_2 & 0 & 0 & N_3 & 0 & 0 & N_4 & 0 & 0 \\
0 & N_1 & 0 & 0 & N_2 & 0 & 0 & N_3 & 0 & 0 & N_4 & 0 \\
0 & 0 & N_1 & 0 & 0 & N_2 & 0 & 0 & N_3 & 0 & 0 & N_4
\end{bmatrix}.
\tag{9.9}
$$

The individual shape functions N_i are defined as

$$N_i = \frac{\alpha_i + x\beta_i + y\gamma_i + z\delta_i}{6V_e} \qquad i = 1, 2, 3, 4 \qquad (9.10)$$

to complete the definition of the displacement field for the basic tetrahedron (compare with equations (8.10) and (8.11) for the basic triangle). The linear shape function N_i is such that it has unit value at node i and zero value at the other three nodes of course.

The strain–displacement equations for three-dimensional elasticity are given by equations (7.2) and can now be used to evaluate matrix $\mathbf{B}$ corresponding to the definition of $\boldsymbol{\varepsilon}$ given by equation (7.4). In the same manner as for the basic triangle in equation (8.12), we have that

$$\mathbf{B} = \begin{bmatrix} \partial/\partial x & 0 & 0 \\ 0 & \partial/\partial y & 0 \\ 0 & 0 & \partial/\partial z \\ \partial/\partial y & \partial/\partial x & 0 \\ 0 & \partial/\partial z & \partial/\partial y \\ \partial/\partial z & 0 & \partial/\partial x \end{bmatrix} \begin{bmatrix} N_1 & 0 & 0 & N_2 & 0 & 0 & N_3 & 0 & 0 & N_4 & 0 & 0 \\ 0 & N_1 & 0 & 0 & N_2 & 0 & 0 & N_3 & 0 & 0 & N_4 & 0 \\ 0 & 0 & N_1 & 0 & 0 & N_2 & 0 & 0 & N_3 & 0 & 0 & N_4 \end{bmatrix} \qquad (9.11)$$

which becomes, using equation (9.10),

$$\mathbf{B} = \frac{1}{6V_e} \begin{bmatrix} \beta_1 & 0 & 0 & \beta_2 & 0 & 0 & \beta_3 & 0 & 0 & \beta_4 & 0 & 0 \\ 0 & \gamma_1 & 0 & 0 & \gamma_2 & 0 & 0 & \gamma_3 & 0 & 0 & \gamma_4 & 0 \\ 0 & 0 & \delta_1 & 0 & 0 & \delta_2 & 0 & 0 & \delta_3 & 0 & 0 & \delta_4 \\ \gamma_1 & \beta_1 & 0 & \gamma_2 & \beta_2 & 0 & \gamma_3 & \beta_3 & 0 & \gamma_4 & \beta_4 & 0 \\ 0 & \delta_1 & \gamma_1 & 0 & \delta_2 & \gamma_2 & 0 & \delta_3 & \gamma_3 & 0 & \delta_4 & \gamma_4 \\ \delta_1 & 0 & \beta_1 & \delta_2 & 0 & \beta_2 & \delta_3 & 0 & \beta_3 & \delta_4 & 0 & \beta_4 \end{bmatrix} \qquad (9.12)$$

It is seen that all individual terms of $\mathbf{B}$ are constants, reflecting the fact that all six strain components have constant value throughout the element volume.

When attention is restricted to a homogeneous isotropic material the elasticity matrix $\mathbf{E}$ linking stresses (equation (7.3)) to strains (equation (7.4)) is given by equation (7.7).

With $\mathbf{B}$ and $\mathbf{E}$ known the stiffness matrix for the basic tetrahedron is given by the general expression (equation (7.21)). Since only constant terms occur under the integration sign of this expression the integration is trivial and the result is

$$\mathbf{k} = V_e \mathbf{B}^t \mathbf{E} \mathbf{B}. \qquad (9.13)$$

The stiffness matrix could be determined explicitly by performing the indicated matrix multiplication although this result will not be given here. Alternatively the matrix multiplication could be carried out numerically within the computer.

The basic tetrahedron will be referred to as element Tet 12 in the presentation of numerical results in Section 9.5.

Mention has not been made yet as to whether the basic tetrahedron is a conforming element or not, although the reader has probably inferred that it is from comparison with the situation that exists for the planar triangles. This is indeed so as can be seen by an extension of the argument given at the end of Section 8.3.1. For the basic tetrahedron each displacement component varies linearly over any plane lying within or on the surface of the element and such a linear variation, in the two dimensions of the plane, is specified by three terms. Since three nodes are available on any outside face of the basic tetrahedron, with values of the displacements used as nodal degrees of freedom, it follows that the variation of any displacement component over an outside face is uniquely defined by the values of the component at the nodes lying on that face. Consequently when the nodal displacements of adjacent basic tetrahedra are connected it follows automatically that displacements are matched over the complete interface. This reasoning can be readily extended to embrace other tetrahedra, described later, which are based on higher-order displacement fields and the use of a greater number of nodal points at which the displacements (alone) are the degrees of freedom.

The stress matrix $\mathbf{S} = \mathbf{EB}$ can easily be determined explicitly for the basic tetrahedron. It is not recorded here but it is emphasized that all six stresses are restricted to uniform values throughout the volume of the tetrahedron.

In three-dimensional analysis any required consistent load column matrices can be derived in similar fashion to that described for planar analysis in Section 8.3.2. Again equation (7.23) (or (7.25)) gives the general statement. In the following example only the case of an applied body force will be briefly looked at, whilst an example of an applied force acting on the surface of a three-dimensional element will be given later in Example 9.2.

Example 9.1. Consider a constant strain tetrahedral element subjected to a body force whose components in the x, y, and z directions are R_x, R_y, and R_z respectively per unit volume. Find the column matrix $\mathbf{Q}$ of nodal loads which is consistent with this body force.

From equations (7.23) and (9.9) the column matrix is

$$\mathbf{Q} = \int_{V_e} \begin{bmatrix} N_1 & 0 & 0 \\ 0 & N_1 & 0 \\ 0 & 0 & N_1 \\ N_2 & 0 & 0 \\ \vdots & \vdots & \vdots \\ 0 & 0 & N_4 \end{bmatrix} \begin{Bmatrix} R_x \\ R_y \\ R_z \end{Bmatrix} dV_e = \int_{V_e} \begin{Bmatrix} N_1 R_x \\ N_1 R_y \\ N_1 R_z \\ N_2 R_x \\ \vdots \\ N_4 R_z \end{Bmatrix} dV_e \qquad (a)$$

where the N_i ($i = 1, 2, 3, 4$) are defined by equation (9.10) and are linear functions of x, y, and z. Assume that each body force component is of uniform intensity throughout the whole element volume. Further, to simplify the required integration take the co-ordinate origin to be at the centroid of the element. Then

$$\iiint_{V_e} x \, dx \, dy \, dz = \iiint_{V_e} y \, dx \, dy \, dz = \iiint_{V_e} z \, dx \, dy \, dz = 0$$

and

$$\alpha_1 = \alpha_2 = \alpha_3 = \alpha_4 = \frac{6V_e}{4}.$$

Using these relationships with the definitions of the N_i in equation (a) gives

$$\mathbf{Q} = \frac{V_e}{4} \{R_x \quad R_y \quad R_z \quad R_x \quad R_y \quad R_z \quad R_x \quad R_y \quad R_z \quad R_x \quad R_y \quad R_z\}.$$

Thus it is seen that consistent use of the potential energy approach demands that the total body force acting in any of the co-ordinate directions is allocated to the four vertex nodes in equal proportions. In this simple case this is what would have been anticipated by physical reasoning.

9.3.2. Higher-order tetrahedra

It has been seen that the basic tetrahedral element for three-dimensional analysis is based on the assumption of each displacement component as a complete linear Cartesian polynomial expression in three dimensions. The relationship between the basic tetrahedron and the basic planar triangle is direct and evident. We have seen that a family of viable higher-order triangular elements can be generated in straightforward fashion by employing various complete two-dimensional Cartesian polynomials in the displacement fields. It is thus no surprise to find that higher-order tetrahedra can be generated in like fashion.

In three dimensions the terms occurring in various levels of complete Cartesian polynomials can be conveniently represented in an extended form of the two-dimensional Pascal triangle of equation (8.36). Such a representation is illustrated in Fig. 9.2 and can be referred to as the three-dimensional Pascal tetrahedron. A family of tetrahedral finite elements can be generated in which each member of the family corresponds to a particular level of complete polynomial displacement representation and has nodal points located in positions which correspond exactly to the positions of the individual terms shown in the Pascal tetrahedron. In theory an unlimited number of element types can be generated but in practice only the first few are of interest; the first three members of the family are shown in Fig. 9.3. At each node the values of u, v, and w are the freedoms and so the total numbers of degrees of freedom for the various elements are 12 (for the basic element already described), 30, 60, and so on. Since strains are given by the first derivatives of displacements

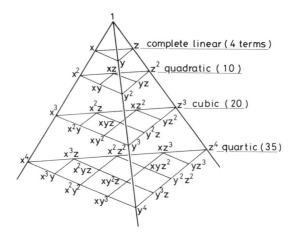

Fig. 9.2. The Pascal tetrahedron, showing terms in complete polynomials in three dimensions.

it follows that the representation of strains within an element is one order of polynomial lower than that of the displacements, i.e. linear displacements give constant strains, quadratic displacements give linear strains, and so on.

As with the members of the planar triangular family these tetrahedral elements are perfectly valid elements with all convergence conditions satisfied: rigid-body motion and constant strain states are present, the elements are conforming, and the displacement field is geometrically invariant. In all cases, if the development of element properties starts with the displacement field being expressed in terms of generalized displacements, the inverse matrix $\mathbf{C}^{-1}$ can be found. Development of element properties when working in terms of Cartesian co-ordinates follows the approach described for the constant strain tetrahedron, except

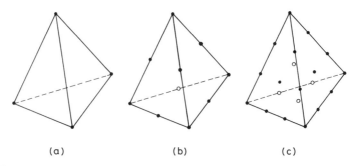

(a)   (b)   (c)

Fig. 9.3. First three members of a family of tetrahedral elements: (a) linear (4 nodes); (b) quadratic (10 nodes); (c) cubic (20 nodes).

that the necessary integrations are made easier if the origin of the co-ordinates is taken to be at the element centroid (as it was in Example 9.1 for calculating the consistent load column matrix corresponding to a body force). However, it should be noted that in the detailed development of the higher-order tetrahedrons the use of a special co-ordinate system which is more 'natural' for the tetrahedron is usually favoured in place of a Cartesian system. Such natural co-ordinate systems are considered in Chapter 10.

The 10-node and 20-node tetrahedra shown in Fig. 9.3 were first introduced by Argyris and co-workers [4–6].

Other tetrahedral elements exist which do not fall within the family of elements just described [7–9]. Such elements are characterized by the use of nodal freedoms other than the displacements alone. Only two such elements will be mentioned here and these are related one to another. The first corresponds again to the adoption of complete cubic polynomials for the three displacement components but, rather than using 20 nodes with three degrees of freedom per node, it is based on the use of just the four vertex nodes with each of the three displacements and all their first derivatives as freedoms, together with nodes at the centre of each face where the three displacements alone are the freedoms. (That is a total of 60 degrees of freedom again of course). The second corresponds to the adoption of an incomplete cubic polynomial in which the centre-face nodes are removed leaving an element with 48 degrees of freedom shared between the four vertices. Thus the element can be represented as in Fig. 9.3(a) but now, instead of just u, v, and w, the degrees of freedom at each node are

$$u, \quad w, \quad v, \quad \partial u/\partial x, \quad \partial u/\partial y, \quad \partial u/\partial z, \quad \partial v/\partial x, \quad \partial v/\partial y, \quad \partial v/\partial z,$$
$$\partial w/\partial x, \quad \partial w/\partial y, \quad \partial w/\partial z.$$

This element will be referred to as Tet 48 in Section 9.5. Use of such elements has some considerable advantages over the use of elements based on the same order of interpolation which have only u, v, and w as nodal degrees of freedom but which have an increased number of nodes. These advantages arise from the connection at the vertex nodes of derivative degrees of freedom. This leads to considerable reduction in the number of structure degrees of freedom and in the bandwidth of the SSM as compared with the situation in which the corresponding elements which use only u, v, and w as freedoms are used. The similar situation that occurs in planar analysis has already been discussed in Section 8.3.4 and Example 8.3. Here, in three-dimensional analysis, it is much more important to improve efficiency as far as is possible by reducing both the structure degrees of freedom and the bandwidth. Of course, a disadvantage of the use of derivative degrees of freedom is that the excess continuity will lead to difficulty in certain problems where the strains are not single valued.

A point of practical difficulty associated with the use of tetrahedra is that it is frequently very difficult to visualize correctly a three-dimensional body made up of such elements. This is particularly so when using the basic tetrahedron since a very considerable number of these elements would often be required to achieve a solution of reasonable accuracy. To overcome this difficulty numbers of tetrahedra can be pre-assembled to form general hexahedra (i.e. brick-type elements) having eight corner nodes, and then the body can be assembled with comparative ease from these 'built-up' elements. Most frequently these latter elements are automatically assembled from five tetrahedra.

9.4. Rectangular hexahedral elements

9.4.1. The basic rectangular hexahedron

The basic rectangular hexahedral element [2] is shown in Fig. 9.4 and is obviously the three-dimensional equivalent of the basic rectangular element of Section 8.4.1. The element has a total of 24 degrees of freedom (the nodal freedoms are u, v, and w at eight corner nodes) and the co-ordinate origin is taken for convenience to be at the centroid of the body.

In assuming a displacement field for the element it is natural to begin by considering expressions for each of u, v, and w as products of linear functions in x, in y, and in z, i.e. as expressions of the form

$$u = (a_1 + b_1 x)(a_2 + b_2 y)(a_3 + b_3 z) \text{ etc.}$$

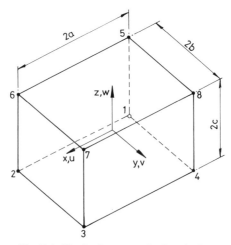

Fig. 9.4. The basic rectangular hexahedron.

where a_1, b_1, etc. are constants. Multiplying this out gives expressions of the form

$$u = A_1 + A_2x + A_3y + A_4z + A_5xy + A_6yz + A_7zx + A_8xyz \quad \text{etc.}$$
$$(9.14)$$

It can be seen (on consulting Fig. 9.2) that such expressions are incomplete three-dimensional polynomials since all first-order terms are present but only three (out of six) second-order terms and one (out of ten) third-order term are present. We could proceed—in much the same way as was done for the basic rectangle in Section 8.4.1—to establish the shape function matrix indirectly from the expressions for u, v, and w written in terms of generalized displacements by determining the matrix product $\boldsymbol{\alpha}\mathbf{C}^{-1}$. However, we need not do this since it is quite clear, following Section 8.4.2, that the individual shape functions can be written directly as triple products of one-dimensional Lagrange interpolation polynomials of the first order in each of the three co-ordinate directions, i.e. a trilinear interpolation is adopted. Thus

$$u = N_1u_1 + N_2u_2 + \ldots N_8u_8 \tag{9.15}$$

where

$$N_1 = N_1(x, y, z) = l_1(x)l_1(y)l_1(z)$$
$$N_2 = N_2(x, y, z) = l_2(x)l_1(y)l_1(z)$$
$$\vdots \tag{9.16}$$
$$N_8 = N_8(x, y, z) = l_1(x)l_2(y)l_2(z).$$

The first-order Lagrange interpolation polynomials $l_1(x)$, $l_2(x)$, $l_1(y)$, and $l_2(y)$ are defined in equations (8.46), and $l_1(z)$ and $l_2(z)$ are defined in like manner in terms of z and c. Expressions of identical form to equation (9.15) apply also for v and w in terms of their nodal values. Thus the complete expression $\mathbf{u} = \mathbf{Nd}$ for the three-component displacement field can be readily established, with

$$\mathbf{N} = \begin{bmatrix} N_1 & 0 & 0 & N_2 & 0 & 0 & N_3 & \ldots & N_8 & 0 & 0 \\ 0 & N_1 & 0 & 0 & N_2 & 0 & 0 & \ldots & 0 & N_8 & 0 \\ 0 & 0 & N_1 & 0 & 0 & N_2 & 0 & \ldots & 0 & 0 & N_8 \end{bmatrix} \tag{9.17}$$

when

$$\mathbf{d} = \{u_1 \quad v_1 \quad w_1 \quad u_2 \quad v_2 \quad w_2 \ldots u_8 \quad v_8 \quad w_8\}. \tag{9.18}$$

The elasticity matrix $\mathbf{E}$ is, for homogeneous isotropic material, again given by equation (7.7) and matrix $\mathbf{B}$ can be established by the appropriate differentiation of $\mathbf{N}$, of similar form to that given in equation (9.11) for the basic tetrahedron, which reflects the form of the strain–displacement equations (7.2).

The 24×24 element stiffness matrix, of the general form of equation (7.21), then becomes

$$\mathbf{k} = \int_{-c/2}^{c/2} \int_{-b/2}^{b/2} \int_{-a/2}^{a/2} \mathbf{B}^t \mathbf{E} \mathbf{B} \, dx \, dy \, dz \tag{9.19}$$

and, given the simplicity of the element geometry, could be evaluated explicitly if desired.

The stress matrix is formed in the usual way as the product of $\mathbf{E}$ and $\mathbf{B}$ and its derivation presents no difficulty (see Problem 9.2). It should be noted that the element stresses are not restricted to uniform values throughout the element, as they were in the case of the basic tetrahedron, since the strains have a restricted linear variation.

The basic hexahedron considered here is referred to as element H24 in Section 9.5.

Consider now the construction of the consistent load column matrix for the hexahedron. If a prescribed body force is acting throughout the element volume V_e the calculation of $\mathbf{Q}$ is quite straightforward and follows the general approach adopted for the basic tetrahedron in Example 9.1. If a surface traction is involved the procedure is a little less straightforward. A specific example will clarify the approach, which extends the ideas of Section 8.3.2.

Example 9.2. Assume that one face of the basic hexahedron of Fig. 9.4, say face 2–3–7–6, forms part of the exterior surface of a three-dimensional body and is subjected to a varying distributed traction acting normal to the face as shown in Fig. E9.2. Determine the consistent load column matrix for this case.

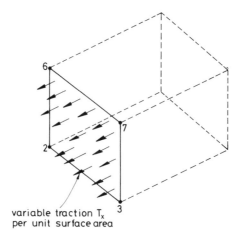

variable traction T_x
per unit surface area

Fig. E9.2

Obviously any consistent replacement of this loading will result in a set of nodal forces acting in the direction normal to the surface—the x direction—at those node points of the element which lie on the pertinent surface, i.e. at nodes 2, 3, 7, and 6. To determine the consistent load column matrix for this case we need concern ourselves only with the way in which the relevant displacement u varies *over the surface 2–3–7–6*.

It is clear that the manner of variation of a displacement component over a face such as 2–3–7–6 of a three-dimensional basic hexahedron is identical with the manner of variation of a displacement component over the two-dimensional basic rectangle shown in Fig. 8.15. This variation is bilinear and could be written down directly in terms of a convenient pair of co-ordinate axes, s_1 and s_2 say, lying in the plane of the face with the origin chosen for convenience to be at the centre of the face. Alternatively the identical expression is obtained by setting $x = a$ and hence $l_1(x) = 0$, $l_2(x) = 1$ in the hexahedral element displacement expression given by equations (9.15) and (9.16). Then the expression for u on surface 2–3–7–6 becomes

$$u^{2376} = \frac{1}{4}\left[\left(1-\frac{y}{b}\right)\left(1-\frac{z}{c}\right) \quad \left(1+\frac{y}{b}\right)\left(1-\frac{z}{c}\right) \quad \left(1+\frac{y}{b}\right)\left(1+\frac{z}{c}\right) \quad \left(1-\frac{y}{b}\right)\left(1+\frac{z}{c}\right)\right]\begin{Bmatrix} u_2 \\ u_3 \\ u_7 \\ u_6 \end{Bmatrix}$$

or

$$u^{2376} = \mathbf{n}_u^{2376}\mathbf{d}_u^{2376} \qquad (a)$$

where the interpretation of the subscripts and superscripts should be clear by extension of the ideas and definitions of Section 8.3.2. The general expression for the consistent load column matrix corresponding to applied traction (equation (7.23))

$$\mathbf{Q} = \int_{S_e} \mathbf{N}^t\mathbf{T}\, dS$$

is now replaced, for the particular case in question, by

$$\mathbf{Q}_u^{2376} = \int_{z=-c}^{c} \int_{y=-b}^{b} (\mathbf{n}_u^{2376})^t T_x\, dy\, dz \qquad (b)$$

(compare with equation (8.24) which concerns an edge of a planar element). Equation (b) can be evaluated quite easily for any reasonable variation of T_x over the face. For example, if T_x varies linearly in the z direction from value zero at edge 2–3 to value $\bar{\sigma}$ at edge 6–7, then

$$T_x = \frac{\bar{\sigma}}{2}\left(1+\frac{z}{c}\right).$$

If this expression for T_x and that for $\mathbf{n}_u^{2376}$ (taken from equation (a)) are substituted into equation (b) and the appropriate integration is performed, the result is

$$\mathbf{Q}_u^{2376} = \bar{\sigma}\frac{bc}{3}\begin{Bmatrix} 1 \\ 1 \\ 2 \\ 2 \end{Bmatrix}. \qquad (c)$$

This result indicates that the given traction acting on the element will be replaced in consistent fashion if loads of one-sixth of the total element surface loading are applied at nodes 2 and 3 and loads of twice this value are applied at nodes 6 and 7.

9.4.2. Higher-order rectangular hexahedra: Lagrange element family

It has been seen that in two-dimensional analysis a family of rectangular elements can be generated with the displacement fields expressed as double products of one-dimensional Lagrange interpolation polynomials in the x and y directions. Clearly, in three-dimensional analysis it is possible to generate similarly a family of rectangular hexahedral elements with the displacement fields expressed as triple products of Lagrange interpolation polynomials in the x, y, and z directions. Obviously the eight-node element H24, shown in Fig. 9.4 and based on trilinear interpolation, is the first member of such a family. Two other members of the Lagrange hexahedral family in which equal-order interpolation is used in all three directions are shown in Figs. 9.5(a) and 9.5(b) [7, 10]. These can be termed the triquadratic and tricubic elements and involve 27 nodes (81 degrees of freedom) and 64 nodes (192 degrees of freedom) respectively, with some nodes positioned within element faces and within the element volume. Again, as with the planar rectangle, unequal interpolation can also be used and one element based on mixed linear and cubic interpolation is illustrated in Fig. 9.5(c).

The Lagrange family of hexahedra provide an obvious and theoretically logical way in which hexahedra can be developed to ever-increasing levels of sophistication. However, in practice there is of course some limit to the number of degrees of freedom that should be associated with a single element and even the tricubic element markedly exceeds this limit in normal operations. Further, the disadvantages discussed earlier (Section 3.4.2) for the Lagrange family of rectangles in two-dimensional analysis have much greater importance for the Lagrange family of hexahedra in

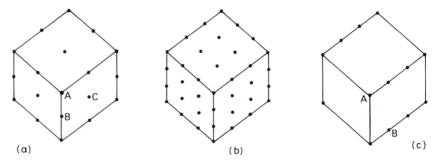

(a)   (b)   (c)

Fig. 9.5. Some Lagrangian hexahedra (hidden nodes not shown): (a) triquadratic (27 nodes); (b) tricubic (64 nodes); (c) mixed (16 nodes).

three-dimensional analysis. The incompleteness of the three-dimensional polynomial displacement fields mitigates against efficiency. As with the rectangle, more efficient elements have been developed, albeit in less direct fashion, which are preferred for three-dimensional analysis. These elements—the so-called serendipity elements—are discussed in the next chapter.

An alternative approach to the generation of hexahedral elements is to employ displacement derivatives as nodal degrees of freedom in the manner mentioned earlier for the tetrahedron and for planar elements. The only such element using a practical number of degrees of freedom [11] is based on the use of cubic Hermitian interpolation polynomials and has 96 degrees of freedom. These are u, v, and w and all their first derivatives with respect to each of the three co-ordinates x, y, and z at the eight corner nodes. For this element, referred to as H96, the element can again be represented by Fig. 9.4 and each component of displacement (u, v, and w) is represented by an expression containing the following terms:

$$1 \quad x \quad y \quad z \quad x^2 \quad y^2 \quad z^2 \quad xy \quad xz \quad yz \quad x^3 \quad y^3 \quad z^3$$
$$x^2y \quad xy^2 \quad x^2z \quad xz^2 \quad y^2z \quad yz^2 \quad xyz \quad x^3y \quad xy^3 \quad x^3z \quad xz^3 \quad y^3z$$
$$yz^3 \quad x^2yz \quad xy^2z \quad xyz^2 \quad x^3yz \quad xy^3z \quad xyz^3. \qquad (9.20)$$

This comprises the complete cubic plus nine quartic terms and three quintic terms.

9.5. Numerical results using three-dimensional elements

There are few studies available which compare the numerical performance of different solid elements in particular structural applications and none which compare the whole range of elements described in Sections 9.3 and 9.4. However, the studies by Fjeld [7] and Clough [12] do provide results which, although necessarily limited in scope, allow certain conclusions to be drawn.

Fjeld [7] compares the performance of two types of tetrahedra and two types of rectangular hexahedra in the solution of two relatively simple test problems involving the bending and torsion of stocky cantilever beams. All the elements have corner nodes only: they are the basic tetrahedron Tet 12, the high-order tetrahedron Tet 48, the basic hexahedron H24, and the high-order hexahedron H96. The manner of convergence of calculated deflection with increase in the total number of degrees of freedom is plotted in Figs. 9.6(a) and 9.6(b). The number of elements involved in the calculation is given at each plotted point. It is clear that in the torsion study (Fig. 9.6(b)) some calculated displacements exceed the

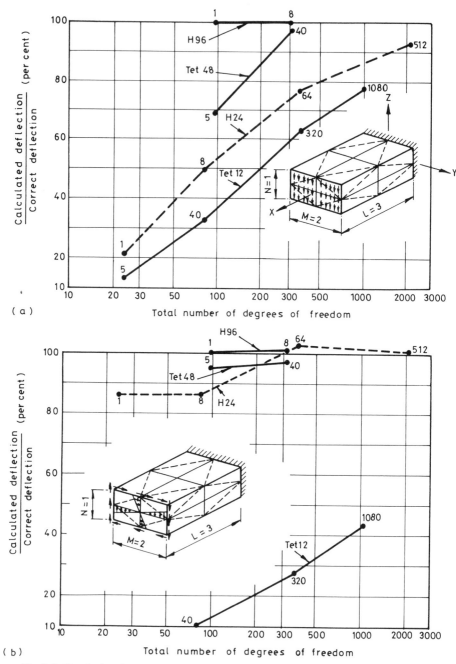

Fig. 9.6. Results for three-dimensional elements (From reference [7]): (a) bending load; (b) torsional load.

correct value; this is stated [7] to be due to some error in the determination of the consistent load column matrix for this case. It is also stated that the accuracy of computed stresses is about the same as that of the computed displacements.

It appears evident from Figs. 9.6(a) and 9.6(b) that (i) higher-order elements are superior to basic elements and (ii) hexahedral elements are superior to tetrahedral elements. It should be mentioned, though, that whilst this conclusion is generally true it can be overturned in some situations. In addition, comparison on the basis of the number of degrees of freedom used does not tell the whole story since other considerations such as size of the bandwidth and the time taken to set up the individual element stiffness matrix need to be taken into account.

Clough [12] has also considered some beam-type problems, using both tetrahedral and hexahedral elements. He also concludes that hexahedral elements are better performers than are tetrahedral elements. The two hexahedral elements considered are the basic eight-node element and a higher-order 20-node element. (The latter element is not one of those considered in Section 9.4. It is in fact a member of the serendipity family which will be described in Section 10.3.) Usually the 20-node element gives superior performance on a solution–time basis although, exceptionally, results presented for a short deep cantilever beam problem show the reverse situation.

9.6. The basic element for the analysis of axisymmetric bodies

Axisymmetric analysis has been defined in Section 7.5 as the analysis of solid bodies of revolution which are subjected to axisymmetric loading only. This means that the displacements, strains, and stresses do not vary in the circumferential (or θ) direction and hence that the problem is effectively reduced to one of two dimensions.

The finite elements used in analysing axisymmetric structures are toroids, i.e. rings of constant cross-section; Fig. 9.7(a) shows such an element of triangular cross-sectional shape. Correspondingly the element nodes are in reality now nodal lines (circles) rather than nodal points. However, provided that we bear in mind the true form of the element when it comes to integrating over the element volume or surface, we can readily represent the element by its cross-section, as in Fig. 9.7(b). The close relationship of axisymmetric finite-element analysis to planar finite-element analysis is now obvious; the element of Fig. 9.7(b)—if it is assumed that the nodal freedoms are u and v at points 1, 2, and 3—is the counterpart of the planar element of Fig. 8.3. The major difference between axisymmetric and planar analysis arises in the number of strain and stress components that are involved. In the axisymmetric case three

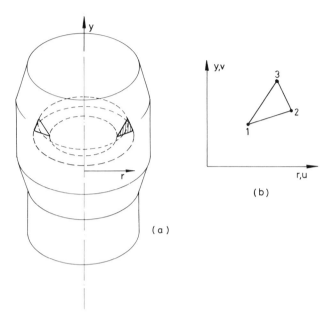

Fig. 9.7. Axisymmetric solid and its basic element.

of the strain components, ε_r, ε_y, and γ_{ry}, have their counterparts in planar analysis but a fourth component, $\varepsilon_\theta = u/r$, also has to be considered (see Section 7.5). This latter component, in the circumferential direction, arises from the radial displacements, and its nature leads to increased difficulty—as compared with the planar case—in carrying out the required integrations.

Consider now the derivation of the properties of the ring finite element shown in Fig. 9.7. This is the simplest ring element, first presented by Clough and Rashid [13] and Wilson [14], and by analogy with the basic planar triangle of Section 8.3 the appropriate displacement field, in terms of generalized displacements, is

$$u = u(r, y) = A_1 + A_2 r + A_3 y$$
$$v = v(r, y) = A_4 + A_5 r + A_6 y. \tag{9.21}$$

This displacement field satisfies all the requirements of Section 6.8.

In regard to the preceding sentence it should be noted that for the axisymmetric body the only possible rigid-body motion (or strain-free state) is a uniform movement in the axial (y) direction and that the only possible state of constant ε_θ strain is when u is proportional to r. These remarks reflect the fact that any radial movement induces a circumferential (or hoop) strain. The above displacement field for the axisymmetric element properly represents the axial rigid-body movement, the state of constant ε_θ strain (in the particular circumstance where it can occur), and states of constant ε_r, ε_y, and γ_{ry} strain. As far as the criteria relating to compatibility are concerned these are clearly satisfied in the same way as they are for the planar triangle.

The above displacement field (equation (9.21)) can be expressed in terms of the nodal displacements

$$\mathbf{d} = \{u_1 \quad v_1 \quad u_2 \quad v_2 \quad u_3 \quad v_3\} \tag{9.22}$$

either indirectly in the form $\mathbf{u} = \alpha\mathbf{C}^{-1}\mathbf{d}$ or directly in the shape function form $\mathbf{u} = \mathbf{Nd}$. The details are exactly as given earlier for the basic planar triangle (equations (8.3)–(8.11)) except that now r replaces x.

The (four) strain–displacement equations for the axisymmetric problem are given by equation (7.39), and if the shape function form of the displacement is used we have that

$$\boldsymbol{\varepsilon} = \left\{ \begin{array}{c} \varepsilon_r \\ \varepsilon_y \\ \varepsilon_\theta \\ \gamma_{ry} \end{array} \right\} = \begin{bmatrix} \partial/\partial r & 0 \\ 0 & \partial/\partial y \\ 1/r & 0 \\ \partial/\partial y & \partial/\partial r \end{bmatrix} \left\{ \begin{array}{c} u \\ v \end{array} \right\} = \begin{bmatrix} \partial/\partial r & 0 \\ 0 & \partial/\partial y \\ 1/r & 0 \\ \partial/\partial y & \partial/\partial r \end{bmatrix} \begin{bmatrix} N_1 & 0 & N_2 & 0 & N_3 & 0 \\ 0 & N_1 & 0 & N_2 & 0 & N_3 \end{bmatrix} \mathbf{d}$$

$$= \mathbf{Bd} \tag{9.23}$$

where

$$N_i = N_i(r, y) = \frac{1}{2\Delta}(\alpha_i + r\beta_i + y\gamma_i) \qquad i = 1, 2, 3 \tag{9.24}$$

and

$$\alpha_1 = r_2 y_3 - r_3 y_2 \qquad \beta_1 = y_2 - y_3 \qquad \gamma_1 = r_3 - r_2 \tag{9.25}$$

with $\alpha_1 \ldots \gamma_3$ defined by cyclic interchange of the subscripts. Also, Δ is the area of the triangular cross-section 1–2–3.

Performing the differentiation indicated in equation (9.23) gives

$$\mathbf{B} = \frac{1}{2\Delta} \begin{bmatrix} \beta_1 & 0 & \beta_2 & 0 & \beta_3 & 0 \\ 0 & \gamma_1 & 0 & \gamma_2 & 0 & \gamma_3 \\ \phi_1 & 0 & \phi_2 & 0 & \phi_3 & 0 \\ \gamma_1 & \beta_1 & \gamma_2 & \beta_2 & \gamma_3 & \beta_3 \end{bmatrix} \tag{9.26}$$

where

$$\phi_i = \frac{1}{r}(\alpha_i + r\beta_i + y\gamma_i) \qquad i = 1, 2, 3. \tag{9.27}$$

Comparison of the $\mathbf{B}$ matrix here with that of the corresponding planar analysis (equation (8.13)) shows that an extra row is of course included here and, more importantly, that the non-zero terms in the extra row, that is the ϕ_i terms associated with the ε_θ strain, are not constants (unlike α_i, β_i, and γ_i). The ϕ_i are functions of both r and y and it is particularly noted that the inverse of r is present in the functions; this complicates the integrations involved in forming the element stiffness matrix, as will be seen shortly.

For the axisymmetric element the elasticity matrix (assuming isotropic material) has been given in equation (7.40) and so the two matrices $\mathbf{B}$ and $\mathbf{E}$ which are the building blocks of the element stiffness matrix $\mathbf{k}$ defined by equation (7.21) are now established. Before this latter equation can be used it is necessary to note that

$$\mathrm{d}V_\mathrm{e} = 2\pi r\, \mathrm{d}r\, \mathrm{d}y.$$

Then the stiffness matrix is

$$\mathbf{k} = \int_{V_\mathrm{e}} \mathbf{B}^\mathrm{t}\mathbf{EB}\, \mathrm{d}V_\mathrm{e} = 2\pi \iint_\Delta \mathbf{B}^\mathrm{t}\mathbf{EB}r\, \mathrm{d}r\, \mathrm{d}y \tag{9.28}$$

where the double integral of the term on the right-hand side is taken over the cross-sectional area 1–2–3 of the element. Performing the triple matrix multiplication involved in equation (9.28) enables the 2×2 sub-matrices $\mathbf{k}_{ij}$ forming $\mathbf{k}$ (see equation (8.15)) to be expressed in the integral form

$$\mathbf{k}_{ij} = \frac{\pi E}{2\Delta^2(1+\nu)(1-2\nu)} \times$$

$$\iint_\Delta \begin{bmatrix} (1-\nu)(\beta_i\beta_j + \phi_i\phi_j) & \nu\gamma_j(\beta_i + \phi_i) \\ + \nu(\beta_i\phi_j + \beta_j\phi_i) & +\tfrac{1}{2}(1-2\nu)\gamma_i\beta_j \\ +\tfrac{1}{2}(1-2\nu)\gamma_i\gamma_j & \\ \hline \nu\gamma_i(\beta_j + \phi_j) & (1-\nu)\gamma_i\gamma_j \\ +\tfrac{1}{2}(1-2\nu)\beta_i\gamma_j & +\tfrac{1}{2}(1-2\nu)\beta_i\beta_j \end{bmatrix} r\, \mathrm{d}r\, \mathrm{d}y. \tag{9.29}$$

Bearing in mind the nature of the ϕ_i (equation (9.27)) it can be seen that the evaluation of $\mathbf{k}$ requires the determination of the following integrals:

$$\iint_\Delta \mathrm{d}r\, \mathrm{d}y \qquad \iint_\Delta r\, \mathrm{d}r\, \mathrm{d}y \qquad \iint_\Delta y\, \mathrm{d}r\, \mathrm{d}y$$

$$\iint_\Delta \frac{1}{r}\, \mathrm{d}r\, \mathrm{d}y \qquad \iint_\Delta \frac{y}{r}\, \mathrm{d}r\, \mathrm{d}y \qquad \iint_\Delta \frac{y^2}{r}\, \mathrm{d}r\, \mathrm{d}y.$$

Explicit evaluation of these integrals is, of course, much more tedious than is the corresponding trivial integration involved in planar analysis. This is particularly so for those integrands with r in the denominator where the integration produces logarithmic expressions. Also, the case when one or more nodes lies on the axis of symmetry requires special treatment. Explicit formulae do exist for all the required integrations [15, 16] but these are rather long and cumbersome even for the simple element shown in Fig. 9.7(b) and will not be detailed here. An alternative to explicit integration is available in the form of numerical integration which evaluates the integrals approximately (usually) by sampling values at a number of discrete points within the element (see Section 10.4.4).

One approximation scheme which has been used and which is effectively a simple numerical integration scheme proceeds as follows. The quantities ϕ_i (equation (9.27)) are evaluated at the centroid c of the triangular area, i.e. at the point (r_c, y_c) where

$$r_c = (r_1 + r_2 + r_3)/3 \qquad y_c = (y_1 + y_2 + y_3)/3.$$

Thus the terms of the matrix **B**, now denoted $\mathbf{B}_c$, are constants—reflecting the fact that the circumferential strain is now uniform throughout the element in common with the other strains—and equation (9.28) can be replaced by

$$\mathbf{k} = 2\pi \mathbf{B}_c^t \mathbf{E} \mathbf{B}_c r_c \Delta. \tag{9.30}$$

Also the definition of the submatrices $\mathbf{k}_{ij}$ now simplifies correspondingly of course. The effect of the approximation is most pronounced close to the axis of the body of revolution where the relative variation of r is greatest. The approximation scheme works well in practice and ultimate convergence to the correct solution is guaranteed if sufficient elements are used. In fact the performance of the element when using the approximation scheme is often superior to its performance when exact integration is employed.

The element stress matrix **S** is easily determined in the standard fashion as the matrix product **EB**. If the element properties are determined exactly the stresses vary throughout the element and these can be evaluated at, say, each of the node points. Alternatively it may well be preferable with this basic axisymmetric element simply to calculate the stresses at the centroid ($\boldsymbol{\sigma}_c = \mathbf{E}\mathbf{B}_c\mathbf{d}$) and take these to be average stresses for the element; this latter approach would be consistent with use of the approximate scheme of evaluation of the element stiffness matrix described in the preceding paragraph. As with the use of the basic triangle in planar analysis, the interpolation of FEM stresses over the whole structure when using meshes of the basic axisymmetric ring element is something of a problem and is often a matter of engineering judgement. Element–centroid values are often used or, alternatively, nodal stresses of elements having common nodes are averaged.

Consider now the generation of the consistent load column matrix **Q**

corresponding to applied body forces and surface tractions. This generation follows the standard pattern exemplified by equation (7.23) (or
equation (7.25)) of course, and presents little extra difficulty in principle
over that associated with the corresponding planar element (Section
8.3.2) as long as it is borne in mind that here the element is actually ring
shaped.

A body force will have components R_r and R_y per unit volume, acting in
the r and y directions. In the axisymmetric problem R_r may be due to
centrifugal action corresponding to the body rotating about the y axis and
would then vary linearly with r. R_y would typically arise owing to gravity.
With a general body force acting the consistent load column matrix is

$$\mathbf{Q} = \int_{V_e} \mathbf{N}^t \mathbf{R} \, dV_e = 2\pi \iint_{\Delta} \mathbf{N}^t \mathbf{R} r \, dr \, dy \qquad (9.31)$$

where $\mathbf{N}$ is defined in equation (8.10), with the N_i given by equation
(9.24), and

$$R = \{R_r \quad R_y\} \qquad (9.32)$$

of course. As with $\mathbf{k}$, integration formulae are available which enable
explicit evaluation of the right-hand side of equation (9.31) or alternatively this evaluation can be made approximately.

A surface traction will generally have components T_r and T_y per unit
surface area, acting in the r and y directions over that part of the surface
S_e of an element which coincides with a loaded region of the exterior
surface of the body. Figure 9.8 shows such a situation, with 2–3 lying on
the surface of the body as a whole. The evaluation of the consistent load
column matrix given in general (see equation (7.23)) by

$$\mathbf{Q} = \int_{S_e} \mathbf{N}^t \mathbf{T} \, dS$$

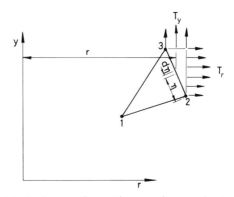

Fig. 9.8. Surface tractions acting on axisymmetric element.

can be carried out for this case along the lines described in Section 8.3.2 for the similar planar situation. The only significant difference is that here the appropriate surface area S_e goes around the circular boundary of the ring so that

$$dS = 2\pi r \, d\eta$$

where η is the length co-ordinate measured along 2–3.

Further to this last remark it should also be realized that in the axisymmetric problem any concentrated external forces $\mathbf{F}$ applied directly to nodes are not applied at points but rather at nodal circles. Thus if the components of applied force per unit length (of circumference) are F_r and F_y in the radial and axial directions respectively, the corresponding applied nodal forces are $2\pi r F_r$ and $2\pi r F_y$.

9.7. Applications of the basic axisymmetric element

Applications of the basic triangular ring element are described in refs 13 and 14. Both sources show that the simple problem of a thick cylinder subjected to pressure loading is solved accurately with the use of a few elements. Here a more difficult problem reported by Clough and Rashid [13] is detailed. This is what is known as the 'Boussinesq problem', a classical elasticity problem [17] concerning a semi-infinite solid subjected to a concentrated load. The element mesh used is shown in Fig. 9.9(a). It should be noted that the region of chief interest is quite local to the load where the stress gradients are steepest and consequently where the mesh is most refined. It should also be noted that the element mesh cannot, of course, extend to infinity to match the classical problem but that the applied roller supports at the mesh boundary are located sufficiently far away from the load point to involve little discrepancy.

The results for the largest direct stress σ_y and the shear stress σ_{ry} are given, for both the classical and the finite-element solutions, in Figs. 9.9(b) and 9.9(c). The agreement between the two sets of results is clearly very good. (The region over which the stresses are plotted is indicated in Fig. 9.9(a).)

9.8. Further remarks on the analysis of axisymmetric bodies

The finite-element analysis of axisymmetric solids under the action of axisymmetric loading has been seen to have a great deal in common with the planar analysis described in Chapter 8. The differences between the two types of analysis are due to the presence of the extra strain and stress component (in the circumferential direction) that occurs in axisymmetric analysis but these differences are so slight that it would be quite easy to

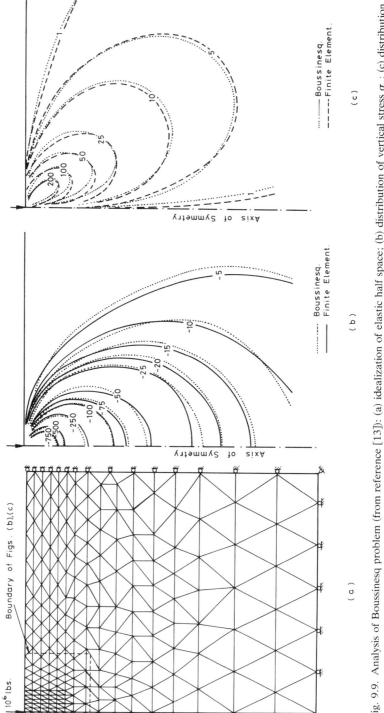

Fig. 9.9. Analysis of Boussinesq problem (from reference [13]): (a) idealization of elastic half space; (b) distribution of vertical stress σ_y; (c) distribution of shear stress τ_{yz}.

accommodate both in a common computer program. We have seen that in planar analysis there exist a wide range of element types based on different displacement assumptions and different geometrical form. It is clear that corresponding types of element can be developed for axisymmetric analysis to supplement the basic element described in Sections 9.6 and 9.7. For example, all the triangles of Figs. 8.9 and 8.10 can be envisaged as the cross-sections of ring elements for axisymmetric analysis. No details need be given here but reference is made to the work of Argyris, Buck, Grieger, and Mareczek [18] for the axisymmetric form of the triangles of Figs. 8.9(b) and 8.9(c) and to Chacour [16] for the axisymmetric form of the triangle of Fig. 8.10. Both references give details of numerical applications to problems of practical interest.

In dealing with the analysis of axisymmetric bodies the concern thus far has been with the response to axisymmetric loading, but obviously many cases arise of non-axisymmetric loading acting on an axisymmetric body. Such cases can be treated by suitable extension of the analysis described above if the circumferential variation of the loading is expressed in a Fourier series. The circumferential component of displacement is now non-zero and all six stress and strain components are involved. The reader is referred to the work of Wilson [14] in which details of this technique were first presented.

References

1 GALLAGHER, R. H., PADLOG, J., and BIJLAARD, P. P. Stress analysis of heated complex shapes. *J. Aerosp. Sci.* **32,** 700–7 (1962).

2 MELOSH, R. J. Structural analysis of solids. *Proc. Am. Soc. Civ. Eng., J. Struct. Div.,* **89** (ST4), 205–23 (1963).

3 ARGYRIS, J. H. Matrix analysis of three-dimensional elastic media, small and large displacements. *AIAA J.* **3,** 45–51 (1965).

4 ARGYRIS, J. H. Continua and discontinua, *Proc. Conf. Matrix Methods in Structural Mechanics, Wright–Patterson Air Force Base,* Dayton, Ohio (1965).

5 ARGYRIS, J. H. Tetrahedron elements with linearly varying strain for the matrix displacement method. *J. R. Aeronaut. Soc.,* **69,** 877–80 (1965).

6 ARGYRIS, J. H., FRIED, I., and SCHARPF, D. W. The TET 20 and the TEA 8 elements for the matrix displacement method, *Aeronaut. J.* **72,** 618–23 (1968).

7 FJELD, S. A. Three dimensional theory of elasticity. In *Finite Element Methods in Stress Analysis,* Chapter 11 (eds. I. Holand and K. Bell). Tapir, Trondheim (1969).

8 HUGHES, T. J. R. and ALLIK, H. Finite elements for compressible and incompressible continua. *Proc. Symp. on Application of Finite Element Methods in Civil Engineering, Vanderbilt University.* American Society of Civil Engineers, Nashville (1969).

9 RASHID, Y. R., SMITH, P. D., and PRINCE, N. On further application of the finite element method to three-dimensional elastic analysis. *Proc. IUTAM*

Symp. on High Speed Computing of Elastic Structures, University of Liege, Belgium, Congrès et Colloques de l'Université de Liège, pp. 433–52 (1971).

10 ARGYRIS, J. H. and FRIED, I. The LUMINA element for the matrix displacement method, *Aeronaut. J.*, **72**, 514–17 (1968).

11 ARGYRIS, J. H., FRIED, I., and SCHARPF, D. W. The HERMES 8 element for the matrix displacement method, *Aeronaut. J.*, **72**, 691 (1968).

12 CLOUGH, R. W. Comparison of three dimensional finite elements. *Proc. Symp. on Application of Finite Element Methods in Civil Engineering, Vanderbilt University*. American Society of Civil Engineers, Nashville (1969).

13 CLOUGH, R. W. and RASHID, Y. Finite element analysis of axisymmetric solids. *Proc. Am. Soc. Civ. Eng., J. Eng. Mech. Div.* **91** (EM1), 71–85 (1965).

14 WILSON, E. L. Structural analysis of axisymmetric solids, *AIAA J* **3**, 2267–74 (1965).

15 UTKU, S. Explicit expressions for triangular torus element stiffness matrix. *AIAA J.* **6**, 1174–5 (1968).

16 CHACOUR, S. A high precision axisymmetric triangular element used in the analysis of hydraulic turbine components, *Trans Am. Soc. Mech. Eng., J. Basic Eng.* **92**, 819–26 (1970).

17 TIMOSHENKO, S. and GOODIER, J. N. Theory of Elasticity (2nd edn.), p. 364. McGraw-Hill, New York (1951).

18 ARGYRIS, J. H., BUCK, K. E., GRIEGER, I., and MARECZEK, G. Application of the matrix displacement method to the analysis of pressure vessels. *Trans. Am. Soc. Mech. Eng., Ser. B*, **92**, 317–29 (1970).

Problems

9.1. Form the stress matrix **S** for the basic tetrahedral element.

9.2. Set up matrix **B** for the basic rectangular hexahedron.

9.3. Form the shape functions for nodes A, B, and C of the triquadratic Lagrangian hexahedral element shown in Fig. 9.5(a). Take the co-ordinate origin and the side lengths to be as shown in Fig. 9.4.

9.4. Form the shape functions for nodes A and B of the Lagrangian hexahedral element shown in Fig. 9.5(c). Take the co-ordinate origin and the side lengths to be as shown in Fig. 9.4.

9.5. For an element of the type referred to in Problem 9.3 derive an expression representing the variation of the u component of displacement over the surface $x = +a$. Use this to determine how the force due to a uniformly distributed surface stress of q per unit area acting over the whole surface $z = +a$, and in the direction normal to it, should be allocated consistently to the nine nodes lying on the surface.

9.6. Figure P9.6 shows the rectangular cross-section of a ring finite element for the analysis of axisymmetric solids. The element has four nodes, with two degrees of freedom (u, and v) at each node. Outline the development of the stiffness matrix for this element, giving details of matrix **B** etc.

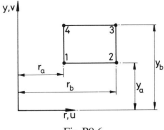

Fig. P9.6

9.7. The ring finite element shown in Fig. P9.7 is based on a biquadratic displacement field. Its surface at $r = R_0$ coincides with the external surface of a body of revolution and is subjected to an applied pressure which varies linearly from p to $2p$ per unit area, as shown in the figure. Determine the loads at nodes 1, 2, and 3 which will consistently represent the applied loading.

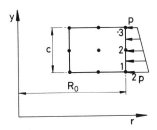

Fig. P9.7

10

MORE-ADVANCED CONCEPTS IN THE FORMULATION OF TWO- AND THREE-DIMENSIONAL ELASTICITY ELEMENTS

10.1. Introduction

FINITE elements for one-, two-, and three-dimensional analysis have been described in Chapters 6, 8, and 9 respectively. Consistent with the aim of simplifying the presentation of material, the developments have in all cases been described within the framework of a conventional Cartesian co-ordinate system. However, such a co-ordinate system is by no means always the most appropriate system to use. Of the elements described earlier, this is particularly so for those of triangular or tetrahedral shape. Fairly early on in the development of the FEM the advantages of working in terms of other co-ordinate systems, which are more appropriate to the particular element shape, were perceived. These co-ordinate systems are intrinsic to the element and as a class are termed 'natural co-ordinates'. Natural co-ordinates are non-dimensional, do not exceed unit value in magnitude, and are related to the global Cartesian co-ordinates by a mapping procedure. Their use often facilitates the direct development of shape functions and simplifies the evaluation of element properties even if the same result could be obtained when using a Cartesian system. The properties of many of the triangular and tetrahedral elements referred to in Chapters 8 and 9 have in fact been developed through the use of natural co-ordinate systems, and certainly anyone concerned with using the FEM to a significant extent will come into contact with such systems. Consequently in this chapter an introduction to natural co-ordinates is given.

In Section 10.2 natural co-ordinates are described for one-, two-, and three-dimensional systems in turn. These co-ordinates are known as length, area (or triangular), and volume (or tetrahedral) co-ordinates respectively. Area and volume co-ordinates are very useful in connection with planar triangular elements and three-dimensional tetrahedral elements.

In Section 10.3 we return to consideration of rectangular elements for planar analysis and rectangular hexahedra for three-dimensional analysis. Again a natural co-ordinate system can be introduced, though here, unlike the situation pertaining to triangular and tetrahedral elements, there is little basic difference between the natural and Cartesian systems. The natural co-ordinates do make it rather easier, though, to develop the shape functions for the important family of serendipity elements. As has been mentioned previously this family of elements is an alternative to the Lagrange family of rectangles and hexahedra, and its elements are generally deemed to be superior since nodes internal to the elements are eliminated.

Finally the concept of isoparametric elements is introduced. In such elements the same form of interpolation is used to define the shape of the element as is used to define its displacement field. Isoparametric co-ordinates are another type of natural co-ordinates. Their use enables basic (or parent) elements of the type considered heretofore, whether these be one-, two-, or three-dimensional, to be mapped into distorted shapes. This is of fundamental importance for it allows the generation of elements having curved sides (in two dimensions) or curved faces (in three dimensions); hence the efficient modelling of complex structures is greatly facilitated. Whilst the isoparametric concept is generally applicable, attention in Section 10.4 is concentrated on distorted forms of the serendipity elements.

10.2. Natural co-ordinates

10.2.1. Length co-ordinates

Length co-ordinates are a system of natural co-ordinates for one-dimensional elements. For such elements the derivation of element properties is in truth simple enough without the introduction of a special co-ordinate system but it will help the understanding of natural co-ordinates in two and three dimensions if some attention is paid to the one-dimensional case.

A line element with ends labelled 1 and 2 is shown in Fig. 10.1(a). Its configuration and geometry are shown in relation to a global co-ordinate

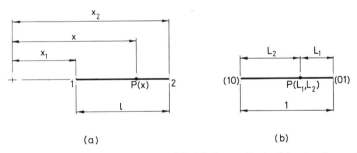

(a)   (b)

Fig. 10.1. Line element coordinate systems: (a) global co-ordinate; (b) natural co-ordinates.

system with an arbitrary origin. P is any general point along the element, located at position x in the global system. We wish to describe the position of such a point in an alternative 'natural' way in terms of suitable co-ordinates associated with the end points of the element. These natural co-ordinates are to be linear functions of the global co-ordinate x over the element length and are to be scaled and non-dimensionalized so that their value does not exceed unity. Figure 10.1(b) shows a suitable pair of natural co-ordinates L_1 and L_2 with point P having co-ordinates (L_1, L_2) in the new system. The end points 1 and 2 have natural co-ordinates $(1, 0)$ and $(0, 1)$ or simply (10) and (01), as in Fig. 10.1(b), if the commas are omitted for convenience. This is akin to saying that natural co-ordinate L_1 varies linearly from zero value at the right-hand end of the element to unit value at the left-hand end, whilst L_2 varies in the opposite fashion.

The global co-ordinate is to be linearly related to the natural co-ordinates by the expression

$$x = L_1 x_1 + L_2 x_2 \tag{10.1}$$

and of course L_1 and L_2 are not independent since

$$L_1 + L_2 = 1. \tag{10.2}$$

Combining these two equations in matrix form gives

$$\begin{Bmatrix} x \\ 1 \end{Bmatrix} = \begin{bmatrix} x_1 & x_2 \\ 1 & 1 \end{bmatrix} \begin{Bmatrix} L_1 \\ L_2 \end{Bmatrix} \tag{10.3}$$

and the solution for L_1 and L_2 as functions of the global co-ordinate is

$$L_1(x) = \frac{x_2 - x}{l} \quad \text{and} \quad L_2(x) = \frac{x - x_1}{l}. \tag{10.4}$$

L_1 and L_2 are clearly ratios of length, hence the name length co-ordinates.

When we come to generate element properties the same basic formulations given earlier apply of course, but now the displacement field will be expressed in terms of the natural co-ordinates L_1 and L_2 rather than the conventional co-ordinate x. It will be necessary, though, to perform differentiations with respect to x (in setting up matrix **B** for example) and to do this the chain rule for differentiation can be used. Thus, if F is a function of L_1 and L_2, then

$$\frac{dF}{dx} = \frac{\partial F}{\partial L_1} \frac{dL_1}{dx} + \frac{\partial F}{\partial L_2} \frac{dL_2}{dx} \tag{10.5}$$

where $dL_1/dx = -1/l$ and $dL_2/dx = 1/l$ from equation (10.4). Also, integrations will be required over the length of the element of various functions of L_1 and L_2. To perform these integrations the following general formula in which i and j are integers is most useful:

$$\int_0^l L_1^i L_2^j \, dx = l \frac{i! \, j!}{(i+j+1)!}. \qquad (10.6)$$

Here, for example,

$$i! = \text{factorial } i = i \times (i-1) \ldots \times 2 \times 1$$

and it should be noted that $0! = 1$.

Consider now the generation of suitable shape functions, written in terms of L_1 and L_2, for the representation of the displacement u in the one-dimensional C_0 continuous problem (i.e. the bar problem).

At the first, or basic, level of interpolation, we seek a linear expression for u corresponding to the presence of two nodes only at the element ends (10) and (01) shown in Fig. 10.1(b). (It will be seen to be convenient to adopt this type of designation of nodal points for ease of specification of the shape function, particularly when extension is made to two- and three-dimensional natural co-ordinates.) Thus we seek expressions for the two shape functions N_{10} and N_{01}, and for this first level of interpolation these take a particularly simple form. The definition of the shape function N_{10} is that it varies linearly from unit value at node (10) to zero at node (01) and this is exactly what the natural co-ordinate L_1 does; a similar relationship exists between N_{01} and L_2. Consequently, for linear interpolation we have the particularly simple relationships

$$N_{10}(L_1, L_2) = L_1 \qquad \text{and} \qquad N_{01}(L_1, L_2) = L_2. \qquad (10.7)$$

(It should be noted that since L_1 and L_2 are not independent (see equation (10.2)) it would also be possible to express N_{10} in terms of L_2 and N_{01} in terms of L_1.) Thus the displacement field is

$$\mathbf{u} = N_{10} u_{10} + N_{01} u_{01} = [L_1 \quad L_2]\{u_{10} \quad u_{01}\} \qquad (10.8)$$

or

$$u = \mathbf{Nd}$$

in the usual fashion.

At higher levels of interpolation the shape functions corresponding to the u freedom at each node can be expressed very systematically if specification of the nodes is again made in the form (e, f) or simply (ef) as was done above for the basic element. Here e is the number of nodes from the node in question to the right-hand end of the element and f is the number of nodes to the left-hand end. To clarify this convention see Fig. 10.2 which shows the labelling of nodes for two bar elements with quadratic and cubic displacement variation. It should be noted that

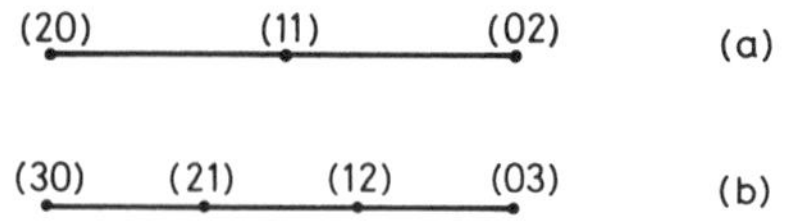

(20)   (11)   (02)   (a)

(30)   (21)   (12)   (03)   (b)

Fig. 10.2. Line elements, showing node numbering convention: (a) quadratic element; (b) cubic element.

$e + f = m$ which is a constant equal to the order of the polynomial curve going through the nodes. The shape function for typical node (ef) is given by

$$N_{ef} = N_e(L_1)N_f(L_2) \tag{10.9}$$

where

$$N_e(L_1) = \prod_{i=1}^{e} \left(\frac{mL_1 - i + 1}{i} \right) \qquad \text{for } e \geqslant 1\dagger$$
$$= 1 \qquad \text{for } e = 0. \tag{10.10}$$

A similar formula holds for $N_f(L_2)$. Equation (10.9) in fact represents Lagrangian interpolation expressed in terms of the natural length co-ordinates.

The recipe for the generation of bar-type elements for any level of interpolation, and within the framework of a natural co-ordinate ap-proach, is now complete. It is emphasized that the end result of applying the natural co-ordinate approach to the development of element proper-ties will be no different to that obtained earlier for bar-type elements in Chapter 6; stiffness matrices and so on will be identical in both ap-proaches for the same type of interpolation. As has been intimated earlier, the advantages of the natural co-ordinate approach are more apparent and more significant in two- and three-dimensional work, and our only concern in dealing with the one-dimensional element is to introduce the concept. With this still in mind an example of the derivation of a bar element, using the natural co-ordinates, is now given to help fix the ideas presented in this section.

Example 10.1. Use the natural co-ordinate approach to determine the stiffness matrix for the bar element of Fig. 10.2(a) based on second-order, or quadratic, interpolation (i.e. $m = 2$). The element length is l.

---

† The symbol $\prod_{i=1}^{e}$ is used to denote the product of all terms for $i = 1$ to $i = e$. If $e = 3$ for example

$$\prod_{i=1}^{e} = (mL_1) \frac{(mL_1 - 1)}{2} \frac{(mL_1 - 2)}{3}.$$

From equation (10.10)

$$N_0(L_1) = 1 \qquad\qquad N_0(L_2) = 1$$
$$N_1(L_1) = 2L_1 \qquad\qquad N_1(L_2) = 2L_2$$
$$N_2(L_1) = L_1(2L_1 - 1) \qquad N_2(L_2) = L_2(2L_2 - 1).$$

The shape functions, using equation (10.9), are then

$$N_{20} = N_2(L_1)N_0(L_2) = 2L_1^2 - L_1$$
$$N_{11} = N_1(L_1)N_1(L_2) = 4L_1L_2$$
$$N_{02} = N_0(L_1)N_2(L_2) = 2L_2^2 - L_2.$$

(If these shape functions are sketched they will be seen to be identical to those shown in Fig. 6.8, as expected.)

The complete displacement field is thus

$$u = N_{20}u_{20} + N_{11}u_{11} + N_{02}u_{02}$$

or

$$u = [2L_1^2 - L_1 \quad 4L_1L_2 \quad 2L_2^2 - L_2]\{u_{20} \quad u_{11} \quad u_{02}\} = \mathbf{Nd}.$$

The matrix $\mathbf{B}$ is

$$\mathbf{B} = \frac{d}{dx}[2L_1^2 - L_1 \quad 4L_1L_2 \quad 2L_2^2 - L_2]$$

and using the chain rule for differentation (equation (10.5)) this becomes

$$\mathbf{B} = \frac{1}{l}[1 - 4L_1 \quad 4(L_1 - L_2) \quad 4L_2 - 1].$$

If the bar is assumed to be uniform, the stiffness matrix is obtained from the general equation (7.21) as

$$\mathbf{k} = A\int_0^l \mathbf{B}^t\mathbf{E}\mathbf{B}\,dx$$

$$= \frac{AE}{l^2}\int_0^l \begin{bmatrix} (1-4L_1)^2 & & \\ 4(L_1-L_2)(1-4L_1) & 16(L_1-L_2)^2 & \text{Symmetric} \\ (4L_2-1)(1-4L_1) & (4L_2-1)4(L_1-L_2) & (4L_2-1)^2 \end{bmatrix} dx.$$

Equation (10.6) yields the required integrals occurring in the stiffness matrix:

$$\int_0^l 1\,dx = l \qquad \int_0^l L_1\,dx = \int_0^l L_2\,dx = \frac{l}{2}$$

$$\int_0^l L_1^2\,dx = \int_0^l L_2^2\,dx = \frac{l}{3} \qquad \int_0^l L_1L_2\,dx = \frac{l}{6}.$$

Then the stiffness matrix of the three-node bar element becomes

$$\mathbf{k} = \frac{AE}{3l}\begin{bmatrix} 7 & -8 & 1 \\ -8 & 16 & -8 \\ 1 & -8 & 7 \end{bmatrix}.$$

This is the same answer as obtained previously without the use of natural co-ordinates (see Example 6.4).

10.2.2. Area or triangular co-ordinates

A natural co-ordinate formulation in two dimensions can be generated using a procedure which is very closely related to that just described for the one-dimensional situation. Our concern here is with plane triangular elements and such an element is shown in a global co-ordinate system in Fig. 10.3(a). By direct extension of the ideas of the preceding section we seek to describe the position of the general point P in a natural way in terms of co-ordinates associated with the corner points of the triangle. This means that three natural co-ordinates L_1, L_2, and L_3 will be involved.

The Cartesian co-ordinates are to be linearly related to the natural co-ordinates by the expressions

$$x = L_1 x_1 + L_2 x_2 + L_3 x_3$$
$$y = L_1 y_1 + L_2 y_2 + L_3 y_3. \tag{10.11}$$

Only two of the natural co-ordinates can be independent (since the general point is described by only two co-ordinates in the Cartesian system) and hence we stipulate in a manner analogous to that used in the preceding section that

$$L_1 + L_2 + L_3 = 1. \tag{10.12}$$

Combining the equations (10.11) and (10.12) in matrix form gives

$$\begin{Bmatrix} x \\ y \\ 1 \end{Bmatrix} = \begin{bmatrix} x_1 & x_2 & x_3 \\ y_1 & y_2 & y_3 \\ 1 & 1 & 1 \end{bmatrix} \begin{Bmatrix} L_1 \\ L_2 \\ L_3 \end{Bmatrix}. \tag{10.13}$$

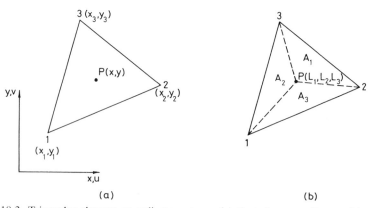

(a)   (b)

Fig. 10.3. Triangular element co-ordinate systems: (a) Cartesian co-ordinates; (b) natural co-ordinates.

The solution for the natural co-ordinates is

$$L_1(x, y) = \frac{1}{2\Delta}(\alpha_1 + \beta_1 x + \gamma_1 y)$$

(10.14)

where

$$\alpha_1 = x_2 y_3 - x_3 y_2 \qquad \beta_1 = y_2 - y_3 \qquad \gamma_1 = x_3 - x_2.$$

(10.15)

The definitions of L_2, L_3, and $\alpha_2 \ldots \gamma_3$ follow directly by cyclic interchange of the subscripts in the order 1, 2, 3. Also

$$2\Delta = \begin{vmatrix} 1 & x_1 & y_1 \\ 1 & x_2 & y_2 \\ 1 & x_3 & y_3 \end{vmatrix} = \alpha_1 + \alpha_2 + \alpha_3.$$

(10.16)

Consider now the physical interpretation of the natural co-ordinates. In the definition of L_1 given in equation (10.14) the quantity

$$\frac{\alpha_1 + \beta_1 x + \gamma_1 y}{2}$$

is in fact the area of the sub-triangle 2–3–P, now denoted as A_1. Further, Δ is the area of the complete triangle 1–2–3. Similar considerations apply in the definitions of L_2 and L_3 so that we see that the natural co-ordinates represent ratios of triangular areas; thus

$$L_1 = A_1/\Delta \qquad L_2 = A_2/\Delta \qquad L_3 = A_3/\Delta$$

(10.17)

The areas A_1, A_2, and A_3 are shown in Fig. 10.3(b) and it is noted that A_1 is the area of the sub-triangle lying opposite vertex 1, and so on. The natural co-ordinates L_1, L_2, and L_3 are termed 'area' or 'triangular' co-ordinates, and clearly these natural co-ordinates in two dimensions are a logical extension to the length co-ordinates in one dimension. (It should be noted that L_1 could also be interpreted as representing the non-dimensional *distance* of a point P measured from the side $L_1 = 0$, with L_2 and L_3 similarly interpreted as *distances* from the sides $L_2 = 0$ and $L_3 = 0$ respectively.) The area co-ordinate L_1 varies linearly from zero at edge 2–3 to unity at vertex 1 so that lines lying parallel to edge 2–3 are lines of constant L_1 value. Similar remarks apply for co-ordinates L_2 and L_3 and Fig. 10.4 illustrates the situation. In terms of the natural co-ordinates the vertices 1, 2, and 3 of the triangle are $(1, 0, 0)$, $(0, 1, 0)$, and $(0, 0, 1)$ respectively. The use of area co-ordinates in generating element properties seems to have been introduced independently by Argyris [1] and by Bazeley, Cheung, Irons, and Zienkiewicz [2] at the same conference.

In generating element properties using the area co-ordinates, functions of L_1, L_2, and L_3 will need to be differentiated with respect to x and y,

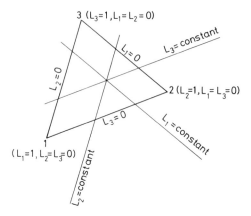

Fig. 10.4. Area co-ordinates.

and other functions will need to be integrated over the region of the triangle. If F is a function of the area co-ordinates then, in a manner akin to equation (10.5), we have for the differentiation process

$$\frac{\partial F}{\partial x} = \frac{\partial F}{\partial L_1}\frac{\partial L_1}{\partial x} + \frac{\partial F}{\partial L_2}\frac{\partial L_2}{\partial x} + \frac{\partial F}{\partial L_3}\frac{\partial L_3}{\partial x}$$

$$\frac{\partial F}{\partial y} = \frac{\partial F}{\partial L_1}\frac{\partial L_1}{\partial y} + \frac{\partial F}{\partial L_2}\frac{\partial L_2}{\partial y} + \frac{\partial F}{\partial L_3}\frac{\partial L_3}{\partial y}$$

(10.18)

where, from equation (10.14),

$$\frac{\partial L_i}{\partial x} = \frac{\beta_i}{2\Delta} \qquad \frac{\partial L_i}{\partial y} = \frac{\gamma_i}{2\Delta} \qquad i = 1, 2, 3.$$

For the integration process the general expression is (compare with equation (10.6))

$$\iint_\Delta L_1^i L_2^j L_3^k \, dx \, dy = \frac{i!\, j!\, k!}{(i+j+k+2)!} 2\Delta.$$

(10.19)

We now turn attention to the specification of displacement fields for a family of triangular elements having various numbers of nodal points with the displacements (u and v) themselves as the only nodal freedoms. Such a family of elements has already been discussed in Chapter 8 but here we are interested in formulations based on the use of area co-ordinates rather than Cartesian co-ordinates. As with the one-dimensional elements of the previous section it is possible to generate the necessary shape functions in a very systematic fashion. To facilitate this a particular form of specification of the nodes is employed, which is an extension of the ideas of the preceding section. Here the nodes are individually (e, f, g) or

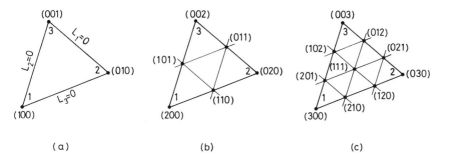

Fig. 10.5. Nodal designations for a family of triangular elements: (a) linear interpolation ($m = 1$); (b) quadratic interpolation ($m = 2$); (c) cubic interpolation ($m = 3$).

simply (efg) where e, f, and g are integers which always sum to the order of polynomial interpolation m used, i.e. $e + f + g = m$. Further, the first integer e represents the number of nodal levels away from the edge $L_1 = 0$ at which the node in question is located; correspondingly f and g represent the number of nodal levels away from edges $L_2 = 0$ and $L_3 = 0$ respectively at which the node is located. Figure 10.5 shows this system of nodal numbering for the first three members of the particular family of triangular elements under consideration, and study of this figure should make clear the method of node specification.

The shape function N_{efg} associated with both the u and v displacement at the typical node (efg) and for mth order interpolation can now be expressed [3, 4] in the form of a triple product of Lagrange interpolation polynomials as

$$N_{efg} = N_e(L_1)N_f(L_2)N_g(L_3) \tag{10.20}$$

where $N_e(L_1)$ is again defined by equation (10.10) and similar formulae hold for $N_f(L_2)$ and $N_g(L_3)$. This direct and systematic generation of the shape functions when working in terms of area co-ordinates is a considerable advantage over the indirect approach adopted in Chapter 8. In adopting equation (10.20) it is assumed that the nodes along any straight line are equally spaced.

In the particular case of linear interpolation of the displacement components over the triangle only the vertex nodes are present and the element is the basic planar triangle with six degrees of freedom shown in Fig. 10.5(a). In this case the shape functions expressed in terms of area co-ordinates have a particularly simple form which is clearly apparent from basic considerations. The area co-ordinates have been defined earlier in precisely the same manner as would be the shape functions corresponding to linear interpolation over the triangle. This is confirmed by noting the equivalence of equations (10.14)–(10.16), which define the area co-ordinates in terms of the Cartesian co-ordinates, with equations (8.11), (8.6), and (8.7), which define the Cartesian shape functions. Thus it is clear that for the basic triangle with six degrees of freedom

$$N_{100}(L_1, L_2, L_3) = L_1$$
$$N_{010}(L_1, L_2, L_3) = L_2 \qquad\qquad (10.21)$$
$$N_{001}(L_1, L_2, L_3) = L_3.$$

It is a relatively simple matter to derive the stiffness matrix for the basic triangle when working in terms of area co-ordinates (see Problem 10.4).

In the above analysis the shape functions N_{efg} are specified directly for the family of triangles which use the displacement components alone as nodal freedoms. It should be realized that the associated displacement fields are in fact complete polynomials in the natural co-ordinates (just as the fields were complete polynomials in the Cartesian co-ordinates in Chapter 8). Thus we could if we wished have assumed displacement fields written in terms of generalized coefficients rather than in shape-function form. For interpolation of order m the typical term of the assumed field is then $L_1^i L_2^j L_3^k$ where i, j, and k are integers which must add up to m and there are $(m+1)(m+2)/2$ terms present. For example the complete quadratic interpolation, corresponding to $m = 2$, requires

$$u \text{ (or } v) = a_1 L_1^2 + a_2 L_2^2 + a_3 L_3^2 + a_4 L_1 L_2 + a_5 L_1 L_3 + a_6 L_2 L_3 \qquad (10.22)$$

and it would be possible to determine the coefficients a_1–a_6 using the 'boundary conditions' at the nodes. This would give the shape functions which would, of course, be identical with those generated rather more easily using equation (10.20). This indirect approach could be used to establish shape functions for other types of element where convenient formulae for such functions are not available. It should be noted that although it is efficient to use complete polynomials to represent element displacements it is not necessary to do so; where incomplete polynomials are employed the use of natural co-ordinates allows geometric invariance to be maintained.

As well as being used in the development of plane stress type elements, area co-ordinates have frequently been employed in deriving plate bending elements (see Chapter 11).

Example 10.2. Determine the shape functions in terms of area coordinates for the plane stress quadratic triangular element shown in Fig. 10.5(b). Hence evaluate the **B** matrix appropriate to the column matrix of nodal displacements

$$\mathbf{d} = \{u_{200} \quad v_{200} \quad u_{020} \quad v_{020} \quad u_{002} \quad v_{002} \quad u_{011} \quad v_{011} \quad u_{101} \quad v_{101} \quad u_{110} \quad v_{110}\}.$$

For quadratic interpolation we have that $m = 2$. The shape functions for both u and v at the nodal point (efg) can be determined using equation (10.20). Prior to this the unidirectional functions $N_e(L_1)$ etc. need to be evaluated using equation (10.10). From this latter equation

$$N_0(L_1) = 1 \qquad N_1(L_1) = 2L_1 \qquad N_2(L_1) = 2L_1^2 - L_1$$
$$N_0(L_2) = 1 \qquad N_1(L_2) = 2L_2 \qquad N_2(L_2) = 2L_2^2 - L_2$$
$$N_0(L_3) = 1 \qquad N_1(L_3) = 2L_3 \qquad N_2(L_3) = 2L_3^3 - L_3.$$

Then

$$N_{200} = N_2(L_1)N_0(L_2)N_0(L_3) = 2L_1^2 - L_1$$
$$N_{011} = N_0(L_1)N_1(L_2)N_1(L_3) = 4L_2L_3$$

and similarly

$$N_{020} = 2L_2^2 - L_2 \qquad N_{002} = 2L_3^2 - L_3 \qquad N_{101} = 4L_1L_3 \qquad N_{110} = 4L_1L_2.$$

(As an exercise the reader should plot the shape functions N_{200} and N_{110} as defined here and verify that their form is precisely that shown (for N_1 and N_6 respectively) in Fig. 8.8.)

Thus the displacement field can be expressed as

$$\begin{Bmatrix} u \\ v \end{Bmatrix} = \begin{bmatrix} 2L_1^2 - L_1 & 0 & 2L_2^2 - L_2 & \cdot & \cdot & \cdot & 4L_2L_3 & 0 & 4L_1L_3 & \cdot & \cdot & 0 \\ 0 & 2L_1^2 - L_1 & 0 & \cdot & \cdot & \cdot & 0 & 4L_2L_3 & 0 & \cdot & \cdot & 4L_1L_2 \end{bmatrix} \mathbf{d}$$

or

$$\mathbf{u} = \mathbf{Nd}.$$

For the plane stress problem we have in general that

$$\mathbf{B} = \begin{bmatrix} \partial/\partial x & 0 \\ 0 & \partial/\partial y \\ \partial/\partial y & \partial/\partial x \end{bmatrix} \mathbf{N}$$

and for $\mathbf{N}$ defined as above, and using equations (10.18), this becomes

$$\mathbf{B} = \frac{1}{2\Delta} \begin{bmatrix} \beta_1(4L_1 - 1) & 0 & \beta_2(4L_2 - 1) & \cdot & \cdot & 4(\beta_2L_3 + \beta_3L_2) \\ 0 & \gamma_1(4L_1 - 1) & 0 & \cdot & \cdot & \cdot & 0 \\ \gamma_1(4L_1 - 1) & \beta_1(4L_1 - 1) & \gamma_2(4L_2 - 1) & \cdot & \cdot & 4(\gamma_2L_3 + \gamma_3L_2) \end{bmatrix}$$

$$\begin{bmatrix} \cdot & \cdot & \cdot & \cdot & 0 \\ \cdot & \cdot & \cdot & \cdot & 4(\gamma_1L_2 + \gamma_2L_1) \\ \cdot & \cdot & \cdot & \cdot & 4(\beta_1L_2 + \beta_2L_1) \end{bmatrix}.$$

The stiffness matrix for the uniform thickness quadratic triangle could now be calculated according to

$$\mathbf{k} = h \int\int_\Delta \mathbf{B}^t\mathbf{EB}\, dx\, dy$$

with appropriate use made of equation (10.19) in performing the integrations.

10.2.3. Volume or tetrahedral co-ordinates

Natural co-ordinate systems have been introduced in the preceding sections for use in deriving the properties of one-dimensional, line elements and of two-dimensional, triangular elements. These co-ordinates have been termed length and area, or triangular, co-ordinates, and certainly in the latter case there is considerable benefit associated with the use of natural co-ordinates with regard to the direct specification of shape functions and the evaluation of the necessary integrals. It will probably seem likely to the reader that the ideas concerning natural co-ordinates could be extended to the three-dimensional situation, and this is indeed so. In particular a system of volume or tetrahedral co-ordinates can be devised for use in deriving the properties of the family of tetrahedral elements described in terms of Cartesian co-ordinates in Section 9.3 (and of other tetrahedral elements). This can be done by straightforward extension of the concepts of the area co-ordinates of two-dimensional analysis.

In brief, the volume co-ordinates are ratios of certain sub-volumes of a tetrahedron to the whole volume. Figure 10.6 shows the location of a general point P within the tetrahedron 1–2–3–4. There are now four natural or volume co-ordinates, L_1, L_2, L_3, and L_4, associated with the four vertices. L_1 is defined as the volume of the sub-tetrahedron P–2–3–4 divided by the volume of the whole tetrahedron 1–2–3–4 and similar definitions apply for L_2, L_3, and L_4. Equations can readily be derived linking the volume co-ordinates to the Cartesian co-ordinates, expressing differentiation and integration of the volume co-ordinates, and giving the shape functions as products of one-dimensional interpolation polynomials for tetrahedral elements based on complete polynomials; these equations are of very similar form to those given for the development of area co-ordinates, with the difference that of course three Cartesian directions and four natural co-ordinates are now involved. Further details will not

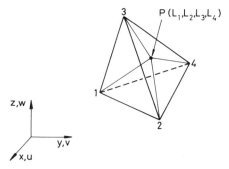

Fig. 10.6. Natural co-ordinates for the tetrahedron.

be given here, largely because tetrahedral elements are not so useful or popular in three-dimensional analysis as are triangular elements in two-dimensional analysis. The reader requiring more information on the use of volume co-ordinates in formulating tetrahedral elements is referred to refs. 3 and 5, and to refs. 7, 8, and 12 of Chapter 9.

10.3. Serendipity rectangles and hexahedra

One family of rectangular elements for planar analysis has already been presented in Section 8.4 whilst a corresponding family of rectangular hexahedra for three-dimensional analysis has been presented in Section 9.4. These elements were referred to as Lagrangian elements because their shape functions could be expressed in very simple and direct manner as products of one-dimensional Lagrange interpolation polynomials. Unfortunately, this simplicity of formulation is paid for by relative inefficiency of performance owing to the presence of interior nodes in the elements and of a large number of terms from incomplete polynomials in the assumed displacement fields. It has already been mentioned that alternative families of elements have been derived for rectangles and rectangular hexahedra. The alternative elements are known as serendipity elements; their shape functions were initially developed by inspection in rather less obvious fashion than are those for the Lagrangian elements but they have improved efficiency. The serendipity elements described in this section are basic rectangles and rectangular hexahedra, but it should be borne in mind that it is possible to transform such simple shapes into much more general geometric forms, as will be seen in the next section.

The serendipity families of elements were introduced in refs. 6 and 7.

10.3.1. Rectangles

In our previous discussion of planar rectangular elements in Section 8.4 the element properties were generated with regard to local x, y co-ordinates and for an element of physical size $2a \times 2b$ (see Fig. 8.15). It will now be more convenient and more natural to use non-dimensional co-ordinates ξ, η with origin at the centre of the rectangle and with a value of ± 1 at all element edges. Figure 10.7 shows such co-ordinates and their relationship to a system of x, y co-ordinates whose origin in general is displaced from that of the ξ, η co-ordinates. Clearly the non-dimensional co-ordinates can be expressed in terms of the Cartesian co-ordinates as

$$\xi = \frac{x - x_c}{a} \qquad \eta = \frac{y - y_c}{b}.$$

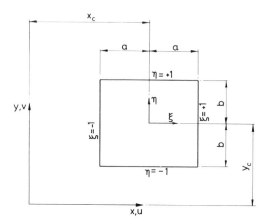

Fig. 10.7. Natural co-ordinates for the rectangle.

The objective is to produce rectangular elements which eliminate the interior nodes associated with the Lagrangian rectangles of quadratic and higher order shown in Fig. 8.17. Thus we seek a family of elements whose first three members are shown in Fig. 10.8. At each node the freedoms are the values of the displacements u and v.

The first member of the serendipity family is based on linear interpolation. The shape functions have to ensure linear variation of displacement at the boundary edges of the rectangle containing the node, so as to meet compatibility requirements, and have to have unit value at the node with which they are associated and zero value at other nodes of course. These conditions are satisfied if the shape function for each of u_i and v_i at node i, where $\xi = \xi_i = \pm 1$ and $\eta = \eta_i = \pm 1$, is defined as

$$N_i = \tfrac{1}{4}(1 + \xi\xi_i)(1 + \eta\eta_i). \tag{10.23}$$

Clearly this shape function is precisely the same as that already found for the basic rectangle, defined by equation (8.41) and illustrated in Fig. 8.16. Thus the lowest-order Lagrangian and serendipity rectangles are one and the same element.

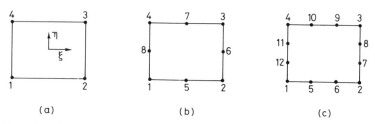

(a)   (b)   (c)

Fig. 10.8. First three serendipity rectangular elements: (a) first order; (b) second order; (c) third order.

For the second-order serendipity rectangle shown in Fig. 10.8(b), the shape functions were originally derived by inspection but the procedure was made more systematic in a manner described by Taylor [8]. The derivation is quite straightforward when account is taken of the basic requirements of the shape function for any particular node: these are that the function has unit value at the particular node and zero value at all other nodes, and that the function varies quadratically on any edge of the rectangle containing the particular node (so as to meet compatibility conditions). For a mid-side node, node 7 say, the pertinent shape function is the product of a second-order Lagrange interpolation function along the edge containing node 7 (edge 4–3) and a first-order Lagrange function in the direction at right angles to this edge. Thus

$$N_7 = (1 - \xi^2)\left(\frac{1+\eta}{2}\right)$$

and this shape function is shown in Fig. 10.9(b). Extending this to embrace all four mid-side nodes gives

$$N_i = \tfrac{1}{2}(1 - \xi^2)(1 + \eta\eta_i) \qquad \text{for} \qquad i = 5 \text{ and } 7$$
$$\text{(with } \eta_i = -1 \text{ and } 1 \text{ respectively)}$$
$$N_i = \tfrac{1}{2}(1 + \xi\xi_i)(1 - \eta^2) \qquad \text{for} \qquad i = 6 \text{ and } 8 \qquad (10.24a)$$
$$\text{(with } \xi_i = 1 \text{ and } -1 \text{ respectively)}.$$

The procedure for deriving the shape function for a corner node is a little more complicated since it involves an addition, in certain proportions, of three other shape functions. This is illustrated for node 4 (of Fig. 10.8(b)) in Fig. 10.10 after the manner of Taylor. The three component shape functions contributing toward N_4 are shown in Fig. 10.10(a): they comprise the *first*-order Lagrange rectangle shape function, denoted by N_4^{L},

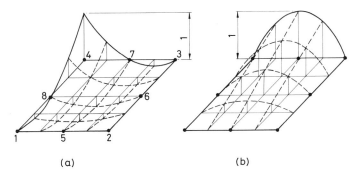

(a)   (b)

Fig. 10.9. Shape functions for the second-order serendipity element: (a) corner-node function N_4; (b) side-node function N_7.

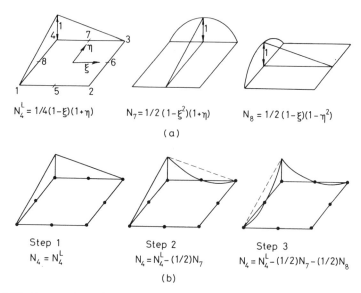

$N_4^L = 1/4(1-\xi)(1+\eta)$ $N_7 = 1/2\,(1-\xi^2)(1+\eta)$ $N_8 = 1/2\,(1-\xi)(1-\eta^2)$

(a)

Step 1   Step 2   Step 3
$N_4 = N_4^L$ $N_4 = N_4^L - (1/2)N_7$ $N_4 = N_4^L - (1/2)N_7 - (1/2)N_8$

(b)

Fig. 10.10. Construction of shape function for corner node 4 of quadratic, serendipity element: (a) component shape functions contributing to N_4; (b) assembly of N_4 from component functions.

and the *second*-order shape functions for the two mid-side nodes adjacent to node 4, denoted by N_7 and N_8 and defined by equation (10.24a). The assembly process leading to the specification of N_4 is shown in Fig. 10.10(b); it starts with N_4^L, subtracts half of N_7 so that zero value is obtained at node 7, and finally subtracts half of N_8 so that zero value is obtained at node 8. The same type of procedure applies, of course, to the shape functions for nodes 1, 2, and 3. Mathematically the corner node shape functions are expressed as

$$N_i = \tfrac{1}{4}(1+\xi\xi_i)(1+\eta\eta_i)(\xi\xi_i + \eta\eta_i - 1) \qquad \text{for} \qquad i = 1, 2, 3, 4. \qquad (10.24b)$$

The shape function N_4 is shown in more detail in Fig. 10.9(a).

It is of interest to compare the quadratic serendipity shape functions for the typical corner and mid-side node, shown in Figs. 10.9(a) and 10.9(b), with the corresponding Lagrangian shape functions shown in Figs. 8.18(a) and 8.18(b). Since both types of shape function vary in quadratic fashion along an edge and have the same value at three points along such an edge, it follows that the functions are identical on the exterior edges of the rectangle. (It also follows, of course, that both elements are conforming.) Obviously there are significant differences in the shape functions of the two types of element throughout the interior, resulting from the removal of the interior node in the serendipity element.

Higher-order serendipity rectangles can be generated in similar fashion to that described for the quadratic element. For an mth-order element the shape functions for nodes on the sides of the element are formed as products of mth-order Lagrange interpolation functions along the nodal edge and first-order Lagrange functions in the direction normal to the nodal edge. The shape function of any corner node is then constructed by subtracting appropriate proportions of the side-node shape functions from the basic first-order Lagrange rectangle shape function for the corner node, so that zero values are obtained at all nodes along the edges joining the corner. For the third-order element shown in Fig. 10.8(c) the procedure yields the following shape functions:

$$N_i = \tfrac{1}{32}(1 + \xi\xi_i)(1 + \eta\eta_i)(9(\xi^2 + \eta^2) - 10) \tag{10.25a}$$

for the corner nodes where $\xi_i = \pm 1$ and $\eta_i = \pm 1$, and

$$N_i = \tfrac{9}{32}(1 + \eta\eta_i)(1 - \xi^2)(1 + 9\xi\xi_i) \tag{10.25b}$$

for side nodes at $\eta_i = \pm 1$, $\xi_i = \pm\tfrac{1}{3}$, with an expression of similar form applying for the remaining side nodes. (To help fix ideas the shape functions for a typical side and a typical corner node of the cubic element are derived in detail in Example 10.3 below.)

By examining the nature of the shape functions for the serendipity rectangles it is easy to establish what polynomial terms are present in the overall displacement fields corresponding to different orders of element. The result is shown symbolically in Fig. 10.11. This shows that the serendipity rectangles are, like the Lagrangian rectangles, based on the use of incomplete two-dimensional polynomial displacement fields. Thus, for example, the second-order serendipity rectangle displacement field contains a complete two-dimensional quadratic polynomial and also contains two cubic terms (the x^2y and xy^2 terms) out of a possible four. As against this the second-order Lagrange rectangle contains the complete quadratic plus the same two cubic terms and a single quartic term

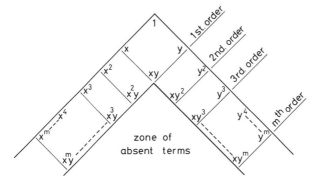

Fig. 10.11. Polynomial terms present in various orders of serendipity rectangles.

(see Fig. 8.19 for the terms present in Lagrange rectangles of various orders). Similarly, if the third-order serendipity and Lagrange elements are compared we see that both elements contain the complete two-dimensional cubic polynomial but that the serendipity element has only two terms $(x^3y$ and $xy^3)$ present belonging to higher-order incomplete polynomials whereas the Lagrange element has six. Thus a smaller number of surplus terms are required for a given complete polynomial displacement field in the serendipity formulation and this points to the greater efficiency of this type of formulation.

From Fig. 10.11 it can also be observed that serendipity rectangles of mth order will lack one or more terms of the complete mth-order two-dimensional polynomial for all m greater than 3. Thus, for example, the displacement field for the fourth-order rectangle does not contain the x^2y^2 term belonging to the complete fourth-order polynomial. This can be remedied by the introduction of an internal node or by a related device [8] but this will not be considered here since it is only the elements of Fig. 10.8 that are used frequently in practice. It is noted that each of these latter elements has an equal number of nodes on all sides but it is perfectly possible using the direct method of constructing shape functions, as outlined above, to derive an element which has different numbers of nodes along each side.

The shape functions for the serendipity rectangles have been constructed here in a direct fashion and this provides a valuable insight into the nature of shape functions as well as being a very useful approach in a practical sense. However, it should be realized that the same shape functions for the elements of Fig. 10.8 could be obtained in the indirect fashion symbolized by the expression $\mathbf{N} = \boldsymbol{\alpha}\mathbf{C}^{-1}$. Thus, for the second-order rectangle for example, we could start by assuming that both u and v are represented by the sum of products of the eight polynomial terms indicated in Fig. 10.11 and generalized coefficients; application of the nodal 'boundary conditions' then allows the determination of the generalized coefficients in the normal way (see Problem 10.9).

10.3.2. Rectangular hexahedra

The rectangular elements shown in Fig. 10.8 have their counterparts in three-dimensional analysis in the form of the rectangular hexahedra shown in Fig. 10.12. For these latter elements a third non-dimensional co-ordinate ζ is involved and the degrees of freedom at each node are now the displacement components u, v, and w in the ξ, η, and ζ directions respectively. The origin of the $\xi\eta\zeta$ co-ordinates is at the centroid of the element and each of the three co-ordinates ranges between -1 and $+1$. In three-dimensional analysis the large numbers of degrees of freedom

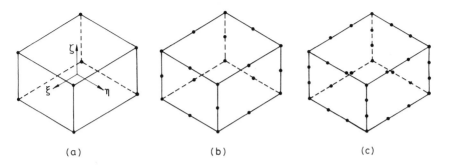

Fig. 10.12. First three serendipity rectangular hexahedra: (a) first order (8 nodes, 24 d.o.f.); (b) second order (20 nodes, 60 d.o.f.); (c) third order (32 nodes, 96 d.o.f.).

usually required to obtain an adequate solution make it especially important to use the most efficient elements, and the serendipity hexahedra again have the advantage over the Lagrange hexahedra (of Section 9.42) of only using external nodes. To derive the shape functions of the hexahedral serendipity family similar techniques can be used as were used in the case of the rectangular family. Only the results will be given here [7].

The first-order hexahedron is precisely the same as the first-order Lagrange element of Section 9.4. The shape functions of the eight corner nodes are defined by

$$N_i = \tfrac{1}{8}(1 + \xi\xi_i)(1 + \eta\eta_i)(1 + \zeta\zeta_i). \tag{10.26}$$

For the second-order element the corner nodes have

$$N_i = \tfrac{1}{8}(1 + \xi\xi_i)(1 + \eta\eta_i)(1 + \zeta\zeta_i)(\xi\xi_i + \eta\eta_i + \zeta\zeta_i - 2) \tag{10.27a}$$

whilst for typical side nodes located at $\xi_i = 0$, $\eta_i = \pm 1$, $\zeta_i = \pm 1$,

$$N_i = \tfrac{1}{4}(1 - \xi^2)(1 + \eta\eta_i)(1 + \zeta\zeta_i). \tag{10.27b}$$

For the third-order element the corner nodes have

$$N_i = \tfrac{1}{64}(1 + \xi\xi_i)(1 + \eta\eta_i)(1 - \zeta\zeta_i)(9(\xi^2 + \eta^2 + \zeta^2) - 19) \tag{10.28a}$$

whilst for typical side nodes located at $\xi_i = \pm\tfrac{1}{3}$, $\eta_i = \pm 1$, $\zeta_i = \pm 1$

$$N_i = \tfrac{9}{64}(1 - \xi^2)(1 + 9\xi\xi_i)(1 + \eta\eta_i)(1 + \zeta\zeta_i). \tag{10.28b}$$

Equations (10.26)–(10.28) for the three-dimensional geometry bear an obvious relationship to equations (10.23)–(10.25) for two-dimensional geometry.

Example 10.3. For the third-order serendipity rectangle shown in Fig. 10.8(c) deduce the shape functions for nodes 4 and 10.

These shape functions will be constructed using first- and third-order Lagrange functions which themselves can be determined using equation (6.72). Working in terms of a non-dimensional co-ordinate s (rather than the x of equation (6.72)) with s ranging from -1 to $+1$ we obtain the following results. (To avoid confusion with the numbering of nodes on the rectangular element (Fig. 10.8(c)) the one-dimensional interpolation functions are written in the form:

$_{-1}^{1}l(s)$ to mean the first-order Lagrangian function for the point at $s = -1$, or

$_{+1}^{3}l(s)$ to mean the third-order Lagrangian function for the point at $s = +1$, and so on.)

For the first-order interpolation

$$_{-1}^{1}l(s) = \frac{1-s}{2} \qquad _{+1}^{1}l(s) = \frac{1+s}{2}.$$

For third-order interpolation, with equispaced points at $s = -1$, $-1/3$, $+1/3$, and $+1$,

$$_{-1/3}^{3}l(s) = \frac{(s+1)(s-1/3)(s-1)}{(2/3)(-2/3)(-4/3)} = \frac{9}{16}(1-s^2)(1-3s)$$

and similarly

$$_{+1/3}^{3}l(s) = \tfrac{9}{16}(1-s^2)(1+3s)$$

($_{+1}^{3}l(s)$ and $_{-1}^{3}l(s)$ are not required in forming the shape functions for the rectangle).

The shape function for side node 10 of the third-order rectangular element is now formed as the product of the third-order function of ξ, with $\xi = -1/3$, and the first-order function of η, with $\eta = +1$. Therefore

$$N_{10} = {}_{+1}^{1}l(\eta) \times {}_{-1/3}^{3}l(\xi) = \left(\frac{1+\eta}{2}\right)\frac{9}{16}(1-\xi^2)(1-3\xi) = \frac{9}{32}(1+\eta)(1-\xi^2)(1-3\xi).$$

This is precisely the result given by equation (10.25b) when $\xi_i = -1/3$, $\eta_i = +1$.

The shape function for corner node 4 of the third-order rectangle is formed along the general lines indicated in Fig. 10.10 for the second-order element. Thus, to determine N_4 will require knowledge of the shape functions of all side nodes along the two sides meeting at corner node 4, i.e. it will require N_{10} as defined above and also N_9, N_{11}, and N_{12}. The latter three functions can be derived in a similar way to N_{10}. Thus

$$N_9 = {}_{+1}^{1}l(\eta) \times {}_{+1/3}^{3}l(\xi) = \tfrac{9}{32}(1+\eta)(1-\xi^2)(1+3\xi)$$
$$N_{11} = {}_{-1}^{1}l(\xi) \times {}_{+1/3}^{3}l(\eta) = \tfrac{9}{32}(1-\xi)(1-\eta^2)(1+3\eta)$$
$$N_{12} = {}_{-1}^{1}l(\xi) \times {}_{+1/3}^{3}l(\eta) = \tfrac{9}{32}(1-\xi)(1-\eta^2)(1-3\eta).$$

The shape function N_4 is now given by

$$N_4 = N_4^{\mathrm{L}} - \tfrac{2}{3}(N_{10} + N_{11}) - \tfrac{1}{3}(N_9 + N_{12})$$

corresponding to the fact that appropriate proportions of N_9–N_{12} are subtracted from the first-order Lagrange rectangle shape function N_4^L (defined in Fig. 10.10(a)) so as to render N_4 zero at nodes 9, 10, 11, and 12. On substituting for N_4^L, N_9, N_{10}, N_{11}, and N_{12} into the preceding equation, the result is

$$N_4 = \tfrac{1}{32}(1 - \xi)(1 + \eta)(9(\xi^2 + \eta^2) - 10)$$

which is the result given by equation (10.25a) when $\xi_i = -1$ and $\eta_i = +1$.
The reader should complete this example by sketching N_4 and N_{10}.

10.4. The isoparametric concept

10.4.1. A linear quadrilateral element

In Fig. 10.13(a) is shown an arbitrary straight-sided quadrilateral element which may form part of a plane stress structure, say. The element lies in the xy plane where x and y are fixed global Cartesian co-ordinates. We wish to derive an 8×8 stiffness matrix for the element, using as degrees of freedom the values of u and v at each of the four corner points. It should be noted that u and v are the displacements measured in the x and y directions respectively. To derive the stiffness matrix we introduce a local natural co-ordinate system $\xi\eta$ whose origin is at the centroid of the quadrilateral element. Consistent with the definition of natural co-ordinates, ξ and η are non-dimensional co-ordinates whose magnitude does not exceed unity. The arbitrary quadrilateral in the global xy co-ordinate system, shown in Fig. 10.13(a), can thus be represented as a square of side length 2 in the local $\xi\eta$ system, as shown in Fig. 10.13(b). The actual element is visualized as a distortion of a 'parent' element and a mapping procedure can be used to relate the two.

The geometry of the actual element is fully defined when the co-ordinate positions x_i, y_i of each of the four corner points are specified.

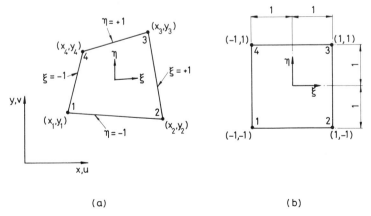

(a)   (b)

Fig. 10.13. The linear isoparametric quadrilateral: (a) actual element; (b) parent element.

Therefore, bearing in mind the rectilinear nature of the element under consideration, we can relate the local and global co-ordinate systems by the equations

$$x = A_1 + A_2\xi + A_3\eta + A_4\xi\eta$$
$$y = A_5 + A_6\xi + A_7\eta + A_8\xi\eta. \quad\quad (10.29)$$

Then the coefficients A_1–A_8 can be determined in terms of the x, y co-ordinates of the four node points i by using the conditions that

$$x = x_i, \ y = y_i \text{ at node } i \text{ where } \xi = \xi_i, \ \eta = \eta_i.$$

Substituting for A_1–A_8 in equations (10.29) then gives

$$x = N_1(\xi, \eta)x_1 + N_2(\xi, \eta)x_2 + \ldots = \sum_1^4 N_i x_i$$
$$y = N_1(\xi, \eta)y_1 + N_2(\xi, \eta)y_2 + \ldots = \sum_1^4 N_i y_i \quad\quad (10.30)$$

where

$$N_i = N_i(\xi, \eta) = \tfrac{1}{4}(1 + \xi\xi_i)(1 + \eta\eta_i). \quad\quad (10.31)$$

Now consider the assumed displacement field. This expresses u and v, the displacement components in the x and y directions, as functions of ξ and η co-ordinates. The parent element shown in Fig. 10.13(b) is recognized as being the first-order Lagrange or first-order serendipity element. The assumed displacement field of this element can be expressed in the indirect form of equation (10.29) (with u, v replacing x, y) or directly in shape-function form as (see Section 8.4 or 10.3),

$$u = N_1(\xi, \eta)u_1 + N_2(\xi, \eta)u_2 + \ldots = \sum_1^4 N_i u_i$$
$$v = N_1(\xi, \eta)v_1 + N_2(\xi, \eta)v_2 + \ldots = \sum_1^4 N_i v_i \quad\quad (10.32)$$

where $N_i(\xi, \eta)$ is precisely as defined in equation (10.31). It is clear from equations (10.30) and (10.32) that the expressions for x and y and for u and v are of identical form. Thus the element geometry and the element displacements are interpolated in precisely the same parametric form. Where this statement applies—whatever the order of the common interpolation or the shape of the element—the element is termed *isoparametric*.

To evaluate the element stiffness matrix the standard form

$$\mathbf{k} = \int_{V_e} \mathbf{B}^t \mathbf{E} \mathbf{B} \, dV_e \quad\quad (7.21)$$

can be used, where $\mathbf{k}$ relates forces to displacements in the Cartesian system and thus corresponds to the following definition of the nodal displacements:

$$\mathbf{d} = \{u_1 \quad v_1 \quad u_2 \quad v_2 \quad u_3 \quad v_3 \quad u_4 \quad v_4\}. \tag{10.33}$$

However, the procedure is complicated by the fact that u and v are expressed in terms of the ξ, η co-ordinates.

For the planar problem the Cartesian components of stress and strain are given by equations (7.32) and (7.33) and the elasticity matrix $\mathbf{E}$, linking the stresses and strains, is given directly by equation (7.35) for plane stress or equation (7.36) for plane strain (assuming isotropic material). To determine the 3×8 matrix $\mathbf{B}$ the strain–displacement equations (7.34) are invoked; thus

$$\boldsymbol{\varepsilon} = \left\{ \begin{array}{c} \varepsilon_x \\ \varepsilon_y \\ \gamma_{xy} \end{array} \right\} = \begin{bmatrix} \partial/\partial x & 0 \\ 0 & \partial/\partial y \\ \partial/\partial y & \partial/\partial x \end{bmatrix} \left\{ \begin{array}{c} u \\ v \end{array} \right\} \tag{10.34}$$

and using equations (10.32)

$$\boldsymbol{\varepsilon} = \begin{bmatrix} \dfrac{\partial N_1}{\partial x} & 0 & \dfrac{\partial N_2}{\partial x} & \cdots & 0 \\[2mm] 0 & \dfrac{\partial N_1}{\partial y} & 0 & \cdots & \dfrac{\partial N_4}{\partial y} \\[2mm] \dfrac{\partial N_1}{\partial y} & \dfrac{\partial N_1}{\partial x} & \dfrac{\partial N_2}{\partial y} & \cdots & \dfrac{\partial N_4}{\partial x} \end{bmatrix} \left\{ \begin{array}{c} u_1 \\ v_1 \\ u_2 \\ \cdot \\ \cdot \\ \cdot \\ v_4 \end{array} \right\} = \mathbf{Bd}. \tag{10.35}$$

Now, the difficulty is that the shape functions N_i (given in equation (10.31)) are functions of the natural co-ordinates ξ and η rather than of the Cartesian co-ordinates x and y. Obviously a relationship needs to be established between the derivatives in the two co-ordinate systems and this can be done by using the chain rule for differentiation. Thus, for function $N_i = N_i(\xi, \eta)$ we have

$$\frac{\partial N_i}{\partial \xi} = \frac{\partial N_i}{\partial x} \frac{\partial x}{\partial \xi} + \frac{\partial N_i}{\partial y} \frac{\partial y}{\partial \xi}$$
$$\frac{\partial N_i}{\partial \eta} = \frac{\partial N_i}{\partial x} \frac{\partial x}{\partial \eta} + \frac{\partial N_i}{\partial y} \frac{\partial y}{\partial \eta} \tag{10.36}$$

which can be expressed in matrix form as

$$\left\{ \begin{array}{c} \dfrac{\partial N_i}{\partial \xi} \\[2mm] \dfrac{\partial N_i}{\partial \eta} \end{array} \right\} = \begin{bmatrix} \dfrac{\partial x}{\partial \xi} & \dfrac{\partial y}{\partial \xi} \\[2mm] \dfrac{\partial x}{\partial \eta} & \dfrac{\partial y}{\partial \eta} \end{bmatrix} \left\{ \begin{array}{c} \dfrac{\partial N_i}{\partial x} \\[2mm] \dfrac{\partial N_i}{\partial y} \end{array} \right\} = \mathbf{J} \left\{ \begin{array}{c} \dfrac{\partial N_i}{\partial x} \\[2mm] \dfrac{\partial N_i}{\partial y} \end{array} \right\}. \tag{10.37}$$

Now the terms such as $\partial x/\partial \xi$ occurring in the 2×2 matrix $\mathbf{J}$ can be obtained from equations (10.30) so that $\mathbf{J}$ becomes

$$\mathbf{J} = \begin{bmatrix} \sum \dfrac{\partial N_i}{\partial \xi} x_i & \sum \dfrac{\partial N_i}{\partial \xi} y_i \\[2ex] \sum \dfrac{\partial N_i}{\partial \eta} x_i & \sum \dfrac{\partial N_i}{\partial \eta} y_i \end{bmatrix} = \begin{bmatrix} J_{11} & J_{12} \\ J_{21} & J_{22} \end{bmatrix} \tag{10.38}$$

where the summations run from 1 to 4. $\mathbf{J}$ is called the *Jacobian* matrix, in the usual terminology of mathematical texts.

The individual terms, J_{11} etc., appearing in $\mathbf{J}$ can be readily evaluated using equation (10.31). For instance

$$J_{11} = \sum \frac{\partial N_i}{\partial \xi} x_i = \sum \tfrac{1}{4} \xi_i (1 + \eta \eta_i) x_i = \tfrac{1}{4}((1 - \eta)(x_2 - x_1) + (1 + \eta)(x_3 - x_4))$$

$$\tag{10.39}$$

and so on. Further, the quantities $\partial N_i/\partial \xi$ and $\partial N_i/\partial \eta$ appearing on the left-hand side of equation (10.37) are available. Thus it is theoretically possible to find $\partial N_i/\partial x$ and $\partial N_i/\partial y$ by the inverse relationship of equation (10.37), that is by the equation

$$\begin{Bmatrix} \dfrac{\partial N_i}{\partial x} \\[2ex] \dfrac{\partial N_i}{\partial y} \end{Bmatrix} = \mathbf{J}^{-1} \begin{Bmatrix} \dfrac{\partial N_i}{\partial \xi} \\[2ex] \dfrac{\partial N_i}{\partial \eta} \end{Bmatrix}. \tag{10.40}$$

This is the required result for the derivatives of the shape functions with respect to x and y, and using this relationship allows the definition of matrix $\mathbf{B}$ (equation (10.35)) to be completed. In detail, by the formal definition of an inverse,

$$\mathbf{J}^{-1} = \frac{1}{|\mathbf{J}|} \begin{bmatrix} J_{22} & -J_{12} \\ -J_{21} & J_{11} \end{bmatrix} \tag{10.41a}$$

where

$$|\mathbf{J}| = J_{11}J_{22} - J_{12}J_{21} \tag{10.41b}$$

is the determinant of $\mathbf{J}$.

Of course, the terms of $\mathbf{B}$ are now functions of ξ and η and so the necessary integration to form $\mathbf{k}$, as in equation (7.21), must be carried out with respect to the ξ and η co-ordinates. The differential areas in the x, y and ξ, η systems are related by

$$dx\, dy = |\mathbf{J}|\, d\xi\, d\eta \tag{10.42}$$

where it is noted that $|\mathbf{J}|$ is in general a function of the ξ, η co-ordinate position. The limits of the integration are simple in the ξ, η system, being $+1$ and -1. The stiffness matrix for an element of uniform thickness h can therefore be written

$$\mathbf{k} = h \int_{-1}^{+1} \int_{-1}^{+1} \mathbf{B^t E B} \, |\mathbf{J}| \, d\xi \, d\eta. \tag{10.43}$$

In general it is not possible to evaluate the stiffness matrix explicitly. Even for the relatively simple case of the rectilinear quadrilateral the inverse of $\mathbf{J}$ is unwieldy since $|\mathbf{J}|$ appears in the denominator (see equation (10.41a)) and contains polynomials in ξ and η. (Only for the rectangle, arbitrarily oriented to the Cartesian axes, is the inverse quite simple (see Example 10.4).) Thus, to evaluate $\mathbf{k}$ use has to be made of (approximate) numerical integration. This need not be viewed as a particular disadvantage since the procedure can be readily systematized and the possibility of algebraic errors is reduced. A brief outline of the numerical integration procedure is given in Section 10.4.4.

The above development of the element stiffness matrix presupposes that matrix $\mathbf{J}$ can be inverted and this will be the case if care is taken in the specification of element geometry. However, it should be realized that it is possible for matrix $\mathbf{J}$ to be singular (i.e. not have an inverse) if the nodes of an (actual) element are incorrectly specified such that the mapping represented by equations (10.30) is not unique. Further (in common with other types of finite element) extremes of element shape should be avoided since the calculation of the inverse of $\mathbf{J}$ may be ultrasensitive in such circumstances.

The basic idea behind the linear isoparametric quadrilateral described here was originally due to Taig [9]. Irons [10, 11] introduced the concept to a wider audience and generalized it to embrace elements of greater complexity having curvilinear geometry which will be described in what follows.

Example 10.4. Determine $\mathbf{J}^{-1}$, appearing in the relationships between derivatives in equation (10.40), for the special case of the rectangular element arbitrarily oriented with respect to the x, y co-ordinates which is shown in Fig. E10.4. Hence determine the submatrix $\mathbf{k}_{11}$ of the element stiffness matrix when used in plane stress analysis.

The Jacobian matrix $\mathbf{J}$ is defined by equations (10.38) and (10.31). The individual components J_{11} etc. are given by equations of the form of equation (10.39). Thus for the arbitrary quadrilateral

$$\mathbf{J} = \frac{1}{4} \begin{bmatrix} (1-\eta)(x_2-x_1)+(1+\eta)(x_3-x_4) & (1-\eta)(y_2-y_1)+(1+\eta)(y_3-y_4) \\ (1-\xi)(x_4-x_1)+(1+\xi)(x_3-x_2) & (1-\xi)(y_4-y_1)+(1+\xi)(y_3-y_2) \end{bmatrix}.$$

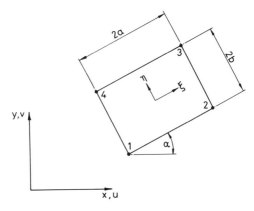

Fig. E 10.4

For the particular case of the rectangle of Fig. E10.4 we have that

$x_2 - x_1 = x_3 - x_4 = 2a \cos \alpha.$

$y_2 - y_1 = y_3 - y_4 = 2a \sin \alpha$

$x_4 - x_1 = x_3 - x_2 = -2b \sin \alpha$

$y_4 - y_1 = y_3 - y_2 = 2b \cos \alpha.$

Thus

$$\mathbf{J} = \begin{bmatrix} a \cos \alpha & a \sin \alpha \\ -b \sin \alpha & b \cos \alpha \end{bmatrix}.$$

Hence (see equations (10.41a) and (10.41b))

$$|\mathbf{J}| = ab(\cos^2 \alpha + \sin^2 \alpha) = ab$$

and

$$\mathbf{J}^{-1} = \frac{1}{ab} \begin{bmatrix} b \cos \alpha & -a \sin \alpha \\ b \sin \alpha & a \cos \alpha \end{bmatrix}.$$

The stiffness matrix for an isoparametric element is given by equation (10.43) and requires the specification of matrices **E** and **B**. The former of these matrices is directly available whilst the latter is defined in equation (10.35) and requires the evaluation of $\partial N_i/\partial x$ and $\partial N_i/\partial y$ ($i = 1, 2, 3, 4$). This evaluation proceeds according to equation (10.40), using the result obtained above for $\mathbf{J}^{-1}$ and the definition of N_i (equation (10.31)). Thus, with $e = \cos \alpha$ and $f = \sin \alpha$,

$$\left\{ \begin{matrix} \dfrac{\partial N_i}{\partial x} \\[2mm] \dfrac{\partial N_i}{\partial y} \end{matrix} \right\} = \frac{1}{ab} \begin{bmatrix} be & -af \\ bf & ae \end{bmatrix} \left\{ \begin{matrix} \frac{1}{4}\xi_i(1 + \eta\eta_i) \\[2mm] \frac{1}{4}\eta_i(1 + \xi\xi_i) \end{matrix} \right\} = \frac{1}{4ab} \begin{bmatrix} be\xi_i(1 + \eta\eta_i) - af\eta_i(1 + \xi\xi_i) \\[2mm] bf\xi_i(1 + \eta\eta_i) + ae\eta_i(1 + \xi\xi_i) \end{bmatrix}.$$

Therefore

$$\mathbf{B} = \frac{1}{4ab} \begin{bmatrix} -be(1-\eta)+af(1-\xi) & 0 & \cdots \\ 0 & -bf(1-\eta)-ae(1-\xi) & \cdots \\ -bf(1-\eta)-ae(1-\xi) & -be(1-\eta)+af(1-\xi) & \cdots \end{bmatrix}.$$

Using this definition of **B** in conjunction with the definitions of $|\mathbf{J}|$ given above and of **E** given by equation (7.35) allows the explicit calculation of **k** (equation (10.43)). In particular the 2×2 submatrix $\mathbf{k}_{11}$ is found to be

$$\mathbf{k}_{11} = \frac{Eh}{12(1-\nu^2)} \begin{bmatrix} \begin{aligned} & 4e^2(r^{-1}+\rho r) \\ & +4f^2(r+\rho r^{-1}) \\ & +6ef(\rho-1) \end{aligned} & \begin{aligned} & 4ef(1-\rho)(r^{-1}-r) \\ & +3(1-\rho)(e^2-f^2) \end{aligned} \\ \begin{aligned} & 4ef(1-\rho)(r^{-1}-r) \\ & +3(1-\rho)(e^2-f^2) \end{aligned} & \begin{aligned} & 4e^2(r+\rho r^{-1}) \\ & +4f^2(r^{-1}+\rho r) \\ & +6ef(1-\rho) \end{aligned} \end{bmatrix}$$

where $r = a/b$ and $\rho = (1-\nu)/2$.

This result is, of course, the same as that found in Example 8.4 by the procedure of transforming the stiffness matrix of a rectangular element from a specialized local co-ordinate system (x, y) to a global system $(\bar{x}, \bar{y})$. (In Example 8.4 the bar convention is used to distinguish the global stiffness matrix from the local matrix whereas here the global stiffness matrix is obtained directly and the bar convention is not needed. Thus $\bar{\mathbf{k}}_{11}$ of Example 8.4 corresponds to $\mathbf{k}_{11}$ here.)

10.4.2. Other isoparametric (and related) elements

The procedure just described for the linear, or first-order, quadrilateral can be extended in a very direct fashion to embrace a considerable range of elements. The linear quadrilateral has been discussed at some length because it provides a relatively simple example with which to introduce the isoparametric concept. However, since its geometry is limited by its straight edges it is not the most useful element; the advantages of the isoparametric concept are heightened when considering higher-order elements in which the edges of the actual element can be curved.

Figure 10.14 shows two possible elements, introduced by Ergatoudis, Irons, and Zienkiewicz [6], in which the actual shapes can be described as 'curvilinear quadrilaterals' whilst the parent elements are again squares of side length 2. The element of Fig. 10.14(a) is the second-order element, with quadratic variation of geometry and displacement along an edge, and that of Fig. 10.14(b) is the third-order element, with cubic variation along an edge. The parent elements are obviously the second- and third-order serendipity rectangles described in Section 10.3. For each element the geometry of the actual element and the displacements are described in the same fashion, the geometry by equations (10.30) and the displacements by equation (10.32), except that the summations are now over

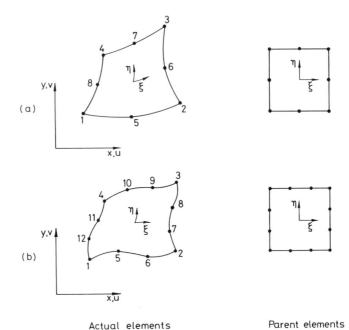

Actual elements　　　　　　　　　　Parent elements

Fig. 10.14. Isoparametric quadrilaterals of the serendipity type: (a) quadratic interpolation; (b) cubic interpolation.

more terms. The quadratic element has i running up to 8 with the $N_i(\xi, \eta)$ defined by equations (10.24) whilst the cubic element has i running up to 12 with the $N_i(\xi, \eta)$ defined by equations (10.25). Aside from the difference in the number and definition of the shape functions the procedure in establishing the stiffness matrix $\mathbf{k}$ for these elements is precisely as that described for the linear quadrilateral.

The isoparametric approach to the formulation of element properties is a very general approach and is not limited to any particular shape or class of element. For two-dimensional analysis, elements of the serendipity type are shown in Fig. 10.14. These are very popular planar elements but the philosphy could equally well embrace elements of the Lagrange type, although in practice these would be less efficient for reasons already discussed. Further, straight-sided planar triangles of the type discussed in earlier chapters can be used as parent elements for a range of curved-sided (or curvilinear) triangles. It should be noted, though, that where the natural co-ordinates L_1, L_2, and L_3 are used to describe the parent triangle it is required (because these co-ordinates are not independent) to express L_3, say, in terms of L_1 and L_2 or it will not be possible to form a square Jacobian matrix. Also the integration limits change, of course, to match the triangular geometry.

The advantages of the isoparametric formulation in dealing with complicated geometry in two-dimensional analysis become much more pronounced in three-dimensional analysis. If advantage is to be taken of the relatively high efficiency of three-dimensional elements with quite sophisticated displacement fields then it is very necessary in most practical problems also to represent the problem geometry in sophisticated fashion by employing curvilinear elements. This can be done using the same philosophy as that employed in two dimensions and again the procedure is not limited to any particular class of element. Curvilinear versions of the serendipity [7, 12] and Lagrange [13] rectangular hexahedra and of the tetrahedral elements, all of which have been described earlier, could be generated. Of these the serendipity family are the most used and so the brief remarks given here on the isoparametric formulation for three-dimensional analysis will relate to this family.

The elements shown earlier in Fig. 10.12 represent the parent elements of first-, second-, and third-order isoparametric solid elements of the serendipity type. The actual elements in three-dimensional Cartesian space have the general shapes shown in Fig. 10.15. Along any edge the geometry and displacements vary in turn in linear, quadratic, and cubic fashion.

As intimated earlier the development of the stiffness matrix for an isoparametric solid (or 'brick') follows very closely the development outlined above for the planar isoparametric element and so only a few points need to be made here. Obviously there are now three co-ordinates and three displacement components to contend with. Consequently the mapping of the geometry has the form

$$x = N_1(\xi, \eta, \zeta)x_1 + N_2(\xi, \eta, \zeta)x_2 + \ldots$$
$$y = N_1(\xi, \eta, \zeta)y_1 + N_2(\xi, \eta, \zeta)y_2 + \ldots \qquad (10.44)$$
$$z = N_1(\xi, \eta, \zeta)z_1 + N_2(\xi, \eta, \zeta)z_2 + \ldots$$

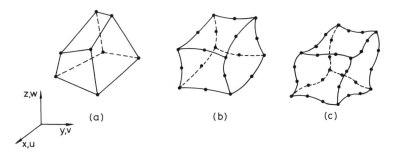

Fig. 10.15. Isoparametric solid elements of the serendipity type: (a) linear interpolation; (b) quadratic interpolation; (c) cubic interpolation.

and the displacements u, v, and w are defined in like manner. The shape functions $N_1(\xi, \eta, \zeta)$ etc. for the elements of Figs. 10.12 and 10.15 are defined in Section 10.3.2. For isotropic material $\mathbf{E}$ is given by equation (7.7), whilst $\mathbf{B}$ is obtained using the strain–displacement equations (7.2) in conjunction with the expressions for u, v, and w, and with the use of the relationship between the derivatives in the two co-ordinate systems; the latter is

$$\begin{Bmatrix} \dfrac{\partial N_i}{\partial \xi} \\[2ex] \dfrac{\partial N_i}{\partial \eta} \\[2ex] \dfrac{\partial N_i}{\partial \zeta} \end{Bmatrix} = \mathbf{J} \begin{Bmatrix} \dfrac{\partial N_i}{\partial x} \\[2ex] \dfrac{\partial N_i}{\partial y} \\[2ex] \dfrac{\partial N_i}{\partial z} \end{Bmatrix} \tag{10.45}$$

where

$$\mathbf{J} = \begin{bmatrix} \dfrac{\partial x}{\partial \xi} & \dfrac{\partial y}{\partial \xi} & \dfrac{\partial z}{\partial \xi} \\[2ex] \dfrac{\partial x}{\partial \eta} & \dfrac{\partial y}{\partial \eta} & \dfrac{\partial z}{\partial \eta} \\[2ex] \dfrac{\partial x}{\partial \zeta} & \dfrac{\partial y}{\partial \zeta} & \dfrac{\partial z}{\partial \zeta} \end{bmatrix} = \begin{bmatrix} \sum \dfrac{\partial N_i}{\partial \xi} x_i & \sum \dfrac{\partial N_i}{\partial \xi} y_i & \sum \dfrac{\partial N_i}{\partial \xi} z_i \\[2ex] \sum \dfrac{\partial N_i}{\partial \eta} x_i & \sum \dfrac{\partial N_i}{\partial \eta} y_i & \sum \dfrac{\partial N_i}{\partial \eta} z_i \\[2ex] \sum \dfrac{\partial N_i}{\partial \zeta} x_i & \sum \dfrac{\partial N_i}{\partial \zeta} y_i & \sum \dfrac{\partial N_i}{\partial \zeta} z_i \end{bmatrix}. \tag{10.46}$$

Finally, the element stiffness matrix is

$$\mathbf{k} = \int_{-1}^{1} \int_{-1}^{1} \int_{-1}^{1} \mathbf{B}^t \mathbf{E} \mathbf{B} \, |\mathbf{J}| \, d\xi \, d\eta \, d\zeta. \tag{10.47}$$

In considering curvilinear two- and three-dimensional elements we have thus far dealt with *isoparametric* elements, with the word isoparametric indicating that the element geometry and the element displacements are defined parametically in the same (iso-) fashion, i.e. the same shape functions are used to express both the Cartesian co-ordinates and the displacement components in terms of values of the co-ordinates and displacements at the same nodes. However, it should be noted that related curvilinear elements exist whose method of formulation is the same as that of the isoparametric elements except that the element geometry and element displacements are defined in different fashion. *Subparametric* elements are elements in which the element geometry is defined by lower-order functions than is the case for the displacement components: *superparametric* elements are elements in which the geometry is defined by higher-order functions than is the case for the displacements. Consequently, for both types of element the number of nodes used to define the geometry is different from that used to define the displacements. An example of a planar subparametric element would be one in which x and y were expressed in terms of four shape functions and four corner-node values (equation (10.30) with N_i defined by equation (10.31)) whilst u and v were expressed in terms of eight shape functions and eight nodal values (as in equation (10.32) but with i running from 1 to 8 and N_i defined by equations (10.24)). An example of a planar superparametric element would be the opposite case. Since isoparametric elements are much more widely used than subparametric or super-parametric elements, these latter elements will not be discussed further here.

10.4.3. Properties of isoparametric elements

As yet the isoparametric concept has not been formally verified with regard to the criteria advanced in Section 6.8 to govern the selection of element displacement fields. It is the purpose of this section to provide such verification by considering first the question of displacement compatibility and then the representation of rigid-body motion and constant strain states [6, 7].

The displacement field within an isoparametric element clearly satisfies compatibility *within* the element and also has the desirable property of invariance. To decide the question of interelement compatibility consider, for example, the eight-noded curved-sided planar element shown in Fig. 10.14(a). The displacement field for this element has been described in Section 10.3.1 and the reader can easily check that along any edge u and v vary as a *quadratic* function in the curvilinear co-ordinate running along the edge, and that u and v at the edge depend only on their values at nodes lying on the edge. Since u and v in adjacent elements are matched at *three* nodal points it follows that they are also matched all along the common edge. In general, for any isoparametric elements we can conclude that if the parent elements are based on interpolation functions that satisfy compatibility, then the actual curvilinear elements will also satisfy compatibility.

By a similar argument to that of the preceding paragraph we can also confirm that isoparametric shape functions define curvilinear element shapes that fit together *before deformation* with no gaps or overlaps and this is obviously most desirable.

Turning to the consideration of constant strain states (which includes rigid-body motions as a special case), it has been seen in Section 8.2 that such states are properly represented in planar finite-element analysis if the element displacement field includes the expressions

$$u = A_1 + A_2 x + A_3 y$$
$$v = A_4 + A_5 x + A_6 y. \tag{10.48}$$

Thus in such states the displacements at each node i will be

$$u_i = A_1 + A_2 x_i + A_3 y_i$$
$$v_i = A_4 + A_5 x_i + A_6 y_i. \tag{10.49}$$

Now the displacement field for a planar isoparametric element is

$$u = \sum N_i u_i \qquad v = \sum N_i v_i \tag{10.50}$$

from equation (10.32), where the summation extends over all nodes i. If the nodal displacements are consistent with equations (10.49) then the displacements at any point in the element are, from equations (10.50),

$$u = \sum (N_i(A_1 + A_2 x_i + A_3 y_i)) = A_1 \sum N_i + A_2 \sum N_i x_i + A_3 \sum N_i y_i$$
$$v = \sum (N_i(A_4 + A_5 x_i + A_6 y_i)) = A_4 \sum N_i + A_5 \sum N_i x_i + A_6 \sum N_i y_i. \tag{10.51}$$

However, if constant strain states are properly represented the displacements u and v are given by equations (10.48). Therefore, from equations (10.48) and (10.51) we have

$$A_1 \sum N_i + A_2 \sum N_i x_i + A_3 \sum N_i y_i = A_1 + A_2 x + A_3 y$$
$$A_4 \sum N_i + A_5 \sum N_i x_i + A_6 \sum N_i y_i = A_4 + A_5 x + A_6 y. \tag{10.52}$$

Equating coefficients of either of these equations gives

$$\sum N_i = 1$$
$$\sum N_i x_i = x \tag{10.53}$$
$$\sum N_i y_i = y$$

and these are the conditions that the element shape functions have to satisfy in order that constant strain (and rigid-body motion) states be properly represented.

The second and third of equations (10.53) are satisfied since they represent the basic philosophy of the representation of element geometry in the isoparametric formulation, as given by equations (10.30). Further, it can easily be shown that the isoparametric shape functions are such that the first of equations (10.53) is also satisfied. To do this, it should be noted that the location of the origin of the global axes is arbitrary. Hence the specification of element geometry, of the form of equation (10.30), could equally be written in terms of co-ordinates X and Y, say, where the position of the co-ordinate origin of the X, Y axes is related to the position of the origin of the x, y axes by equations of the type

$$X = a + x \qquad Y = b + y \qquad (10.54)$$

where a and b are constants and are the distances between the two origins in the x (or X) and y (or Y) direction. This being so we have (from equations of the form of (10.30)) that

$$X = \sum N_i X_i$$

or $\qquad\qquad\qquad\qquad\qquad\qquad\qquad\qquad\qquad\qquad (10.55)$

$$a + x = \sum N_i(a + x_i) = a \sum N_i + \sum N_i x_i$$

with a similar expression for Y. However, it has already been seen that $x = \sum N_i x_i$ and so, from equation (10.55), we have that

$$a = a \sum N_i$$

and this can only mean that $\sum N_i = 1$ as required. Thus all the conditions of equation (10.53) are satisfied and it has been verified that constant strain states (and rigid-body motion states) are properly represented in isoparametric formulations.

In view of the above arguments we can conclude that if the compatibility and completeness criteria are satisfied in the parent element then they are also satisfied in the curvilinear element. Thus the validity of isoparametric elements is theoretically proven, but it should be remembered that in calculating the element stiffness matrix approximations are introduced through the use of an inexact numerical integration scheme. Such approximations are justifiable and perhaps even beneficial, but do mean that bound conditions are destroyed.

The arguments given above in regard to planar analysis can be extended quite readily to encompass three-dimensional analysis.

10.4.4. Numerical integration

As has been mentioned earlier it is generally not possible to perform analytically the integrations necessary to set up the stiffness matrices (and some other properties such as consistent load column matrices) of the isoparametric family of elements. Resort has to be made to some suitable scheme of numerical integration (otherwise known as quadrature). There are a number of schemes available for numerically evaluating definite integrals; for example the reader probably has had some contact with Simpson's rule in evaluating areas as integrals under curves. In finite-element work one particular scheme has been found to be very useful and this scheme is known as Gauss (or Gauss–Legendre) numerical integration. This scheme alone is briefly described here, and the reader requiring

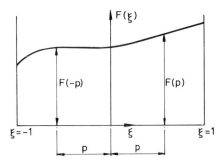

Fig. 10.16. One-dimensional numerical integration.

more information on numerical integration in general is referred to books dealing with numerical analysis, amongst which may be mentioned that of Kopal [14]. For a given accuracy the Gauss method requires the least number of evaluations of the function which is being integrated.

Let us begin by considering the one-dimensional situation in which it is required to evaluate a definite integral I of the type

$$I = \int_{-1}^{+1} F(\xi)\, d\xi, \tag{10.56}$$

i.e. the integral of a function F of ξ with respect to ξ in the range -1 to $+1$. The variation of F with ξ is illustrated in Fig. 10.16; I of course is the area under the curve. In the Gauss quadrature procedure we aim to evaluate I as a summation of a series of n products of values of the function at particular 'sampling' points, i.e. values of $F(\xi)$, and appropriate weighting factors W_i. Thus we calculate

$$I = \sum_{i=1}^{n} W_i F(\xi_i). \tag{10.57}$$

In the Gauss procedure the sampling points are symmetrically located about the centre of the interval of interest (that is about $\xi = 0$) and the magnitudes of the weighting factors are the same at points $\xi = -\xi_i$ and $\xi = +\xi_i$. With these facts in mind the optimum position of the sampling points (at which the function is evaluated) and the values of the weighting factors for a particular degree of polynomial curve can be established.

For a linear function $F(\xi) = A + B\xi$, for instance, it is easy to find analytically that $I = \int_{-1}^{+1} F(\xi)\, d\xi = 2A$ where A must be $F(0)$, that is the value of F at $\xi = 0$. Thus in this case only one sampling point is required to evaluate the linear function exactly. This point is at $\xi = 0$ and the corresponding weighting factor is 2. Thus, in equation (10.57) when $n = 1$, then $\xi_i = 0$ and $W_i = 2$.

For a cubic polynomial function $F = A + B\xi + C\xi^2 + D\xi^3$, analytic evaluation of the definite integral I, according to equation (10.56), gives $I = 2A + 2C/3$. We seek to determine the definite integral numerically by using two sampling (or integrating) points only (i.e. $n = 2$). These are the points shown in Fig. 10.16 at locations $\xi = +p$ and $\xi = -p$. If we remember that the weighting factors are to have equal value at both points, and we denote this common value as W, the numerical integration method estimates the integral as $I = WF(p) + WF(-p)$ and for the given cubic function of F this becomes $I = 2W(A + Cp^2)$. The difference between the numerical and analytical values of I is $2A(W-1) + 2C(Wp^2 - 1/3)$. This difference vanishes for any values of A and C if $W = 1$ and $p = 1/\sqrt{3}$. This result tells us that the exact value of the definite integral of any cubic (or lower-order) function will be obtained numerically by simply summing the two values of the function evaluated at $\xi = +1/\sqrt{3}$ and at $\xi = -1/\sqrt{3}$.

In a similar way it is possible to determine the optimum positions ξ_i and weighting factors W_i for higher-order polynomial functions with the use of more sampling points (that is with greater n). The results are given in Table 10.1 for n up to 4.

It should be noted that with n integration points any polynomial function of order $2n - 1$ or less is evaluated exactly.

TABLE 10.1

Details of positions of sampling points and associated weighting factors for Gauss quadrature

No. of sampling points n	Position of points ξ_i	Weighting factors W_i
1	0	2
2	$+1/\sqrt{3} = 0.577340269\ldots$	1
	$-1/\sqrt{3}$	1
3	0	$8/9 = 0.88888888\ldots$
	$+\sqrt{(3/5)} = 0.774596669\ldots$	$5/9 = 0.55555555\ldots$
	$-\sqrt{(3/5)}$	$5/9$
4	$+\left(\dfrac{3+\sqrt{4.8}}{7}\right)^{1/2} = 0.861136311\ldots$	$\dfrac{18-\sqrt{30}}{36} = 0.347854845\ldots$
	$-\left(\dfrac{3+\sqrt{4.8}}{7}\right)^{1/2}$	$\dfrac{18-\sqrt{30}}{36}$
	$+\left(\dfrac{3-\sqrt{4.8}}{7}\right)^{1/2} = 0.339981043\ldots$	$\dfrac{18+\sqrt{30}}{36} = 0.652145154\ldots$
	$-\left(\dfrac{3-\sqrt{4.8}}{7}\right)^{1/2}$	$\dfrac{18+\sqrt{30}}{36}$

The above has been concerned with the one-dimensional situation but is also of direct (and more important) concern in two- and three-dimensional situations, as will now be demonstrated.

In the two-dimensional situation—in determining the stiffness matrix of an isoparametric quadrilateral—we are concerned with evaluating definite integrals of the form

$$I = \int_{-1}^{+1} \int_{-1}^{+1} F(\xi, \eta) \, d\xi \, d\eta. \tag{10.58}$$

This can be achieved numerically by first evaluating the inner integral with η kept constant and then evaluating the outer integral. Thus

$$I = \int_{-1}^{+1} \left(\int_{-1}^{+1} F(\xi, \eta) \, d\xi \right) d\eta = \int_{-1}^{+1} \left(\sum_{j=1}^{nj} W_j(\xi_j, \eta) \right) d\eta$$

$$= \sum_{i=1}^{ni} W_i \left(\sum_{j=1}^{nj} W_j F(\xi_j, \eta_i) \right) = \sum_{i=1}^{ni} \sum_{j=1}^{nj} W_i W_j F(\xi_j, \eta_i). \tag{10.59}$$

In this expression it is assumed that there are nj sampling points in the ξ direction and ni in the η direction; there are consequently $ni \times nj$ sampling points in all. Usually the same number of sampling points are used in each direction, so that $ni = nj = n$, but this does not have to be so. In Fig. 10.17 are shown the positions of the integrating (or Gauss) points on the parent element for $n = 2$ and $n = 3$, that is for 2×2 and 3×3 integration.

The ideas just described can be readily extended to the three-dimensional situation in which the properties of an isoparametric brick element are to be evaluated. Then integrals of the form

$$I = \int_{-1}^{+1} \int_{-1}^{+1} \int_{-1}^{+1} F(\xi, \eta, \zeta) \, d\xi \, d\eta \, d\zeta \tag{10.60}$$

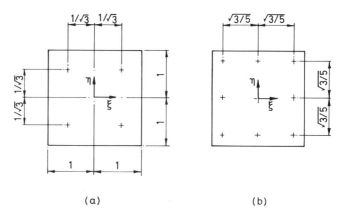

(a)   (b)

Fig. 10.17. Positions of integrating points: (a) 2×2 integration; (b) 3×3 integration.

arise and can be calculated numerically as

$$I = \sum_{i=1}^{ni} \sum_{j=1}^{nj} \sum_{k=1}^{nk} W_i W_j W_k F(\xi_k, \eta_j, \zeta_i). \tag{10.61}$$

The question naturally arises as to what level of numerical integration could or should be used in evaluating element properties. It is recalled that in one-dimensional analysis the use of n integration points in the Gaussian quadrature procedure provides exact evaluation of the integral of any polynomial function of degree $2n-1$ or less. In two- and three-dimensional analysis this remark applies in each co-ordinate direction; in Fig. 10.17(a) for instance numerical integration using the indicated points would be exact for a polynomial of up to third order in each of the ξ and η directions. Now, although it might appear at first sight that the integration procedure should be carried out exactly, or at least as close to exactly as possible, this in fact is rarely the aim. The reasons for this are that use of an excessive number of integration points is computationally expensive and that the use of a relatively low .order (and therefore inexact) scheme of integration can be beneficial in practice as it tends to result in reduced stiffness which compensates for the normal overstiffness of the displacement-based finite element. Of course, if an inexact integration scheme is used in evaluating element properties there can no longer be any guarantee as to the over-stiffness of the whole structure; thus bound conditions no longer apply. However, this is not particularly important so long as ultimate convergence to the correct solution can be assured with mesh refinement.

It can be reasoned that convergence to the correct solution will occur as the mesh of numerically integrated elements is refined if the numerical integration scheme used is sufficient to evaluate the volume of the element exactly. This reasoning is linked with satisfaction of the constant strain condition, for if in the limit the element strains are constant this means that the result of the triple product $\mathbf{B^t EB}$, appearing in equations (10.43) or (10.47), will then be a matrix of constants. Consequently the required integrals in evaluating the element stiffness matrix will be simply integrals of differential volumes (i.e. $h\,|\mathbf{J}|\,d\xi\,d\eta$ for the two-dimensional case and $|\mathbf{J}|\,d\xi\,d\eta\,d\zeta$ for the three-dimensional case). This reasoning leads to the conclusion that the minimum orders of (Gauss) numerical integration to be used are as follows: for the planar constant-thickness isoparametric quadrilaterals, one-point integration for the linear element and 2×2 integration for the quadratic element; for the isoparametric bricks, $2\times2\times2$ integration for the linear element and $3\times3\times3$ integration for the quadratic element. Higher order numerical integration schemes than are given by these rules, may, however, be used in practice with the most appropriate scheme being decided upon perhaps on the basis of test studies. There are sometimes dangers associated with the use of low-order integration rules since this can lead to zero-energy deformation states and consequent singular stiffness matrices.

The above description of the numerical integration procedure is necessarily brief owing to limitations of space. It should be noted that various special schemes of numerical integration, other than the above, are used on occasion. Further, numerical integration formulae exist for triangular and tetrahedral elements, couched in terms of natural (area or volume) co-ordinates. Fuller descriptions of various aspects of the numerical integration procedure are available elsewhere [15, 16].

10.4.5. Numerical applications using the eight-node curvilinear quadrilateral

Here, as some illustration of the general performance and applicability of isoparametric elements, results are presented of the application of the eight-node plane stress quadrilateral which is illustrated in Fig. 10.14(b). The results pertain to the two problems previously discussed in Section 8.3.6 where solutions were documented based on the use of the CST and LST straight-sided triangular elements. In the first problem subsidiary use is also made of the four-node element (i.e. the basic rectangle).

Uniformly loaded deep beam

The beam is shown in Fig. 8.11. Here, for the eight-node element, the finite-element analysis is made using the three uniform meshes shown in Fig. 10.18 in which the individual elements are in fact straight-sided. Some results of the application of the eight-node element are shown in

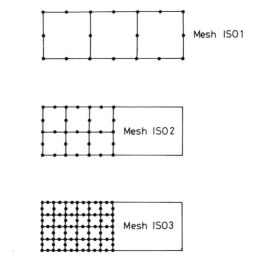

Fig. 10.18. Deep beam problem: meshes of 8-node rectangular elements.

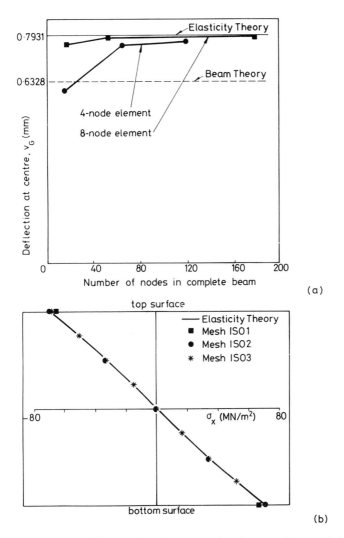

Fig. 10.19. Deep beam problem: results using rectangular elements: (a) central deflection; (b) σ_x at centre line.

Fig. 10.19 and these are concerned with the same quantities as were considered in our earlier look at this problem in Section 8.3.6. Comparison of Fig. 10.19 with Fig. 8.12 shows that on a degree-of-freedom basis the performance of the eight-node element is far superior to that of the CST element and slightly superior to that of the LST element. This is particularly so in the prediction of the central vertical displacement of the beam.

In Fig. 10.19(a) are also shown the displacement results calculated using the basic four-node rectangle and employing three meshes which have nodes in exactly the same locations as depicted for the meshes of triangles shown in Fig. 8.11. The performance of the basic rectangle in predicting the value of the deflection at the beam centre is significantly superior to that of the basic triangle, i.e. the CST element. Similar remarks apply to the calculated values of σ_x at the central section which are not shown here.

Rectangular plate with central hole

This problem is as illustrated in Fig. 8.13. Now the finite-element analysis is based on the two meshes shown in Fig. 10.20. The curvilinear eight-node quadrilaterals obviously allow a much better representation of the region of the plate in the vicinity of the hole that do the straight-sided triangular elements used earlier. It is noted that some of the elements in the finer mesh are rather longer and narrower than is really desirable but this appears to have no ill effect in this application.

Point values of the stress concentration factor at A are given in the table shown in Fig. 10.20 for applied stress corresponding to $f_1 = f_2$, $f_1 = 2f_2$, and $f_2 = 0$, and these can be compared with the tabulated values given in Fig. 8.13 for the triangular elements. Additionally the distribution of σ_x along the line AA' for the case in which $f_2 = 0$ is shown in Fig. 10.21. This figure for the eight-node curvilinear quadrilateral can be compared with Fig. 8.14 which relates to the CST and LST elements.

In this example the isoparametric element is again seen to be superior to the earlier triangular elements.

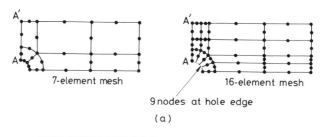

7-element mesh   16-element mesh

9 nodes at hole edge

(a)

Mesh	No. of nodes	Values of σ_x^A / f_1		
		$f_1 = f_2$	$f_1 = 2f_2$	$f_2 = 0$
7-element	34	2·07	2·34	2·62
16-element	67	2·09	2·62	2·96
Exact solution		2·00	2·50	3·00

(b)

Fig. 10.20. Hole-in-plate problem: (a) meshes of curvilinear quadrilateral elements; (b) results for stress concentration factor σ_x^A / f_1.

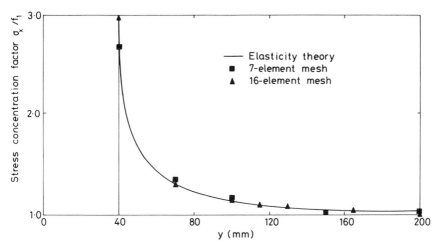

Fig. 10.21. Hole-in-plate problem: variation of σ_x along AA′ when using curvilinear quadrilateral elements.

References

1 ARGYRIS, J. H. Continua and discontinua. *Proc. Conf. on Matrix Methods in Structural Mechanics, Wright–Patterson Air Force Base, Ohio* (1965).

2 BAZELEY, G. P., CHEUNG, Y. K., IRONS, B. M., and ZIENKIEWICZ, O. C. Triangular elements in bending-conforming and non-conforming solutions. *Proc. Conf. on Matrix Methods in Structural Mechanics, Wright–Patterson Air Force Base, Ohio* (1965).

3 ARGYRIS, J. H., FRIED, I., and SCHARPF, D. W. The TET20 and the TEA8 elements for the matrix displacement method *Aeronaut. J.* **72,** 618–25 (1968).

4 SILVESTER, P. Higher-order polynomial triangular finite elements for potential problems. *Int. J. Eng. Sci.* **7,** 849–61 (1969).

5 SILVESTER, P. Timoshenko polynomial finite elements for the Helmholtz equation. *Int. J. Numer. Methods Eng.* **4,** 405–13 (1972).

6 ERGATOUDIS, J. G., IRONS, B. M., and ZIENKIEWICZ, O. C. Curved isoparametric quadrilateral elements for finite element analysis. *Int. J. Solids Struct.* **4,** 31–42 (1968).

7 ZIENKIEWICZ, O. C., IRONS, B. M., ERGATOUDIS, J., AHMAD, S., and SCOTT, F. C. Isoparametric and associated element families for two and three dimensional analysis, in *Finite element methods in stress analysis* (eds. I. Holand and K. Bell), Chapter 13. Tapir, Trondheim (1969).

8 TAYLOR, R. L. On completeness of shape functions for finite element analysis. *Int. J. Numer. Methods Eng.* **4,** 17–22 (1972).

9 TAIG, I. C. Structural analysis by the matrix displacement method. *English Electric Aviation Rep. SO17* (1961).

10 IRONS, B. M. Numerical integration applied to finite element methods. *Conf. on the Use of Electronic Digital Computers in Structural Engineering, University of Newcastle upon Tyne,* Paper 19 (1966).

11   IRONS, B. M. Engineering applications of numerical integration in stiffness methods, *AIAA J.* **4,** 2035–7 (1966).
12   ERGATOUDIS, J., IRONS, B. M., and ZIENKIEWICZ, O. C. Three-dimensional analysis of arch dams and their foundations. *Symp. on Arch Dams and their Foundations, Institution of Civil Engineers,* London (1968).
13   ARGYRIS, J. H. The LUMINA element for the matrix displacement method. *Aeronaut. J.* **72,** 514–17 (1968).
14   KOPAL, Z. *Numerical analysis* (2nd edn.). Chapman and Hall, London (1961).
15   ZIENKIEWICZ, O. C. *The finite element method* (3rd edn.). McGraw-Hill, London (1977).
16   TONG, P. and ROSETTOS, J. N. *Finite element method: basic techniques and implementation.* MIT Press, Cambridge, Mass. (1977).

Problems

10.1. Continue Example 10.1 to derive the consistent load column matrix **Q** corresponding to a uniformly distributed axial loading of intensity p per unit length acting along the length of the three-node quadratic bar element; verify that the result is identical to that recorded in part (b) of Example 6.4.

10.2 Using the natural co-ordinate approach determine the matrix **N** of shape functions for the bar element of Fig. 10.2(b) based on cubic polynomial interpolation ($m = 3$) with element length l.

10.3. Determine the stiffness matrix for the prismatic bar element of Problem 10.2. Also determine column matrix **Q** corresponding to a distributed axial loading varying linearly from intensity f per unit length at node (30) to λf per unit length at node (03).

10.4. For the basic plane stress triangular element the shape functions when working in terms of area co-ordinates are defined in equations (10.21). Proceed to derive the 6×6 element stiffness matrix and compare with the derivation given in Section 8.3.1.

10.5. Show that applying the nodal 'boundary conditions' to equation (10.22) leads to the shape functions recorded in Example 10.2 for the quadratic plane stress triangular element.

10.6. Imagine that the quadratic plane stress triangular element (of uniform thickness) shown in Fig. 10.5(b) and considered in Example 10.2 is subjected to a body force in the u direction which is uniformly distributed over the whole volume. Show that consistent representation of this force corresponds to loads in the u direction of one-third of the total at each mid-side node, and zero at each vertex node.

10.7. Continue Example 10.2 by evaluating the four terms k_{31}, k_{32}, k_{41}, and k_{42} of the 12×12 stiffness matrix **k**.

10.8. Construct the shape functions in terms of area co-ordinates for the cubic plane stress triangular element shown in Fig. 10.5(c). Sketch typical shape functions.

10.9. Consider the second-order serendipity rectangle (Fig. 10.8(b)) and show that its shape functions, defined by equations (10.24), can be produced by first assuming a field of the form

$$A_1 + A_2\xi + A_3\eta + A_4\xi^2 + A_5\xi\eta + A_6\eta^2 + A_7\xi^2\eta + A_8\xi\eta^2$$

and applying the element 'boundary conditions'.

10.10. The second-order serendipity rectangle of uniform thickness, shown in Fig. 10.8(b), is subjected to a uniformly distributed body force in the u direction. Determine the nodal loads acting in the u direction which represent this distributed body force in consistent fashion, expressing your answer as proportions of the total body force.

10.11. Repeat Problem 10.10 for the third-order serendipity rectangle shown in Fig. 10.8(c).

10.12. Construct the shape functions for all nodes along edge 2–3 of both of the mixed-order serendipity elements shown in Fig. P10.12.

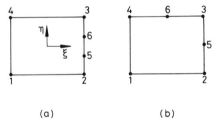

(a)   (b)

Fig. P10.12.

10.13. A uniform body force acts in the ξ direction throughout the volume of the second-order rectangular hexahedron shown in Fig. 10.12(b). Determine what loads should act in the ξ direction at a typical corner node and a typical side node so as to represent the body force in consistent fashion, expressing the answer as proportions of this force.

10.14. The stiffness matrix for the basic four-node rectangular plane stress element of uniform thickness is recorded in equation (8.44). Consider the calculation of some typical stiffness terms, say the four of those belonging to the $\mathbf{k}_{11}$ submatrix, using numerical integration. Obtain solutions for the terms of this submatrix by using (a) one integrating point (at the element centre) and (b) 2×2 integration. Show that the result of (a) is approximate whilst that of (b) gives the correct solution. (Note that Example 10.4 also provides the solution for $\mathbf{k}_{11}$, when $e = 1$, $f = 0$).

11
FINITE ELEMENTS FOR PLATE BENDING ANALYSIS

11.1. Introduction

THE development of displacement-based finite elements within the framework of the classical plate theory described in Section 7.6 poses an extra level of difficulty compared with the situations described in previous chapters. This extra difficulty arises from the assumptions which are introduced in order to convert the equations of three-dimensional elasticity into a theoretical (but approximate) model of plate behaviour in which deformation throughout the plate is expressed solely in terms of the middle surface deflection w. Chief amongst these assumptions is the Kirchhoff hypothesis that straight lines which are initially normal to the plate middle surface remain straight and normal to the middle surface in the loaded configuration. The fact that classical plate behaviour is characterized by the single variable w has some considerable advantages but it does mean that the nature of classical plate theory differs significantly from that of two- or three-dimensional elasticity theory. For instance the governing differential equations of equilibrium—when written in terms of displacements—are of second order for elasticity theory whereas the corresponding equation in the plate theory has been seen to be of fourth order. More pertinently so far as the finite-element displacement approach is concerned, the assumptions of classical plate theory mean that C_1 continuity, i.e. continuity of w and its first derivatives, is required for strict admissibility. This will be discussed further in Section 11.2 when the general requirements of classical plate theory elements will be debated, but for the moment it is simply noted that the requirement for C_1 continuity does considerably complicate element development. It will be seen in what follows that the strict requirement for C_1 continuity can be relaxed in certain circumstances whilst still producing useful (non-conforming) elements, though at the cost of nullifying bound conditions.

The simplest element shape to consider in the plate bending problem is the rectangle. Whilst the shape of the rectangle obviously puts some limit on its applicability, there is, nevertheless, good reason for considering the rectangle in some detail since such consideration will clarify the difficulties associated with classical plate bending elements and will outline the means by which these difficulties can—to a considerable extent at least—be overcome. Rectangular elements are considered in Sections 11.3–11.5.

Elements of less restricted shape, particularly those of triangular shape, are then detailed in Sections 11.6–11.8. It will be seen that by and large the viable elements that have emerged belong to categories that have their counterparts in the realm of rectangular elements.

Sections 11.2–11.8 concern elements whose properties are based on the use of classical plate theory in conjunction with the usual PMPE (or assumed-displacement) approach. A large number of element types are contained within this classification and the elements mentioned in these sections are by no means the only ones belonging to this classification which are available. Furthermore it must be made clear that in plate (and shell) finite-element analysis the problems associated with the C_1 continuity requirement in the 'conventional' classical plate theory–PMPE approach have led to the development of a wide range of alternative approaches to the problem and consequently to the presentation of a greater variety of individual element types than in any other category of finite-element analysis. Prolonged discussion of the alternative approaches to plate analysis is not warranted in a book of this nature but some remarks are appropriate and these are given in Section 11.9. Finally, notwithstanding the title of this chapter, some mention should be made of the important topic of the analysis of shells using the FEM and a brief overview of this complicated topic is given in Section 11.10.

11.2. Requirements for element displacement fields and element formulation

The requirements that element displacement fields should strictly satisfy have been discussed in general terms in Section 6.8. These requirements apply to plate bending elements just as to other element types, but we now need to interpret them in detail for the specific case of elements based on the classical plate theory. In doing this we restrict attention at this stage to consideration of displacement fields expressed in terms of Cartesian or xy co-ordinates. (Area co-ordinates, described in Section 10.2, are a convenient alternative to Cartesian co-ordinates when developing the properties of triangular plate bending elements.)

It has been shown in Section 7.6 that the deformation of a plate based on the classical theory is expressed in terms of the single variable w, this being the lateral deflection of the middle surface. Thus, when talking of a displacement field for plate bending elements we mean simply an expression for the lateral deflection w alone. Further, it has been seen that the measures of strain in plate bending analysis are the bending and twisting curvatures. These pseudo-strains depend on the second derivatives of the deflection, as does the curvature in the bending of beams; specifically, as defined in equation (7.53), the pseudo-strains are

$$\boldsymbol{\varepsilon}_p = \{\chi_x \quad \chi_y \quad 2\chi_{xy}\} = -\left\{\frac{\partial^2 w}{\partial x^2} \quad \frac{\partial^2 w}{\partial y^2} \quad 2\frac{\partial^2 \omega}{\partial x\,\partial y}\right\}. \qquad (11.1)$$

It is also recalled from Section 7.6 that the stress-type quantities of plate theory are the pseudo-stresses or moments (per unit length) defined as

$$\boldsymbol{\sigma}_p = \{M_x \quad M_y \quad M_{xy}\} \qquad (11.2)$$

and that the moments are related to the curvatures by the expression

$$\boldsymbol{\sigma}_p = \mathbf{E}_p \boldsymbol{\varepsilon}_p \qquad (11.3)$$

where the special elasticity matrix is (assuming isotropic material) defined as

$$\mathbf{E}_p = D\begin{bmatrix} 1 & \nu & 0 \\ \nu & 1 & 0 \\ 0 & 0 & \frac{1}{2}(1-\nu) \end{bmatrix} \qquad (11.4)$$

with

$$D = \frac{Eh^3}{12(1-\nu^2)}. \qquad (11.5)$$

The first requirement itemized in Section 6.8 concerns the inclusion in a displacement field of any possible rigid-body motion. For the plate bending element the rigid-body motions that can occur are

(i) a rigid-body translation in the z direction (see Fig. 11.1) of the form $w = \text{constant}$

(ii) a rigid body rotation about the x axis of the form $w = (\text{constant})y$

(iii) a rigid-body rotation about the y axis of the form $w = (\text{constant})x$.

Thus the complete rigid-body motion representation has the form

$$w_r = A_1 + A_2 x + A_3 y. \qquad (11.6)$$

The second requirement, of which the first is really only a special case, concerns the representation of any possible states of constant strain. In plate bending this effectively means the representation of states of constant curvature, i.e. of constant second derivatives of w. This is clearly achieved if the expression for w contains x^2, xy and y^2 terms. Therefore, to satisfy both the first and second requirements the assumed deflection must contain the following terms:

$$w = A_1 + A_2 x + A_3 y + A_4 x^2 + A_5 xy + A_6 y^2. \qquad (11.7)$$

The third requirement of Section 6.8 is satisfied by simply assuming an expression for the deflection which is smoothly continuous over the whole element. It should be noted, though, that elements have been derived for plate bending applications in which different deflection expressions are assumed for different subregions of the individual element; this point will be taken up in Section 11.4.

The fourth requirement concerns the compatibility condition at interelement boundaries and here increased difficulty is met in plate analysis as compared with what has gone before. The requirement is for continuity of derivatives of one order less than the highest-order derivative appearing in the potential energy expression (see Section 6.8). The highest-order derivatives are in fact the second derivatives of w, or the curvatures, appearing in the strain energy (see equation (7.56))

$$U_p = \tfrac{1}{2} \int \int \boldsymbol{\varepsilon}_p^t \mathbf{E}_p \boldsymbol{\varepsilon}_p \, dx \, dy \qquad (11.8)$$

Thus continuity of w and its two first derivatives $\partial w/\partial x$ and $\partial w/\partial y$ (the slopes of the deformed middle surface) is required over an element assemblage, including the interelement boundaries. As has been mentioned before this is referred to as a C_1 continuity requirement.

We have, of course, already considered one C_1 continuity problem in the form of the beam problem and there no special difficulty was met since the beam elements are connected one to another at nodal *points* only. In the plate bending problem, however, the C_1 continuity condition should strictly be met at all points along all interelement boundaries so that no gaps or kinks appear under load in the assemblage of elements. It will be seen in this chapter that this causes considerable difficulty but also that useful elements can be generated which relax this condition subject to the satisfaction of an extended form of the second requirement of Section 6.8.

The procedure used in evaluating the stiffness matrix and any consistent load column matrices for a plate bending element follows the general line described in Section 7.3 but with one or two obvious specialist modifications to account for the fact that the plate behaviour is referred to the lateral deflection of the middle surface. Thus, given that the strain energy for the element is of the form of equation (11.8), with $\boldsymbol{\varepsilon}_p$ defined by equation (11.1), it is easy to see, following Section 7.3, that the expression for a plate element stiffness matrix will be a modified form of equation (7.21), i.e. will be

$$\mathbf{k} = \int_{A_e} \mathbf{B}_p^t \mathbf{E}_p \mathbf{B}_p \, dA_e. \qquad (11.9)$$

Here A_e is the area of the middle surface of the element and $\mathbf{E}_p$ is defined in equation (11.4). The matrix $\mathbf{B}_p$ relates the pseudo-strains to the nodal freedoms of the element and is obtainable by appropriate differentiation of the shape function matrix $\mathbf{N}$ in the usual fashion. The alternative form of $\mathbf{k}$ if the derivation is based on a displacement field expressed in terms of generalized displacements is (see equation (7.24)),

$$\mathbf{k} = (\mathbf{C}^{-1})^t \left(\int_{A_e} \boldsymbol{\beta}_p^t \mathbf{E}_p \boldsymbol{\beta}_p \, dA_e \right) \mathbf{C}^{-1} \tag{11.10}$$

where the definitions of $\mathbf{C}$ and $\boldsymbol{\beta}_p$ should be obvious from what has gone before.

It has been mentioned in Section 7.6 that in plate theory lateral loads of both the surface and body types are assumed to act at the plate middle surface and hence both types of load can be represented as a distribution of intensity $q(x, y)$ per unit area at the middle surface. The potential energy of the load acting over the middle surface of the elements is of the form of equation (7.57). The expression for the consistent load column matrix for a plate bending element then becomes, by similar arguments to those of Section 7.6,

$$\mathbf{Q} = \int_{A_e} \mathbf{N}^t q \, dA_e \tag{11.11}$$

or, in the alternative form when generalized displacements are used in the expression for w,

$$\mathbf{Q} = (\mathbf{C}^{-1})^t \int_{A_e} \boldsymbol{\alpha}^t q \, dA_e. \tag{11.12}$$

These two equations are modified forms of equations (7.23) and (7.25). The loading $q(x, y)$ can vary over the middle surface and can act over only part of the middle surface, of course. In particular it should be noted that in the limit we can consider a concentrated lateral load acting at a single point. For instance, if a lateral load W_c acts at the point whose co-ordinate position is x_c, y_c then its contribution to the consistent load column matrix becomes, by modifying equation (11.11) or (11.12),

$$\mathbf{Q} = \mathbf{N}_c^t W_c \qquad \text{or} \qquad \mathbf{Q} = (\mathbf{C}^{-1})^t \boldsymbol{\alpha}_c^t W_c \tag{11.13}$$

where $\mathbf{N}_c$ and $\boldsymbol{\alpha}_c$ are $\mathbf{N}$ and $\boldsymbol{\alpha}$ evaluated at x_c, y_c. Furthermore, we can extend this argument to include the possibility of a concentrated couple acting at a specific point such as x_c, y_c as was done for the beam in Section 6.5. For instance, if a couple M_c acts in the x direction (i.e. about the y axis) at this point its contribution to the consistent load column matrix is

$$\mathbf{Q} = \frac{\partial}{\partial x} (\mathbf{N}^t)_c M_c = (\mathbf{C}^{-1})^t \frac{\partial}{\partial x} (\boldsymbol{\alpha}^t)_c M_c \tag{11.14}$$

since the partial differential of $\mathbf{N}$ with respect to x is the rotation (or slope of the middle surface) in the x direction.

11.3. A 12-degree-of-freedom rectangular element (R1)

A rectangular plate bending element is shown in Fig. 11.1; the origin of co-ordinates is located at the centroid of the element but could, of course, be located elsewhere—at a corner, say. In formulating an element of this type it seems clear that we could assume that the corners of the rectangle are nodal points and that the degrees of freedom at each node must include the two components of rotation, about the x and y axes respectively, as well as the lateral deflection. This would be in line with the situation existing in the basic beam element where the deflection and the rotation in the direction along the element are the nodal degrees of freedom.

The rotations shown in Fig. 11.1 are identically the slopes of the middle surface and are defined as

$$\phi = \frac{\partial w}{\partial y} \qquad \text{and} \qquad \theta = \frac{\partial w}{\partial x} \qquad (11.15)$$

where ϕ is the rotation about the x axis and θ that about the y axis. The 12 degrees of freedom of the element are defined as

$$\mathbf{d} = \{w_1 \ \phi_1 \ \theta_1 \ w_2 \ \phi_2 \ \theta_2 \ w_3 \ \phi_3 \ \theta_3 \ w_4 \ \phi_4 \ \theta_4\}. \qquad (11.16)$$

In seeking a suitable deflection field for the element we bear in mind that if we are expressing the deflection in generalized displacement form the six terms given by equation (11.7) have to be included. Further, if we consider interelement continuity of w itself (which is only part of the requirement for full interelement compatibility) it is clear that w should vary as a cubic polynomial along any element edge since four nodal degrees of freedom (the values of w and its first derivative in the direction of the edge at each of two nodes) are available to define w along the edge. Thus the product of a cubic polynomial in x and a cubic polynomial in y is of interest; effectively this corresponds to the product of the

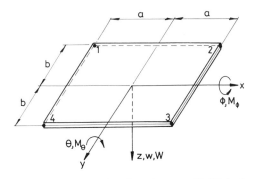

Fig. 11.1. Rectangular plate element with 12 d.o.f.

displacement fields of two mutually perpendicular basic beam elements. The terms of the bicubic expansion are all the 16 terms shown in equation (11.17):

$$
\begin{matrix}
 & & 1 & & & \\
 & & x & & y & \\
 & x^2 & & xy & & y^2 \\
x^3 & & x^2y & & xy^2 & & y^3 \\
 & x^3y & & x^2y^2 & & xy^3 \\
 & & x^3y^2 & & x^2y^3 \\
 & & & x^3y^3 & &
\end{matrix}
\tag{11.17}
$$

However, there are of course four too many terms in this expansion for our present purpose. Thus four terms must be removed and it makes sense to omit higher-order terms and at the same time retain symmetry of the deflection field. Therefore the four terms shown beneath the broken lines in equation (11.17) are to be removed. Finally, the assumed 12-term deflection expression, which includes proper representation of rigid-body motion and constant strain terms, is

$$
w = [1 \ x \ y \ x^2 \ xy \ y^2 \ x^3 \ x^2y \ xy^2 \ y^3 \ x^3y \ xy^3]
\begin{Bmatrix} A_1 \\ A_2 \\ \vdots \\ A_{12} \end{Bmatrix}
\tag{11.18}
$$

or

$$
w = \alpha \mathbf{A}
$$

and this expression includes a complete two-dimensional cubic polynomial field. It is noted that this deflection expression satisfies the governing differential equation of an unloaded part of the plate (i.e. satisfies equation (7.60) with $q = 0$) though this does not concern us directly when generating element properties.

Development of the 12-degree-of-freedom rectangular plate element described here, based on the deflection expression of equation (11.18), was first introduced by Adini [1] and independently rediscovered by Melosh [2], Zienkiewicz and Cheung [3], and Dawe [4]. This element will be referred to as R1 in what follows. The derivation of the element stiffness will now be detailed and equation (11.10) will be used for this purpose; the steps of the derivation follow the familiar path outlined previously for a number of types of element. It is noted that corresponding to the definition of the 12 nodal displacements the element nodal forces are

$$
\mathbf{P} = \{W_1 \ M_{\phi 1} \ M_{\theta 1} \ W_2 \ M_{\phi 2} \ M_{\theta 2} \ W_3 \ M_{\phi 3} \ M_{\theta 3} \ W_4 \ M_{\phi 4} \ M_{\theta 4}\},
\tag{11.19}
$$

that is a direct lateral force and two couples in the ϕ and θ directions at each corner node. It should be noted further that these couples are concentrated at a point, unlike the quantities M_x, M_y, and M_{xy} which are moments per unit length.

Matrix $\mathbf{C}$, relating the nodal displacements $\mathbf{d}$ to the generalized displacements $\mathbf{A}$, is obtained by applying the nodal boundary conditions that

$$w = w_i \qquad \phi = \frac{\partial w}{\partial y} = \phi_i \qquad \theta = \frac{\partial w}{\partial x} = \theta_i$$

at node i $(i = 1\text{--}4)$ where

$$x = x_i \text{ (i.e. } x = \pm a) \qquad y = y_i \text{ (i.e. } y = \pm b).$$

The differentials $\partial w/\partial y$ and $\partial w/\partial x$ are simply obtained by straightforward differentiation of equation (11.18). Thus $\mathbf{C}$, a 12×12 matrix, becomes

$$\mathbf{C} =$$

$$\begin{bmatrix} 1 & -a & -b & a^2 & ab & b^2 & -a^3 & -a^2b & -ab^2 & -b^3 & a^3b & ab^3 \\ 0 & 0 & 1 & 0 & -a & -2b & 0 & a^2 & 2ab & 3b^2 & -a^3 & -3ab^2 \\ 0 & 1 & 0 & -2a & -b & 0 & 3a^2 & 2ab & b^2 & 0 & -3a^2b & -b^3 \\ 1 & a & -b & a^2 & -ab & b^2 & a^3 & -a^2b & ab^2 & -b^3 & -a^3b & -ab^3 \end{bmatrix}$$

$$\text{etc.}$$

$$(11.20)$$

The inverse of $\mathbf{C}$ always exists and in fact can be found by hand (by solving the 12 equations for the generalized displacements) without overwhelming difficulty. The inverse is recorded in Table 11.1.

Matrix $\boldsymbol{\beta}_\mathrm{p}$, which occurs in equation (11.10), relates the pseudo-strains $\boldsymbol{\varepsilon}_\mathrm{p}$ to the generalized displacements $\mathbf{A}$. Given the definition of $\boldsymbol{\varepsilon}_\mathrm{p}$ (equation (11.1)) and the expression for the plate deflection (equation (11.18)) the matrix $\boldsymbol{\beta}_\mathrm{p}$ is readily established. Thus

$$\boldsymbol{\varepsilon}_\mathrm{p} = -\begin{Bmatrix} \dfrac{\partial^2 w}{\partial x^2} \\[2mm] \dfrac{\partial^2 w}{\partial y^2} \\[2mm] \dfrac{2\partial^2 w}{\partial x\, \partial y} \end{Bmatrix}$$

$$= -\begin{bmatrix} 0 & 0 & 0 & 2 & 0 & 0 & 6x & 2y & 0 & 0 & 6xy & 0 \\ 0 & 0 & 0 & 0 & 0 & 2 & 0 & 0 & 2x & 6y & 0 & 6xy \\ 0 & 0 & 0 & 0 & 2 & 0 & 0 & 4x & 4y & 0 & 6x^2 & 6y^2 \end{bmatrix} \begin{Bmatrix} A_1 \\ A_2 \\ \vdots \\ A_{12} \end{Bmatrix} \qquad (11.21)$$

TABLE 11.1
Matrix $\mathbf{C}^{-1}$ *for element R1*

$$\mathbf{C}^{-1} = \frac{1}{8a^3b^3}$$

$$
\begin{bmatrix}
2a^3b^3 & a^3b^4 & a^4b^3 & 2a^3b^3 & a^3b^4 & a^4b^3 & -a^3b^4 & -a^4b^3 & -a^3b^4 & -a^4b^3 & -a^3b^4 & a^4b^3 \\
-3a^2b^3 & -a^2b^4 & -a^3b^3 & 3a^2b^3 & a^2b^4 & -a^3b^3 & -a^2b^4 & -a^3b^3 & a^2b^4 & -a^3b^3 & a^2b^4 & -a^3b^3 \\
-3a^3b^2 & -a^3b^3 & -a^4b^2 & -3a^3b^2 & -a^3b^3 & a^4b^2 & -a^3b^3 & -a^4b^2 & -a^3b^3 & a^4b^2 & -a^3b^3 & a^4b^2 \\
0 & 0 & -a^2b^3 & 0 & 0 & a^2b^3 & 0 & a^2b^3 & 0 & a^2b^3 & 0 & -a^2b^3 \\
4a^2b^2 & a^2b^3 & a^3b^2 & -4a^2b^2 & -a^2b^3 & a^3b^2 & -a^2b^3 & -a^3b^2 & -a^2b^3 & a^3b^2 & a^2b^3 & -a^3b^2 \\
0 & -a^3b^2 & 0 & 0 & -a^3b^2 & 0 & a^3b^2 & 0 & a^3b^2 & 0 & a^3b^2 & 0 \\
b^3 & 0 & ab^3 & -b^3 & 0 & ab^3 & 0 & ab^3 & 0 & ab^3 & 0 & ab^3 \\
0 & 0 & a^2b^2 & 0 & 0 & -a^2b^2 & 0 & a^2b^2 & 0 & -a^2b^2 & 0 & -a^2b^2 \\
0 & a^2b^2 & 0 & 0 & a^2b^2 & 0 & a^2b^2 & 0 & -a^2b^2 & 0 & -a^2b^2 & 0 \\
a^3 & a^3b & 0 & a^3 & a^3b & 0 & a^3b & 0 & a^3b & 0 & a^3b & 0 \\
-b^2 & 0 & -ab^2 & b^2 & 0 & -ab^2 & 0 & ab^2 & 0 & ab^2 & 0 & ab^2 \\
-a^2 & -a^2b & 0 & a^2 & a^2b & 0 & a^2b & 0 & a^2b & 0 & -a^2b & 0
\end{bmatrix}
$$

or

$$\boldsymbol{\varepsilon}_p = \boldsymbol{\beta}_p \mathbf{A}$$

It should be noted that matrix $\boldsymbol{\beta}_p$ is the 3×12 matrix of equation (11.21) with the accompanying minus sign. However, as far as constructing the element stiffness matrix is concerned (equation (11.10)) the minus sign can be ignored since $\boldsymbol{\beta}_p$ and its transpose appear in the required matrix multiplication.

With $\mathbf{E}_p$ defined by equation (11.4)—assuming isotropic material—the necessary ingredients for the formulation of $\mathbf{k}$ according to equation (11.10) are now available. The stiffness matrix can be evaluated explicitly, with $\mathrm{d}A_e = \mathrm{d}x\,\mathrm{d}y$ of course and the integration being performed over the range $x = -a$ to $+a$ and $y = -b$ to $+b$; the result for a uniform thickness element is recorded in Table 11.2. As with all other elements considered in this text this stiffness matrix satisfies the general properties itemized in Section 3.6. In particular the second property, concerning overall equilibrium of the element, requires that for each column s of the stiffness matrix

$$k_{1s} + k_{4s} + k_{7s} + k_{10s} = 0$$
$$k_{2s} + k_{5s} + k_{8s} + k_{11s} + b(-k_{1s} - k_{4s} + k_{7s} + k_{10s}) = 0$$
$$k_{3s} + k_{6s} + k_{9s} + k_{12s} + a(-k_{1s} + k_{4s} + k_{7s} - k_{10s}) = 0.$$

These equations are expressions of equilibrium of vertical forces, of moments about the x axis, and of moments about the y axis. The fact that the individual terms of the stiffness matrix do satisfy these equations provides some check on the validity of the stiffness matrix.

If the element is subjected to any loading other than directly at the nodes the effect of this loading can be accommodated in consistent fashion by calculating the column matrix $\mathbf{Q}$ according to equation (11.12) (since we are here working in terms of generalized displacements) or according to equations such as (11.13) and (11.14), depending upon the particular form of the loading. For the case of a uniform transverse loading q per unit area over the whole middle-surface area of the element the column matrix $\mathbf{Q}$ can be evaluated quite readily by substituting into equation (11.12) the expression for $\boldsymbol{\alpha}$ given in equation (11.18) and the definition of $\mathbf{C}^{-1}$ given in Table 11.1. The result is

$$\mathbf{Q} = qab \left\{ 1 \quad \frac{b}{3} \quad \frac{a}{3} \quad 1 \quad \frac{b}{3} \quad -\frac{a}{3} \quad 1 \quad -\frac{b}{3} \quad -\frac{a}{3} \quad 1 \quad -\frac{b}{3} \quad \frac{a}{3} \right\}. \quad (11.22)$$

This reveals the fact that the replacement of a lateral loading by nodal quantities in a manner consistent with the PMPE results in both nodal transverse loads and nodal couples. This result would not have been

TABLE 11.2
Stiffness matrix for element R1

$$\mathbf{k} = \frac{D}{60ab} \qquad\qquad p = \left(\frac{a}{b}\right)^2 \qquad \text{Symmetric}$$

	w_1	ϕ_1	θ_1	w_2	ϕ_2	θ_2	w_3	ϕ_3	θ_3	w_4	ϕ_4	θ_4
	$60p+60p^{-1}+42-12\nu$											
	$b(60p+6+24\nu)$	$b^2(80p+16-16\nu)$										
	$a(60p^{-1}+6+24\nu)$	$60\nu ab$	$a^2(80p^{-1}+16-16\nu)$									
	$k_{4,2}$	$b(30p-6-24\nu)$	$a(-60p^{-1}-6+6\nu)$	$k_{1,1}$								
	$-k_{4,3}$	$b^2(40p-16+16\nu)$	0	$k_{2,1}$	$k_{2,2}$							
	$30p-60p^{-1}+42+12\nu$	$-k_{5,3}$	$a^2(40p^{-1}-4+4\nu)$	$-k_{3,1}$	$-k_{3,2}$	$k_{3,3}$						
	$-k_{7,2}$	$b(-30p+6-6\nu)$	$a(-30p^{-1}+6-6\nu)$	$k_{10,1}$	$k_{10,2}$	$-k_{10,3}$	$k_{1,1}$					
	$-k_{7,3}$	$b^2(20p+4-4\nu)$	0	$k_{11,1}$	$k_{11,2}$	$-k_{11,3}$	$-k_{2,1}$	$k_{2,2}$				
	$-60p+30p^{-1}-42+12\nu$	$k_{8,3}$	$a^2(20p^{-1}+4-4\nu)$	$-k_{12,1}$	$-k_{12,2}$	$k_{12,3}$	$-k_{3,1}$	$k_{3,2}$	$k_{3,3}$			
	$-k_{10,2}$	$b(-60p-6+6\nu)$	$a(30p^{-1}-6-24\nu)$	$k_{7,1}$	$k_{7,2}$	$-k_{7,3}$	$k_{4,1}$	$-k_{4,2}$	$-k_{4,3}$	$k_{1,1}$		
	$k_{10,3}$	$b^2(40p-4+4\nu)$	0	$k_{8,1}$	$k_{8,2}$	$-k_{8,3}$	$-k_{5,1}$	$k_{5,2}$	$k_{5,3}$	$-k_{2,1}$	$k_{2,2}$	
		$-k_{11,3}$	$a^2(40p^{-1}-16+16\nu)$	$-k_{9,1}$	$-k_{9,2}$	$k_{9,3}$	$-k_{6,1}$	$k_{6,2}$	$k_{6,3}$	$k_{3,1}$	$-k_{3,2}$	$k_{3,3}$

forecast by a crude lumping scheme which would have missed the contribution of the couples.

With element stiffness matrices and consistent load column matrices established the assembly of plate elements and subsequent solution of the set of plate structure equations follows the usual lines. Once the nodal displacements are calculated the analyst will probably require a knowledge of the distribution of stress-type quantities, and in plate bending analysis the important stress-type quantities are the bending and twisting moments, referred to above as the pseudo-stresses σ_p. Now, from equations (11.2), (11.3), and (11.21), and remembering the general definition $\mathbf{A} = \mathbf{C}^{-1}\mathbf{d}$, the pseudo-stresses are expressed in terms of the element nodal displacements as

$$\sigma_p = \mathbf{E}_p\boldsymbol{\beta}_p\mathbf{C}^{-1}\mathbf{d} = \mathbf{S}_p\mathbf{d}. \tag{11.23}$$

Performing the indicated triple matrix product yields the (pseudo-) stress matrix $\mathbf{S}_p$ which is recorded in Table 11.3. It is noted from this table that the bending moments M_x and M_y vary linearly along any line running parallel to the x axis or parallel to the y axis whilst the twisting moment varies quadratically along such a line. Using Table 11.3 the moments can of course be calculated at any point in the element but normally calculations would be made at the node points. In general, discontinuities of moments will occur at interelement boundaries.

The development of element properties is now complete and the element is ready for use. However, before doing so let us consider the question of interelement compatibility to decide whether the element is conforming or not. It has already been stated earlier in this section that interelement continuity of the lateral deflection w is achieved using the deflection expression (equation (11.18)) since w varies as a cubic polynomial along any element edge and there are four freedoms available at the ends of the edge which uniquely define this variation. This though is only part of the compatibility requirement since continuity of the slopes is also required across element boundaries. The slope in the direction along an edge presents no problem since if w is matched between adjacent elements at all points along their common edge then clearly the slope of w along the edge is also matched. It is the slope in the direction normal to an edge which presents the difficulty.

Consider the edge $x = a$, say, in Fig. 11.1. The slope in the direction normal to this edge is $\partial w/\partial x$, and by differentiating equation (11.18) it can be seen that this slope varies along the edge as a cubic polynomial function of y. However, there are only two nodal freedoms, and not four, which are available at nodes 2 and 3 to specify the cubic variation; these are the values of the normal slopes θ_2 and θ_3. Hence the cubic variation cannot be uniquely defined by nodal freedoms associated solely with the

TABLE 11.3
Stress matrix for element R1

$$\begin{Bmatrix} M_x \\ M_y \\ M_{xy} \end{Bmatrix} = \frac{D}{4ab}\,[\ \cdots\ ]$$

	w_1	ϕ_1	θ_1	w_2	ϕ_2	θ_2
M_x	$3(r^{-1}X(Y-1)+vrY(X-1))$	$va(1-X)(1-3Y)$	$b(1-Y)(1-3X)$	$3(r^{-1}X(1-Y)-vrY(1+X))$	$va(1+X)(1-3Y)$	$b(Y-1)(1+3X)$
M_y	$3(vr^{-1}X(Y-1)+rY(X-1))$	$a(1-X)(1-3Y)$	$vb(1-Y)(1-3X)$	$3(vr^{-1}X(1-Y)-rY(1+X))$	$a(1+X)(1-3Y)$	$vb(Y-1)(1+3X)$
M_{xy}	$\rho(-4+3(X^2+Y^2))$	$\rho b(3Y^2-2Y-1)$	$\rho a(3X^2-2X-1)$	$\rho(4-3(X^2+Y^2))$	$\rho b(1+2Y-3Y^2)$	$\rho a(3X^2+2X-1)$

	w_3	ϕ_3	θ_3	w_4	ϕ_4	θ_4
M_x	$3(r^{-1}X(1+Y)+vrY(1+X))$	$-va(1+X)(1+3Y)$	$-b(1+Y)(1+3X)$	$3(-r^{-1}X(1+Y)+vrY(1-X))$	$-va(1-X)(1+3Y)$	$b(1+Y)(1-3X)$
M_y	$3(vr^{-1}X(1+Y)+rY(1+X))$	$-a(1+X)(1+3Y)$	$-vb(1+Y)(1+3X)$	$3(-vr^{-1}X(1+Y)+rY(1-X))$	$-a(1-X)(1+3Y)$	$vb(1+Y)(1-3X)$
M_{xy}	$\rho(-4+3(X^2+Y^2))$	$\rho b(1-2Y-3Y^2)$	$\rho a(1-2X-3X^2)$	$\rho(4-3(X^2+Y^2))$	$\rho b(3Y^2+2Y-1)$	$\rho a(1+2X-3X^2)$

$r=a/b$, $\rho=\tfrac{3}{2}(1-v)$, $X=x/a$, and $Y=y/b$.

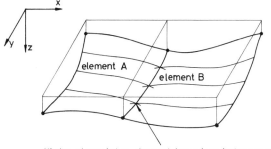

Fig. 11.2. Non-conformity of 12 d.o.f. rectangular element.

edge in question and therefore discontinuities of the normal slope, or kinks, will generally occur along interelement boundaries between the nodes. (The discontinuities in the normal slope value mean that the moments are theoretically infinite at interelement boundaries, but these infinities do not contribute to the potential energy summation.) The situation is shown in somewhat exaggerated form for a two-element assemblage in Fig. 11.2 and means that the element is non-conforming or, in other words, that the overall structure displacement field formed by combining the element fields is, strictly speaking, inadmissible in the Rayleigh–Ritz sense. It follows that the total potential energy of an assemblage of these elements will not necessarily be an upper bound on the true minimum potential energy and that the finite-element solution obtained may therefore be over- or under-stiff.

Despite its non-conformity the rectangular element described here is a useful one and does provide results which converge to the true results, albeit without any bound condition applying. The present element is one of a number of non-conforming but viable plate bending elements. It should be recognized that non-conformity is not necessarily a disadvantage numerically; the presence of kinks, say, at interelement boundaries make the structure more flexible than it otherwise would be and hence tends to compensate for the over-stiff representation within the individual elements.

The following question arises: Under what circumstances are non-conforming elements acceptable, in the sense that results obtained using them will converge to the correct solution with mesh refinement? The answer generally accepted nowadays is that the element displacement field should properly represent rigid-body motion and constant strain states (or constant curvature states in the plate case) and additionally that the element should pass the *patch test*. The patch test, introduced by Bazeley, Cheung, Irons, and Zienkiewicz [5], is a numerical test which

Fig. 11.3. A patch of elements.

extends the philosophy of the constant strain requirement from the individual element to a group of elements. In this test an arbitrary group (or patch) of elements, such as is shown in Fig. 11.3, has applied to it at its boundary a set of nodal displacements corresponding to a constant strain state throughout the patch. The calculated displacements and stresses in the interior of the patch should then be consistent with the constant strain state and if so the element is viable; if not the element is suspect and the results obtained using it may not converge correctly. The present non-conforming 12-degree-of-freedom rectangular element does satisfy the patch test so that convergence to the correct solution is guaranteed when using it.

Example 11.1 Using element R1 determine the central deflection of the square plate shown in Fig. E11.1(a) when the plate edges are clamped and a vertical (downward) load P acts at the plate centre. Assume $\nu = 0.3$.

Clearly, for the given geometry and loading the lines XX' and YY' shown in Fig. E11.1(a) are both axes of symmetry. Therefore only a quarter of the plate need be considered and this (top left) quadrant can be represented by a single element as shown in Fig. E11.1(b).

The deflection and both slopes (rotations) are zero at all points on the clamped boundary, and both slopes have zero value at the plate centre from symmetry considerations. Hence w_3 is the only non-zero degree of freedom and the corresponding externally applied load is $P/4$. From Table 11.2 we have that the

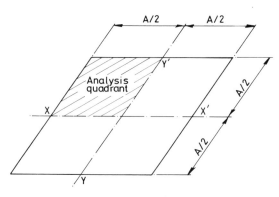

Fig. E11.1(a)

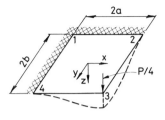

Fig. E11.1(b)

sole stiffness equation is

$$F_{w3} = \frac{P}{4} = \frac{D}{60ab}(60p + 60p^{-1} + 42 - 12\nu)w_3.$$

Now $a = b = A/4$, $p = 1$, and $\nu = 0.3$. Then the central deflection becomes

$$w_3 = 0.005919PA^2/D.$$

This compares with a solution [6] of $0.00560PA^2/D$ which may be considered to be exact for our purposes.

The FEM solution is thus seen to be about 6 per cent in error as far as central deflection is concerned and it is noted that the deflection is overestimated, i.e. the finite-element model of the plate is under-stiff. This is a consequence of the non-conforming nature of the element. It will be seen later (Fig. 11.8) that the accuracy of the result given here is a little fortuitous since a mesh with four elements in the plate quadrant gives a less accurate result than that found here.

Example 11.2. Consider again the square plate shown in Fig. E11.1(a) but now assume that all the plate edges are simply supported and that a uniformly distributed load of intensity q per unit area acts over the whole plate surface. Using a single R1 element in the plate quadrant calculate the central deflection and the rotation at the middle of an edge, and hence determine the distribution of moment M_y along YY′

Now that the edges are simply supported the rotations ϕ_2 and θ_4 are non-zero in addition to the deflection w_3, as indicated in Fig. E11.2.

The consistent load column matrix **Q** for the single element is given by equation (11.22). From this the column matrix $\mathbf{F}_{0r}$, whose entries correspond to the

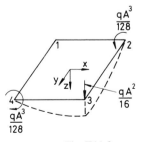

Fig. E11.2

freedoms ϕ_2, w_3, and θ_4, is obtained directly as

$$F_{0r} = qab\{b/3 \quad 1 \quad a/3\}.$$

The stiffness matrix $\mathbf{K}_r$ is obtained from the element stiffness matrix given in Table 11.2. Thus the general expression $\mathbf{F}_{0r} = \mathbf{K}_r\mathbf{D}_r$ becomes

$$qab\begin{Bmatrix} b/3 \\ 1 \\ a/3 \end{Bmatrix} = \frac{D}{60ab}\begin{bmatrix} b^2(80p+16-16\nu) & b(-60p-6+6\nu) \\ b(-60p-6+6\nu) & 60p+60p^{-1}+42-12\nu \\ 0 & a(-60p^{-1}-6+6\nu) \end{bmatrix}$$

$$\begin{bmatrix} 0 \\ a(-60p^{-1}-6+6\nu) \\ a^2(80p^{-1}+16-16\nu) \end{bmatrix}\begin{Bmatrix} \phi_2 \\ w_3 \\ \theta_4 \end{Bmatrix}.$$

For a hand solution the number of equations can be further reduced to 2 by noting the extra symmetry condition that $\theta_4 = \phi_2$. Then, with $a = b = A/4$, $p = 1$, and $\nu = 0.3$ the first two of the above equations become

$$\frac{qA^2}{16}\begin{Bmatrix} A/12 \\ 1 \end{Bmatrix} = \frac{16D}{60A^2}\begin{bmatrix} 5.7A^2 & -16.05A \\ -32.1A & 158.4 \end{bmatrix}\begin{Bmatrix} \phi_2 \\ w_3 \end{Bmatrix}$$

and the third equation can be ignored since it merely repeats the information given by the first equation. The solution of these equations gives the following values of the mid-side rotation and the central deflection:

$$\phi_2 = 0.01768qA^3/D = \theta_4 \qquad\qquad w_3 = 0.005063qA^4/D.$$

The accurate comparative solution for the central deflection is $0.00406qA^4/D$ [6] so that the FEM solution is about 25 per cent in error in predicting this quantity (again on the under-stiff side).

The moments at any point in the element can be calculated using the relationships given in Table 11.3. For the calculation of M_y along element edge 2–3 we have that

$$M_y = \frac{D}{4ab}[a(1+X)(1-3Y) \qquad 3\nu r^{-1}X(1+Y)+3rY(1+X)$$

$$\nu b(1+Y)(1-3X)]\begin{Bmatrix} \phi_2 \\ w_3 \\ \theta_4 \end{Bmatrix}$$

where $X = 1$, $r = 1$, and a, b, ν, ϕ_2, w_3, and θ_4 are as stated above. Clearly M_y varies linearly between points 2 and 3. Its values at these points become

$$M_{y2} = 0.01995qA^2 \qquad\qquad M_{y3} = 0.06602qA^2.$$

Obviously the correct value of M_{y2} is zero since this is the moment acting about a simply supported edge. An accurate value of M_{y3} is $0.0479qA^2$. Thus point values of the FEM moments are seen to be in considerable error when using a single element, and this is not surprising. With the use of more R1 elements the errors would be expected to reduce of course; some other numerical results for moments which show this are presented later in Fig. 11.9.

Before leaving element R1 it is noted that the displacement field for this element can be expressed in shape function form as

$$w_{R1} = \sum_{i=1}^{4} (f_i^{I} w_i + f_i^{II} \phi_i + f_i^{III} \theta_i) \tag{11.24}$$

where f_i^{I}, f_i^{II}, and f_i^{III} are (shape) functions of the co-ordinates defined as

$$f_i^{I} = \tfrac{1}{8}(1 + X_i X)(1 + Y_i Y)(2 + X_i X(1 - X_i X) + Y_i Y(1 - Y_i Y))$$

$$f_i^{II} = -\frac{bY_i}{8}(1 + X_i X)(1 + Y_i Y)(1 + Y_i Y)(1 - Y_i Y) \tag{11.25}$$

$$f_i^{III} = -\frac{aX_i}{8}(1 + X_i X)(1 + X_i X)(1 - X_i X)(1 + Y_i Y).$$

Here X and Y are dimensionless natural co-ordinates defined as

$$X = x/a \qquad Y = y/b. \tag{11.26}$$

and X_i, Y_i are their values at node i.

It should be noted that these dimensionless co-ordinates X and Y have been introduced here purely for convenience and that the displacement field of equation (11.24) is precisely that obtained by evaluating $w = \alpha C^{-1} d$ where α, C^{-1}, and d are defined in equations (11.16) and (11.18) and Table 11.1. To complete the picture as far as the displacement field of element R1 is concerned the shape functions for w_1 and ϕ_1 are illustrated in Fig. 11.4. (The other shape functions can easily be visualized from these two.) It should be noted that the shape function for ϕ_1 is restricted to a linear variation in the x direction. It should also be noted that at an edge the derivative of the shape functions with respect to the co-ordinate normal to the edge is generally not zero (except that $\partial N_{\phi 1}/\partial y = 0$ at the edge $Y = -1$); this reflects the normal slope nonconformity of the element.

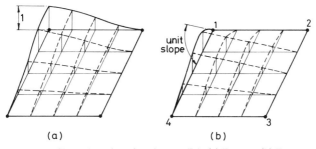

(a)   (b)

Fig. 11.4. Shape functions for element R1: (a) For w_1; (b) For ϕ_1.

11.4. Other rectangular elements

Element R1 has been considered in some detail in Section 11.3 since it is a relatively straightforward element with which to introduce plate finite-element analysis and at the same time it is quite a useful element which provides results of reasonable accuracy; its lack of complete compatibility need not be regarded as a particular disadvantage since ultimate convergence of results to the correct level when using it is guaranteed. However, other rectangular elements are available and some of these yield improved accuracy. Also the development of their displacement fields highlights some interesting points about classical plate finite-element analysis. For these reasons the alternative rectangular elements are described in this section.

11.4.1. An invalid conforming element (R2)

Now we know that the displacement field given by equations (11.24) and (11.25) is such that the lateral deflection is continuous across interelement boundaries whilst the slope normal to the boundary, i.e. $\partial w/\partial n$ where n is the direction normal to the boundary, is not. If extra terms are added to w_{R1} such that they vanish at the element boundary then the continuity of the normal slope can be altered whilst not interfering with the continuity of the lateral deflection. Consider, then, displacement fields of the form

$$w = w_{R1} + (1+X)(1-X)(1+Y)(1-Y)(P+QX+RY+SXY) \quad (11.27)$$

where P, Q, R, and S are coefficients.

A detailed examination of the variation of normal slope at the edges of the original R1 element reveals that complete continuity of lateral deflection *and normal slope* is achieved if in equation (11.27) the coefficients are defined as

$$
\begin{aligned}
P &= 0 \\
Q &= b(-\phi_1 + \phi_2 - \phi_3 + \phi_4)/16 \\
R &= a(-\theta_1 + \theta_2 - \theta_3 + \theta_4)/16 \\
S &= (w_1 - w_2 + w_3 - w_4 + b\phi_1 - b\phi_2 - b\phi_3 + b\phi_4 \\
&\quad + a\theta_1 + a\theta_2 - a\theta_3 - a\theta_4)/16.
\end{aligned}
\quad (11.28)
$$

However, although this displacement field is fully conforming it does have a very serious drawback. The problem is that the adjustment of the displacement field to achieve interelement continuity of normal slope has resulted in a situation in which the twist $\partial^2 w/\partial x\,\partial y$ is zero at all corner nodes. Thus no state of non-zero constant twist can be represented in the conforming element. This is clearly at variance with the basic requirement of completeness of the displacement field, as stipulated by the constant

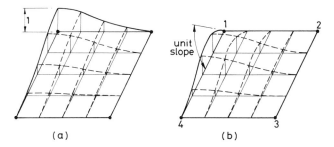

Fig. 11.5. Shape functions for element R2: (a) for w_1; (b) for ϕ_1.

strain condition. The result is that values obtained using this conforming, but invalid, element will not converge to the correct level however many elements are used; rather, convergence will be to a solution which is somewhat under-stiff.

This element is referred to as element R2 and its shape functions for w_1 and ϕ_1 are shown in Fig. 11.5. Comparing these with the functions for element R1 (shown in Fig. 11.4) clarifies the differences discussed above between the R1 and R2 element displacement fields. These differences appear small for the w_1 shape function, though there are differences in the interior of the element, but are more pronounced for the ϕ_1 function. In particular it is noted that now, for the conforming displacement field, the derivative of the shape functions with respect to the co-ordinate normal to an edge are zero at the edge, except that $\partial N_{\phi 1}/\partial y$ is, of course, necessarily non-zero at the edge $Y = -1$.

11.4.2. An improved non-conforming 12-degree-of-freedom element (R3)

Proceeding further, if the extra coefficients are modified to become [7]

$$
\begin{aligned}
P &= 0 \\
Q &= Cb(-\phi_1 \; +\phi_2 \; -\phi_3 \; +\phi_4)/8 \\
R &= Ca(-\theta_1 \; +\theta_2 \; -\theta_3 \; +\theta_4)/8 \\
S &= C(2w_1 \; -2w_2 \; +2w_3 \; -2w_4 \; +b\phi_1 \; -b\phi_2 \; -b\phi_3 \; +b\phi_4 \\
&\quad + a\theta_1 \; +a\theta_2 \; -a\theta_3 \; -a\theta_4)/8
\end{aligned}
\tag{11.29}
$$

where C is a constant, then it will be found that the constant strain condition is satisfied once more. Discontinuities of normal slope are, of course, re-introduced at element boundaries but their magnitude may now be controlled through selection of different values of the constant C, leading to a series of 12-degree-of-freedom rectangles. The complete displacement field can be expressed in the shape function form of equation (11.24) where now, on using equations (11.27) and (11.29), the new shape functions are

$$f_i^{\mathrm{I}} = \tfrac{1}{8}[(1+X_iX)(1+Y_iY)(2+X_iX(1-X_iX)+Y_iY(1-Y_iY))$$
$$+2\mu X_iY_iXY)]$$

$$f_i^{\mathrm{II}} = \frac{b}{8}[-Y_i((1+X_iX)(1+Y_iY)(1+Y_iY)(1-Y_iY)+\mu X_iX(1+Y_iY))]$$

$$f_i^{\mathrm{III}} = \frac{a}{8}[-X_i((1+X_iX)(1+X_iX)(1-X_iX)(1+Y_iY)+\mu Y_iY(1+X_iX))]$$

and
$$(11.30)$$

$$\mu = C(1+X)(1-X)(1+Y)(1-Y). \tag{11.31}$$

The basis for selection of the value of C is discussed in reference 7 where it is concluded that the value $C = \tfrac{1}{3}$ is particularly appropriate in reducing the magnitude of normal slope discontinuities at interelement boundaries.

TABLE 11.4

Additional stiffness terms for element R3 (see equation (11.32))

i, j	k_{ij}^{*}	k_{ij}^{**}
1,1	$24p + 24p^{-1} + 48\nu$	$252 + 180p + 180p^{-1}$
2,1	$b(18p + 36p^{-1} + 30\nu)$	$b(126 + 90p + 90p^{-1})$
3,1	$a(36p + 18p^{-1} + 30\nu)$	$a(126 + 90p + 90p^{-1})$
4,1	$-24p - 24p^{-1} - 48\nu$	$-252 - 180p - 180p^{-1}$
7,1	$24p + 24p^{-1} + 48\nu$	$252 + 180p + 180p^{-1}$
10,1	$-24p - 24p^{-1} - 48\nu$	$-252 - 180p - 180p^{-1}$
2,2	$b^2(16p + 32\nu)$	$b^2(168 + 60p + 360p^{-1})$
3,2	$ab(36p + 36p^{-1} + 24\nu)$	$ab(63 + 45p + 45p^{-1})$
4,2	$b(-18p - 36p^{-1} - 30\nu)$	$b(-126 - 90p - 90p^{-1})$
5,2	$b^2(-16p - 32\nu)$	$b^2(-168 - 60p - 360p^{-1})$
7,2	$b(18p - 24p^{-1} + 18\nu)$	$b(126 + 90p + 90p^{-1})$
8,2	$b^2(-8p + 8\nu)$	$b^2(42 - 30p + 270p^{-1})$
10,2	$b(-18p + 24p^{-1} - 18\nu)$	$b(-126 - 90p - 90p^{-1})$
11,2	$b^2(8p - 8\nu)$	$b^2(-42 + 30p - 270p^{-1})$
3,3	$a^2(16p^{-1} + 32\nu)$	$a^2(168 + 360p + 60p^{-1})$
4,3	$a(24p - 18p^{-1} - 18\nu)$	$a(-126 - 90p - 90p^{-1})$
5,3	$ab(24p - 36p^{-1} - 12\nu)$	$ab(-63 - 45p - 45p^{-1})$
6,3	$a^2(8p^{-1} - 8\nu)$	$a^2(-42 - 270p + 30p^{-1})$
7,3	$a(-24p + 18p^{-1} + 18\nu)$	$a(126 + 90p + 90p^{-1})$
8,3	$ab(24p + 24p^{-1})$	$ab(-63 - 45p - 45p^{-1})$
9,3	$a^2(-8p^{-1} + 8\nu)$	$a^2(42 + 270p - 30p^{-1})$
10,3	$a(-36p - 18p^{-1} - 30\nu)$	$a(-126 - 90p - 90p^{-1})$
11,3	$ab(36p - 24p^{-1} + 12\nu)$	$ab(63 + 45p + 45p^{-1})$
12,3	$a^2(-16p^{-1} - 32\nu)$	$a^2(-168 - 360p - 60p^{-1})$·

The 12-degree-of-freedom element based on the use of the shape functions of equation (11.30), with C having the value of $\frac{1}{3}$, will be referred to as R3. Drawings of the shape functions for this element will not be presented here since they look little different from those of element R2 shown in Fig. 11.5. However, it is emphasized that the differences, although appearing small in an illustration, are of vital significance since with the R3 functions $\partial w/\partial n$ is generally non-zero at element edges as is the twist at the corners.

The stiffness matrix for element R3 can be expressed (with $C = \frac{1}{3}$)

$$\mathbf{k} = \mathbf{k}_{R1} + \frac{D}{60ab}\left(C\mathbf{k}^* + \frac{8}{105}C^2\mathbf{k}^{**}\right) \tag{11.32}$$

where $\mathbf{k}_{R1}$ is the stiffness matrix for element R1, given in Table 11.2, and the additional matrices $\mathbf{k}^*$ and $\mathbf{k}^{**}$ are defined in Table 11.4. (Only the same 24 terms of the stiffness matrix as were given in Table 11.2 are recorded in Table 11.4 since all other terms are again related to the recorded ones in the manner shown in Table 11.2.)

11.4.3. Compatibility and twist

So far we have seen that in considering the category of bending elements which have a single displacement field over the whole element and which use only values of lateral deflection and its two first derivatives (the slopes) as nodal freedoms, it is possible to generate displacement fields which either satisfy the constant strain condition (including the patch test) but are non-conforming, or are conforming but violate the constant strain condition (with regard to the twist). The suspicion arises that it will not be possible to derive a displacement field belonging to this category of elements which simultaneously is fully conforming and satisfies the constant strain condition. This is indeed the case, as the following argument [8] shows.

Assume that an element of the type shown in Fig. 11.1 is fully compatible. Consider, for example, the adjacent edges 1–2 and 1–4 meeting at node 1. With the assumption of compatibility the slope $\partial w/\partial y$ all along edge 1–2 must depend only upon values of the degrees of freedom at nodes 1 and 2. Therefore

$$\frac{\partial^2 w}{\partial x\,\partial y} = \frac{\partial}{\partial x}\left(\frac{\partial w}{\partial y}\right)$$

along edge 1–2 must also depend only on the degrees of freedom at nodes 1 and 2. However, by a similar argument $\partial w/\partial x$ and

$$\frac{\partial^2 w}{\partial y\,\partial x} = \frac{\partial}{\partial y}\left(\frac{\partial w}{\partial x}\right)$$

along edge 1–4 must depend only on the degrees of freedom at nodes 1 and 4 if full compatibility is to be attained. Since the values of the nodal degrees of freedom at 2 and 4 are generally different it follows that in these circumstances the values of $\partial^2 w/\partial x\,\partial y$ and $\partial^2 w/\partial y\,\partial x$ at node 1 will need to be different; this is obviously contrary to what can occur when using a single continuous displacement field except where, as in the case of element R2, the specialized condition

$$\frac{\partial^2 w}{\partial x\,\partial y} = \frac{\partial^2 w}{\partial y\,\partial x} = 0$$

applies at the nodes. Thus it is confirmed that it will not be possible to find a displacement field which satisfies the constant strain condition (with particular regard to the twist) and which completely satisfies the compatibility condition if we restrict attention to single displacement fields applying over the whole element and to the use of only w and the two slopes as nodal degrees of freedom. We shall now consider the removal of these restrictions in two different ways: (a) using extra types of nodal freedoms, and (b) using a displacement field which envisages the element as being sub-divided into separate sub-regions.

11.4.4. Elements based on the use of extra nodal freedoms (R4)

Consider again the bicubic expansion shown in equation (11.17). This expansion formed the basis of the non-conforming element R1 when used in incomplete (12-term) form. The complete (16-term) expansion can be used if one extra degree of freedom is provided at each corner node. We recall that along any edge of a rectangular element the deflection and the slope in the direction normal to the edge vary as a cubic polynomial when using the bicubic expansion. Additionally, the continuity of w itself presented no difficulty in element R1 but continuity of the normal slope was not achieved since only two nodal freedoms were available at an edge to specify a cubic variation. With this in mind it is clear that if an extra degree of freedom is to be used at each corner then it must be the twist $\partial^2 w/\partial x\,\partial y$. (Then at an edge $x = \text{constant}$, say, there would be available the values of $\partial w/\partial x$ and

$$\frac{\partial}{\partial y}\left(\frac{\partial w}{\partial x}\right) = \frac{\partial^2 w}{\partial x\,\partial y}$$

at both nodes at the ends of the edge, giving four values with which to specify uniquely the cubic variation of $\partial w/\partial x$ along the edge.)

The 16-degree-of-freedom rectangular plate bending element, using w, $\partial w/\partial x$, $\partial w/\partial y$, and $\partial^2 w/\partial x\,\partial y$ as freedoms at the four corner nodes, was presented by Bogner, Fox, and Schmit [9]; it will be referred to as element R4 here. The element is fully conforming and satisfies the constant strain condition, and whilst the use of $\partial^2 w/\partial x\, \partial y$ as a nodal freedom is a little inconvenient it does not preclude the use of the element in complicated circumstances of step changes in plate thickness or material properties. Since the displacement field (equation (11.17)) is a product of cubic polynomials in x and y, it is not difficult to see that it will correspond to appropriate products of cubic Hermitian polynomials. These one-dimensional polynomials have been described in Section 6.9 and their shapes are shown in Fig. 6.5. It is recalled that the four cubic Hermitian polynomials, designated H_{01}^1, H_{11}^1, H_{02}^1, and H_{12}^1 correspond precisely to the shape functions of the basic beam finite element. For the rectangular plate element illustrated in Fig. 11.1, with the co-ordinate origin at the element centroid, the following definitions of Hermitian cubic polynomials in the x and y directions apply:

$$H_{01}^1(x) = \frac{1}{4}(2 - 3X + X^3) \qquad H_{01}^1(y) = \frac{1}{4}(2 - 3Y + Y^3)$$

$$H_{11}^1(x) = \frac{a}{4}(1 - X - X^2 + X^3) \qquad H_{11}^1(y) = \frac{b}{4}(1 - Y - Y^2 + Y^3)$$

$$(11.33)$$

$$H_{02}^1(x) = \frac{1}{4}(2 + 3X - X^3) \qquad H_{02}^1(y) = \frac{1}{4}(2 + 3Y - Y^3)$$

$$H_{12}^1(x) = \frac{a}{4}(-1 - X + X^2 + X^3) \qquad H_{12}^1(y) = \frac{b}{4}(-1 - Y + Y^2 + Y^3).$$

Then the displacement field for the 16-degree-of-freedom element is

$$w = \sum_{i=1}^{4} (f_i^I w_i + f_i^{II}\phi_i + f_i^{III}\theta_i + f_i^{IV}\Omega_i) \qquad (11.34)$$

where $\phi = \partial w/\partial y$, $\theta = \partial w/\partial x$ (as before), and $\Omega = \partial^2 w/\partial x\, \partial y$. The functions f_i^I etc. are shape functions defined as

$$
\begin{aligned}
f_1^I &= H_{01}^1(x)H_{01}^1(y) & f_2^I &= H_{02}^1(x)H_{01}^1(y) \\
f_3^I &= H_{02}^1(x)H_{02}^1(y) & f_4^I &= H_{01}^1(x)H_{02}^1(y) \\
f_1^{II} &= H_{01}^1(x)H_{11}^1(y) & f_2^{II} &= H_{02}^1(x)H_{11}^1(y) \\
f_3^{II} &= H_{02}^1(x)H_{12}^1(y) & f_4^{II} &= H_{01}^1(x)H_{12}^1(y) \\
f_1^{III} &= H_{11}^1(x)H_{01}^1(y) & f_2^{III} &= H_{12}^1(x)H_{01}^1(y) \\
f_3^{III} &= H_{12}^1(x)H_{02}^1(y) & f_4^{III} &= H_{11}^1(x)H_{02}^1(y) \\
f_1^{IV} &= H_{11}^1(x)H_{11}^1(y) & f_2^{IV} &= H_{12}^1(x)H_{11}^1(y) \\
f_3^{IV} &= H_{12}^1(x)H_{12}^1(y) & f_4^{IV} &= H_{11}^1(x)H_{12}^1(y).
\end{aligned}
$$

$$(11.35)$$

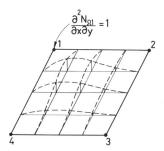

Fig. 11.6. Shape function corresponding to Ω_1 for element R4.

That these definitions of the shape functions are logical should become clear on referring to Fig. 6.5 (in conjunction with Fig. 11.1) and bearing in mind the general definition of a shape function.

Now, although not obvious at first, it transpires that the displacement field for the 16-degree-of-freedom element R4 is a very direct extension of the field for the conforming, but invalid, 12-degree-of-freedom element R2. In fact the field for R2, given in total by equations (11.24)–(11.28), can be abstracted from the field for R4, given by equations (11.33)–(11.35), if the twist Ω_i and associated shape functions $f_i^{IV}(i = 1\text{--}4)$ are ignored in the latter equations. Thus the shape functions for the 12 freedoms w_i, ϕ_i, and θ_i are exactly the same for element R4 as for element R2 and hence the functions illustrated in Fig. 11.5 apply for R4 as well as for R2. The shape functions for the extra four freedoms Ω_i of the R4 element are typified by that corresponding to Ω_1 which is illustrated in Fig. 11.6.

The procedure of introducing extra nodal degrees of freedom need not stop with the 16-degree-of-freedom element R4 of course. Bogner *et al.* [9] have also described refined rectangular elements with 24 and 36 degrees of freedom respectively. In the first of these the degrees of freedom at the corner nodes are w, $\partial w/\partial x$, $\partial w/\partial y$, $\partial^2 w/\partial x^2$, $\partial^2 w/\partial x\,\partial y$, and $\partial^2 w/\partial y^2$; in the second the quantities $\partial^3 w/\partial x\,\partial y^2$, $\partial^3 w/\partial x^2\,\partial y$, and $\partial^4 w/\partial y^2\,\partial x^2$ are used in addition. Refined elements such as these are conforming and can give highly accurate results but the excessive nodal continuity can lead to difficulty in conditions where step changes in plate thickness or material property occur.

11.4.5. An element based on the sub-region approach (R5)

A second way of generating a valid conforming element, without the use of extra nodal degrees of freedom, is to imagine subdividing the individual element into a number of sub-regions with separate displacement fields assumed for each sub-region. Such a rectangular element was

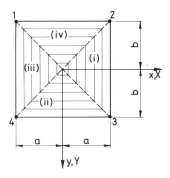

Fig. 11.7. Sub-regional element R5.

presented by Deak and Pian [10] and is illustrated in Fig. 11.7; the element degrees of freedom are the basic ones (i.e. w, ϕ, and θ) at each of the four corner nodes. The sub-regions are the triangular regions numbered (i)–(iv) in the figure and the suggested displacement field is

$$w = \sum_{j=1}^{12} A_j F_j(x, y) \tag{11.36}$$

where the A_j are coefficients. Some of the functions F_j are specified differently in different sub-regions, as follows ($X = x/a$ and $Y = y/b$ again):

$$F_1 = 1,\ F_2 = X^2,\ F_3 = Y^2,\ F_4 = X,\ F_5 = X^3,\ F_7 = Y,\ F_8 = Y^3,\ F_{10} = XY,$$

and

$$F_{11} = 3X^3Y + 3Y^3X - X^3Y^3 - 5XY \text{ in regions (i), (ii), (iii), and (iv)}$$

$$F_6 = \begin{cases} X^2 - 2X + Y^2 \text{ in region (i)} \\ 2XY - 2X \text{ in region (ii)} \\ -X^2 - 2X - Y^2 \text{ in region (iii)} \\ -2XY - 2X \text{ in region (iv)} \end{cases}$$

$$F_9(X, Y) = F_6(Y, X)$$

$$F_{12} = \begin{cases} \tfrac{1}{4}(X^3Y^3 - YX^5 - 3XY^3 + 3YX^3) \text{ in region (i) and (iii)} \\ \tfrac{1}{4}(XY^5 - X^3Y^3 - 3XY^3 + 3YX^3) \text{ in region (ii) and (iv).} \end{cases} \tag{11.37}$$

It can be shown that C_1-type continuity is achieved across both interior and exterior boundaries and that the constant strain condition is satisfied. It is noted that at the exterior boundaries w varies as a cubic function whilst the normal slope is restricted to a linear variation. A further consequence of requiring full compatibility whilst only using the basic nodal freedoms is that the twist $\partial^2 w/\partial x\, \partial y$ is generally not uniquely defined at the element nodes. This is expected, for reasons given earlier, and means that at node 1, say, $\partial^2 w/\partial x\,\partial y$ will have different values in sub-regions (iii) and (iv). This does not invalidate this element (designated element R5 in what follows).

11.5. Numerical results for rectangular elements

The most readily available numerical results for the rectangular plate bending elements are those for the deflection under a central point load of a square plate. The convergence curves for this deflection for simply supported and clamped edges are shown in Figs. 11.8(a) and 11.8(b) respectively. The ordinate is the percentage error in the calculated central deflection and, since this deflection is directly proportional to the potential energy in the point-load cases, this provides a convenient measure of overall accuracy of solution. The abscissa measures on a logarithmic scale the number of degrees of freedom in a symmetric plate quadrant before applying any boundary conditions; the plate quadrant is idealized with 1, 4, 9, and 16 equal square elements in turn.

It is noted that results for the 12-degree-of-freedom conforming element R2 are apparently converging to an incorrect solution in Fig. 11.8(a), and this is not unexpected for the reasons given earlier. The results for all other elements do converge to the correct solution, though for some of them further refinement of the element mesh would be necessary to achieve very high accuracy. The conforming elements R4 (using the twist as a nodal freedom) and R5 (using the subregion

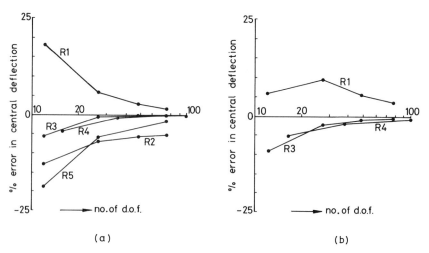

Fig. 11.8. Convergence of central deflection when using rectangular elements for a square plate with central load: (a) simply supported edges; (b) clamped edges.

approach) demonstrate monotonic convergence from the over-stiff side, as expected. The non-conforming elements R1 and R3 may yield over- or under-stiff results, though for the two examples chosen R1 results are always under-stiff and R3 results are always over-stiff.

Elements R3 and R4 are the most successful elements and their performance is very similar in these (and other) problems. The 16-degree-of-freedom conforming element R4 has the advantage of being known to yield bounded solutions. On the other hand the improved 12-degree-of-freedom non-conforming element R3 has the considerable advantage of only using w, θ, and ϕ as nodal degrees of freedom; this element is considerably more efficient than is the other 12-degree-of-freedom non-conforming element R1.

In static analysis we are often more concerned with calculating moment values rather than displacements and since the moments depend upon the second derivatives of displacement we expect calculated moments to be less accurate than are the calculated displacements. Values of moment appear not to be available for all the rectangular elements considered here and so no direct comparison can be made between all five element types. However, comparison can be made between element types R1 and R4 and Fig. 11.9 shows this for the variation of moment M_y along the centre line YY' of Fig. E11.1(a). With both types of element discontinuities occur at interelement boundaries but these reduce with refinement of the mesh. (It should be noted that for element R4 the variation of moment between the nodes of an element is not strictly linear, although it is shown as such in Fig. 11.9).

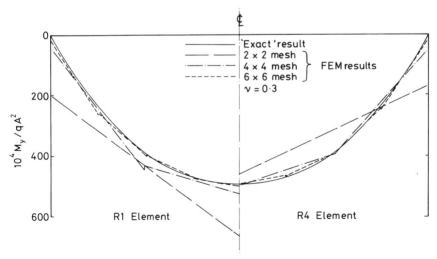

Fig. 11.9. Distribution of M_y along YY' (see Fig. E11.1(a)) for a simply supported plate carrying a uniformly distributed load.

11.6. Elements of more general shape

Considerable attention has been given thus far in this chapter to plate bending elements of rectangular shape. This has been done for the reasons that in plate analysis rectangular elements are the simplest to deal with (using Cartesian co-ordinates) and that the description of the displacement fields for the five rectangular elements considered does highlight the difficulties associated with the C_1 continuity problem and some ways in which these can be overcome. Clearly, the rectangular elements are of somewhat limited application and more general element shapes must be considered.

As a start we might consider whether the described rectangles can be successfully transformed into general quadrilaterals along the lines, say, of the transformations described for planar and three-dimensional elements in Chapter 10. Unfortunately this approach is not generally successful in the C_1-continuity problem since it leads to violation of the constant strain condition and to interelement compatibility problems of increased significance. (For the non-conforming elements R1 and R3, for instance, the lateral deflection, as well as the normal slope, becomes discontinuous at element edges if extension to the general quadrilateral is attempted.) Only in the case of the element of parallelogrammic shape can the transformation be successfully achieved such that the parallelogram has the same basic properties as the 'parent' rectangle. Parallelogrammic plate elements are of use in the analysis of skew plate problems but again are not applicable to general problems and so will not be considered further here. It is noted, though, that full details of the derivation of the parallelogrammic element equivalent of the R1 rectangular element and numerical studies are given by Dawe [11]. (Plate elements of sector shape are also possible as transformations of rectangles when working in terms of polar co-ordinates.)

For the analysis of plates of general shape it is clear that triangular and/or quadrilateral elements are needed and a wide range of such elements have been developed as a consequence. The elements are almost exclusively straight-sided (rectilinear) since transformation to curvilinear versions is impractible in the C_1 continuity problem. The triangular elements are more widely used than are quadrilateral elements and often the quadrilateral elements are in any case assembled from triangles. Therefore attention will be restricted here to triangular elements.

11.7. Triangular elements

It has been seen that it is impossible to achieve complete compatibility in classical plate bending elements by using a single simple displacement field if only w and its first derivatives are prescribed at the nodes. In considering rectangular elements three types of element have been identified.

(a) Elements which use a single displacement field in conjunction with basic degrees of freedom only, and in which complete compatibility is relaxed. Such elements can be accurate and practical, and their validity is assured if the element displacement field contains the constant strain representation and if the patch test is satisfied. With these elements the twist is, of course, single valued at the nodes.

(b) Elements whose displacement fields are constructed by dividing the element into sub-regions. It is possible to derive conforming elements in this way since the twist (and other second- and higher-order derivatives of w for that matter) is non-unique (or multi-valued) at the nodes.

(c) Elements which use a single field but use quantities other than just w and its first derivatives at the nodes. Such elements can be conforming and satisfy the constant strain condition but the use of the higher derivatives of w as nodal freedoms is somewhat inconvenient.

In moving on to consider non-rectangular elements it will be seen that these same element types occur.

For triangular elements the use of Cartesian co-ordinates is not usually particularly appropriate, although they have been successfully employed from time to time. An alternative system of co-ordinates which is more natural for the triangle is the area co-ordinate system which has been described in Section 10.2.2. The use of area co-ordinates simplifies the specification of a geometrically invariant displacement field and often eases the detailed development of element properties. When using Cartesian co-ordinates it is necessary to use complete polynomial expressions to ensure invariance.

A considerable number of triangular plate elements has been developed over the years—whether based on Cartesian or area co-ordinates—and it is not appropriate here to attempt to describe the full range of these elements or to go into great detail. Instead, brief descriptions of a limited number of elements—including some of the most popular ones—are given and it is left to the interested reader to consult the given references for more detail.

11.7.1. Non-conforming elements using basic freedoms only

In the search for a viable triangular plate bending element an obvious starting point is to consider an element with nodes at the vertices only and with w and two slopes as nodal degrees of freedom. Such an element has nine degrees of freedom and this implies a nine-term displacement field. If we wish to use a Cartesian co-ordinate formulation it is immediately obvious from the Pascal triangle (equation (8.36)) that a point of difficulty arises since a complete cubic expansion in two dimensions involves 10 terms. Somehow one term of the expansion has to be effectively eliminated.

In the first attempt to do this the xy term was deleted so as to leave a symmetric but incomplete expansion. This means that a state of non-zero constant twist cannot be represented and this defect means that convergence to correct energy levels will generally not occur, regardless of how fine the element mesh is made. This type of behaviour has already been seen for the rectangular element R2.

In a second attempt two of the cubic terms were combined such that the displacement field has the form

$$w = A_1 + A_2x + A_3y + A_4x^2 + A_5xy + A_6y^2 + A_7x^3 + A_8(x^2y + xy^2) + A_9y^3. \quad (11.38)$$

Here symmetry has been maintained but the element is not invariant and for some orientations of the element sides with regard to the x, y co-ordinate axes the transformation matrix $\mathbf{C}$ is singular. Another approach used the complete 10-term polynomial expression and included as a tenth degree of freedom the deflection at the centroid of the triangular element; this degree of freedom could be eliminated by static condensation.

The above mentioned triangular elements are all non-conforming and, more importantly, have such poor convergence properties that they will not be considered further here; details of their performance are given in reference 12.

A more successful triangular element with nine degrees of freedom is that due to Bazeley *et al.* [5] which is formulated in terms of area co-ordinates and illustrated in Fig. 11.10. The element displacement field is constructed as the sum of two parts in the form

$$w = w^1 + w_r \qquad (11.39)$$

Fig. 11.10. Element T1.

where w_r is the rigid-body deflection and w^1 is the relative deflection of the element when it is regarded as simply supported at the nodes. The procedure will not be detailed here but is similar to that described for a related element in Example 11.3. In reference 5 direct formation of the shape functions is made with the requirements of the constant strain (curvature) condition being satisfied. The resulting total displacement field is an incomplete cubic polynomial function of the area co-ordinates which can be written in terms of generalized coefficients a_i as

$$w = a_1 L_1 + a_2 L_2 + a_3 L_3 + a_4(L_2^2 L_1 + \tfrac{1}{2} L_1 L_2 L_3) + a_5(L_3^2 L_2 + \tfrac{1}{2} L_1 L_2 L_3)$$
$$+ a_6(L_1^2 L_3 + \tfrac{1}{2} L_1 L_2 L_3) + a_7(L_1^2 L_2 + \tfrac{1}{2} L_1 L_2 L_3) + a_8(L_2^2 L_3 + \tfrac{1}{2} L_1 L_2 L_3)$$
$$+ a_9(L_3^2 L_1 + \tfrac{1}{2} L_1 L_2 L_3). \tag{11.40}$$

It is noted that the $L_1 L_2 L_3$ term vanishes at the element edges and also has zero slopes at the element nodes; this term therefore has similar characteristics to the extra terms shown in equation (11.27) which are added to the original rectangular element displacement field. The deflection varies along any side of the triangle as a cubic polynomial function and will match that of an adjacent element. On the other hand the normal slope varies quadratically on an edge and since only two end values are available to define this slope it follows that discontinuities of normal slope will occur between adjacent elements.

The element just described will be referred to as T1. Without for the moment presenting any numerical results for this element it should be noted that the validity of the results obtained depends upon the geometry of the element mesh used. Convergence to correct energy levels occurs when the mesh is generated by three sets of equally spaced parallel lines but convergence to incorrect levels can occur in other cases (in, say, a 'union jack' type mesh) where the patch test is violated. This is an unusual circumstance but fortunately the numerical errors involved in the converged results are so small that the element is still regarded as useful. Modifications of this element are available which will be discussed a little later.

Our consideration of triangular elements has started with what might appear to be the lowest-order viable element—with nine degrees of freedom—and this type of element was the first to be considered historically. However, an element with only six degrees of freedom was later derived by Morley [13] and this represents the simplest possible plate element. The element is shown in Fig. 11.11 from which it is seen that the degrees of freedom are w alone at the vertices and the normal slope $\partial w / \partial n$ at the mid-points of each side. As with the T1 element the shape functions for the 6-degree-of-freedom element were originally constructed directly in terms of area co-ordinates by considering the total

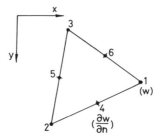

Fig. 11.11. Element T2.

deflection as the sum of a rigid-body deflection and a relative deflection; the procedure is described in Example 11.3. Essentially, however, the displacement field is a complete quadratic polynomial field and consequently the curvatures and moments can only have constant value throughout the element. This element is designated T2 and it shatters all the usual conceptions regarding compatibility since discontinuities of deflection as well as of normal slope occur between adjacent elements. However, since strains are constant within the element and the patch test is satisfied the element will yield results which converge to the correct solution. Furthermore, the element does in fact correspond exactly to an earlier *equilibrium-type* formulation. This need not concern us here except to note that it means that solutions obtained using the element will be under-stiff.

11.7.2. Conforming elements based on sub-region or related approach

Two approaches are considered in this category.

In the first of these Clough and Tocher [12] developed the stiffness matrix for the triangular element shown in Fig. 11.12(a) by using a Cartesian coordinate approach along the lines described earlier for rectangle R5. The triangular element is divided into three sub-regions ((i),

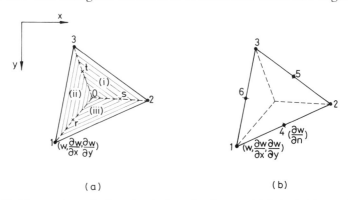

(a)   (b)

Fig. 11.12. Two versions of sub-regional triangular element: (a) 9 d.o.f. (element T3); (b) 12 d.o.f. (element T4).

(ii), and (iii)) with a common vertex at point O which is taken to be at the centroid of the element. In each subregion a different incomplete nine-term cubic polynomial expression is assumed for the deflection; for example in sub-region (i) it is assumed that

$$w = A_1 + A_2\bar{x} + A_3\bar{y} + A_4\bar{x}^2 + A_5\bar{x}\bar{y} + A_6\bar{y}^2 + A_7\bar{x}^3 + A_8\bar{x}\bar{y}^2 + A_9\bar{y}^3$$

$$(11.41)$$

where the $\bar{x}$ and $\bar{y}$ axes are directed parallel and perpendicular respectively to the exterior edge of the subregion. The $\bar{x}^2\bar{y}$ term is excluded so that the normal slope varies linearly along the exterior boundary and this ensures continuity of this slope with any adjacent element. There are a total of 27 displacement coefficients but 18 of these are used in satisfying internal compatibility requirements between adjacent sub-regions (by matching displacements and Cartesian slopes at nodal points 1, 2, and 3 and normal slopes at points r, s, and t). The remaining nine are then related to the nine degrees of freedom of the complete element (w, $\partial w/\partial x$, and $\partial w/\partial y$ at the vertices). The element, designated T3, satisfies all monotonic convergence requirements but it may be suspected that it is rather over-stiff owing to the restriction of the normal slope to a linear variation at the element boundary.

An improvement on this formulation is possible if the introduction of extra nodal points is accepted. Then a 12-degree-of-freedom triangular element can be formed by using values of the normal slope at the mid-points of the exterior sides as extra degrees of freedom. Correspondingly a complete 10-term cubic polynomial is assumed for the displacement in each subregion and the normal slope varies quadratically along element boundaries. The resulting element, which is still fully conforming, is designated T4 and is shown in Fig. 11.12(b)).

The second approach is described in the paper by Bazeley *et al.* [5] which was the source of the non-conforming element T1. To produce a conforming triangle special displacement functions are devised which ensure a *linear* variation of normal slope at the element edges when added to the original displacement field of element T1. The effect of this, as in the sub-region approach, is to produce non-unique values of the second derivatives of w at the vertex nodes. This does not prevent the 9-degree-of-freedom conforming element satisfying all convergence requirements. Again, though, the restriction of a linear variation of normal slope along element edges results in an over-stiff element which is numerically much less efficient than is its non-conforming counterpart T1.

11.7.3. Conforming elements using higher-order degrees of freedom

Consider now the generation of a triangular element based on the use of a single displacement field (with consequent single-valued second derivatives everywhere) and satisfying all convergence conditions including full compatibility. We know that such an element will have to employ as nodal degrees of freedom quantities other than just w and its first derivatives and that in the realm of rectangular elements the element R4 which uses the twist $\partial^2 w/\partial x\, \partial y$ as a nodal freedom is efficient, as are other related elements using higher-order degrees of freedom. In Section 11.4 the need for the presence of $\partial^2 w/\partial x\, \partial y$ as a nodal freedom of a conforming rectangular element was explained. Now, when using triangular elements, the element edges lie in arbitrary directions and it is the presence of $\partial^2 w/\partial n\, \partial s$ which is required to maintain continuity of the normal slope; here s and n are the directions along and normal to the boundary between any two adjacent elements. To ensure the nodal continuity of $\partial^2 w/\partial n\, \partial s$ for arbitrary direction of s and n requires the nodal continuity of all second derivatives of w with respect to x and y; thus $\partial^2 w/\partial x^2$, $\partial^2 w/\partial x\, \partial y$, and $\partial^2 w/\partial y^2$ must be used as nodal degrees of freedom if a conforming element is sought. Consequently, since w, $\partial w/\partial x$, and $\partial w/\partial y$ are also present as nodal freedoms, it is clear that a conforming triangular plate element will have at least 18 degrees of freedom.

The complete two-dimensional quintic polynomial contains 21 terms, as can be seen for Cartesian co-ordinates in the Pascal triangle given by equation (8.36). If this polynomial is used as an element displacement field, then along any straight line, and particularly along any element edge, the deflection w will vary as a quintic polynomial in the co-ordinate s running along the edge and the normal slope $\partial w/\partial n$ as a quartic polynomial in s. Thus six nodal quantities are required at the edge to specify w and five to specify $\partial w/\partial n$. If w and all its first and second derivatives are used as degrees of freedom at the vertices, then values of w, $\partial w/\partial s$ and $\partial^2 w/\partial s^2$ are available at two nodes at the ends of the edge to specify w uniquely. On the other hand, to specify $\partial w/\partial n$ along the edge there will be available at the two vertex nodes only values of $\partial w/\partial n$ and $\partial^2 w/\partial s\, \partial n$. Thus one more condition is required at each element edge to specify $\partial w/\partial n$ uniquely and the obvious thing to do is to introduce mid-side nodes at which the value of $\partial w/\partial n$ is the sole degree of freedom. This completes the specification of the 21-degree-of-freedom conforming triangular element which is based on a complete two-dimensional quintic polynomial displacement field and which is shown in Fig. 11.13(a).

Clearly it would be advantageous if the single degree of freedom at each mid-side node could be eliminated without destroying the compatibility of the element or introducing any other adverse effect. This can be done by constraining the normal slope to vary as a cubic, rather than quartic, function along an edge. The result is an 18-degree-of-freedom (constrained–quintic) element with vertex nodes only, as in Fig. 11.13(b). This element, which will be referred to as T5, is generally considered a more useful element than is the corresponding 21-degree-of-freedom

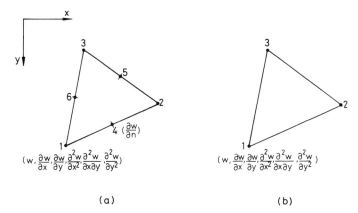

Fig. 11.13. Two versions of a high-order triangular element: (a) 21 d.o.f.; (b) 18 d.o.f. (element T5).

element since little loss in accuracy is experienced in eliminating the inconvenient mid-side nodes.

The quintic elements (full and constrained) were presented at much the same time in a number of different formulations amongst which were references 14–17; some formulations are based on the exclusive use of Cartesian co-ordinates whilst others employ area co-ordinates. Even more-refined triangular elements can be generated based on complete sixth- or higher-order polynomial displacement fields [14] but in practical use the quintic elements would probably be the most refined elements used.

Example 11.3. Deduce the shape functions, in terms of area co-ordinates, for the 6-degree-of-freedom constant-moment triangular plate element T2.

The element and its degrees of freedom are shown in Fig. 11.11. The manner of the construction of the shape functions detailed here is based on the original work of Morley [13]. The reader's attention is directed to Section 10.2.2 for a description of area co-ordinates.

The total deflection w is considered to be made up of two parts in the form

$$w = w_r + w^1. \tag{a}$$

Here w_r is the rigid-body deflection which in terms of area co-ordinates is the linear field (see equation (11.6) for the expression for w_r in terms of Cartesian co-ordinates):

$$w_r = w_1 L_1 + w_2 L_2 + w_3 L_3 \tag{b}$$

The other part, w^1, is the relative deflection of the element when it is regarded as simply supported at its vertices. Figure E11.3(a) gives a diagrammatic representation of equation (a). It should be noted that since the element remains flat in the w_r rigid-body movement it is, of course, only w^1 which contributes to the curvatures and hence to the element strain energy.

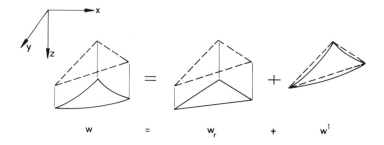

Fig. E11.3(a)

An expression is now sought for the relative deflection w in the shape function form

$$w^1 = N_4^1 \frac{\partial w_4^1}{\partial n} + N_5^1 \frac{\partial w_5^1}{\partial n} + N_6^1 \frac{\partial w_6^1}{\partial n} \tag{c}$$

since $\partial w_4^1/\partial n$, $\partial w_5^1/\partial n$, and $\partial w_6^1/\partial n$ (the slopes of the relative, not total, deflection) are the only non-zero degrees of freedom in this deflection. The requirements of the shape functions are typified by those for N_4^1 which are that

$N_4^1 = 0$ at the vertex points 1, 2, and 3

$\dfrac{\partial N_4^1}{\partial n} = 0$ at the mid-side points 5 and 6

$\dfrac{\partial N_4^1}{\partial n} = 1$ at the mid-side point 4.

The shape function

$$N_4^1 = -\frac{2\Delta}{s_{12}} L_3(1 - L_3) \tag{d}$$

satisfies the stated requirements, where s_{12} is the length of side 1–2 as shown in Fig. E11.3(b) and Δ has its usual meaning as the area of the triangular element. That the given definition for N_4^1 is suitable will now be demonstrated.

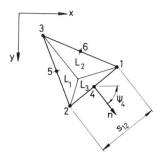

Fig. E11.3(b)

Since $L_3 = 0$ at vertices 1 and 2 and $L_3 = 1$ at vertex 3, it can be seen straight away that the shape function N_4^1, given by equation (d), does indeed satisfy the condition of having zero value at the three vertices. The stated requirements on $\partial N_4^1/\partial n$ are less easy to verify but the process is started by noting that the derivative of any function $f(x, y)$ in a direction n is related to the derivatives in the x and y directions by the expresion

$$\frac{\partial f}{\partial n} = \cos \psi \frac{\partial f}{\partial x} + \sin \psi \frac{\partial f}{\partial y} \tag{e}$$

where ψ is the angle that n makes with the x axis as shown in Fig. E11.3(b). Now the function that we are interested in here is N_4^1 which is expressed in equation (d) in terms of the area co-ordinate L_3 rather than in terms of x and y. However, by the chain rule for differentiation we have

$$\frac{\partial N_4^1}{\partial x} = \frac{\partial N_4^1}{\partial L_1}\frac{\partial L_1}{\partial x} + \frac{\partial N_4^1}{\partial L_2}\frac{\partial L_2}{\partial x} + \frac{\partial N_4^1}{\partial L_3}\frac{\partial L_3}{\partial x} = \frac{2\Delta}{s_{12}}(2L_3 - 1)\frac{\beta_3}{2\Delta} = \frac{\beta_3}{s_{12}}(2L_3 - 1) \tag{f}$$

where equation (10.18) has been used. Similarly,

$$\frac{\partial N_4^1}{\partial y} = \frac{\gamma_3}{s_{12}}(2L_3 - 1). \tag{g}$$

Now, from these last two equations it can be seen that $\partial N_4^1/\partial x$ and $\partial N_4^1/\partial y$ are both zero at points where $L_3 = \frac{1}{2}$ and that therefore, through use of equation (e), $\partial N_4^1/\partial n = 0$ at the mid-side points 5 and 6 as required. For the mid-side point 4, $\psi = \psi_4$ and substitution of equations (f) and (g) into an equation of the form of equation (e) gives

$$\frac{\partial N_4^1}{\partial n} = \frac{\beta_3}{s_{12}}(2L_3 - 1)\cos \psi_4 + \frac{\gamma_3}{s_{12}}(2L_3 - 1)\sin \psi_4 \tag{h}$$

where β_3 and γ_3 are defined (see equation (10.15)) as

$$\beta_3 = y_1 - y_2 \qquad \gamma_3 = x_2 - x_1.$$

Now

$$\sin \psi_4 = -\frac{\gamma_3}{s_{12}} \qquad \cos \psi_4 = -\frac{\beta_3}{s_{12}} \tag{i}$$

and so, from equation (h),

$$\frac{\partial N_4^1}{\partial n} = -(2L_3 - 1)\frac{(\beta_3^2 + \gamma_3^2)}{s_{12}^2} = 1 - 2L_3. \tag{j}$$

Since $L_3 = 0$ at the mid-side point 4 (or indeed at any point along side 1–2) then $\partial N_4^1/\partial n = 1$ as required. This completes the check on N_4^1.

The shape functions corresponding to the element degrees of freedom shown in Fig. 11.11 are not yet established since N_4^1 etc. refer to the relative slopes $\partial w_4^1/\partial n$ etc, and not to the total slopes $\partial w_4/\partial n$ etc. The two types of slope are related by the expression (from equation (a))

$$\frac{\partial w^1}{\partial n} = \frac{\partial w}{\partial n} - \frac{\partial w_r}{\partial n} \tag{k}$$

Now, using equation (e),

$$\frac{\partial w_r}{\partial n} = \cos \psi \frac{\partial w_r}{\partial x} + \sin \psi \frac{\partial w_r}{\partial y} \tag{l}$$

and, with w_r given by equation (b),

$$\frac{\partial w_r}{\partial x} = \frac{\partial w_r}{\partial L_1}\frac{\partial L_1}{\partial x} + \frac{\partial w_r}{\partial L_2}\frac{\partial L_2}{\partial x} + \frac{\partial w_r}{\partial L_3}\frac{\partial L_3}{\partial x} = \frac{1}{2\Delta}(\beta_1 w_1 + \beta_2 w_2 + \beta_3 w_3)$$

$$\frac{\partial w_r}{\partial y} = \frac{\partial w_r}{\partial L_1}\frac{\partial L_1}{\partial y} + \frac{\partial w_r}{\partial L_2}\frac{\partial L_2}{\partial y} + \frac{\partial w_r}{\partial L_3}\frac{\partial L_3}{\partial y} = \frac{1}{2\Delta}(\gamma_1 w_1 + \gamma_2 w_2 + \gamma_3 w_3). \tag{m}$$

Therefore

$$\frac{\partial w^1}{\partial n} = \frac{\partial w}{\partial n} - \frac{1}{2\Delta}\Big((\beta_1 \cos \psi + \gamma_1 \sin \psi)w_1$$

$$+ (\beta_2 \cos \psi + \gamma_2 \sin \psi)w_2 + (\beta_3 \cos \psi + \gamma_3 \sin \psi)w_3\Big). \tag{n}$$

This relationship applies for the mid-point of each side; for point 4, for instance, $\psi = \psi_4$ and $\sin \psi_4$ and $\cos \psi_4$ are defined as in equations (i) so that the relationship becomes

$$\frac{\partial w_4^1}{\partial n} = \frac{\partial w_4}{\partial n} + \frac{1}{2\Delta}\left(\frac{\beta_1\beta_3 + \gamma_1\gamma_3}{s_{12}}w_1 + \frac{\beta_2\beta_3 + \gamma_2\gamma_3}{s_{12}}w_2 + (s_{12})w_3\right) \tag{o}$$

with similar relationships applying for points 5 and 6.

The total deflection w is given by equations (a)–(c). With N_4^1 etc. and $\partial w_4^1/\partial n$ etc. defined by relationships of the type given by equations (d) and (o) respectively the total deflection can now be expressed in the desired form

$$w = N_1 w_1 + N_2 w_2 + N_3 w_3 + N_4 \frac{\partial w_4}{\partial n} + N_5 \frac{\partial w_5}{\partial n} + N_6 \frac{\partial w_6}{\partial n}. \tag{p}$$

The shape functions corresponding to the true element degrees of freedom are found to be

$$N_1 = L_1^2 - \frac{(\beta_2\beta_1 + \gamma_2\gamma_1)}{s_{31}^2}(L_2 - L_2^2) - \frac{(\beta_3\beta_1 + \gamma_3\gamma_1)}{s_{12}^2}(L_3 - L_3^2) \tag{q}$$

and

$$N_4 = -\frac{2\Delta}{s_{12}}L_3(1 - L_3) = N_4^1 \tag{r}$$

with the other shape functions obtained by cyclic interchange of the subscripts. The functions N_1 and N_4 are sketched in Fig. E11.3(c).

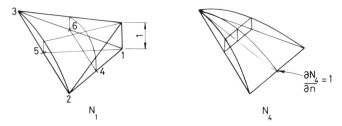

Fig. E11.3(c)

The direct construction of the element shape functions in the manner detailed here is an interesting and useful approach. Expressing w in the form of equation (a) ensures that rigid-body motions are properly included in the displacement field from the beginning and allows attention to be concentrated on the curvature-inducing part of the field (the w^1 part). Further, once this part of the field is established the detailed calculation of the element stiffness matrix is eased since only w^1 contributes to the plate curvatures. (For details of the procedure the reader is referred to reference 13 and to the related earlier reference 5.) However, it should be emphasized that the displacement field for the constant-curvature element can be established without the use of the above approach. Rather, the same result as given by equations (p), (q), and (r) can be obtained by assuming initially that the total deflection w is a complete quadratic polynomial function of the area co-ordinates, written in terms of generalized displacements (see Problem 11.11).

11.8. Numerical results for triangular elements

Figure 11.14 shows convergence curves for the five triangular elements T1–T5 described in Section 11.7 when applied to the solution of the problem of the square plate simply supported on all edges and carrying a central concentrated load. (Results for the same problem when using rectangular elements are given in Fig. 11.8(a).) The results are obtained using regular meshes of triangles (of the sort shown in the inserts to Fig. 11.15 and 11.16). We expect the conforming models T3, T4, and T5 to demonstrate monotonic convergence from the over-stiff side and this is indeed so. Further, model T2 shows convergence from the under-stiff side and this is in line with the fact that it is an equilibrium-type element,

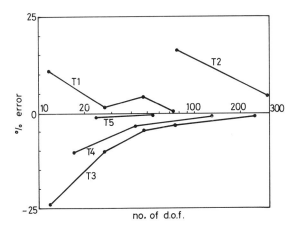

Fig. 11.14. Convergence of central deflection when using triangular elements for simply supported, square plate with central load.

although presented in this text in the guise of a non-conforming displacement element. Model T1 satisfies no bound conditions but does demonstrate convergence for meshes generated by three sets of equally spaced parallel lines as used here.

On a degree-of-freedom basis the most efficient element in this typical application is clearly the refined model T5. (The number of degrees of freedom are those in a symmetric quadrant of the plate, before applying any boundary or symmetry conditions. Remember that the number of degrees of freedom used in obtaining a solution is only a rough measure of the computing effort used since various other factors should be taken into account, such as the effort used in setting up individual element stiffness matrices and the bandwidth of the system of structure stiffness equations.) Model T5 is generally very accurate and reliable but, in common with all other plate elements, it does have some disadvantages. These are related primarily to the excessive nodal continuity associated with the use of all second derivatives of w as nodal degrees of freedom. This leads to difficulty in situations where step changes of thickness or material property occur and some curvatures are therefore not continuous in the real structure. However, Cowper, Lindberg, and Olson [18] have shown how a step change of thickness can be accommodated when using the refined element. Of the two conforming Clough–Tocher triangles T3 and T4 the latter, with its quadratic variation of normal slope at an element edge, is the more accurate but is also the less convenient because of the presence of its mid-side nodes. The simple triangle T2 does not show up at all well in the comparison of calculated deflection, but bear in mind that since this element is in reality an equilibrium element the moments rather than the displacements are primary variables.

To give some idea of the accuracy of calculated bending moment distributions (or of the related curvature quantity) some results are given in Figs. 11.15 and 11.16. In Fig. 11.15 the 'exact' bending moment distribution for a square plate with uniformly distributed loading is shown together with the calculated distributions using models T1 and T2 for the element mesh shown in the inset. Clearly, large step discontinuities are present at interelement boundaries for both models; this would have to be so, of course, for the constant-moment element T2. In interpreting the moment distributions in practice it is probably wisest to plot the values at element centroids, whether using T1 or T2. In Fig. 11.16 a comparison is made between the 'exact' distribution of a curvature quantity and that calculated using coarse meshes of model T5 elements for a simply supported square plate with central concentrated load. The accuracy of the FEM results is high with continuity of curvature (and moment) exhibited at the nodes.

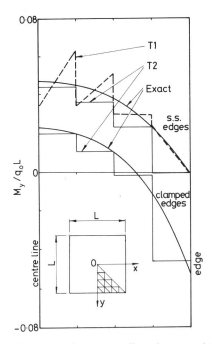

Fig. 11.15. Distribution of moment along centre line of square plate carrying a uniformly distributed load. (From reference [13].)

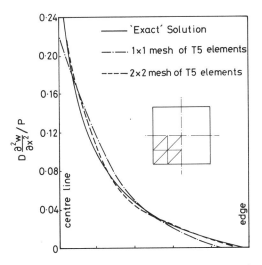

Fig. 11.16. Distribution of curvature along centre line of square plate carrying central load P. (From reference [16(b)].)

11.9. Alternative philosophies in the generation of plate finite elements

Thus far in this chapter the finite element analysis of flat plates based on the classical theory–minimum potential energy approach has been described. It has been demonstrated that the analysis is relatively complicated owing to the Kirchhoff normalcy condition and the resulting requirement for C_1 compatibility. Fully compatible plate models of this type can only be generated by either using a sub-region approach, in which the twist is not single valued at the corners, or using higher-order derivatives at the corners. The difficulty associated with the imposition of C_1 continuity has led to the development of alternative approaches to finite-element plate bending analysis. The alternative approaches generally aim to replace the C_1 continuity requirement with a C_0 continuity requirement.

Two broad avenues of approach can be identified.

In the first the conventional classical plate theory is retained and the PMPE is replaced by one or other alternative variational principles. We have seen in considering the finite-element displacement method, based on the PMPE, that the field variables are displacements, that the compatibility (and elasticity) condition is satisfied everywhere (in a rigorous formulation), and that satisfaction of the equilibrium condition is obtained approximately through application of the variational principle. The opposite side of the coin is the force method, based on the principle of minimum complementary energy, where the field variables are stresses which satisfy the equilibrium condition everywhere at the outset and satisfy the compatibility condition only approximately through the variational procedure. Between the displacement and stress approaches lies the mixed method based on a mixed variational principle, such as the Reissner principle, in which the field variables are both displacements and stresses and equal weight is given to the satisfaction of the equilibrium and compatibility conditions. (It is noted in passing that in fact the three principles that have been mentioned are themselves specializations of a very general variational principle, known as the Hu–Washizu principle, in which displacements, strains, and stresses are all field variables.) Elements for plate (and shell) analysis have been developed directly on the basis of the complementary energy principle and the Reissner principle but alternative elements also exist which are based on modified forms of the more conventional variational principles. These alternatives are known as hybrid elements and in such elements one form of field is assumed for the element interior and an independent field is defined on the element boundary. Two classes of hybrid formulation can be identified which are referred to as the hybrid-displacement formulation and the hybrid-stress formulation respectively. In the former a displacement field is chosen, written in terms of generalized parameters, to represent behaviour in the interior of the element and a second expression is assumed in terms of nodal displacements—completely independently of the interior field—to represent an interelement-compatible displacement at the element boundary. In the hybrid-stress approach an equilibrium stress field is assumed in the element interior, written in terms of generalized parameters, whilst an interelement-compatible displacement field, expressed in terms of nodal displacements, is assumed at the element boundary. The hybrid-stress method in particular has been used quite frequently in the development of plate elements. We shall not pursue further here the question of elements based on variational principles other than the PMPE except to note that a given finite element can sometimes be derived from a number of different variational principles and that some non-conforming displacement elements can be shown to be equivalent to mixed or hybrid elements. For further information on the variational principles of structural mechanics the reader is referred to the text of Washizu [19].

In the second avenue of approach the PMPE is retained whilst the constraints of the classical plate theory are relaxed. It is recalled (see Section 7.6) that the employment of the Kirchhoff hypothesis in the classical theory implies that the effect of transverse shear deformation is ignored. (Transverse shear stresses are τ_{yz} and τ_{zx}; transverse shear strains are γ_{yz} and γ_{zx}). This means, of course, that the classical plate theory is an approximate theory, although for truly thin plates the errors involved are very small. For moderately thick plates, though, the effects of transverse shear can be physically significant and should be included. Also, it has been implied earlier that relaxation of the Kirchhoff hypothesis by including transverse shear effects will have benefits related to interelement compatibility whether a plate is moderately thick or truly thin.

One obvious way of bypassing the assumptions of the classical plate theory is to treat a plate directly as a three-dimensional body, which it obviously is in actuality. This could be envisaged through the use of solid finite elements; in particular use of the serendipity isoparametric bricks shown in Fig. 10.15 might be contemplated (for shell as well as plate analysis) with the consequent advantage of their curvilinear geometry. However, there are disadvantages associated with the direct use of solid isoparametric elements in situations of thin geometry. Firstly, if the element dimension in one direction is relatively very small, as it would be in the ζ direction (see Fig. 11.17(a) which shows a 20-node quadratic element), the stiffness in this direction becomes much greater than do the stiffnesses in the other directions and this can lead to ill-conditioned equations and inaccurate results. Secondly, the element will be un-

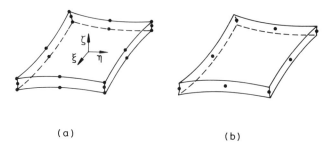

Fig. 11.17. Isoparametric element for plate analysis: (a) 60 d.o.f. solid element (nodal freedoms are u, v, w); (b) 24 d. o. f. specialized plate element (nodal freedoms are w, θ, ϕ).

economical owing to the unnecessary use of a considerable number of degrees of freedom along lines through the thickness. Some specialization of the isoparametric solid element has to be made to turn it into a satisfactory plate (and shell) element. Ahmad, Irons, and Zienkiewicz [20] tackled this problem in such a way that the total degrees of freedom of the quadratic solid element are reduced from the usual 60 to 24 for the plate (or to 40 for the shell). In this approach the direct strain in the (normal) ζ direction is ignored and the direct ξ and η strains are assumed to vary linearly through the thickness. It follows that lines which are originally normal to the mid-surface are constrained to remain straight and unchanged in length after loading. It should be noted particularly, though, that such lines are not constrained to remain normal to the mid-surface and so the element can accommodate transverse shear deformation. These assumptions allow the deformation of the quadratic element, when used for plate analysis, to be completely defined by the mid-surface nodal displacement w and by the rotations θ and ϕ of a line originally normal to the mid-surface about two mutually orthogonal axes. The element thus becomes as shown in Fig. 11.17(b). In determining element properties it should be noted that, since the strain quantities vary only linearly in the ζ direction, the integration in this direction can be performed analytically and then numerical integration is required only in the ξ and η directions over the mid-surface. When 3×3 numerical integration is used over the mid-surface the quadratic element gives good results when applied to the solution of moderately thick plate problems but fails comprehensively owing to excessive stiffness when thin geometry is considered (3×3 integration for the quadratic element is termed 'full' integration since this level of integration is sufficient to evaluate the stiffness matrix exactly if the element is of rectangular or parallelogrammic shape). However it is found that this situation can usually be very considerably improved if, paradoxically, the order of the numerical integration used in calculating the stiffness matrix is lowered. This can be done either in a general way by using 2×2 integration for calculating all components of the stiffness matrix (when the term 'reduced' integration is used) or by using 2×2 integration for the transverse shear contribution only (when the term 'selective' integration is used). The use of this simple technique, which was introduced in references 21 and 22, produces a useful and versatile element for plate bending problems; of course no bound conditions apply because of the inaccurate integration.

Another way of producing a plate element based on the PMPE but without the encumbrance of the C_1 continuity requirement of the classical theory is to use directly an existing 'improved' or 'thick' plate theory which includes the effects of transverse shear deformation. Two such theories are those of Reissner [23] and Mindlin [24], with the latter being used to a somewhat greater extent than the former. In Mindlin plate theory any line which is originally normal to the plate median surface is assumed during deformation to remain straight but not generally normal to the median surface. The effect of this is that the rotations θ and ϕ of such lines about orthogonal axes can no longer be expressed solely in terms of derivatives of the deflection w. Thus these rotations have to be interpolated independently of w in a finite-element approach and so three reference quantities (w, θ, and ϕ) are involved (requiring C_0 continuity) instead of the one (requiring C_1 continuity) of the classical theory. For those readers familiar with Timoshenko beam theory it is noted that Mindlin plate theory is the two-dimensional equivalent of this beam theory. (A range of Timoshenko beam finite elements has been produced in the literature; one of these is presented, and most of the others are discussed, in reference 25.) Early work concerning finite elements based on improved plate theories related to straight-sided rectangles and triangles (see for example references 26 and 27), whereas more recently a range of curved-sided elements has been developed based on the isoparametric concept and employing both serendipity and Lagrangian formulations as described by Hinton and Bicanic [28]. The eight-node quadratic serendipity element is obviously very closely related to the specialized solid element described above and illustrated in Fig. 11.17(b). Efficient performance in solving thin-plate problems using the Mindlin plate elements depends again on the use of either reduced or selective numerical integration. Surprisingly the Lagrangian elements appear superior to the serendipity elements in the plate applications. There are some difficulties which can occur which are associated with mechanisms or near mechanisms, that is zero or low energy modes of deformation, but these appear to be avoided in so-called 'heterosis' elements by judicious use of both serendipity and Lagrange shape functions [29].

In this section only the briefest and most general descriptions have been given of some (but not all) of the possible approaches which are available for the analysis of plate bending problems as alternatives to the conventional classical plate theory–PMPE approach described at greater length earlier in this chapter. The reader interested in pursuing these alternative approaches will find much further information concerning them in the texts of Zienkiewicz [30], Gallagher [31], and Cook [32].

11.10. Brief remarks on shell analysis

Thin curved shell structures constitute possibly the most difficult class of structure to analyse by the finite-element method and the difficulties involved have led to the development of a considerable variety of approaches to the problem and of a large number of element types. A comprehensive description of the subject of shell finite-element analysis clearly lies outside the scope of the present volume. However, some very brief remarks are appropriate and will be made here in very general terms to give the reader some small idea of the work that has been done in this area of finite-element analysis.

The difficulties involved in shell analysis can be readily imagined if some thought is given to the range of possible shell geometries and possible shell responses to loading. The shell geometry will generally be doubly curved with the initial radii of curvature varying over the shell surface and with a thickness which may be only moderately small or may be extremely small compared to these radii. The response of the shell to loading usually involves a combination of bending and membrane actions and these actions are coupled together throughout the shell owing to its curvature. Depending upon the shape of the shell, upon its boundary conditions, and upon the nature of its loading, the response may be dominated by bending behaviour or by membrane behaviour, or each of these types of behaviour may in turn be dominant in different regions of the shell.

Three broad categories of approach to finite-element shell analysis will now be outlined.

One approach, in which the complexities of curved shell theory can be avoided, is what is known as the 'facet' approach. This was the first approach investigated in finite-element shell analysis, and in it the actual curved shell is modelled with an assemblage of flat elements (or facets) located such that their vertices lie in the middle surface of the actual shell, as is shown in Fig. 11.18. Clearly the most useful element shape is the triangle. The fundamental advantage of the facet approach, which follows from the fact that the element is flat, is that bending and membrane behaviour are completely uncoupled at the element level. This means that a facet element is simply a combination of a plate bending element and a membrane element; representation of rigid-body motion and constant

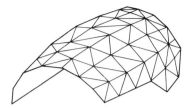

Fig. 11.18. Facet element representation of a shell.

strain states presents no difficulty since if these are present in both the bending element and the membrane element they will automatically be present in the flat facet element. The coupling between bending and membrane actions which occurs continuously over the middle surface of the actual curved shell only occurs in the facet approach when adjacent elements lying in different planes are joined in the usual element assembly procedure. The question now becomes: Which bending and membrane elements should be combined together to form facet elements? One consideration which might be thought to affect the choice is that of interelement compatibility but, surprisingly, this is not really so. Compatibility is almost always violated in an assembly of facet elements and usually discontinuities of displacement as well as of normal slope are present. Despite this blatant disregard of the accepted rules of admissibility, convergence to correct results does always seem to occur, presumably because the discontinuities progressively reduce with mesh refinement. If we accept that convergence will occur despite the discontinuities then we have a fairly free hand in choosing how to combine available assumed displacement bending and membrane elements. A range of facet elements has been developed but generally it pays to use fairly low order elements, so that a large number of elements can be used to approximate the curved shell geometry better. As an example the simplest facet element [33], which has only 12 degrees of freedom, is formed by combining the CST membrane element and the T2 bending element, as shown in Fig. 11.19.

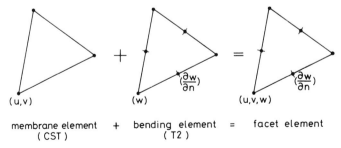

Fig. 11.19. A simple facet shell element.

Of course the stiffness equations of all facets have to be transformed from their local configuration to a common global configuration before assembly of the structure stiffness equations. With some facet elements which use rotations as degrees of freedom at the element vertices a difficulty can occur when all elements meeting at a particular nodal point are coplanar since a null stiffness is then associated with the rotation about the axis normal to the common plane. (This is avoided in the element shown in Fig. 11.19.) Despite its obvious shortcomings the facet approach retains considerable popularity, mainly because of the difficulties associated with other approaches.

A second approach is to use elements of curved geometry whose properties are based directly on thin-shell theory. This certainly is a superior approach in principle since it avoids the major physical idealization errors which are present in the facet approach. However, the better physical idealization of the shell obtained by using curved elements will not necessarily yield a better structural model since success is very dependent upon the nature of the assumed displacement field and the approach is inevitably much more complicated than is the facet approach. The complexity of curved shell finite-element analysis starts with the background shell theory and, in fact, there exists a variety of shell theories rather than a single generally accepted one (Kraus [34] gives a description of some of these). Usually the shell behaviour is described with reference to a system of curvilinear co-ordinates α and β embedded in the middle surface of the shell (see Fig. 11.20) and to 'surface' displacement components u, v, and w which act tangentially to the surface (u and v) and normal to it (w). Also the assumption that shell normals remain normal during deformation is commonly employed, with consequent requirement in a PMPE approach for C_1 continuity as far as w is concerned; thus at interelement boundaries the requirement is for continuity of u, v, w, and $\partial w/\partial n$ and this poses difficulties which clearly are at least as great as those described earlier in discussing plate elements.

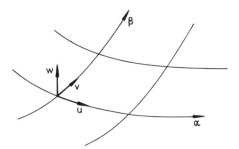

Fig. 11.20. Curvilinear co-ordinates and surface displacements.

In curved elements coupling exists between bending and membrane behaviour throughout the element and the initial curvature of the element leads to expressions for rigid-body motion states, for example, which are extremely complicated, at least for arbitrary geometry, and which conflict with interelement compatibility requirements. Such conflict can be avoided by seeking to approximate closely, rather than represent precisely, the rigid-body motion states. For deep thin geometry efficient solution requires the use of high-order polynomial representations for each of u, v, and w, typically of quintic order and using higher derivatives as nodal degrees of freedom, as in the representation of w in the triangular plate elements shown in Fig. 11.13. In view of the complications of shell theory it is no surprise to find that most of the many shell elements described in the literature are for shells of specialized geometrics form, such as cylindrical shells, shells of revolution, shallow shells, etc., although there do exist some shell elements of more general geometric form. The remarks given here have implied use of the assumed displacement–PMPE approach and this indeed is much the most popular approach. However, some elements have also been derived based on the use of shell theory in conjunction with other variational principles, as in the case of plate elements.

A third approach is the extension to shell analysis of the approach described in the previous section for plate analysis whereby the specialized quadratic isoparametric solid element, or Ahmad element, is used. The quadratic shell element looks like the element depicted in Fig. 11.17(b) but now the mid-surface is curved rather than flat, of course, and 40 rather than 24 degrees of freedom are involved. The degrees of freedom at each node are the three Cartesian components of displacement u, v, and w and the two rotations θ and ϕ. The remarks made earlier for the plate version of this element with regard to numerical integration of element properties still broadly apply and reduced integration is almost always used. The type of formulation described here (which need not only apply to the quadratic element) provides a popular approach to the solution of shell problems because of its versatile geometric form. It should be borne in mind, though, that the effect of reduced integration is going to be particularly problem dependent for shells since membrane and bending effects might be relatively of vastly different orders of importance in moving from one situation to another. Where membrane effects dominate it might well be that full integration will sometimes provide superior results to reduced integration.

Finally, it is noted that there exists a considerable amount of literature concerning finite-element shell analysis which it is not possible to detail here. For the reader wishing to explore the matter further references 35 and 36 give reviews of early work, reference 37 is a later specialist book on the topic (produced from the proceedings of a conference), and some general finite-element texts such as references 30, 32, and 38 also contain considerable information on the topic.

References

1 ADINI, A. and CLOUGH, R. W. Analysis of plate bending by the finite element method. *Report, Grant G 7337*. National Science Foundation, U.S.A. (1961).
2 MELOSH, R. J. Basis for derivation of matrices for the direct stiffness method. *AIAA J.* **1,** 1631–7 (1963).
3 ZIENKIEWICZ, O. C. and CHEUNG, Y. K. The finite element method for analysis of elastic isotropic and orthotropic slabs. *Proc. Inst. Civ. Eng.* **28,** 471–88 (1964).
4 DAWE, D. J. A finite element approach to plate vibration problems, *J. Mech. Eng. Sci.* **7,** 28–32 (1965).
5 BAZELEY, G. P., CHEUNG, Y. K., IRONS, B. M., and ZIENKIEWICZ, O. C. Triangular elements in bending-conforming and non-conforming solutions. *Proc. Conf. on Matrix Methods in Structural Mechanics, Air Force Institute of Technology, Wright–Patterson Air Force Base*, Dayton, Ohio (1965).
6 TIMOSHENKO, S. and WOINOWSKY-KRIEGER, S. *Theory of plates and shells*, 2nd edn. McGraw-Hill, New York (1959).
7 DAWE, D. J. On assumed displacements for the rectangular plate bending element, *J. R. Aeronaut. Soc.* **71,** 722–4 (1967).
8 IRONS, B. M. and DRAPER, K. J. Inadequacy of nodal connections in a stiffness solution for plate bending. *AIAA J.* **3,** 961–3 (1965).
9 BOGNER, F. K., FOX, R. L., and SCHMIT, L. A. The generation of interelement-compatible stiffness and mass matrices by the use of interpolation formulas. *Proc. Conf. on Matrix Methods in Structural Mechanics, Air Force Institute of Technology, Wright–Patterson Air Force Base*, Dayton, Ohio (1965).
10 DEAK, A. L. and PIAN, T. H. Application of the smooth surface interpolation to the finite element analysis. *AIAA J.* **5,** 187–9 (1967).
11 DAWE, D. J. Parallelogrammic elements in the solution of rhombic cantilever plate problems. *J. Strain Anal.* **1,** 223–30 (1966).
12 CLOUGH, R. W. and TOCHER, J. L. Finite element stiffness matrices for analysis of plate bending. *Proc. Conf. on Matrix Methods in Structural Mechanics, Air Force Institute of Technology, Wright–Patterson Air Force Base*, Dayton, Ohio (1965).
13 MORLEY, L. S. D. On the constant moment plate bending element. *J. Strain Anal.* **6,** 20–4 (1971).
14 ARGYRIS, J. H., FRIED, I., and SCHARPF, D. W. The TUBA family of plate elements for the matrix displacement method. *J. R. Aeronaut. Soc.* **72,** 701–9 (1968).
15 BELL, K. A refined triangular plate bending finite element, *Int. J. Num. Meth. Eng.* **1,** 101–22 (1969).
16 (a) COWPER, G. R., KOSKO, E., LINDBERG, G. M. and OLSON, M. D. Static and dynamic applications of a high-precision triangular plate element. *AIAA J.* **7,** 1957–65 (1969).
   (b) COWPER, G. R., KOSKO, E., LINDBERG, G. M., and OLSON, M. D. A

high precision triangular plate bending element. *Aeronautical Rept.* LR-514, National Research Council of Canada (1968).

17 BUTLIN, G. and FORD, R. A compatible triangular plate bending finite element. *Int. J. Solids Struct.* **6,** 323–32 (1970).

18 COWPER, G. R., LINDBERG, G. M., and OLSON, M. D. *Int. J. Num. Meth. Eng.* **2,** 453–4 (1970).

19 WASHIZU, K. *Variational methods in elasticity and plasticity,* 2nd edn. Pergamon, Oxford (1975).

20 AHMAD, S., IRONS, B. M., and ZIENKIEWICZ, O. C. Analysis of thick and thin shell structures by curved elements. *Int. J. Num. Meth. Eng.* **2,** 419–51 (1970).

21 ZIENKIEWICZ, O. C., TAYLOR, R. L., and TOO, J. M. Reduced integration technique in general analysis of plates and shells. *Int. J. Num. Meth. Eng.* **3,** 275–90 (1971).

22 PAWSEY, S. and CLOUGH, R. W. Improved numerical integration of thick shell finite elements, *Int. J. Num. Meth. Eng.* **3,** 575–86 (1971).

23 REISSNER, E. The effect of transverse shear deformation on the bending of elastic plates. *J. Appl. Mech.* **12,** A69–A77 (1945).

24 MINDLIN, R. D. Influence of rotary inertia and shear on flexural motions of isotropic elastic plates. *J. Appl. Mech.* **18,** 31–8 (1951).

25 DAWE, D. J. A finite element for the vibration analysis of Timoshenko beams, *J. Sound Vib.,* **60,** 11–20 (1978).

26 PRYOR, C. W., BARKER, R. M., and FREDERICK, D. Finite element bending analysis of Reissner plates, *Proc. Am. Soc. Civ. Eng.* **96** (EM6), 967–83 (1970).

27 GREIMANN, L. F. and LYNN, P. P. Finite element analysis of plate bending with transverse shear deformation. *Nucl. Eng. Des.,* **14,** 223–30 (1970).

28 HINTON, E. and BICANIC, N. A comparison of Lagrangian and serendipity Mindlin plate elements for free vibration analysis. *Comput. Struct.,* **10,** 483–93 (1979).

29 HUGHES, T. J. R. and COHEN, M. The 'heterosis' finite element for plate bending. *Comput. Struct.* **9,** 445–50 (1978).

30 ZIENKIEWICZ, O. C. *The finite element method,* 3rd edn. McGraw-Hill, New York (1977).

31 GALLAGHER, R. H. *Finite element analysis: Fundamentals.* Prentice-Hall, Englewood Cliffs, NJ. (1975).

32 COOK, R. D. *Concepts and applications of finite element analysis.* Wiley, New York (1974).

33 DAWE, D. J. Shell analysis using a simple facet element. *J. Strain Anal.* **7,** 266–70 (1972).

34 KRAUS, H. *Thin elastic shells.* Wiley, New York (1967).

35 GALLAGHER, R. H. Analysis of plate and shell structures. *Proc. Conf. on Application of Finite Element Method in Civil Engineering, Vanderbilt University* (1969).

36 DAWE, D. J. Curved finite elements in the analysis of shell structures. *Proc. 1st Conf. on Structural Mechanics in Reactor Technology, Berlin,* Paper J1/4. Commission of the European Communities, Luxembourg (1971).

37 ASHWELL, D. G. and GALLAGHER, R. H. (eds.) *Finite elements for thin shells and curved members.* Wiley, New York (1976).

38 BREBBIA, C. A. and CONNOR, J. J. *Fundamentals of finite element techniques.* Butterworths, London (1973).

Problems

11.1. Show that for element R3 the consistent load column matrix $\mathbf{Q}$ for a uniformly distributed loading is the same as that given for element R1 in equation (11.22).

11.2. Solve Example 11.1 using element R3.

11.3. Solve Example 11.2 using element R3.

11.4. For element R1 (see Fig. 11.1) derive column matrix $\mathbf{Q}$ for the following loading cases;
  (i) a distributed lateral loading acting over the whole element whose intensity varies linearly with x from the value q per unit surface area at edge 1–4 to $2q$ at edge 2–3;
  (ii) a concentrated lateral load W acting at the point $x = +a/2$, $y = +b/2$.

11.5. Consider a typical edge of rectangular element R1, say edge 2–3 of the element shown in Fig. 11.1. Determine explicitly, in terms of the element nodal freedoms, how w and θ vary along this edge.

11.6. Repeat Problem 11.5 for element R4.

11.7. A uniform-thickness square cantilever plate (i.e. with one edge clamped and the others free) of side length A is subjected to a uniform normal loading of intensity q per unit surface area. Model the whole plate using a single R1 element and calculate the displacements at the free corners. It should be noted that by taking advantage of the problem symmetry the number of equations to be solved is three. Take $\nu = 0.3$.

11.8. A square plate of side length A has three edges clamped and the remaining edge free. It is subjected to a uniform normal loading of intensity q per unit surface area. Model a symmetric half of the plate using a single rectangular plate element and determine the lateral deflection at the centre of the free edge. Use the R1 and R3 elements in turn. Take $\nu = 0.3$.

11.9. Determine the column matrix $\mathbf{Q}$ for a uniformly distributed lateral load acting over the surface area of element R4.

11.10. In rectangular element R5 show that C_1 continuity is achieved at the boundaries of the triangular subregions by considering a typical boundary, say that between subregions (i) and (ii) where the boundary is defined by the line $X = Y$. Show also that the twist has different values in subregions (i) and (ii) at node 3, and that at an exterior edge w and the normal slope vary as cubic and linear polynomials respectively.

11.11. In Example 11.3 the shape functions were deduced for the constant moment triangular plate element T2. Confirm that these shape functions can be obtained, albeit not without considerable difficulty, by assuming initially that w is the following complete quadratic polynomial function of the area co-ordinates:

$$w = a_1 L_1^2 + a_2 L_2^2 + a_3 L_3^2 + a_4 L_1 L_2 + a_5 L_1 L_3 + a_6 L_2 L_3.$$

12

VIBRATION AND BUCKLING:
THE EIGENVALUE PROBLEM

12.1. Introduction

THE predominant concern in preceding chapters has been with the linear elastic analysis of structures subjected to a static dead loading whose nature is such as not to lead to structural instability. This linear static type of analysis is much the most common category in the realm of structural analysis, but there frequently occur situations where a different type of analysis is needed and where the versatile FEM can again be used to good effect. The full range of such situations is considerable, including non-linear behaviour of the geometric (large deflection) and physical (plasticity) kinds, and will not be explored here. Rather, attention in this chapter will be directed to two specific types of problem which retain the essentially linear nature which characterizes the rest of this text.

This first type of problem comes from the realm of structural dynamics. This subject is one of considerable breadth and clearly cannot be treated in other than a limited fashion here. In fact, in the early part of this chapter, attention is restricted to the problem of calculating the natural frequencies and corresponding mode shapes of free undamped vibration of common structural components and forms. This requires the development for each finite element in the model structure of a mass matrix which will represent the effect of the dynamic loading (proportional to the square of the natural frequency) which is set up during vibration. The mass matrix is used in conjunction with the conventional stiffness matrix in formulating a set of n algebraic equations which govern the structural behaviour. These equations are of what is known as the eigenvalue type; their solution yields n values of the square of natural frequency (the eigenvalues or characteristic values) and n corresponding sets of nodal degrees of freedom (the eigenvectors) which define the deformed shape of the structure when vibrating at a particular natural frequency.

The second type of problem comes from the realm of structural stability. Again this subject is very broad and only a very limited aspect of it can be considered in this chapter. In general, structural instability problems are non-linear but an important category of linear problem exists of the kind typified by the problem of the lateral buckling under axial compression of a perfect pin-ended column or strut at the so-called Euler load. It is this category of problem which is considered in the latter part of this chapter, with relation to flat plates as well as to struts. The linear stability problem is accommodated in the FEM by the introduction of what is known as a geometric stiffness matrix, which accounts for the effect that an axial (for the strut) or in-plane (for the plate) loading has on conventional bending stiffness: the effective bending stiffness vanishes at a buckling load. In common with the free vibration problem, the linear stability problem is an eigenvalue problem, with the eigenvalues now being the critical values of loading magnitude at which buckling occurs; usually only the lowest of these is of practical interest.

The reader is assumed to have at least some slight acquaintance with the two subject problems, and some introductory material is presented for both before considering the finite-element approach. However, the reader may require more background reading and, if so, is recommended to consult relevant texts such as references 1–4 for the vibration problem and references 5–7 for the buckling problem.

12.2. Free undamped vibration of a spring–mass system

We begin our study of the vibration problem by considering the very simple one-degree-of-freedom mathematical model shown in Fig. 12.1(a). This system comprises an elastic spring of stiffness k, which is assumed to have negligible mass, to which is attached a rigid body of mass m resting on a frictionless surface. The system is in static equilibrium in its datum position and then a disturbance is applied by displacing the mass horizontally and releasing it, whereafter the system vibrates indefinitely in the absence of any damping.

Consider an instant during the vibration when the displacement of the body is u and its acceleration is $a = \mathrm{d}^2u/\mathrm{d}t^2 = \ddot{u}$. (The dots represent differentiation with respect to time t.) At this instant the spring force is ku. Newton's second law of motion for a particle, or for a rigid body of

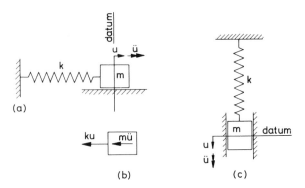

Fig. 12.1. Simple spring-mass system.

the type considered here, states that force equals mass times acceleration, or

$$F = ma. \tag{12.1}$$

Acceleration, like force, is a vector quantity of course.

For the system shown in Fig. 12.1(a), in which only horizontal motion takes place, application of Newton's Law gives

$$-ku = m\ddot{u} \tag{12.2}$$

where the negative sign associated with the spring force is present since the force acts in the direction opposite to that of the assumed motion.

It is more convenient for our purpose to adopt the modified view of equation (12.2) which is known as *d'Alembert's principle*. This principle states that a system is in a state of *dynamic equilibrium* under the action both of external forces and of so-called inertia (or inertial) forces which are products of mass and acceleration. An inertia force acts in the direction opposite to that of the relevant assumed acceleration, that is opposite to the relevant co-ordinate direction. For the system shown in Fig. 12.1(a) we apply d'Alembert's principle by drawing a free-body diagram for the mass, as in Fig. 12.1(b), on which we show acting the spring force ku and the inertia force $m\ddot{u}$, the latter directed in the negative u sense. The inertia force is now treated as just another static force, and the equation of equilibrium in the horizontal direction is therefore

$$m\ddot{u} + ku = 0 \tag{12.3}$$

which is identical with equation (12.2) of course.

Incidentally, it is noted that the system shown in Fig. 12.1(c), in which the spring is vertical, is effectively identical with that shown in Fig. 12.1(a). Equation (12.3) still applies if u is measured from the datum position of static equilibrium in which the weight of the body is balanced by an initial spring force.

Equation (12.3) is a homogeneous second-order linear differential equation with constant coefficients which has the simple standard solution

$$u = A \sin(\omega t + \phi). \tag{12.4}$$

This solution shows that in free undamped vibration the motion is a *simple harmonic motion* (SHM) in which u varies with t as shown in Fig. 12.2. The quantity A is the amplitude of vibration, that is the maximum displacement from the datum position. The quantity ω is the circular natural frequency of vibration of the system, which is measured in radians

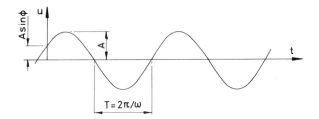

Fig. 12.2. Harmonic motion in the free vibration problem.

per second and has the value, for the system shown in Fig. 12.1, of

$$\omega = \sqrt{\frac{k}{m}}.$$ (12.5)

The quantity ϕ is a phase angle whose value depends on the conditions pertaining at the onset of vibration. This angle need not concern us here and can be eliminated from consideration if we assume that $u = 0$ at $t = 0$; then

$$u = A \sin \omega t.$$ (12.6)

It should be noted that in a SHM defined by equation (12.6) the velocity at time t is

$$\dot{u} = A\omega \cos \omega t$$ (12.7)

and the acceleration is

$$\ddot{u} = -A\omega^2 \sin \omega t = -\omega^2 u.$$ (12.8)

Two other quantities which are characteristic of the free vibration problem are recalled in passing. These are the period T, which is the time required for one complete cycle of vibration, and the frequency f, which is the number of cycles of vibration occurring in unit time (f is usually given in hertz (Hz) where $1 \text{ Hz} = 1 \text{ cycle s}^{-1}$). In terms of ω

$$T = \frac{2\pi}{\omega} \qquad f = \frac{1}{T} = \frac{\omega}{2\pi}.$$ (12.9)

In the one-degree-of-freedom problem considered here it has been a simple matter to determine the (single) circular natural frequency of vibration (equation (12.5)). For more complicated situations we shall be working with mathematical models of structures which have distributed elastic and inertia properties, based on the use of a finite number of degrees of freedom. In the model structure there will exist as many frequencies of vibration, each with an associated (mode) shape of vibration, as there are degrees of freedom. (In the real structure there are an infinite number of natural frequencies.) As with the static problem, to obtain (approximate) results for multi-degree-of-freedom problems recourse has to be made to work or energy principles. There are a number of related ways in which the problem can be set up and some of these will now be outlined for the one-degree-of-freedom system; all will give exactly the same result for the natural frequency.

Consider again the problem shown in Fig. 12.1. We have seen that d'Alembert's principle effectively converts the dynamic problem into a static problem by incorporating the inertia force into an equilibrium equation. Clearly, use of d'Alembert's principle will allow a dynamic version of the PVD, discussed in Section 5.3, to be established by the simple expedient of including the virtual work of inertia forces. In the given problem we assume from the outset that the body is vibrating in SHM with a circular frequency ω which we wish to calculate. At any displacement u from the datum position the forces acting on the body when it is in dynamic equilibrium are the spring force ku and the inertia force $m\ddot{u}$, both acting from right to left. However, with the assumption of SHM ($\ddot{u} = -\omega^2 u$) the inertia force becomes $m\omega^2 u$ acting from left to right. Now impose a virtual displacement δu (in the positive sense of u) from this equilibrium position. Applying the PVD means setting the resulting work done to zero; that is

$$\delta u(m\omega^2 u - ku) = 0$$

and since δu is arbitrary

$$m\omega^2 u - ku = 0$$

or

$$(k - \omega^2 m)u = 0 \tag{12.10}$$

Hence $\omega = \sqrt{(k/m)}$ as before.

The PMPE described in Section 5.4 and following sections in relation to static problems can also be extended to deal with the free vibration problem [8]. The change in total potential energy of a system in moving from a datum position is given in equation (5.25) as the summation of the changes of the strain energy U_p and of the potential energy of the loads V_p. For static loading V_p is the negative of the sum of all products of external load and corresponding displacement at the load position measured in the direction that the load acts (see equation (5.26)). For dynamic loading the potential energy of the inertia loads is not given in this way since these loads do not now have constant value throughout the displacement. For the system shown in Fig. 12.1 the change in potential energy of the inertia load in moving during free vibration from the datum position with $u = 0$ through intermediate values u_1 to the position u is

$$V_{pv} = -\int_{u_1=0}^{u} (-m\ddot{u}_1)\, du_1 = -m\omega^2 \int_{u_1=0}^{u} u_1\, du_1 = -\tfrac{1}{2}m\omega^2 u^2. \qquad (12.11)$$

Now the change in strain energy in moving from the datum position to the position u is $U_p = \tfrac{1}{2}ku^2$ and therefore the change in total potential energy is

$$\begin{aligned}
\Pi_p &= U_p + V_{pv}\\
&= \tfrac{1}{2}ku^2 - \tfrac{1}{2}m\omega^2 u^2.
\end{aligned} \qquad (12.12)$$

Application of the minimizing procedure (with ω not subject to variation of course) gives

$$\Pi_p = \text{minimum} \qquad (12.13)$$

or

$$\frac{\partial \Pi_p}{\partial u} = \frac{\partial U_p}{\partial u} + \frac{\partial V_{pv}}{\partial u} = ku - \omega^2 mu = 0$$

which gives the same result for ω as obtained above. (It should be noted that u is itself a sinusoidal function of time, from equation (12.6), but that the result given here does apply for any finite value of u, including in particular the case when u equals the amplitude of vibration A.)

In what follows the emphasis in solving free vibration problems will be placed on the extended PMPE approach just referred to. It is noted, though, that the concept of kinetic energy could equally well be incorporated in the analysis. The expression for V_{pv} (equation (12.11)) can be further expressed using equations (12.6) and (12.7) as

$$\begin{aligned}
V_{pv} &= -\tfrac{1}{2}m\omega^2 u^2 = -\tfrac{1}{2}m\omega^2(A^2 \sin^2\omega t)\\
&= -\tfrac{1}{2}m(\omega A)^2 + \tfrac{1}{2}m(\omega A \cos \omega t)^2 = -\tfrac{1}{2}m\dot{u}_{max}^2 + \tfrac{1}{2}m\dot{u}^2
\end{aligned} \qquad (12.14)$$

Here $\dot{u}$ is the velocity of the body at general position u and $\dot{u}_{max}$ is the maximum velocity at the datum position $u = 0$. Thus it is seen that V_{pv} represents the difference in kinetic energy values between the general and datum positions. (If the general position is specialized to be the extreme position where u has the value of the amplitude A, then V_{pv} becomes simply the negative of the maximum kinetic energy.) It is also noted that the well-known Hamilton (variational) principle, or the associated Lagrange equations [1, 2, 8], could also be used as the basis for the calculation of the natural frequencies; for free vibration the Lagrangian function is the kinetic energy minus the strain energy.

12.3. Finite-element formulation of the free vibration problem

In the general problem, unlike the situation considered in Section 12.2, a real structure and (usually) the mathematical model used to represent it will have distributed elasticity and inertia. In the finite-element approach the problem will of course be formulated finally in terms of a finite number of structure degrees of freedom. We choose to use the extended PMPE approach described in Section 12.2 to study the free vibration problem; equations (12.12) and (12.13) still apply for the multi-degree-of-freedom situation, with U_p and V_{pv} expressed as functions of the structure nodal degrees of freedom.

For the general three-dimensional body, any point in the body will experience harmonic movement during free vibration in each of the three orthogonal directions x, y, and z. Let the mass per unit volume at a point be ρ and consider a finite element of volume V_e which forms part of such a body. The strain energy U_p is unchanged from static analysis whilst the new energy term V_{pv} becomes, by analogy with equation (12.11),

$$-V_{pv} = \tfrac{1}{2}\omega^2 \int_{V_e} (u^2 + v^2 + w^2)\rho \, dV_e \tag{12.15}$$

or

$$-V_{pv} = \frac{1}{2}\omega^2 \int_{V_e} [u \quad v \quad w] \begin{Bmatrix} u \\ v \\ w \end{Bmatrix} \rho \, dV_e$$

$$= \frac{1}{2}\omega^2 \int_{V_e} \mathbf{u}^t \mathbf{u} \rho \, dV_e. \tag{12.16}$$

Since the displacement field of a finite element is defined spatially at any time t by

$$\mathbf{u} = \mathbf{Nd}$$

this becomes

$$-V_{pv} = \tfrac{1}{2}\omega^2 \mathbf{d}^t \mathbf{md} \tag{12.17}$$

where

$$\mathbf{m} = \int_{V_e} \mathbf{N}^t \mathbf{N}\rho \, dV_e. \tag{12.18}$$

The matrix $\mathbf{m}$ is known as the element *consistent mass matrix* (or sometimes as the *inertia matrix*). Like the element stiffness matrix it is a square symmetric matrix. The energy V_{pv} (equation (12.17)) is seen to be a quadratic function of the nodal displacements, in much the same way as is U_p (see equation (6.10)).

If the displacement field is expressed in terms of generalized displacements, rather than directly in shape function form, then in equation (12.18) $\mathbf{N}$ can be replaced by $\boldsymbol{\alpha}\mathbf{C}^{-1}$ and an alternative expression for the element consistent mass matrix is

$$\mathbf{m} = (\mathbf{C}^{-1})^{t}\left(\int_{V_e} \boldsymbol{\alpha}^{t}\boldsymbol{\alpha}\rho \; \mathrm{d}V_e \right)\mathbf{C}^{-1}. \tag{12.19}$$

The consistent mass matrix, as defined by either of equations (12.18) and (12.19), replaces the effects of distributed mass or inertia throughout an element by a set of masses or inertias at the nodes in a manner which is 'consistent' with the variational principle: the same shape functions are used to derive the mass matrix as are used to derive the stiffness matrix. This is the usual procedure for finite-element analysis but it is noted that a more intuitive approach was used in early work and is still used to a considerable extent today. This intuitive approach is the *lumped mass* approach whereby the distributed mass is 'lumped' at the nodes in a manner which seems physically reasonable. When using simple elements this approach is quite reasonable and has some advantages with regard to economy of computational effort. Broadly speaking it is the consistent mass approach which will be used here although some further consideration is given to the lumped mass approach later in this chapter.

Consider now an assemblage of E elements and use the appellation e to refer to a typical element, as in Chapter 6, Section 6.6. The total energy terms for the whole structure, denoted $\bar{\bar{U}}_{p}$ and $\bar{\bar{V}}_{pv}$, are simply summations of the element terms, and thus

$$\bar{\bar{\Pi}}_{p} = \bar{\bar{U}}_{p} + \bar{\bar{V}}_{pv} = \sum_{e=1}^{E} U_{p}^{e} + \sum_{e=1}^{E} V_{pv}^{e}. \tag{12.20}$$

In the manner of Section 6.6, and employing the same notation, this can be expressed as

$$\bar{\bar{\Pi}}_{p} = \tfrac{1}{2}\mathbf{D}^{t}\left(\sum_{e=1}^{E} \mathbf{k}_{e}^{*} \right)\mathbf{D} - \tfrac{1}{2}\omega^{2}\mathbf{D}^{t}\left(\sum_{e=1}^{E} m_{e}^{*} \right)\mathbf{D} \tag{12.21}$$

where expansion of element matrices to structure size is indicated. The stiffness and consistent mass matrices for the complete structure are symbolically represented as

$$\mathbf{K} = \sum_{e=1}^{E} \mathbf{k}_{e}^{*} \qquad \mathbf{M} = \sum_{e=1}^{E} \mathbf{m}_{e}^{*}. \tag{12.22}$$

What this means in practice is simply that the structure consistent mass matrix is assembled from the element matrices by exactly the same direct procedure used to assemble the stiffness matrix. The structure consistent mass matrix will therefore have a similar arrangement of terms as does the structure stiffness matrix.

Performing the minimizing procedure (equation (12.13)) whilst at the same time taking account of prescribed geometric boundary conditions (that is setting certain nodal degrees of freedom to zero) gives the reduced set of n equations:

$$(\mathbf{K}_r - \omega^2 \mathbf{M}_r)\mathbf{D}_r = \mathbf{0}. \tag{12.23}$$

This set of equations has the form of a *characteristic value* or *eigenvalue* problem. If $\mathbf{D}_r$ is non-zero, as it would be if vibration is taking place, the set of equations is only satisfied if the determinant of the term within parentheses in equation (12.2) is zero, i.e. if

$$|\mathbf{K}_r - \omega^2 \mathbf{M}_r| = 0. \tag{12.24}$$

This is called the *characteristic equation*. When expanded the given determinant yields in general a polynomial equation of degree n in ω^2 whose n roots are the eigenvalues, that is the squares of the natural frequencies ω_i^2 ($i = 1, 2, \ldots$). Associated with each particular eigenvalue will be a corresponding *eigenvector* $\mathbf{D}_{ri}$. This is a particular set of nodal displacements which describes the shape that the structure takes up when vibrating at the particular (ith) natural frequency, i.e. describes the particular *natural mode shape of vibration*. For any particular calculated natural frequency ω_i the set of equations (12.23) can be used to determine the components of the particular mode shape $\mathbf{D}_{ri}$. These components of displacement in a natural mode of vibration will not be determined as absolute values but rather will be determined simply in proportion to one another. Thus each eigenvector contains an arbitrary constant whose magnitude may be chosen so that the eigenvector has some desired numerical property; when this is done the eigenvector is said to be *normalized* and the modes are called *normal* modes. The desired numerical property might be simply that the component of the eigenvector with the largest magnitude is given the value unity. Another commonly used method of normalization is to choose the constant such that the condition $\mathbf{D}_{ri}^t \mathbf{M}_r \mathbf{D}_{ri} = 1$ holds.

The type of analysis described here can be used to determine the natural frequencies and modes of vibration of any kind of structure, with the use of mass matrices developed, through use of equation (12.18) or equation (12.19) for any of the range of finite elements discussed in previous chapters. In general the vibration of thin or slender structures is of greatest practical interest and thus it is beam, framework, plate, and shell finite elements which are perhaps of particular interest in free vibration analysis. In what follows attention will be paid to the first three

of these categories of elements. The usual purpose of the analysis is to
determine a certain number of the lowest natural frequencies and associated mode shapes since the lowest frequencies are most readily
excited and are thus of most importance.

12.4. Lateral vibration of beams

The basic prismatic Bernoulli–Euler beam element was described in
Chapter 6, Section 6.3, and is illustrated in Fig. 6.2. The consistent mass
matrix corresponding to lateral vibration can be determined in
straightforward fashion using equation (12.18) or equation (12.19), with
the matrices $\boldsymbol{\alpha}$, $\mathbf{C}^{-1}$, and $\mathbf{N}$ occurring in the expression for the lateral
deflection v given for this element in equations (6.19), (6.22), and (6.24).
If equation (12.19) is chosen as the basis for the development of $\mathbf{m}$ then
with $dV_e = A\,dx$, where A is the uniform cross-sectional area, we have

$$\mathbf{m} = (\mathbf{C}^{-1})^{t}\rho A \left(\int_0^l \begin{Bmatrix} 1 \\ x \\ x^2 \\ x^3 \end{Bmatrix} [1 \quad x \quad x^2 \quad x^3]\,dx \right)\mathbf{C}^{-1} \tag{12.25}$$

or

$$\mathbf{m} = \rho A\,(\mathbf{C}^{-1})^{t} \begin{bmatrix} l & & \text{Symmetric} \\ l^2/2 & l^3/3 & & \\ l^3/3 & l^4/4 & l^5/5 & \\ l^4/4 & l^5/5 & l^6/6 & l^7/7 \end{bmatrix}\mathbf{C}^{-1}.$$

With the aforementioned definition of $\mathbf{C}^{-1}$, as in equation (6.22), the
consistent mass matrix for the basic beam element becomes

$$\mathbf{m} = \frac{\rho A l}{420} \begin{matrix} \begin{matrix} v_1 \quad\;\; \theta_{z1} \quad\;\; v_2 \quad\;\; \theta_{z2} \end{matrix} \\ \begin{bmatrix} 156 & & \text{Symmetric} & \\ 22l & 4l^2 & & \\ 54 & 13l & 156 & \\ -13l & -3l^2 & -22l & 4l^2 \end{bmatrix} \end{matrix} \tag{12.26}$$

and, of course, relates to the same column matrix of nodal displacements
as does the stiffness matrix $\mathbf{k}$ defined in equation (6.32). The matrix $\mathbf{m}$
given by equation (12.26) was presented independently at much the same
time by Leckie and Lindberg [9] and by Archer [10]; it represents the
first development of a consistent mass matrix.

The $\mathbf{k}$ and $\mathbf{m}$ matrices discussed here are not exact matrices for the
dynamic problem since the deflected shape of a vibrating beam element is
not a cubic polynomial. Because we are using a consistent potential

energy approach the finite-element structure will be over-stiff (except in the limit when sufficient elements are used to give the exact solution within the context of the chosen mathematical model) and hence calculated natural frequencies will be higher than 'exact' frequencies.

It is noted that in developing the consistent mass matrix for the Bernoulli–Euler beam element, only the effects of transverse inertia, namely that inertia associated with the deflection v of the beam centroidal axis, are taken into account. The inertia associated with the rotation of cross-sections—the rotary inertia—is excluded from consideration in this mathematical model. This is consistent with the exclusion of the effects of transverse shear deformation in developing the stiffness matrix and leads to little error for very slender beams. For deeper beams the effects of transverse shear and rotary inertia are represented in the alternative mathematical model of beam behaviour known as Timoshenko beam theory. The finite-element vibration analysis of Timoshenko beams was discussed at an early stage by Archer [11] and later developments were described by Dawe [12] but the topic will not be considered here.

With regard to the basic Bernoulli–Euler beam element it is sometimes useful in obtaining solutions manually to combine together the stiffness and consistent mass matrices to form what is known as a *dynamic stiffness matrix*, of the form exhibited for the complete structure in equations (12.23) and (12.24). The element dynamic stiffness matrix $\mathbf{k}_D$ is simply

$$\mathbf{k}_D = \mathbf{k} - \omega^2 \mathbf{m} \tag{12.27}$$

and for the uniform basic beam element becomes, using equations (6.32) and (12.26),

$$\mathbf{k}_D = \frac{EI_z}{l^3} \begin{bmatrix} 12-156\psi & & \text{Symmetric} & \\ l(6-22\psi) & l^2(4-4\psi) & & \\ -(12+54\psi) & -l(6+13\psi) & 12-156\psi & \\ l(6+13\psi) & l^2(2+3\psi) & -l(6-22\psi) & l^2(4-4\psi) \end{bmatrix} \tag{12.28}$$

with columns v_1, θ_{z1}, v_2, θ_{z2}

where

$$\psi = \frac{\rho A l^4 \omega^2}{420 E I_z}. \tag{12.29}$$

Assembly of the element matrices $\mathbf{k}_D$ by the usual direct stiffness procedure will result in the structure dynamic stiffness matrix

$$\mathbf{K}_{Dr} = \mathbf{K}_r - \omega^2 \mathbf{M}_r \tag{12.30}$$

the determinant of which is zero for any natural frequency, as stated earlier.

Example 12.1. Within Bernoulli–Euler beam theory the exact values of the infinite number of natural frequencies of lateral vibration of a simply supported prismatic beam of length L are given by the expression $\omega_m = m^2 \pi^2 \sqrt{(EI_z/\rho AL^4)}$ where $m = 1, 2, 3, \ldots$. Determine the lowest two natural frequencies using the FEM with one and two basic elements modelling the beam in turn.

One-element solution. The beam is shown in Fig. E12.1(a) and is represented by a single element of length $l = L$. The boundary conditions are that $v_1 = v_2 = 0$ and so the problem degrees of freedom are only θ_{z1} and θ_{z2}. For this problem $\mathbf{K}_r$ and $\mathbf{M}_r$ can easily be obtained, or $\mathbf{K}_{Dr}$ can be obtained directly from the expression for $\mathbf{K}_D$ (equation (12.28)) simply by eliminating the rows and columns corresponding to v_1 and v_2. Thus

$$\mathbf{K}_{Dr}\mathbf{D}_r = (\mathbf{K}_r - \omega^2\mathbf{M}_r)\mathbf{D}_r = \frac{EI_z}{L^3}\begin{bmatrix} L^2(4-4\psi) & L^2(2+3\psi) \\ L^2(2+3\psi) & L^2(4-4\psi) \end{bmatrix}\begin{Bmatrix} \theta_{z1} \\ \theta_{z2} \end{Bmatrix} = \begin{Bmatrix} 0 \\ 0 \end{Bmatrix}. \tag{a}$$

The condition governing free vibration is

$$|\mathbf{K}_{Dr}| = \frac{EI_z}{L}\begin{vmatrix} 4-4\psi & 2+3\psi \\ 2+3\psi & 4-4\psi \end{vmatrix} = 0$$

or

$$(4-4\psi)^2 - (2+3\psi)^2 = 0$$

or

$$(7\psi - 2)(\psi - 6) = 0.$$

From this equation two values of ψ are obtained corresponding to the lowest two natural frequencies of vibration. These are $\psi_1 = 2/7$ and $\psi_2 = 6$ where ψ is defined in equation (12.29). Hence

$$\omega_1 = 10.9544\sqrt{\left(\frac{EI_z}{\rho AL^4}\right)} \qquad \omega_2 = 50.200\sqrt{\left(\frac{EI_z}{\rho AL^4}\right)}.$$

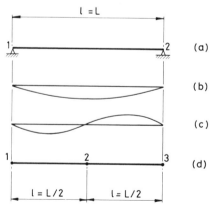

Fig. E12.1

The exact solutions calculated from the given expression are

$$\omega_1 = 9.8696 \sqrt{\left(\frac{EI_z}{\rho AL^4}\right)} \qquad \omega_2 = 39.478 \sqrt{\left(\frac{EI_z}{\rho AL^4}\right)}.$$

The relative values of the nodal degrees of freedom in the mode shapes corresponding to these two frequencies can easily be determined by substituting $\psi = 2/7$ and $\psi = 6$ in turn in either of equations (a). The result is

$$\theta_{z1} = -\theta_{z2} \quad \text{for mode 1}$$

$$\theta_{z1} = \theta_{z2} \quad \text{for mode 2}.$$

Between nodes v varies as a cubic polynomial, of course, in the FEM solution and the overall deflected form can easily be constructed. (The 'exact' deflected form is sinusoidal.) Natural modes 1 and 2 are shown in Figs. E12.1(b) and (c) respectively.

Incidentally, an important general property of natural modes of vibration that is commonly referred to in texts dealing with structural dynamics is the *orthogonality* property. This states that any two natural modes, the ith and jth modes $\mathbf{D}_{ri}$ and $\mathbf{D}_{rj}$ say in the present context, must satisfy the relationships

$$\mathbf{D}_{ri}^t \mathbf{M}_r \mathbf{D}_{rj} = 0 \qquad \text{and} \qquad \mathbf{D}_{ri}^t \mathbf{K}_r \mathbf{D}_{rj} = 0.$$

The ith and jth modes are said to be orthogonal with respect to the mass matrix and to the stiffness matrix. In this particular problem we have only two natural modes which are

$$\mathbf{D}_{r1} = \{1 \quad -1\} \qquad \mathbf{D}_{r2} = \{1 \quad 1\}$$

and with $\mathbf{K}_r$ and $\mathbf{M}_r$ as defined within equation (a) it can be verified easily that the orthogonality condition holds, which serves as a check on $\mathbf{D}_{r1}$ and $\mathbf{D}_{r2}$.

Two-element solution. For a two-element modelling of the whole beam, as shown in Fig. E12.1(d), there would be four structure degrees of freedom, namely $\mathbf{D}_r = \{\theta_{z1} \ v_2 \ \theta_{z2} \ \theta_{z3}\}$. However, rather than consider the whole beam it is better to take advantage of the obvious symmetry of the problem. Thus, by considering the use of one element in, say, the left-hand half of the beam, denoted 1–2 in Fig. E12.1(d), symmetric and anti-symmetric modes can be considered in turn.

For symmetric modes, as shown in Fig. E12.1(b), use the boundary conditions that $v_1 = \theta_{z2} = 0$, so that the degrees of freedom are θ_{z1} and v_2. Then, using equation (12.28) again, with $l = L/2$, the condition governing vibration is

$$\begin{vmatrix} L^2(4-4\psi)/4 & -L(6+13\psi)/2 \\ -L(6+13\psi)/2 & 12-156\psi \end{vmatrix} = 0.$$

Hence

$$(4-4\psi)(12-156\psi) - (6+13\psi)^2 = 0$$

or

$$455\psi^2 - 828\psi + 12 = 0.$$

The lowest value of ψ satisfying this equation is $\psi = 0.014\,610$ and the corresponding lowest frequency of symmetric vibration is

$$\omega_1 = \sqrt{\left(\frac{420EI_z\psi}{\rho A(L/2)^4}\right)} = 9.9086\ \sqrt{\left(\frac{EI_z}{\rho AL^4}\right)}.$$

For antisymmetric modes, as in Fig. E12.1(c), use the boundary conditions that $v_1 = v_2 = 0$ (in Fig. E12.1(d)), so that the degrees of freedom are θ_{z1} and θ_{z2}. Then equations (a) apply again, except that L is now replaced by $L/2$ ($=l$) in these equations and the definition of ψ is

$$\psi = \frac{\rho A(L/2)^4\omega^2}{420EI_z}.$$

Hence the lowest frequency of antisymmetric vibration—which is the second-lowest frequency overall—is

$$\omega_2 = 4\left(10.9544\ \sqrt{\left(\frac{EI_z}{\rho AL^4}\right)}\right) = 43.818\ \sqrt{\left(\frac{EI_z}{\rho AL^4}\right)}.$$

These predicted values of ω_1 and ω_2 are an improvement over those obtained earlier in this example, as would be expected.

In determining the natural frequencies of vibration by hand in this example (and in those that follow) the mathematical condition that the determinant of the structure dynamic stiffness matrix is zero has been used in direct fashion to set up a polynomial equation of degree n in ω^2 (or ψ) whose roots can readily be determined and yield the n natural frequencies. This approach is suitable for very-small-order problems but it should be appreciated that it is not suitable for problems in which more than a few degrees of freedom are involved. For such problems a variety of numerical approaches is available to determine the eigenvalues and some discussion of this is made in Appendix A.

Example 12.2. Determine the fundamental natural frequency of lateral vibration of a prismatic cantilever beam using a single basic element in a FEM solution. Then determine again the fundamental frequency when a concentrated mass of magnitude equal to that of the beam itself is attached at the free end of the beam.

The beam is shown in Fig. E12.2 and it can be seen that the problem degrees of freedom are v_2 and θ_{z2}. Ignoring the concentrated mass and using equation (12.28) the condition governing free vibration is

$$\begin{vmatrix} (12-156\psi) & -L(6-22\psi) \\ -L(6-22\psi) & L^2(4-4\psi) \end{vmatrix} = 0.$$

Fig. E12.2

Thus

$$(12 - 156\psi)(4 - 4\psi) - (6 - 22\psi)^2 = 0$$

or

$$140\psi^2 - 408\psi + 12 = 0$$

The lowest root is $\psi_1 = 0.029\,715 = \rho A L^4 \omega_1^2 / 420 E I_z$. Thus

$$\omega_1 = 3.5327 \sqrt{\left(\frac{E I_z}{\rho A L^4}\right)}$$

is the fundamental frequency of the beam without the concentrated mass. (The exact solution is $\omega_1 = 3.5160\sqrt{(E I_z/\rho A L^4)}$.)

The presence of the concentrated mass at node 2 is readily included in the analysis if equation (12.11), giving the change in potential energy of a point inertia load, is recalled. For the present situation we have that the change in potential energy of the mass of magnitude $\rho A L$ at node 2 is

$$-V_{pv} = \tfrac{1}{2}(\rho A L)\omega^2 v_2^2 = \tfrac{1}{2}\omega^2 v_2(\rho A L)v_2.$$

By analogy with equation (12.17) this indicates simply the addition of $\rho A L$, the magnitude of the mass, into the structure mass matrix in location on the leading diagonal corresponding to the degree of freedom v_2. If the dynamic stiffness matrix is used, as it is here, it means correspondingly the addition of $(-\rho A L \omega^2)$ in the appropriate location. Since $\rho A L \omega^2 = 420\psi E I_z/L^3$, with ψ defined in equation (12.29), the governing condition becomes

$$\begin{vmatrix} (12 - 156\psi - 420\psi) & -L(6 - 22\psi) \\ -L(6 - 22\psi) & L^2(4 - 4\psi) \end{vmatrix} = 0$$

or

$$(12 - 576\psi)(4 - 4\psi) - (6 - 22\psi)^2 = 0$$

or

$$1820\psi^2 - 2088\psi + 12 = 0.$$

The lowest root is $\psi_1 = 0.0\,057\,762$ and hence

$$\omega_1 = 1.5576 \sqrt{\left(\frac{E I_z}{\rho A L^4}\right)}$$

is the fundamental frequency when the concentrated mass is present.

In the examples just given some results have been calculated for the Bernoulli–Euler beam vibration problem using the coarsest of finite-element meshes of basic elements. We now follow on from this with the presentation of more exhaustive results [9, 10] for the same two problems already considered in preliminary fashion, namely the simply supported beam and the cantilever beam.

The calculated natural frequencies for the first four modes of vibration of the two types of beam are presented in Tables 12.1 and 12.2. In these tables N_E is the number of elements in the complete beam and n is the

TABLE 12.1

Natural frequencies of a uniform simply supported beam using the consistent mass matrix

Element type	N_E	n	Values of $\omega\sqrt{(\rho AL^4/EI_z)}$			
			Mode 1	Mode 2	Mode 3	Mode 4
Cubic	1	2	10.954	50.200		
	2	4	9.9086	43.818	110.14	200.80
	3	6	9.8776	39.945	98.592	183.32
	4	8	9.8720	39.634	90.450	175.27
	5	10	9.8707	39.544	89.530	161.55
Exact solution			9.8696	39.478	88.826	157.91
Quintic	1	4	9.8725	39.646	131.81	276.02
	2	8	9.8696	39.490	89.089	158.58

number of degrees of freedom after appropriate boundary conditions have been applied. For the moment the results of interest are those for the basic beam element, i.e. for the cubic element type. It can be seen from the tables that, as expected, use of the basic element in the FEM approach gives overestimates of natural frequencies and that progressive convergence towards the exact results occurs with increase in the number of elements used. (This latter point is not absolutely guaranteed unless each succeeding mesh contains within it all the freedoms of the preceding meshes.) It can also be seen that the accuracy of solution reduces as the mode number increases. This is simply because the mode shapes, i.e. the variations of v along the beam when vibrating at the natural frequencies, become increasingly complicated with increase in mode number. The mode shapes for modes 1–4 of the two beams are illustrated in Fig. 12.3.

TABLE 12.2

Natural frequencies of a uniform cantilever beam using the consistent mass matrix

Element type	N_E	n	Values of $\omega\sqrt{(\rho AL^4/EI_z)}$			
			Mode 1	Mode 2	Mode 3	Mode 4
Cubic	1	2	3.5327	34.807		
	2	4	3.5177	22.221	75.157	218.14
	3	6	3.5164	22.107	62.466	140.67
	4	8	3.5162	22.060	62.175	122.66
	5	10	3.5161	22.046	61.919	122.32
Exact solution			3.5160	22.034	61.697	120.90
Quintic	1	4	3.5160	22.158	63.347	281.60
	2	8	3.5160	22.035	61.781	122.59

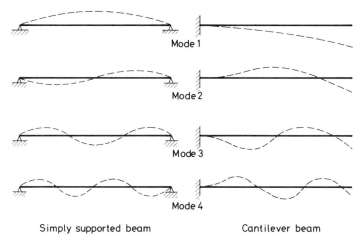

Mode 1

Mode 2

Mode 3

Mode 4

Simply supported beam   Cantilever beam

Fig. 12.3. Mode shapes of two beams.

In predicting beam natural frequencies using the FEM we are not restricted to the use of only the basic cubic beam element of course. It is perfectly possible to use more-refined element types, with the hope of increasing the efficiency of the solution. In Section 6.10 mention was made, for instance, of two versions of a beam element whose stiffness properties are based on representing v as a quintic polynomial (equation (6.77)). These two element versions are illustrated in Fig. 6.10; both have six degrees of freedom but differ one from another in the allocation of the nodal degrees of freedom. Consistent mass matrices can be derived for either element without undue effort and the elements can then be incorporated into a dynamic analysis in straightforward manner. Pestel [13] considered the element type shown in Fig. 6.10(b) and other related elements based on displacement fields represented as Hermitian polynomials and having higher-order derivatives as degrees of freedom at the end nodes. Carnegie, Thomas, and Dokumaci [14] considered the use of the element type shown in Fig. 6.10(a) in vibration applications. For uniform beams the alternative types of refined element with a common level of interpolation yield frequency estimates whose levels of accuracy are similar. With quintic interpolation it is generally preferable to use the three-node element of Fig. 6.10(a) rather than the two-node element of Fig. 6.10(b) so as to avoid excess continuity between adjacent elements. Some results obtained using the three-node quintic element are recorded in Tables 12.1 and 12.2. By and large, for the same number of structure degrees of freedom the quintic element provides superior results to the cubic element; this conclusion does not necessarily apply, however, in circumstances where very large errors are involved with the use of either type of element.

The above has been concerned with prismatic beam elements but clearly a cross-sectional area varying with x could be accommodated quite readily by moving the quantity A under the integral sign in equation (12.25). Linearly tapered beam elements based on both cubic [15] and quintic [16] displacement interpolation have been used in free vibration analysis. The three-node quintic element appears to be significantly superior to the two-node quintic element and to the cubic element in documented numerical studies of tapered beams [16].

12.5. Vibration of bars and frameworks

The derivation of the stiffness matrix for the basic element of a bar subjected only to axial deformation u has been presented in Section 6.2 (with **k** given by equation (6.12)). We begin this section by deriving the consistent mass matrix for this element (illustrated in Fig. 6.1). This is a simple matter since equation (12.18) can be used with (see equation (6.5))

$$\mathbf{N} = \left[\left(1 - \frac{x}{l} \right) \quad \frac{x}{l} \right] \tag{12.31}$$

and $dV_e = A\,dx$. Then, for a prismatic bar of cross-sectional area A,

$$\mathbf{m} = \rho A \int_0^l \mathbf{N}^t\mathbf{N}\,dx = \frac{\rho Al}{6} \begin{matrix} u_1 & u_2 \\ \begin{bmatrix} 2 & 1 \\ 1 & 2 \end{bmatrix} \end{matrix} \tag{12.32}$$

is the consistent mass matrix for the basic bar element with axial (i.e. extensional) motion.

In a very similar way we can generate the consistent mass matrix for a basic bar element with torsional motion. The analogy between the axial and torsional behaviour of a bar is so close that the derivation of the stiffness matrix for the bar in torsion by way of the FEM was not directly considered in Chapter 6. However, this stiffness matrix can be derived readily, based on the assumption that the rotation θ_x of the element shown in Fig. 2.4 is expressed in terms of nodal rotational displacements θ_{x1} and θ_{x2} by way of the same shape function matrix **N** as defined in equation (12.31).† The stiffness matrix that results is exactly that derived by a direct approach in Section 2.7 and defined by equation (2.17). The consistent mass matrix for the basic prismatic bar element undergoing

---

† It should be noted that in the early chapters the nodes of line elements were designated i, j rather than 1, 2.

torsional motion is, by analogy with the axial motion case,

$$\mathbf{m} = \frac{\rho I_p l}{6} \begin{matrix} \theta_{x1} & \theta_{x2} \\ \begin{bmatrix} 2 & 1 \\ 1 & 2 \end{bmatrix} \end{matrix} \tag{12.33}$$

where I_p is the polar second moment of area of the cross-section. (For a circular cross-section $J = I_p$.)

In Chapters 3 and 4 the static analysis of a number of different types of frame was considered. These frames included planar pin-jointed and rigid-jointed frames, grillages, and space ball-jointed and rigid-jointed frames. Stiffness matrices were established for the elements of such frames by applying a standard transformation procedure to convert from a local to a global co-ordinate system. The element stiffness matrices occurring in these chapters were in fact derived by direct application of basic structural principles but the reader will be aware that precisely the same results would follow from straightforward application of the finite-element displacement approach (with the basic displacement fields). Now that consistent mass matrices are established in local co-ordinate systems for the beam, the axial bar, and the torsional bar (in equations (12.26), (12.32), and (12.33)) it is clearly possible to develop mass matrices for any type of frame element in a global co-ordinate system by exactly the same procedures as used for stiffness matrices. The assumptions implicit in this with regard to the assumed uncoupling of bending, axial, and torsional behaviour remain as discussed in the earlier chapters.

Here we give just one particular example of the development of a consistent mass matrix, in the global co-ordinate system, of a frame element by considering the prismatic element of a planar rigid-jointed frame. This type of element has been discussed in relation to static behaviour in Section 3.4, is illustrated in Fig. 3.1, and has a stiffness matrix in the global system given by equation (3.28). (Again i, j are equivalent to 1, 2.) In considering the consistent mass matrix we proceed in the same way as for the stiffness matrix by firstly setting up the 6×6 matrix in the local system by combining together bending behaviour (see equation (12.26)) and axial behaviour (equation (12.32)). The result is

$$\mathbf{m} = \frac{\rho A l}{420} \begin{matrix} u_1 \quad\; v_1 \quad\;\; \theta_{z1} \quad\; u_2 \quad\; v_2 \quad\; \theta_{z2} \\ \begin{bmatrix} 140 & & & & & \\ 0 & 156 & & & & \\ 0 & 22l & 4l^2 & \text{Symmetric} & & \\ 70 & 0 & 0 & 140 & & \\ 0 & 54 & 13l & 0 & 156 & \\ 0 & -13l & -3l^2 & 0 & -22l & 4l^2 \end{bmatrix} \end{matrix}. \tag{12.34}$$

This mass matrix relates, of course, to vibration in the plane of the frame.

In the global system the mass matrix is denoted $\bar{\mathbf{m}}$ where, in an analogous way to that for transformed stiffness (as in equation (3.18) say),

$$\bar{\mathbf{m}} = \boldsymbol{\tau}^{t}\mathbf{m}\boldsymbol{\tau}. \tag{12.35}$$

Here, as in Section 3.4,

$$\boldsymbol{\tau} = \begin{bmatrix} \mathbf{T} & \mathbf{0} \\ \mathbf{0} & \mathbf{T} \end{bmatrix} \quad \text{where} \quad \mathbf{T} = \begin{bmatrix} e & f & 0 \\ -f & e & 0 \\ 0 & 0 & 1 \end{bmatrix} \tag{12.36}$$

and e and f are defined in equation (3.13). Performing the matrix multiplication indicated in equation (12.35) is not difficult in this case and gives the global element mass matrix $\bar{\mathbf{m}}$ as in equation (12.37); this mass matrix is presented for use in conjunction with the global element stiffness matrix $\bar{\mathbf{k}}$ given by equation (3.28);

$$\bar{\mathbf{m}} = \frac{\rho Al}{420} \begin{bmatrix} 140e^2 + 156f^2 & & & & & \\ -16ef & 140f^2 + 156e^2 & & & \text{Symmetric} & \\ -22lf & 22le & 4l^2 & & & \\ 70e^2 + 54f^2 & 16ef & -13lf & 140e^2 + 156f^2 & & \\ 16ef & 70f^2 + 54e^2 & 13le & -16ef & 140f^2 + 156e^2 & \\ 13lf & -13le & -3l^2 & 22lf & -22le & 4l^2 \end{bmatrix}$$

with column headings $\bar{u}_1$, $\bar{v}_1$, $\theta_{\bar{z}1}$, $\bar{u}_2$, $\bar{v}_2$, $\theta_{\bar{z}2}$.

$$\tag{12.37}$$

The element for plane frame vibration analysis considered above is of the basic type, with minimum displacement fields, but there is of course no reason why other types of frame elements based on more-refined displacement fields should not be used in an attempt to improve solution efficiency. For instance a refined element of a planar rigid-jointed frame could be based on quintic v displacement, cubic u displacement with three nodes (at the ends and the centre) with three degrees of freedom (u, v, and θ_z) at each node.

Example 12.3. Set up the system of equations governing the free vibration in its own plane of the portal frame shown in Fig. E12.3 using a single basic finite element to represent each frame member. The frame cross-section is the same throughout, with $A = 1.85187 \times 10^{-5}$ m^2, $I_z = 2.85785 \times 10^{-11}$ m^4, $E = 2.06829 \times 10^{11}$ N m^{-2}, $\rho = 25613.5$ kg m^{-3}, and $L = 0.2413$ m.

For each element the stiffness and consistent mass matrices in the global system are calculated using the definitions of $\bar{\mathbf{k}}$ and $\bar{\mathbf{m}}$ given in equations (3.28) and (12.37) respectively; for the horizontal element 1–2 $e = 1$ and $f = 0$ whilst for the vertical elements 1–3 and 2–4 $e = 0$ and $f = 1$. The non-zero structure freedoms are clearly $\bar{u}_1$, $\bar{v}_1$, $\theta_{\bar{z}1}$, $\bar{u}_2$, $\bar{v}_2$, and $\theta_{\bar{z}2}$.

The structure stiffness and mass matrices are assembled from the element matrices in the usual manner and are

$$
\mathbf{K}_r =
\begin{array}{c}
\begin{array}{cccccc}
\bar{u}_1 & \bar{v}_1 & \theta_{\bar{z}1} & \bar{u}_2 & \bar{v}_2 & \theta_{\bar{z}2}
\end{array}\\
\left[
\begin{array}{cccccc}
\begin{matrix}AE/L\\+12EI_z/L^3\\+0\end{matrix} & & & & & \\[2em]
\begin{matrix}0\\+0\\+0\end{matrix} & \begin{matrix}12EI_z/L^3\\+AE/L\\+0\end{matrix} & & \text{Symmetric} & & \\[2em]
\begin{matrix}0\\-6EI_z/L^2\\+0\end{matrix} & \begin{matrix}6EI_z/L^2\\+0\\+0\end{matrix} & \begin{matrix}4EI_z/L\\+4EI_z/L\\+0\end{matrix} & & & \\[2em]
\begin{matrix}-AE/L\\+0\\+0\end{matrix} & \begin{matrix}0\\+0\\+0\end{matrix} & \begin{matrix}0\\+0\\+0\end{matrix} & \begin{matrix}AE/L\\+0\\+12EI_z/L^3\end{matrix} & & \\[2em]
\begin{matrix}0\\+0\\+0\end{matrix} & \begin{matrix}-12EI_z/L^3\\+0\\+0\end{matrix} & \begin{matrix}-6EI_z/L^2\\+0\\+0\end{matrix} & \begin{matrix}0\\+0\\+0\end{matrix} & \begin{matrix}12EI_z/L^3\\+0\\+AE/L\end{matrix} & \\[2em]
\begin{matrix}0\\+0\\+0\end{matrix} & \begin{matrix}6EI_z/L^2\\+0\\+0\end{matrix} & \begin{matrix}2EI_z/L\\+0\\+0\end{matrix} & \begin{matrix}0\\+0\\-6EI_z/L^2\end{matrix} & \begin{matrix}-6EI_z/L^2\\+0\\+0\end{matrix} & \begin{matrix}4EI_z/L\\+0\\+4EI_z/L\end{matrix}
\end{array}
\right]
\end{array}
$$

$$
\mathbf{M}_r = \frac{\rho AL}{420}
\begin{array}{c}
\begin{array}{cccccc}
\bar{u}_1 & \bar{v}_1 & \theta_{\bar{z}1} & \bar{u}_2 & \bar{v}_2 & \theta_{\bar{z}2}
\end{array}\\
\left[
\begin{array}{cccccc}
\begin{matrix}140\\+156\\+0\end{matrix} & & & & & \\[2em]
\begin{matrix}0\\+0\\+0\end{matrix} & \begin{matrix}156\\+140\\+0\end{matrix} & & \text{Symmetric} & & \\[2em]
\begin{matrix}0\\-22L\\+0\end{matrix} & \begin{matrix}22L\\+0\\+0\end{matrix} & \begin{matrix}4L^2\\+4L^2\\+0\end{matrix} & & & \\[2em]
\begin{matrix}70\\+0\\+0\end{matrix} & \begin{matrix}0\\+0\\+0\end{matrix} & \begin{matrix}0\\+0\\+0\end{matrix} & \begin{matrix}140\\+0\\+156\end{matrix} & & \\[2em]
\begin{matrix}0\\+0\\+0\end{matrix} & \begin{matrix}54\\+0\\+0\end{matrix} & \begin{matrix}13L\\+0\\+0\end{matrix} & \begin{matrix}0\\+0\\+0\end{matrix} & \begin{matrix}156\\+0\\+140\end{matrix} & \\[2em]
\begin{matrix}0\\+0\\+0\end{matrix} & \begin{matrix}-13L\\+0\\+0\end{matrix} & \begin{matrix}-3L^2\\+0\\+0\end{matrix} & \begin{matrix}0\\+0\\-22L\end{matrix} & \begin{matrix}-22L\\+0\\+0\end{matrix} & \begin{matrix}4L^2\\+0\\+4L^2\end{matrix}
\end{array}
\right]
\end{array}
$$

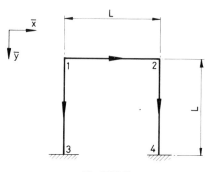

Fig. E12.3

In the above the three elemental contributions to each of the individual terms of $\mathbf{K}_r$ and $\mathbf{M}_r$ are given in the order of elements 1–2, 1–3, and 2–4. For a portal of arbitrary geometry the quantities E, A, I_z, L, and ρ would vary from element to element but in this particular case do not. Substituting in the given numerical values gives the following set of equations:

$$
\left(
10^9
\begin{bmatrix}
1587.8253 & & & & & \\
0 & 1587.8253 & & \text{Symmetric} & & \\
-0.060910 & 0.060910 & 0.019597 & & & \\
-1587.3204 & 0 & 0 & 1587.8253 & & \\
0 & -0.504847 & -0.060910 & 0 & 1587.8253 & \\
0 & 0.060910 & 0.004899 & -0.060910 & -0.060910 & 0.019597
\end{bmatrix}
\right.
$$

$$
\left.
-\omega^2
\begin{bmatrix}
8066.38 & & & & & \\
0 & 8066.38 & & \text{Symmetric} & & \\
-144.666 & 144.666 & 12.6938 & & & \\
1907.59 & 0 & 0 & 8066.38 & & \\
0 & 1471.57 & 85.4846 & 0 & 8066.38 & \\
0 & -85.4846 & -4.76018 & -144.666 & -144.666 & 12.6938
\end{bmatrix}
\right)
\begin{Bmatrix}
\bar{u}_1 \\
\bar{v}_1 \\
\theta_{\bar{z}1} \\
\bar{u}_2 \\
\bar{v}_2 \\
\theta_{\bar{z}2}
\end{Bmatrix}
=
\begin{Bmatrix}
0 \\
0 \\
0 \\
0 \\
0 \\
0
\end{Bmatrix}.
$$

This is the required system of equations which governs the free vibration problem and which has the form of equation (12.23). The calculation of the frame natural frequencies and mode shapes is too difficult to attempt manually but these frequencies and mode shapes have been calculated on the computer and are recorded in what follows.

We proceed now to extend Example 12.3 by presenting some numerical results for precisely the problem discussed in the example, but using in turn one, two, three, and four equal-length basic elements per frame member. Calculated FEM frequencies are recorded for the first six modes in Fig. 12.4 together with, for the first four modes, comparative values documented elsewhere [17] and obtained using a force method; the mode shapes are also shown in Fig. 12.4. The FEM results demonstrate

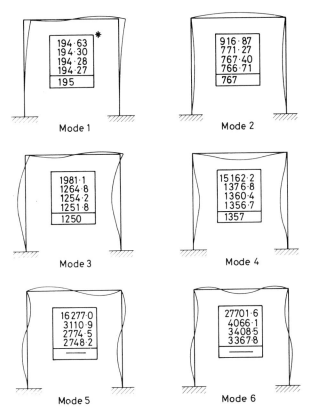

194·63 *
194·30
194·28
194·27
195

Mode 1

916·87
771·27
767·40
766·71
767

Mode 2

1981·1
1264·8
1254·2
1251·8
1250

Mode 3

15162·2
1376·8
1360·4
1356·7
1357

Mode 4

16277·0
3110·9
2774·5
2748·2

Mode 5

27701·6
4066·1
3408·5
3367·8

Mode 6

* Values quoted are frequencies (rads/sec) for, in turn,
1,2,3 and 4 elements per member in FEM meshes, and
a force method result for modes 1 to 4.

Fig. 12.4. Natural frequencies and mode shapes of a simple portal frame.

monotonic convergence from the high side, with the accuracy of solution very dependent on the relative complexity of the mode shape of course.

For this and other more complicated frame vibration problems a significant improvement in solution efficiency is often achieved by using a refined rather than a basic element type. With the simple portal frame, for instance, if a three-node nine-degree-of-freedom element type (with quintic v, cubic u displacement) is used the coarsest mesh of one element per member has the same number of nodes and degrees of freedom as when the basic element type is used with two elements per member. Accuracy is improved for all of the first six natural frequencies when using the refined element, the improvement being most significant for modes 5 and 6 when the 14 per cent and 22 per cent errors involved with the use of two basic elements per member are reduced to around 1 per cent and 2 per cent respectively when using one refined element per member.

Returning consideration briefly to Example 12.3 it is noted in passing (but not as a suggestion for general use) that we can obtain solutions manually for this very simple geometry if we invoke assumptions of member inextensibility to reduce the number of effective degrees of freedom in a similar way to that described for static problems in Section 3.6. Thus we can set $\bar{v}_1 = \bar{v}_2 = 0$ and either $\bar{u}_1 = \bar{u}_2 = 0$ and $\theta_{\bar{z}1} = -\theta_{\bar{z}2}$ for a symmetric mode or $\bar{u}_1 = \bar{u}_2 \neq 0$ and $\theta_{\bar{z}1} = \theta_{\bar{z}2}$ for an antisymmetric mode. The result is $\omega_1 = 194.64$ rad s^{-1} and $\omega_2 = 917.65$ rad s^{-1} which compares very closely indeed with the values of $\omega_1 = 194.63$ rad s^{-1} and $\omega_2 = 916.87$ rad s^{-1} obtained in the 'full' solution without invoking inextensibility; this demonstrates the very predominant bending nature of the early modes.

12.6. Lateral vibration of plates

Finite elements for the analysis of plate bending problems based, in the main, on the classical plate theory, have been introduced in Chapter 11. In the classical theory the plate behaviour is expressed solely in terms of the lateral, or transverse deflection w of the middle surface. Consequently in considering the development of a consistent mass matrix for the lateral vibration of a plate bending element it is only terms associated with w on the right-hand side of equations (12.15) and (12.16) which are pertinent. (Rotary inertia is ignored, as it was for the Bernoulli–Euler beam.)

Consistent mass matrices can be derived for any type of plate bending element, of course, but consider first in a little detail a particular plate element which was one of the very first for which such a matrix was generated; this is the 12-degree-of-freedom non-conforming (in slope) rectangular element designated R1 in Chapter 11, Section 11.3, and illustrated in Fig. 11.1. Since we have for this element that

$$\boldsymbol{\alpha} = [1 \quad x \quad y \quad x^2 \quad xy \quad y^2 \quad x^3 \quad x^2y \quad xy^2 \quad y^3 \quad x^3y \quad xy^3] \quad (12.38)$$

(from equation (11.18)) and that $\mathbf{C}^{-1}$ is as defined in Table 11.1, it is convenient to use equation (12.19) to form the consistent mass matrix. For the plate the infinitesimal volume dV_e becomes

$$dV_e = h\, dA_e = h\, dx\, dy \quad (12.39)$$

where dA_e is the infinitesimal area of the middle surface and h is the plate thickness which in general will vary over the middle surface. The expression for the consistent mass matrix becomes

$$\mathbf{m} = \rho(\mathbf{C}^{-1})^t \left(\int_{-b}^{b} \int_{-a}^{a} h\boldsymbol{\alpha}^t \boldsymbol{\alpha}\, dx\, dy \right) \mathbf{C}^{-1} \quad (12.40)$$

TABLE 12.3

Consistent mass matrix for element R1

$$\mathbf{m} = \frac{\rho hab}{6300}$$

	w_1	ϕ_1	θ_1	w_2	ϕ_2	θ_2	w_3	ϕ_3	θ_3	w_4	ϕ_4	θ_4
w_1	3454											
ϕ_1	$922b$	$320b^2$										
θ_1	$922a$	$252ab$	$320a^2$									
w_2	1226	$398b$	$548a$	$m_{1,1}$								
ϕ_2	$m_{4,2}$	$160b^2$	$168ab$	$m_{2,1}$	$m_{2,2}$			Symmetric				
θ_2	$-m_{4,3}$	$-m_{5,3}$	$-240a^2$	$-m_{3,1}$	$-m_{3,2}$	$m_{3,3}$						
w_3	394	$232b$	$232a$	$m_{10,1}$	$m_{10,2}$	$m_{10,3}$	$m_{1,1}$					
ϕ_3	$-m_{7,2}$	$-120b^2$	$-112ab$	$m_{11,1}$	$m_{11,2}$	$-m_{11,3}$	$-m_{2,1}$	$m_{2,2}$				
θ_3	$-m_{7,3}$	$m_{8,3}$	$-120a^2$	$-m_{12,1}$	$-m_{12,2}$	$m_{12,3}$	$-m_{3,1}$	$m_{3,2}$	$m_{3,3}$			
w_4	1226	$548b$	$398a$	$m_{7,1}$	$m_{7,2}$	$-m_{7,3}$	$m_{4,1}$	$m_{3,2}$	$-m_{4,3}$	$m_{1,1}$		
ϕ_4	$-m_{10,2}$	$-240b^2$	$-168ab$	$m_{8,1}$	$m_{8,2}$	$-m_{8,3}$	$-m_{5,1}$	$m_{5,2}$	$m_{5,3}$	$-m_{2,1}$	$m_{2,2}$	
θ_4	$m_{10,3}$	$-m_{11,3}$	$160a^2$	$-m_{9,1}$	$-m_{9,2}$	$m_{9,3}$	$-m_{6,1}$	$m_{6,2}$	$m_{6,3}$	$m_{3,1}$	$-m_{3,2}$	$m_{3,3}$

and can be evaluated explicitly for uniform value of h or for simple variations thereof. Mass (and stiffness) matrices for uniform thickness and for linearly varying thickness in one co-ordinate direction have been presented by Dawe [18, 19]. The consistent mass matrix for the uniform thickness case is recorded in Table 12.3; this mass matrix corresponds to the stiffness matrix given in Table 11.2.

Example 12.4. Determine the lowest (or fundamental) frequency of flexural vibration of a uniform thickness square plate of side A and thickness h with (a) all edges clamped and (b) all edges simply supported. Model one quadrant of the plate with a single R1 element and take $\nu = 0.3$.

(a) With the edges clamped the problem is as illustrated in Fig. E11.1 except that in place of the static load $P/4$ shown there, we now have a distributed inertia loading. The boundary conditions are (as in Example 11.1) that the deflection and both slopes are zero at all points on the clamped boundary and (for the doubly symmetric first mode of vibration) that both slopes have zero value at the plate centre. Hence w_3 is the only non-zero degree of freedom and

$$\mathbf{K}_r = k_{7,7} = \frac{D}{60ab}(60p + 60p^{-1} + 42 - 12\nu) \quad \text{(from Table 11.2)}$$

$$\mathbf{M}_r = m_{7,7} = \frac{\rho hab}{6300} 3454 \quad \text{(from Table 12.3)}.$$

Thus, using equation (12.24) for this single-degree-of-freedom system, with $a = b = A/4$, $p = 1$, and $\nu = 0.3$, gives

$$\frac{16D}{60A^2} 158.4 - \omega^2 \frac{\rho h A^2}{16} \frac{3454}{6300} = 0.$$

Hence

$$\omega = 35.11 \sqrt{\left(\frac{D}{\rho h A^4}\right)}$$

is the fundamental frequency of vibration (in radians per second). This compares with the exact value of $\omega = 35.99\sqrt{(D/\rho h A^4)}$, corresponding to an error of 2.4 per cent on the low side (owing to the non-conforming nature of the element).

(b) With the edges simply supported the dynamic problem is closely related to the static problem considered in Example 11.2 and illustrated in Fig. E11.2. After applying the usual boundary and symmetry conditions the non-zero degrees of freedom for the doubly symmetric mode are ϕ_2, w_3, and θ_4 but we have further that $\phi_2 = \theta_4$ and can therefore work in terms of ϕ_2 and w_3 only. This has been done in the related static example (Example 11.2) where it is shown that

$$\mathbf{K}_r = \frac{16D}{60A^2} \begin{bmatrix} \overset{\phi_2}{5.7A^2} & \overset{w_3}{-16.05A} \\ -32.1A & 158.4 \end{bmatrix}.$$

In a similar way the corresponding $\mathbf{M}_r$, obtained using Table 12.3, is

$$\mathbf{M}_r = \frac{\rho hab}{6300} \begin{array}{cc} \phi_2 & w_3 \end{array} \\ \begin{bmatrix} (320b^2 + 112ab) & 548b \\ (548b + 548a) & 3454 \end{bmatrix} = \frac{\rho hA^2}{16 \times 6300} \begin{array}{cc} \phi_2 & w_3 \end{array} \\ \begin{bmatrix} 27 & 137A \\ 274A & 3454 \end{bmatrix}.$$

Thus, using equation (12.25), the condition governing free vibration is

$$\begin{vmatrix} A^2(5.7 - 27\Omega) & A(-16.05 - 137\Omega) \\ A(-32.1 - 274\Omega) & (158.4 - 3454\Omega) \end{vmatrix} = 0$$

where $\Omega = 60\rho hA^4\omega^2/(256 \times 6300D)$. Evaluating the determinant gives the characteristic equation

$$A^2(5.7 - 27\Omega)(158.4 - 3454\Omega) - A^2(16.05 + 137\Omega)(32.1 + 274\Omega) = 0$$

or

$$55720\Omega^2 - 32760\Omega + 387.675 = 0.$$

There are, of course, two values of Ω which satisfy this quadratic equation, the lowest of which gives the value of Ω corresponding to the fundamental frequency. This value is $\Omega = 0.0120821$ and hence the fundamental frequency of flexural vibration is

$$\omega = 18.02 \sqrt{\left(\frac{D}{\rho hA^4}\right)}.$$

An accurate value for this frequency is $\omega = 19.74\sqrt{(D/\rho hA^4)}$; hence the FEM solution is in error by 8.7 per cent, again on the low side.

(Further results for this problem obtained using element R1 and other types of element are given below.)

Element R1 is a quite useful and popular element but its accuracy is not as high as that of the other 12-degree-of-freedom non-conforming rectangle [20] designated R3 in Chapter 11. The displacement field for this latter element is given in shape function form by equations (11.24), (11.30), and (11.31). The consistent mass matrix for element R3 can be evaluated in the standard fashion by use of equation (12.18) and is defined in Table 12.4. It corresponds to the stiffness matrix recorded in Section 11.4.2 (see equation (11.32) and Table 11.4).

Another successful rectangular element described in Chapter 11 which has been used in dynamic applications is element R4 of Bogner, Fox, and Schmit [21] which has 16 degrees of freedom including the use of the twist degree of freedom at the corner nodes. This element has also been discussed and used in relation to the free vibration problem by Mason [22] who, in the same reference, also considers a further rectangular element having 24 degrees of freedom with w, $\partial w/\partial x$, $\partial w/\partial y$, $\partial^2 w/\partial x^2$, $\partial^2 w/\partial x\,\partial y$, and $\partial^2 w/\partial y^2$ as nodal freedoms. An element closely related to this latter one, with the same degrees of freedom, is presented by

TABLE 12.4

Consistent mass matrix for element R3

$$
\mathbf{m} = \frac{\rho hab}{6300}
\begin{bmatrix}
3484.349 & & & \\
962.508b & 343.365b^2 & & \\
962.508a & 268.254ab & 343.365a^2 & \\
1195.651 & 357.492b & 567.492a & \\
m_{4,2} & 136.635b^2 & 165.079ab & \\
-m_{4,3} & -m_{5,3} & -257.524a^2 & \\
424.349 & 212.508b & 212.508a & \\
-m_{7,2} & -102.476b^2 & -101.587ab & \\
-m_{7,3} & m_{8,3} & -102.476a^2 & \\
1195.651 & 567.492b & 357.492a & \\
-m_{10,2} & -257.524b^2 & -165.079ab & \\
m_{10,3} & -m_{11,3} & 136.635a^2 &
\end{bmatrix}
$$

with column headings $w_1 \quad \phi_1 \quad \theta_1 \quad w_2 \ \cdots \ \theta_4$

The terms in the nine columns which are not itemized, corresponding to degrees of freedom w_2 through to θ_4, are defined with respect to the given terms in the same way as for the **k** or **m** matrices in Tables 11.2 and 12.3 respectively.

Popplewell and McDonald [23]. All the elements mentioned in this paragraph are fully conforming and are capable of yielding high accuracy but they also have excessive nodal continuity of course.

Amongst non-rectangular plate bending elements that have been used in predicting plate frequencies are the triangular elements referred to as T1 (nine degrees of freedom non-conforming), T3 (nine degrees of freedom conforming), and T5 (18 degrees of freedom conforming) in Chapter 11, and the parallelogrammic element equivalent of the R1 rectangle which was mentioned in Section 11.6. Applications to vibration problems using these elements are reported in references 24–27. Results for plate, and other, problems are also included in the comprehensive work of Argyris [28].

One or two results obtained using some of the above mentioned plate elements are now detailed.

One 'standard' problem that has been considered by a number of investigators is that of a square plate with all edges simply supported for which 'exact' solutions are known. In Table 12.5 results are presented for the first five modes of vibration using elements R1, R3, R4, and T5 in regular meshes of 4 through to 64 squares in the complete plate. (For the triangular element T5 each square is further sub-divided by a diagonal into two triangular elements.) The presentation of numerical results in this way, on the basis of mesh size, is clearly not at all fair on elements R1

TABLE 12.5
Natural frequencies of a simply supported square plate

Mode	Element type	Finite-element mesh[a]					Exact solution[a]
		2×2	3×3	4×4	6×6	8×8	
1	R1	18.02	18.78	19.15	19.46	19.58	19.74
	R3	19.73	—	19.69	19.71	19.72	
	R4	19.78	19.75	19.74	19.74	19.74	
	T5	19.74	19.74	19.74	19.74	—	
2	R1	50.91	—	47.40	48.32	48.73	49.35
	R3	55.17	—	49.34	49.27	49.30	
	R4	52.98	—	49.47	—	—	
	T5	50.22	49.42	49.36	49.35	—	
3	R1	—	—	72.08	75.11	76.60	78.96
	R3	—	—	78.91	78.69	78.75	
	R4	83.93	—	79.12	—	—	
	T5	80.88	79.10	78.99	78.96	—	
4	R1	124.42	—	96.21	96.87	97.49	98.70
	R3	—	—	100.10	98.84	98.67	
	R4	118.71	—	100.18	—	—	
	T5[b]	100.76	100.81	98.95	98.72	—	
		100.42	99.26	98.85	98.71	—	
5	R1	—	—	116.50	120.67	123.37	128.30
	R3	—	—	129.77	128.10	127.98	
	R4	146.94	—	129.56	—	—	
	T5	142.80	130.87	128.66	128.34	—	

[a] The values are in units of $\omega(D/\rho hA^4)^{-1/2}$.
[b] Two values of frequency are given for mode 4 in reference 26.

and R3 which have less degrees of freedom per element than do R4 and T5. A better comparison, though still not the ideal one, is on the basis of the number of degrees of freedom after applying the boundary conditions; for this plate the number of degrees of freedom is approximately in the ratio 9:6:4 for equal meshes of T5, R4, and R1/R3 elements.

The vibration of variable-thickness rectangular cantilever plates has been considered in reference 19 where the natural frequencies and mode shapes were measured experimentally as well as being calculated using the FEM with element type R1 incorporating the true linear variation of thickness. One example considered, illustrated in Fig. 12.5, is a plate of aspect ratio 2 with linear taper across the width of the plate. In the analysis the plate is modelled with uniform meshes of, in turn, eight elements (four lengthwise by two crosswise) and 24 elements (six lengthwise by four crosswise). Calculated and measured frequencies for the first 12 modes are recorded in Table 12.6. The experimental and fine-mesh calculated mode patterns are shown in the form of plots of lines

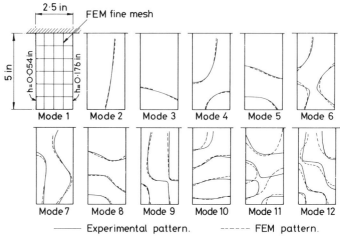

Fig. 12.5. Mode shapes for a variable thickness, rectangular, cantilever plate.

of zero deflection in Fig. 12.5. Good agreement is demonstrated between the experimental and calculated results for both natural frequencies and mode shapes. As an alternative to the use of variable-thickness elements it is possible to obtain solutions by using constant-thickness elements so that the cross-section of the model plate is 'stepped', but this procedure is not very efficient [19].

In this brief look at the analysis of the free lateral vibration of plates consideration has been restricted to classical thin-plate theory. It has been mentioned in Section 11.9 that other approaches to FEM plate analysis

TABLE 12.6
Natural frequencies of a tapered plate

	Dimensionless frequency $\omega(D/\rho hL^4)^{-1/2}$ [a]		
	Finite-element method using R1		
Mode	8 elements	24 elements	Experimental
1	2.525	2.525	2.536
2	10.78	10.76	10.80
3	15.77	15.54	15.50
4	31.53	31.49	31.92
5	42.42	40.13	39.86
6	58.81	58.63	59.35
7	60.59	61.45	62.57
8	71.40	69.82	68.90
9	—	79.54	81.43
10	—	101.54	99.29
11	—	105.45	105.26
12	—	112.18	113.96

[a] h is the *maximum* plate thickness and L is the plate length.

are possible, amongst them being the adoption of an 'improved' plate theory such as the Mindlin plate theory. Plate finite elements based on Mindlin theory include the effects of transverse shear deformation and of rotary inertia; these effects become significant for plates of moderate thickness and lead to lower predictions of natural frequencies. Reference 29 provides an example of the approach.

12.7. Lumped mass matrices

Thus far we have considered the solution of free vibration problems using an analysis which is perfectly consistent with the underlying variational principle. The adjective 'consistent' is included in the phrase 'consistent mass matrix' to indicate that this is so and that, in particular, precisely the same assumed displacement form is used in constructing this mass matrix as is used in constructing the elastic stiffness matrix. The result is that the element mass matrix is fully populated in the same way as is the element stiffness matrix; this reflects the fact that there is dynamic coupling between any particular nodal inertia force and all nodal displacements. In an assemblage of finite elements with efficient node numbering the structure consistent mass matrix will have the same banded form as does the structure stiffness matrix.

Mass matrices of other than the consistent type have been, and still are, used to a considerable extent in finite-element vibration analysis. Element mass matrices can be generated through the use of a simpler form of assumed displacement field than is used in generating the corresponding stiffness matrix. The simplest form of mass matrix is obtained by placing concentrated masses at the node points associated with translational degrees of freedom and perhaps with rotational degrees of freedom as well. The magnitude of a particular concentrated or lumped mass is calculated effectively on the basis of the assumption that a certain volume of the element local to the node in question moves as a rigid body whilst the rest of the element does not move. The resulting *lumped* element mass matrix does not reflect any dynamic coupling and is thus simply a *diagonal* matrix. A few examples of lumped mass matrices for simple elements will now be given.

For the basic prismatic bar element the lumped mass matrix is obtained very easily by associating half the total mass of the element with each node. Thus†

$$\mathbf{m} = \frac{\rho A l}{2} \begin{matrix} u_1 & u_2 \\ \begin{bmatrix} 1 & 0 \\ 0 & 1 \end{bmatrix} \end{matrix} \quad \text{or} \quad \mathbf{m} = \frac{\rho A l}{2} \begin{matrix} u_1 & u_2 \\ \lceil 1 & 1 \rfloor \end{matrix} \qquad (12.41)$$

---

† It is recalled that the brackets $\lceil \ \rfloor$ denote a diagonal matrix.

and this lumped mass matrix could be used in place of the consistent mass matrix recorded in equation (12.32).

For the basic prismatic beam element both translational and rotational degrees of freedom are present in the element stiffness matrix. In using the lumped mass approach it is usually assumed that only translational inertia effects need be considered. In this case the lumped mass matrix is

$$\mathbf{m} = \frac{\rho A l}{2} \begin{matrix} v_1 & \theta_{z1} & v_2 & \theta_{z2} \\ \lceil 1 & 0 & 1 & 0 \rfloor \end{matrix} \qquad (12.42)$$

and takes the place of the consistent mass matrix recorded in equation (12.26). Where a lumping scheme of this type is used, i.e. where the diagonal mass matrix contains zero contributions associated with the rotational degrees of freedom it is usual to eliminate the rotational freedoms from the set of equations governing the eigenvalue problem by a process of condensation; this is discussed in the next section. On the other hand it is perfectly possible to include in the lumped mass matrix for the beam values associated with the rotational freedoms. These values correspond to the mass moment of inertia about the end nodes of half the beam length and the diagonal mass matrix would then become

$$\mathbf{m} = \rho \frac{A l}{2} \begin{matrix} v_1 & \theta_{z1} & v_2 & \theta_{z2} \\ \lceil 1 & l^2/12 & 1 & l^2/12 \rfloor \end{matrix} \qquad (12.43)$$

but, as stated earlier, this is not normally done.

For the basic rectangular membrane element described in Section 8.4.1 and illustrated in Fig. 8.15 (with side lengths $2a$ and $2b$) the lumped mass matrix corresponds simply to lumping one-quarter of the total element mass at each corner node and is

$$\mathbf{m} = \rho a b h \begin{matrix} u_1 & v_1 & u_2 & v_2 & u_3 & v_3 & u_4 & v_4 \\ \lceil 1 & 1 & 1 & 1 & 1 & 1 & 1 & 1 \rfloor \end{matrix}. \qquad (12.44)$$

The consistent mass matrix for this element is given by the solution to Problem 12.14.

For the rectangular plate bending element R1 (or R3) described in Sections 11.3 and 12.6 (again with side lengths $2a$ and $2b$) the lumped mass matrix is simply

$$\mathbf{m} = \rho h a b \begin{matrix} w_1 & \phi_1 & \theta_1 & w_2 & \phi_2 & \theta_2 & w_3 & \phi_3 & \theta_3 & w_4 & \phi_4 & \theta_4 \\ \lceil 1 & 0 & 0 & 1 & 0 & 0 & 1 & 0 & 0 & 1 & 0 & 0 \rfloor \end{matrix}$$

$$(12.45)$$

if contributions associated with rotational degrees of freedom are ignored, or

$$\mathbf{m} = \rho h a b \left\lceil 1 \quad \frac{b^2}{3} \quad \frac{a^2}{3} \quad 1 \quad \frac{b^2}{3} \quad \frac{a^2}{3} \quad 1 \quad \frac{b^2}{3} \quad \frac{a^2}{3} \quad 1 \quad \frac{b^2}{3} \quad \frac{a^2}{3} \right\rfloor \qquad (12.46)$$

if such contributions are included. The consistent mass matrix is given by Table 12.3.

The above are examples of lumped or diagonal mass matrices for relatively simple elements. For more complicated elements with more than the minimum number of nodes and/or number of freedoms per node it is not so straightforward to decide in what proportion the mass of an element should be allocated to the individual nodal freedoms. Different procedures have been used to do this [30] but these will not be discussed here.

Obviously the question arises as to which procedure is the better one to adopt: the consistent mass approach or the lumped mass approach? The consistent mass approach appears to be built on the sounder theoretical foundation and does yield a bounded solution. However, the element consistent mass matrix is fully populated and is more expensive, in terms of computing requirements, to generate than is the diagonal mass matrix; there are also other significant computational advantages associated with a diagonal structure mass matrix. It is quite possible to obtain numerically more accurate results by using the lumped mass approach than by using the consistent one. This might happen since use of lumped masses generally tends to reduce the value of calculated frequencies whilst, for a mesh of compatible elements, the structure stiffness is overestimated which tends to increase the value of frequencies; the two effects can counteract one another to yield accurate frequency estimates. Some studies have been conducted which compare results obtained using the two approaches (and some comparison for line elements is included below). For instance, Tong, Pian, and Bucciarelli [31] considered bar, beam, and membrane finite elements in considerable theoretical detail. They concluded that for simple bar and membrane elements a proper lumped mass formulation will not suffer any loss of rate of convergence of results as compared with the consistent mass formulation; however, for beam and plate elements, and for refined bar and membrane elements based on high-order interpolation, their analysis shows that a consistent mass formulation often will provide a better rate of convergence. In general the weight of evidence does not seem to be sufficiently clear cut to enable dogmatic statements to be made regarding the relative efficiency of the consistent mass and lumped mass approaches.

Example 12.5. Using one and two basic elements in turn, calculate the natural frequencies of extensional vibration of a prismatic bar held at one end and free at the other (see Fig. E12.5). Compare results obtained using the consistent mass and lumped mass approaches.

The exact results for the first two natural frequencies are

$$\omega_1 = \frac{\pi}{2} \sqrt{\left(\frac{E}{\rho L^2}\right)} \qquad \omega_2 = \frac{3\pi}{2} \sqrt{\left(\frac{E}{\rho L^2}\right)}.$$

Fig. E12.5

It is recalled that the stiffness and mass matrices for a basic element are

$$\mathbf{k} = \frac{AE}{l}\begin{bmatrix} 1 & -1 \\ -1 & 1 \end{bmatrix} \qquad \mathbf{m}_{\text{consistent}} = \rho\frac{Al}{6}\begin{bmatrix} 2 & 1 \\ 1 & 2 \end{bmatrix} \qquad \mathbf{m}_{\text{lumped}} = \rho\frac{Al}{2}\begin{bmatrix} 1 & 0 \\ 0 & 1 \end{bmatrix}.$$

One-element solutions

Consistent mass: $\dfrac{AE}{l} - \rho\dfrac{AL}{6}\times 2\omega^2 = 0$; therefore $\omega \equiv \omega_1 = 1.7321\sqrt{\left(\dfrac{E}{\rho L^2}\right)}$.

Lumped mass: $\dfrac{AE}{L} - \rho\dfrac{Al}{2}\times 1\omega^2 = 0$; therefore $\omega \equiv \omega_1 = 1.4142\sqrt{\left(\dfrac{E}{\rho L^2}\right)}$.

Two-element solutions (see Fig. E12.5 for node numbering)

$$\mathbf{K}_r = \frac{AE}{L/2}\begin{matrix} u_1 & u_2 \\ \begin{bmatrix} 2 & -1 \\ -1 & 1 \end{bmatrix} \end{matrix}$$

$$\mathbf{M}_{r\,(\text{consistent})} = \rho\frac{A(L/2)}{6}\begin{matrix} u_1 & u_2 \\ \begin{bmatrix} 4 & 1 \\ 1 & 2 \end{bmatrix} \end{matrix}$$

$$\mathbf{M}_{r\,(\text{lumped})} = \rho\frac{A(L/2)}{2}\begin{matrix} u_1 & u_2 \\ \begin{bmatrix} 2 & 0 \\ 0 & 1 \end{bmatrix} \end{matrix}.$$

Consistent mass:

$$\begin{vmatrix} 2-4\mu_c & -1-\mu_c \\ -1-\mu_c & 1-2\mu_c \end{vmatrix} = 0 \qquad \text{where} \qquad \mu_c = \frac{\rho L^2}{24E}\omega^2.$$

The solution is

$$\omega_1 = 1.6114\sqrt{\left(\frac{E}{\rho L^2}\right)} \qquad\qquad \omega_2 = 5.6293\sqrt{\left(\frac{E}{\rho L^2}\right)}.$$

Lumped mass:

$$\begin{vmatrix} 2-2\mu_l & -1 \\ -1 & 1-\mu_l \end{vmatrix} = 0 \qquad \text{where} \qquad \mu_l = \frac{\rho L^2}{8E}\omega^2$$

The solution is

$$\omega_1 = 1.5307\sqrt{\left(\frac{E}{\rho L^2}\right)} \qquad\qquad \omega_2 = 3.6955\sqrt{\left(\frac{E}{\rho L^2}\right)}.$$

When the results are collected together the percentage errors of the FEM solutions are seen to be as follows:

	Percentage error		
	ω_1		ω_2
	1 element	2 elements	2 elements
Consistent mass	+10.27	+2.59	+19.46
Lumped mass	−9.97	−2.55	−21.58

The percentage errors associated with the two approaches are of very similar magnitude but opposite sign in this example of extensional vibration.

12.8. Eigenvalue economization

It frequently becomes imperative to reduce the effective number of equations of a large-order eigenvalue problem so as either to obtain a solution more economically than otherwise or to achieve a meaningful solution at all. The way in which this is done is to specify a certain number of structure degrees of freedom as independent or 'master' freedoms, with the rest of the freedoms becoming dependent or 'slave' freedoms. The procedure follows directly on from the static condensation procedure described in Section 6.11 and is applied at the structure level rather than the element level. The procedure to reduce the effective size of the dynamic problem is referred to variously as mass reduction, mass condensation, dynamic condensation, or eigenvalue economization; it seems to have been first considered in the FEM by Irons [32] and Guyan [33] and is consequently sometimes also referred to as Guyan reduction.

Consider that we have established the stiffness matrix and mass matrix (whether of the consistent or lumped kind) corresponding to all the degrees of freedom of a finite-element model of a structure other than those set to zero to represent the appropriate structure boundary conditions. Denote these matrices **K** and **M**, omitting the subscript r which has earlier been used to indicate that boundary conditions have been applied, so as to avoid proliferation of subscripts in what follows. Let the column matrix of nodal displacements corresponding to **K** and **M** be **D** and let this be partitioned in the form

$$\mathbf{D} = \begin{Bmatrix} \mathbf{D}_m \\ \mathbf{D}_s \end{Bmatrix} \tag{12.47}$$

where $\mathbf{D}_m$ is the column matrix of selected master degrees of freedom and $\mathbf{D}_s$ is the column matrix of slave degrees of freedom.

Now, for the moment imagine that the problem under consideration is *static* and that there are no loads acting which correspond to any of the $\mathbf{D}_s$ freedoms. The system of static structure equations would be

$$\mathbf{F} = \mathbf{KD} \quad \text{or} \quad \begin{Bmatrix} \mathbf{F}_m \\ \mathbf{F}_s \end{Bmatrix} = \begin{Bmatrix} \mathbf{F}_m \\ \mathbf{0} \end{Bmatrix} = \begin{bmatrix} \mathbf{K}_{mm} & \mathbf{K}_{ms} \\ \mathbf{K}_{sm} & \mathbf{K}_{ss} \end{bmatrix} \begin{Bmatrix} \mathbf{D}_m \\ \mathbf{D}_s \end{Bmatrix}. \tag{12.48}$$

Solving the lower set of equations gives a relationship between the slave and master degrees of freedom: thus

$$\mathbf{D}_s = -\mathbf{K}_{ss}^{-1}\mathbf{K}_{sm}\mathbf{D}_m. \tag{12.49}$$

The upper set of equations now gives

$$\mathbf{F}_m = \mathbf{K}_{mm}\mathbf{D}_m + \mathbf{K}_{ms}\mathbf{D}_s \tag{12.50}$$

or

$$\mathbf{F}_m = (\mathbf{K}_{mm} - \mathbf{K}_{ms}\mathbf{K}_{ss}^{-1}\mathbf{K}_{sm})\mathbf{D}_m \tag{12.51}$$

where $\mathbf{K}_{ms} = \mathbf{K}_{sm}^t$ from the symmetry of the stiffness matrix. The term in parentheses in equation (12.51) is the condensed stiffness matrix

$$\mathbf{K}_c = \mathbf{K}_{mm} - \mathbf{K}_{ms}\mathbf{K}_{ss}^{-1}\mathbf{K}_{sm}. \tag{12.52}$$

The procedure described in this paragraph is the process of static condensation discussed in Section 6.11.

To encompass the dynamic problem we now note that the production of the condensed stiffness matrix $\mathbf{K}_c$ from the full matrix $\mathbf{K}$ can in fact be envisaged as resulting from a co-ordinate transformation of the form

$$\mathbf{D} = \begin{Bmatrix} \mathbf{D}_m \\ \mathbf{D}_s \end{Bmatrix} = \begin{bmatrix} \mathbf{I}_m \\ -\mathbf{K}_{ss}^{-1}\mathbf{K}_{sm} \end{bmatrix} \mathbf{D}_m = \mathbf{T}\mathbf{D}_m \tag{12.53}$$

where $\mathbf{I}_m$ is a unit matrix of the same order as $\mathbf{D}_m$. This is to say that $\mathbf{K}_c$ given by equation (12.52) can be alternatively expressed as

$$\mathbf{K}_c = \mathbf{T}^t\mathbf{K}\mathbf{T}. \tag{12.54}$$

For the free vibration problem we now assume that the static relationship between total degrees of freedom and master degrees of freedom, expressed by equation (12.53), still applies and can be used to produce a condensed mass matrix. The full mass matrix is, in partitioned form,

$$\mathbf{M} = \begin{bmatrix} \mathbf{M}_{mm} & \mathbf{M}_{ms} \\ \mathbf{M}_{sm} & \mathbf{M}_{ss} \end{bmatrix} \tag{12.55}$$

(with $\mathbf{M}_{ms} = \mathbf{M}_{sm}^t$) and the condensed mass matrix becomes

$$\mathbf{M}_c = \mathbf{T}^t\mathbf{M}\mathbf{T} \tag{12.56}$$

with $\mathbf{T}$ defined in equation (12.53). It is noted that $\mathbf{T}$ is dependent upon stiffness terms and hence that the condensed mass matrix contains contributions from both the full structure stiffness matrix and the full mass matrix. Expanding equation (12.56) in fact yields

$$\mathbf{M}_c = \mathbf{M}_{mm} \quad -\mathbf{M}_{ms}\mathbf{K}_{ss}^{-1}\mathbf{K}_{sm} \quad -\mathbf{K}_{ms}\mathbf{K}_{ss}^{-1}\mathbf{M}_{sm} + \mathbf{K}_{ms}\mathbf{K}_{ss}^{-1}\mathbf{M}_{ss}\mathbf{K}_{ss}^{-1}\mathbf{K}_{sm}$$

(12.57)

which is a considerably more complicated expression than that for the condensed stiffness matrix (equation (12.52)).

With $\mathbf{K}_c$ and $\mathbf{M}_c$ established as above the free vibration problem is now governed by the set of equations

$$(\mathbf{K}_c - \omega^2\mathbf{M}_c)\mathbf{D}_c = \mathbf{0}$$

(12.58)

which is, of course, a statement of an eigenvalue problem of lower order than the original one.

The above procedure to produce a condensed eigenvalue problem relies on the assumption that equations (12.49) and (12.53), which were developed through static considerations, apply to the dynamic problem. In fact this will be exactly so only in certain circumstances, though it will be approximately so in others. A particular circumstance in which the assumption is exact occurs when $\mathbf{M}$ is a lumped or diagonal mass matrix with only the $\mathbf{D}_m$ degrees of freedom having associated inertia (as would occur, for instance, when considering beam or plate vibration and ignoring inertia associated with rotational degrees of freedom (see Example 12.6)). Then $\mathbf{M}_{ms}$, $\mathbf{M}_{sm}$, and $\mathbf{M}_{ss}$ are null matrices and $\mathbf{M}_c = \mathbf{M}_{mm}$ with $\mathbf{K}_c$ defined by equation (12.52). The condensed system of equations (equation (12.58)) will in these circumstances be exactly equivalent to the original, full system of equations and will provide precisely the same eigenvalues. In more general circumstances where both the $\mathbf{D}_m$ and $\mathbf{D}_s$ degrees of freedom have associated inertia the assumed relationship between the degrees of freedom is not precise in the vibration problem. The aim then is to select for the slave freedoms $\mathbf{D}_s$ only those freedoms associated with small inertia contributions and to retain as master freedoms $\mathbf{D}_m$ only sufficient freedoms to represent adequately those vibration modes which are of interest. There is a considerable element of judgement in this, though there are some pointers to success. For beam, plate, and shell problems, for instance, the lower modes of vibration are usually the ones of interest and are of a flexural nature; thus it makes sense to retain many or all of the transverse nodal displacements as master degrees of freedom and to select many of the in-plane nodal displacements and the rotational nodal displacements as slave degrees of freedom. It should be clear that in using the eigenvalue economizer technique we shall not be able to determine the higher frequencies of vibration of the original model structure because some of the degrees of freedom have been discarded. The frequencies for the lower modes of vibration that are determined from the equation set (12.58) will generally be higher than the frequencies determined from the full uncondensed equation set because of the constraints that are imposed on the displacements.

Example 12.6. A uniform cantilever beam is modelled with two equal basic beam finite elements as shown in Fig. E12.6. Compare solutions for the lowest two natural frequencies obtained by
    (a) use of the full, consistent mass matrix
    (b) use of a lumped mass matrix, taking into account only translational inertia
    (c) use of a condensed form of the consistent mass matrix obtained by eliminating the rotational degrees of freedom using the eigenvalue economizer method.

(a) The structure degrees of freedom are v_1, θ_{z1}, v_2, θ_{z2}, and the governing set of four equations of the form of equation (12.23) are assembled in the standard way from the contributions of the two elements to give

$$\left(EI_z \begin{bmatrix} 192/L^3 & & \text{Symmetric} & \\ 0 & 16/L & & \\ -96/L^3 & -24/L^2 & 96/L^3 & \\ 24/L^2 & 4/L & -24/L^2 & 8/L \end{bmatrix} \right.$$

$$\left. - \frac{\rho A L \omega^2}{840} \begin{bmatrix} 312 & & \text{Symmetric} & \\ 0 & 2L^2 & & \\ 54 & 13L/2 & 156 & \\ -13L/2 & -3L^2/4 & -11L & L^2 \end{bmatrix} \right) \begin{Bmatrix} v_1 \\ \theta_{z1} \\ v_2 \\ \theta_{z2} \end{Bmatrix} = \begin{Bmatrix} 0 \\ 0 \\ 0 \\ 0 \end{Bmatrix}.$$

The first two natural frequencies obtained by solving this eigenvalue problem are (see Table 12.2)

$$\omega_1 = 3.5177 \sqrt{\left(\frac{EI_z}{\rho A L^4}\right)} \qquad \omega_2 = 22.221 \sqrt{\left(\frac{EI_z}{\rho A L^4}\right)}.$$

(b) The master and slave degrees of freedom are $\mathbf{D}_m = \{v_1 \; v_2\}$, $\mathbf{D}_s = \{\theta_{z1} \; \theta_{z2}\}$. We rearrange rows and columns of the stiffness matrix of part (a) to correspond to these definitions of $\mathbf{D}_m$ and $\mathbf{D}_s$. Thus

$$\mathbf{K} = \begin{bmatrix} \mathbf{K}_{mm} & \mathbf{K}_{ms} \\ \mathbf{K}_{sm} & \mathbf{K}_{ss} \end{bmatrix} = \begin{array}{cccc} & v_1 & v_2 & \theta_{z1} & \theta_{z2} \\ & \left[\begin{array}{cc:cc} 192/L^3 & -96/L^3 & 0 & 24/L^2 \\ -96/L^3 & 96/L^3 & -24/L^2 & -24/L^2 \\ \hdashline 0 & -24/L^2 & 16/L & 4/L \\ 24/L^2 & -24/L^2 & 4/L & 8/L \end{array} \right] \end{array}$$

           L/2   L/2

Fig. E12.6

The condensed stiffness matrix $\mathbf{K}_c$ can be calculated using equation (12.52) or using equation (12.54) with $\mathbf{T}$ defined in equation (12.53). Choosing the latter form we can readily determine that

$$\mathbf{T} = \left[\begin{array}{c} \mathbf{I}_m \\ \hline -\mathbf{K}_{ss}^{-1}\mathbf{K}_{sm} \end{array}\right] = \frac{1}{7L} \left[\begin{array}{cc} 7L & 0 \\ 0 & 7L \\ \hline 6 & 6 \\ -24 & 18 \end{array}\right].$$

Hence

$$\begin{array}{cc} v_1 & v_2 \end{array}$$
$$\mathbf{K}_c = \mathbf{T}^t\mathbf{K}\mathbf{T} = \frac{EI_z}{L^3} \left[\begin{array}{cc} 109.714 & -34.286 \\ -34.286 & 13.714 \end{array}\right].$$

For the lumped mass approach with only translational inertia taken into account we use the element matrix given by equation (12.42) but simply ignore the null θ_z entries. The condensed structure mass matrix is then assembled from the two element contributions as

$$\begin{array}{cc} v_1 & v_2 \end{array}$$
$$\mathbf{M}_c = \frac{\rho AL}{4} \left[\begin{array}{cc} 2 & 0 \\ 0 & 1 \end{array}\right].$$

With these definitions of $\mathbf{K}_c$ and $\mathbf{M}_c$ the statement of the eigenvalue problem becomes

$$\left|\begin{array}{cc} 109.714 - 2\gamma_l & -34.286 \\ -34.286 & 13.714 - \gamma_l \end{array}\right| = 0$$

where $\gamma_l = \rho AL^4\omega^2/4EI_z$. The solution is $\omega_1 = 3.1562\sqrt{(EI_z/\rho AL^4)}$, $\omega_2 = 16.258\sqrt{(EI_z/\rho AL^4)}$.

(c) Rearrange the consistent mass matrix of part (a) to correspond to the $\mathbf{K}$ matrix of part (b). Thus

$$\begin{array}{cccc} v_1 & v_2 & \theta_{z1} & \theta_{z2} \end{array}$$
$$\mathbf{M} = \left[\begin{array}{cc} \mathbf{M}_{mm} & \mathbf{M}_{ms} \\ \mathbf{M}_{sm} & \mathbf{M}_{ss} \end{array}\right] = \frac{\rho AL}{840} \left[\begin{array}{cc|cc} 312 & 54 & 0 & -13L/2 \\ 54 & 156 & 13L/2 & -11L \\ \hline 0 & 13L/2 & 2L^2 & -3L^2/4 \\ -13L/2 & -11L & -3L^2/4 & L^2 \end{array}\right].$$

The condensed mass matrix can be calculated using equation (12.56) with $\mathbf{T}$ as defined in part (b) (or using equation (12.57)) and becomes

$$\begin{array}{cc} v_1 & v_2 \end{array}$$
$$\mathbf{M}_c = \frac{\rho AL}{840 \times 49} \left[\begin{array}{cc} 18336 & 3615 \\ 3615 & 5652 \end{array}\right].$$

The condensed stiffness matrix is the same as in part (b) and thus the statement of the eigenvalue problem in this case is

$$\left|\begin{array}{cc} 109.714 - 18336\gamma_e & -34.286 - 3615\gamma_e \\ -34.286 - 3615\gamma_e & 13.714 - 5652\gamma_e \end{array}\right| = 0$$

where $\gamma_e = \rho A L^4 \omega^2 / (840 \times 49 E I_z)$. The solution is $\omega_1 = 3.5220\sqrt{(E I_z/\rho A L^4)}$, $\omega_2 = 22.279\sqrt{(E I_z/\rho A L^4)}$.

For convenience the results are collected together and compared with the exact solution in the following table.

| | Values of $\omega\sqrt{(\rho A L^4/E I_z)}$ | | | |
Mode	Full system, 4 degrees of freedom as (a)	Lumped mass, 2 degrees of freedom as (b)	Eigenvalue economization, 2 degrees of freedom as (c)	Exact solution
1	3.5177	3.1562	3.5220	3.5160
2	22.221	16.258	22.279	22.034

It is seen that the eigenvalue economization results (c) are only very slightly higher than the full-system results (a) and the exact results (and are superior to the full-system results with two degrees of freedom quoted in Table 12.2). The results (b) obtained using the lumped mass approach are very considerably lower than the exact results.

Following on from Example 12.6, further results obtained using the three kinds of approach described therein are presented graphically in Fig. 12.6. It should be noted that the abscissa shows the number of

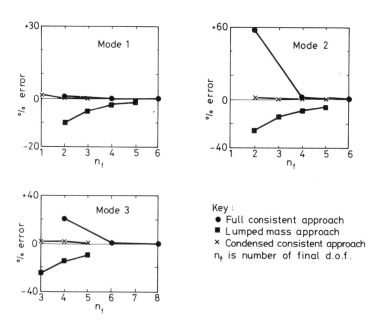

Fig. 12.6. Error in calculated frequencies using different approaches for a cantilever beam.

degrees of freedom in the final eigenvalue problem rather than the number of elements used. Neither of these measures is a perfect base for estimating relative performance; it should be remembered, particularly with regard to using the number of degrees of freedom as a base, that significant extra matrix operations are required in the eigenvalue economization approach, as compared with the full approach, before the eigenvalue solution stage is reached. Nevertheless, the results obtained using the eigenvalue economizer approach in conjunction with the consistent mass matrix are impressive. Convergence is slow in this problem when using the lumped mass approach. Related further results for beam problems are documented in reference 10.

We now turn to a larger-order problem; frequency results for a square cantilever plate based on the use of a regular 4×4 mesh of type R1

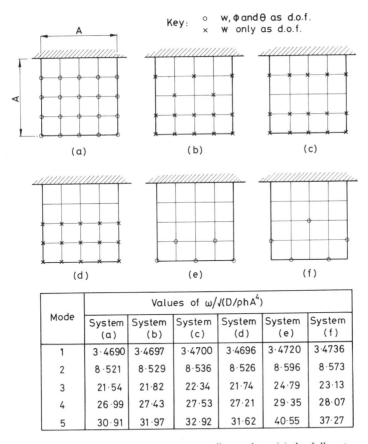

Mode	Values of $\omega/\sqrt{(D/\rho hA^4)}$					
	System (a)	System (b)	System (c)	System (d)	System (e)	System (f)
1	3·4690	3·4697	3·4700	3·4696	3·4720	3·4736
2	8·521	8·529	8·536	8·526	8·596	8·573
3	21·54	21·82	22·34	21·74	24·79	23·13
4	26·99	27·43	27·53	27·21	29·35	28·07
5	30·91	31·97	32·92	31·62	40·55	37·27

Fig. 12.7. Finite element results for a square cantilever plate: (a) the full system with 60 d.o.f.; (b)–(f) condensed systems with 15 d.o.f.

elements (with stiffness and consistent mass matrices given in Tables 11.2 and 12.3 respectively) and using different selections of 15 master degrees of freedom in the eigenvalue economizer approach have been presented by Jennings [34]. Figure 12.7 shows the arrangement of master degrees of freedom for the full problem, designated (a), and five condensed problems, designated (b)–(f), with the corresponding calculated frequencies for the first five modes of vibration (The results for the full problem are taken from reference 27). Of the condensed systems that of (d), which uses the lateral displacements only at nodes furthest from the clamped edge as master freedoms, gives the closest comparison of results with the full system.

The same cantilever plate has also been considered by Anderson et al. [24] who used triangular plate elements and various patterns of condensation. This reference also gives further examples of the use of eigenvalue economization, including one rectangular plate problem in which the initial problem contains 936 degrees of freedom which are condensed to 32 degrees of freedom, where the results compare well with known solutions. Amongst other works dealing with eigenvalue economization are those of Ramsden and Stoker [35] and Henshell and Ong [36], the latter giving a method for the automatic selection of the master degrees of freedom by computer.

12.9. Buckling of a strut

In this and the following two sections the use of the FEM in solving problems of structural instability or buckling is described. Only a brief treatment of a limited aspect of the problem is given with attention restricted to the problem of linear elastic instability (of the flexural type); the only geometric forms considered are the slender (Bernoulli–Euler) beam and the thin (classical) plate.

In linear elastic instability analysis the stresses within a structure have a known set form of distribution and increase linearly with increase in the magnitude of the applied loading until a critical level is reached at which the structure becomes unstable and buckling, characterized by a sudden change in the form of the deformation, occurs. The aim of linear elastic instability analysis is to determine the critical magnitude of the applied loading at which buckling occurs and also the corresponding form that the buckled structure takes up, i.e. the buckled shape or buckled mode. The problem is another example of an eigenvalue problem, as will be demonstrated.

Consider the slender strut shown in Fig. 12.8(a) which is subjected to a compressive axial force P increasing progressively from zero value. (For convenience the strut is shown with pinned ends but this need not be so in

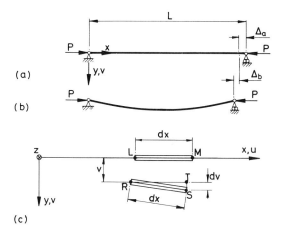

Fig. 12.8. Buckling of a strut: (a) prebuckling deformation; (b) buckling deformation; (c) movement of an infinitesimal length of strut during buckling when end movement is allowed.

the analysis that follows.) The strut is assumed to have a doubly symmetrical cross-section with I_z as the least principal second moment of area, so that any bending will take place about the z axis or, in other words, in the xy plane. The usual assumptions are made, which include the statements that the strut is perfectly straight and unstressed prior to loading, the load acts longitudinally through the centroid of the cross-section, and material behaviour is linearly elastic.

As P increases from zero the strut will initially simply be compressed axially, the amount of compression being directly proportional to the magnitude of P, and the strut remains straight in a state of stable equilibrium. The compressive force effectively reduces the bending resistance of the strut until at a critical magnitude of P this resistance vanishes altogether, the straight configuration becomes one of unstable equilibrium, and buckling occurs such that a new state of stable equilibrium is taken up in which the strut deflects laterally to the bent configuration shown in Fig. 12.8(b). Immediately prior to buckling the strut in its straight configuration will of course have shortened by an amount Δ_a say (where $\Delta_a = PL/AE$). During the actual process of buckling the load P does not change in magnitude and it is assumed that no further change in the length of the centroidal axis occurs, i.e. the centroidal axis is regarded as inextensible during buckling. However, as shown in Fig. 12.8(b) the ends of the strut do approach each other of course, by an amount Δ_b, owing to the bending that takes place.

We wish to examine the buckling problem on the basis of use of the PMPE. To do this we take as the datum position from which to measure the change in total potential energy that position immediately prior to buckling. This allows us to eliminate from consideration the potential energy of the load P with regard to the pre-buckling axial deformation Δ_a and the corresponding strain energy of compression. It leaves for consideration the potential energy of P with regard to the axial shortening Δ_b and the strain energy associated with the bending of the strut. The change in total potential energy between configurations immediately prior to and immediately after buckling is therefore

$$\Pi_p = U_{p\,(bend)} + V_{pg} \tag{12.59}$$

where

$$U_{p\,(bend)} = \int_0^L \frac{EI_z}{2} \left(\frac{d^2v}{dx^2}\right)^2 dx \tag{12.60}$$

is the usual bending strain energy (within Bernoulli–Euler theory) and

$$V_{pg} = -P\Delta_b \tag{12.61}$$

represents the loss of potential energy of the end thrust as the strut bends and the ends approach each other. We now need to establish an expression for Δ_b in terms of the lateral deflection.

Figure 12.8(c) shows an infinitesimally short length of strut lying horizontally, in position LM, before buckling and inclined to the horizontal, in position RS, after buckling. Since the centroidal axis is assumed to be inextensible during buckling, the lengths LM and RS are the same and are equal to dx. After buckling the infinitesimally short length of strut has a horizontal projection RT given by

$$RT = \sqrt{(dx^2 - dv^2)} = dx\left(1 - \left(\frac{dv}{dx}\right)^2\right)^{1/2} = dx\left(1 - \frac{1}{2}\left(\frac{dv}{dx}\right)^2\right)$$

where the binomial theorem has been used to establish the last term of this equation, with higher powers of $(dv/dx)^2$ neglected in view of the assumed smallness of the deformations. The apparent shortening of the length LM during buckling is therefore $\frac{1}{2}(dv/dx)^2$ and the total apparent shortening Δ_b is this quantity integrated with respect to x between 0 and L. Thus

$$V_{pg} = -\frac{1}{2}\int_0^L P\left(\frac{dv}{dx}\right)^2 dx \tag{12.62}$$

is the change in potential energy of the axial force. The total potential energy (change) is now

$$\Pi_p = U_{p\,(bend)} + V_{pg} = \int_0^L \frac{EI_z}{2}\left(\frac{d^2v}{dx^2}\right)^2 dx - \frac{1}{2}\int_0^L P\left(\frac{dv}{dx}\right)^2 dx \tag{12.63}$$

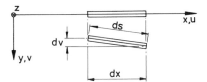

Fig. 12.9. Movement of an infinitesimal length of strut during buckling when end movement is prevented.

This same expression for Π_p can be derived in another equally valid way by making a different assumption pertaining to the process of buckling. Specifically we now assume that at the critical level of P lateral buckling occurs with the ends of the strut prevented from having any u movement. Thus in Fig. 12.8(b) the displacement Δ_b does not occur and instead the centroidal axis of the strut is stretched owing solely to the rotation occurring in the lateral deflection associated with the buckled configuration. An infinitesimally short length of strut dx lying horizontally before buckling becomes inclined to the horizontal after buckling with stretched length ds, as shown in Fig. 12.9. Now

$$ds = dx\left(1+\left(\frac{dv}{dx}\right)^2\right)^{1/2} \approx dx\left(1+\frac{1}{2}\left(\frac{dv}{dx}\right)^2\right).$$

Thus the axial strain of the centroidal axis *owing solely to the rotation of the length dx* is

$$\frac{ds-dx}{dx} = \frac{1}{2}\left(\frac{dv}{dx}\right)^2. \tag{12.64}$$

It is noted that this forms part of the expression for total axial strain at the centroidal axis, defined as

$$\varepsilon = \frac{du}{dx}+\frac{1}{2}\left(\frac{dv}{dx}\right)^2,$$

or of the expression for total axial strain at a fibre at distance y from the centroid, defined as

$$\varepsilon = \frac{du}{dx} - y\frac{d^2v}{dx^2}+\frac{1}{2}\left(\frac{dv}{dx}\right)^2. \tag{12.65}$$

In this equation du/dx is the usual linear axial strain and $-y\,d^2v/dx^2$ is the usual linear bending strain; in the present problem the first of these terms can be ignored since it affects only pre-buckling deformation whilst the second term will give rise to the usual strain energy of bending $U_{p(bend)}$. The non-linear part of the strain, given by equation (12.64), is usually ignored in most circumstances as being a very small quantity but must be included in the present analysis because it is associated with a large axial force P. (The present analysis is an example of what is termed a geometrically non-linear problem.) The change in strain energy released due to stretching of the centroidal axis during buckling (when P remains unchanged in magnitude) is

$$(U_p)_{\text{stretch}} = -\frac{1}{2} \int_0^L P \left(\frac{\mathrm{d}v}{\mathrm{d}x} \right)^2 \mathrm{d}x. \tag{12.66}$$

In this view of the buckling process there is no change in potential energy of the end forces (since Δ_b is zero) and thus the total potential energy change is

$$\Pi_p = U_{p(\text{bend})} + U_{p(\text{stretch})} \tag{12.67}$$

which gives exactly the same result as equation (12.63).

Either of the two views of the buckling process described above can be taken and it does not really matter which is adopted since the result is identical. In what follows, though, we shall adopt the second philosophy and refer to strain energy of stretching $U_{p(\text{stretch})}$ rather than to change in potential energy of external forces V_{pg}.

The total potential energy Π_p (equation (12.63)) depends on the form of the lateral deflection $v(x)$. We could examine *all* possible forms of $v(x)$ that satisfy the end conditions and make Π_p a minimum corresponding to the equilibrium state of the buckled strut. This could be done through the calculus of variations and would result in the governing differential equation of the strut buckling problem which is

$$\frac{\mathrm{d}^2}{\mathrm{d}x^2} \left(EI_z \frac{\mathrm{d}^2 w}{\mathrm{d}x^2} \right) + P \frac{\mathrm{d}^2 w}{\mathrm{d}x^2} = 0 \tag{12.68}$$

together with statements of the natural boundary conditions (those of zero bending moment for the pinned-end case) at the two ends; it would still remain to solve these equations, if possible, to give the exact result for particular cases. This is not our purpose here, of course, since we are concerned with examining a restricted range of admissible possible forms of $v(x)$ and obtaining approximate solutions through the finite-element approach.

12.10. Buckling of struts using the FEM

Consider now that a beam finite element forms part (or all) of a strut and that when buckling occurs the deflected form of the element is represented in the familiar way

$$\mathbf{u} \equiv v = \mathbf{Nd}.$$

The bending strain energy of the element U_p is unchanged from static analysis and can be expressed as

$$U_{p(bend)} = \tfrac{1}{2}\mathbf{d}^t\mathbf{kd} \tag{12.69}$$

where $\mathbf{k}$ is the conventional flexural stiffness matrix for an element.

To evaluate the new stretching strain energy term (equation (12.66)) for the element we require an expression for dv/dx; this can be written

$$\frac{dv}{dx} = \mathbf{Hd} \qquad \text{where} \qquad \mathbf{H} = \frac{d}{dx}\mathbf{N}. \tag{12.70}$$

Then

$$U_{p(stretch)} = -\tfrac{1}{2}\mathbf{d}^t\mathbf{gd} \tag{12.71}$$

where

$$\mathbf{g} = \int_0^l P\mathbf{H}^t\mathbf{H}\,dx \tag{12.72}$$

for an element of length l with co-ordinate origin at one end. The energy $U_{p(stretch)}$ is clearly a quadratic function of the nodal displacements and the matrix $\mathbf{g}$ is clearly a symmetric matrix of the same size as the conventional beam stiffness matrix $\mathbf{k}$. Matrix $\mathbf{g}$ can be used in association with $\mathbf{k}$ in a range of geometrically non-linear problems, and not just in the linear instability problem considered here. It is used to take account of the effect of in-plane loading (i.e. axial loading for the beam) on bending stiffness; a compressive loading reduces the effective bending stiffness (to zero at the buckling load) and a tensile loading increases the bending stiffness. Matrix $\mathbf{g}$ is commonly referred to as the *geometric stiffness matrix* or alternatively as the initial stress stiffness matrix or the stability coefficient matrix.

If v were expressed in terms of generalized displacements rather than directly in shape function form then $\mathbf{N}$ could be replaced by $\boldsymbol{\alpha}\mathbf{C}^{-1}$ and correspondingly $\mathbf{H}$ could be replaced by $\mathbf{LC}^{-1}$ where $\mathbf{L} = d\boldsymbol{\alpha}/dx$. Then $\mathbf{g}$ could be alternatively expressed as

$$\mathbf{g} = (\mathbf{C}^{-1})^t\left(\int_0^l P\mathbf{L}^t\mathbf{L}\,dx\right)\mathbf{C}^{-1}. \tag{12.73}$$

Before considering in detail the evaluation of matrix $\mathbf{g}$ for any specific type of element we proceed to establish symbolically the set of equations pertaining to an assemblage of E elements, the typical one of which is element e. This follows the now familiar pattern established in Sections 6.6, for the static problem, and 12.3 for the dynamic problem. The total potential energy of the whole structure is

$$\bar{\bar{\Pi}}_p = \sum_{e=1}^{E} \Pi_p^e = \sum_{e=1}^{E} U_{p(\text{bend})}^e + \sum_{e=1}^{E} U_{p(\text{stretch})}^e \tag{12.74}$$

which can be expressed

$$\bar{\bar{\Pi}}_p = \tfrac{1}{2}\mathbf{D}^t \left(\sum_{e=1}^{E} \mathbf{k}_e^* \right)\mathbf{D} - \tfrac{1}{2}\mathbf{D}^t \left(\sum_{e=1}^{E} \mathbf{g}_e^* \right)\mathbf{D}$$
$$= \tfrac{1}{2}\mathbf{D}^t\mathbf{K}\mathbf{D} - \tfrac{1}{2}\mathbf{D}^t\mathbf{G}\mathbf{D}. \tag{12.75}$$

The structure matrix $\mathbf{G}$ is assembled from element matrices $\mathbf{g}_e$ in precisely the same way that structure matrix $\mathbf{K}$ is assembled from element matrices $\mathbf{k}_e$, i.e. by the direct stiffness procedure. $\mathbf{G}$ has a similar arrangement of terms as does $\mathbf{K}$ (provided that the same displacement field is used in deriving the corresponding element matrices). Performing the minimizing procedure and at the same time applying the boundary conditions by setting appropriate degrees of freedom to zero gives the set of n equations

$$(\mathbf{K}_r - \mathbf{G}_r)\mathbf{D}_r = \mathbf{0}. \tag{12.76}$$

Now $\mathbf{G}_r$, assembled from various element $\mathbf{g}$ matrices, depends of course upon the magnitude and distribution of axial force (in the case of the beam) or of membrane stress (in the case of the plate, as will be seen) in each element. The distribution of axial force or membrane stress throughout the structure is assumed to be known and fixed. What has to be found is the intensity of loading which causes buckling.

This is usually done by setting $\mathbf{G}_r = f\bar{\mathbf{G}}_r$ where $\bar{\mathbf{G}}_r$ is $\mathbf{G}_r$ evaluated at some arbitrarily chosen level of loading and f is a *load factor*. The set of n equations governing the buckling problem then becomes

$$(\mathbf{K}_r - f\bar{\mathbf{G}}_r)\mathbf{D}_r = \mathbf{0} \tag{12.77}$$

and we seek the critical values of load factor f and the corresponding forms of $\mathbf{D}_r$ that satisfy these equations. These critical values correspond to situations when the effect of compressive axial force on bending behaviour reduces the total structure bending stiffness to zero and hence leads to buckling. Normally our interest is confined to the lowest of the critical load factors, corresponding to initial elastic buckling; the higher critical load factors are of little practical interest.

Equations (12.77) are of the same form as equations (12.23), which govern the free vibration problem, and thus define an eigenvalue problem. The calculation of critical values of f follows the same procedures as those outlined earlier for the calculation of values of the squares of

natural frequencies and generally there is complete correspondence between f and ω^2. In particular the characteristic equation governing
buckling is

$$|\mathbf{K_r} - f\bar{\mathbf{G}}_r| = 0 \tag{12.78}$$

and expansion of the determinant provides a polynomial equation of
degree n in f which can form the basis of a manual solution when n is
small. For larger n some other method of solving the eigenvalue problem
has to be selected. Methods of solution of eigenvalue problems are briefly
discussed in broad terms in Appendix A. For the buckling problem it
should be borne in mind that usually the interest is in determining only
one eigenvalue/vector corresponding to the lowest critical load. Thus a
popular approach is to recast equation (12.77) in the form (by premultiplying both terms in the parentheses by $\mathbf{K}_r^{-1}$)

$$\frac{1}{f}\mathbf{D_r} = \mathbf{K}_r^{-1}\bar{\mathbf{G}}_r\mathbf{D_r}. \tag{12.79}$$

If an iterative method is used to obtain a solution this form has the virtue
that the largest eigenvalue (value of $1/f$), which is found first in the
iterative method, corresponds to the lowest value of the load factor f.

We now concentrate attention on the basic beam finite element (illustrated in Fig. 6.2) to provide a specific example of the evaluation and use
of an element geometric stiffness matrix. We choose to use the form given
by equation (12.73) to evaluate the geometric stiffness matrix and note
that for the basic beam element

$$\mathbf{L} = \frac{\mathrm{d}}{\mathrm{d}x}\boldsymbol{\alpha} = \frac{\mathrm{d}}{\mathrm{d}x}[1 \quad x \quad x^2 \quad x^3] = [0 \quad 1 \quad 2x \quad 3x^2] \tag{12.80}$$

and $\mathbf{C}^{-1}$ is recorded in equation (6.22). It is a straightforward procedure
to evaluate $\mathbf{g}$ for this element which becomes

$$\mathbf{g} = \frac{P}{30l}\begin{bmatrix} \overset{v_1}{36} & \overset{\theta_{z1}}{} & \overset{v_2}{} & \overset{\theta_{z2}}{} \\ 3l & 4l^2 & \text{Symmetric} & \\ -36 & -3l & 36 & \\ 3l & -l^2 & -3l & 4l^2 \end{bmatrix}. \tag{12.81}$$

This geometric stiffness matrix corresponds in consistent fashion (since
the same deflection expression is used) to the usual elastic stiffness matrix
$\mathbf{k}$ defined by equation (6.32); it was originally introduced by Gallagher
and Padlog [37]. Buckling loads calculated through the use of the element
matrices defined in equations (12.81) and (6.32) will be upper bounds to
the exact buckling loads.

Example 12.7. Calculate the buckling load of a uniform pin-ended strut of length L using a single element 1–2 in a symmetric half of the strut, as indicated in Fig. E12.7.

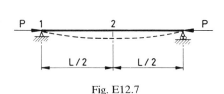

Fig. E12.7

The FEM solution of this type of problem proceeds in very similar fashion to the free vibration problem considered earlier (Example 12.1) with the geometric stiffness matrix replacing the mass matrix of course.

The boundary conditions for the symmetric half of the structure are that $v_1 = \theta_{z2} = 0$, so that the problem degrees of freedom are θ_{z1} and v_2 only.

The **k** and **g** matrices for the single element 1–2 are given by equations (6.32) and (12.81) respectively. From these, having applied the displacement boundary conditions, we obtain (with $l = L/2$)

$$\mathbf{K}_r = 8\frac{EI_z}{L^3}\begin{bmatrix} L^2 & -3L \\ -3L & 12 \end{bmatrix} \qquad \mathbf{G}_r = \frac{P}{15L}\begin{bmatrix} L^2 & -3L/2 \\ -3L/2 & 36 \end{bmatrix} = f\bar{\mathbf{G}}_r.$$

where the columns are labelled θ_{z1}, v_2 for both matrices.

If $\bar{\mathbf{G}}_r$ is evaluated when P has unit value then the load factor f at which buckling occurs is simply identical with P_{cr}, the critical magnitude of P, in this quite trivial example. Thus the condition governing buckling (equation (12.78)) becomes

$$|\mathbf{K}_r - f\bar{\mathbf{G}}_r| = \begin{vmatrix} L^2(1-\beta) & L(-3+3\beta/2) \\ L(-3+3\beta/2) & 12-36\beta \end{vmatrix} = 0$$

where $\beta = fL^2/120EI_z$. It follows that

$$135\beta^2 - 156\beta + 12 = 0.$$

The lowest root of this equation is $\beta = 0.082\,865$ and therefore

$$f = P_{cr} = (0.082\,865)120\frac{EI_z}{L^2} = 9.9438\frac{EI_z}{L^2} = 1.00752\pi^2\frac{EI_z}{L^2}.$$

The well-known 'exact' solution is $\pi^2 EI_z/L^2$ and so the error in the FEM solution is only 0.752 per cent.

To give some idea of the rate of convergence of results in buckling situations when using the basic beam element the above example of the pin-ended strut is extended by considering other regular element meshes. The results are given in Table 12.7 where it is seen that convergence is quite rapid from the over-stiff side. Some other numerical results for the use of the prismatic basic element in beam and frame structures are given by Hartz [38] whilst tapered beams have been considered by Gallagher and Lee [39]. Refined beam elements could also be used in instability

TABLE 12.7
Buckling of pin-ended strut

N_E^{*a}	1	2	3	4	5	6
Percentage error[b]	21.5	0.75	0.16	0.05	0.02	0.01

[a] Number of elements in complete beam.
[b] Percentage error in FEM result as compared with the exact solution of $\pi^2 EI_z/L^2$.

analysis, of course, but results for this do not appear to be available in the literature. Beam elements based on Timoshenko theory rather than Bernoulli–Euler theory were discussed at an early stage of development of finite-element instability analysis by Rodden *et al.* [40] and by Archer [11].

The geometric stiffness matrix of equation (12.81) can be termed 'consistent' since it is based on the same displacement field as is the conventional stiffness matrix. Just as is the case with the mass matrix, it should be mentioned that it is also possible to develop inconsistent geometric stiffness matrices by using a lower-order displacement field for their development than for the development of the associated conventional stiffness matrix. For example, the basic beam element could, for the purpose of calculating $\mathbf{g}$ (only), be assumed to have the linear displacement variation

$$v = \left(1 - \frac{x}{l}\right)v_1 + \frac{x}{l}v_2.$$

Then the (inconsistent) geometric stiffness matrix becomes

$$\mathbf{g} = \frac{P}{l} \begin{matrix} & \begin{matrix} v_1 & \theta_{z1} & v_2 & \theta_{z2} \end{matrix} \\ & \begin{bmatrix} 1 & 0 & -1 & 0 \\ 0 & 0 & 0 & 0 \\ -1 & 0 & 1 & 0 \\ 0 & 0 & 0 & 0 \end{bmatrix} \end{matrix} \qquad (12.82)$$

with only translational displacements now making a contribution. The rotational displacements are easily eliminated from this $\mathbf{g}$ matrix and correspondingly $\mathbf{k}$ could be condensed to correspond to only v_1 and v_2 by the same process of eigenvalue economization described in Section 12.8 in relation to the free vibration problem. Use of the inconsistent $\mathbf{g}$ matrix of equation (12.82) would destroy any bound conditions on a solution and would be of questionable value when compared with use of the consistent $\mathbf{g}$ matrix of equation (12.81).

The process of eigenvalue economization can generally be applied to the linear buckling problem just as it can to the free vibration problem.

12.11. Buckling of plates

The method presented in the preceding two sections for the FEM analysis of the buckling of struts can be extended quite directly to embrace the corresponding plate problem. In this problem the middle plane of a flat plate is subjected to applied membrane stresses of known distribution whose magnitude, which is governed by a load factor f, is increased until a critical level is reached at which flexural buckling occurs. In general the plate may be of arbitrary shape and the membrane stress distribution may also be arbitrary, but many of the available results are for rectangular plates and uniform applied stresses σ_x, σ_y, and τ_{xy}, as illustrated in Fig. 12.10.

Plate bending elements, based on the use of the classical plate theory, have been considered in detail in Chapter 11 and the derivation of conventional flexural stiffness matrices (the $\mathbf{k}$ matrices) has been dealt with there. The derivation of the geometric stiffness matrix $\mathbf{g}$ for a plate element is based on the definition of the stretching strain energy arising from the membrane stresses acting on the second-order strains induced by the deflection w of the middle surface occurring in the buckling process. These second-order strains are the direct strains $\frac{1}{2}(\partial w/\partial x)^2$ and $\frac{1}{2}(\partial w/\partial y)^2$ in the x and y directions respectively (akin to equation (12.64) for the beam), and the shear strain $(\partial w/\partial x)(\partial w/\partial y)$ [5–7]. The stretching strain energy is

$$U_{\text{p(stretch)}} = -\frac{1}{2} \iint \left(\sigma_x \left(\frac{\partial w}{\partial x}\right)^2 + \sigma_y \left(\frac{\partial w}{\partial y}\right)^2 + 2\tau_{xy} \frac{\partial w}{\partial x} \frac{\partial w}{\partial y} \right) h \, dx \, dy \qquad (12.83)$$

In matrix form this can be expressed as

$$U_{\text{p(stretch)}} = -\frac{1}{2} \iint \begin{bmatrix} \dfrac{\partial w}{\partial x} & \dfrac{\partial w}{\partial y} \end{bmatrix} \begin{bmatrix} \sigma_x & \tau_{xy} \\ \tau_{xy} & \sigma_y \end{bmatrix} \begin{Bmatrix} \partial w/\partial x \\ \partial w/\partial y \end{Bmatrix} h \, dx \, dy$$

$$= -\frac{1}{2} \iint \mathbf{R}^{\mathrm{t}} \boldsymbol{\sigma} \mathbf{R} h \, dx \, dy. \qquad (12.84)$$

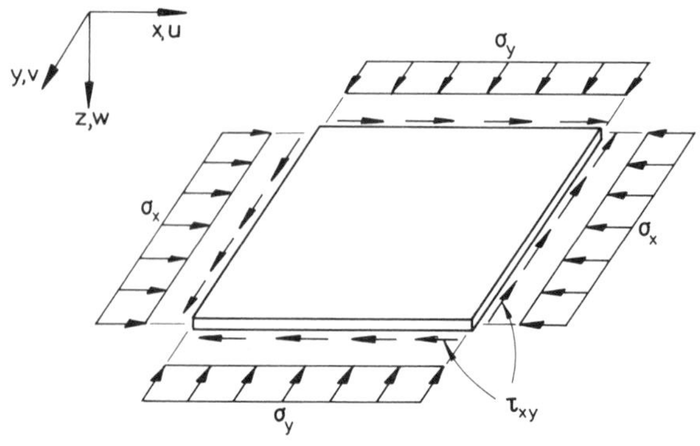

Fig. 12.10. Membrane stresses acting on a rectangular plate.

For any plate finite element the matrix $\mathbf{R}$ can be determined from the assumed displacement field. If this field is expressed in the shape function form $w = \mathbf{Nd}$ then

$$\mathbf{R} = \begin{Bmatrix} \dfrac{\partial w}{\partial x} \\ \dfrac{\partial w}{\partial y} \end{Bmatrix} = \begin{bmatrix} \dfrac{\partial \mathbf{N}}{\partial x} \\ \dfrac{\partial \mathbf{N}}{\partial y} \end{bmatrix} \mathbf{d} = \mathbf{Hd} \qquad (12.85)$$

and

$$U_{\text{p(stretch)}} = -\tfrac{1}{2}\mathbf{d}^{\text{t}}\left(\int\int \mathbf{H}^{\text{t}}\boldsymbol{\sigma}\mathbf{H}h \, dx \, dy\right)\mathbf{d}$$

where the integration is carried out over the middle surface area of the element. Hence the consistent geometric stiffness matrix for the plate element is

$$\mathbf{g} = \int\int \mathbf{H}^{\text{t}}\boldsymbol{\sigma}\mathbf{H}h \, dx \, dy. \qquad (12.86)$$

Alternatively, if the displacement field is expressed in the indirect form $w = \boldsymbol{\alpha}\mathbf{C}^{-1}\mathbf{d}$ then

$$\mathbf{R} = \mathbf{LC}^{-1}\mathbf{d} \qquad \text{where} \qquad \mathbf{L} = \begin{bmatrix} \dfrac{\partial \boldsymbol{\alpha}}{\partial x} \\ \dfrac{\partial \boldsymbol{\alpha}}{\partial y} \end{bmatrix}$$

and

$$\mathbf{g} = (\mathbf{C}^{-1})^{\text{t}}\left(\int\int \mathbf{L}^{\text{t}}\boldsymbol{\sigma}\mathbf{L}h \, dx \, dy\right)\mathbf{C}^{-1}. \qquad (12.87)$$

As a particular and relatively simple illustration of the derivation of $\mathbf{g}$ for a plate element, consider the rectangular element R1 (described in Section 11.3) and use the definition of $\mathbf{g}$ given in equation (12.87). For this element, matrix $\mathbf{C}^{-1}$ is defined in Table 11.1. With $\boldsymbol{\alpha}$ given by equation (11.18) matrix $\mathbf{L}$ becomes

$$\mathbf{L} = \begin{bmatrix} 0 & 1 & 0 & 2x & y & 0 & 3x^2 & 2xy & y^2 & 0 & 3x^2y & y^3 \\ 0 & 0 & 1 & x^2 & x & 2y & 0 & x^2 & 2xy & 3y^2 & x^3 & 3xy^2 \end{bmatrix} \quad (12.88)$$

The 2×2 matrix $\boldsymbol{\sigma}$ contains the individual membrane stresses σ_x, σ_y, and τ_{xy}. In general each of these stresses may be present and may vary with the x and y co-ordinates, leading to a complicated integral within the parentheses of equation (12.87). Kapur and Hartz [41] have given explicit details of the geometric stiffness matrix for element R1 for uniform values of all three stresses and for linear variations thereof. Here, as an example, Table 12.8 records $\mathbf{g}$ for the case of uniform value of σ_x only. For the other non-conforming 12-degree-of-freedom rectangular element R3 explicit details of the geometric stiffness matrix corresponding to linear variations of the membrane stresses are available elsewhere [42].

TABLE 12.8

Geometric stiffness matrix for element R1 for uniform compressive stress σ_x

	w_1	ϕ_1	θ_1	w_2	$\cdots$	θ_4
	276					
	$66b$	$24b^2$				
	$42a$	0	$112a^2$			
	-276	$-66b$	$-42a$			
	$-66b$	$-24b^2$	0			
$\mathbf{g} = -\dfrac{\sigma_x h\, b}{630\, a}$	$42a$	0	$-28a^2$			
	-102	$-39b$	$-21a$			
	$39b$	$18b^2$	0			
	$21a$	0	$-14a^2$			
	102	$39b$	$21a$			
	$-39b$	$-18b^2$	0			
	$21a$	0	$56a^2$			

Terms in the nine columns not itemized, corresponding to degrees of freedom w_2 through to θ_4, are defined with respect to the given terms in the same way as for the **k** or **m** matrices in Tables 11.2 and 12.3 respectively.

Quite a number of different types of plate finite element have been used in the analysis of plate buckling problems and references 24, 28, and 41–46 provide examples of these.

The case of the square plate with all edges simply supported subjected to a uniform uniaxial stress is the most well documented plate buckling problem. Solutions for this problem have been obtained from the FEM displacement approach using uniform square meshes of the element types defined in Chapter 11 as R1 [41], R3 [42, 43], R4 [44, 45], R5 [46], and T1 [24]. The relative accuracy of these solutions is illustrated in Fig. 12.11. It can be seen that the conforming elements R4 and R5 provide upper bound solutions to the buckling stress, as expected; results for R1, R3, and T1 happen to converge from below but this would not be guaranteed. Element R4 provides very accurate results in this problem but has an extra nodal degree of freedom as compared with all the other elements. Element R3 also performs very well and its performance is much the best of the elements with three degrees of freedom per node. It is noted that reference 24 also provides results for the problem under consideration when using element T1 and employing the eigenvalue economization procedure to eliminate a large proportion of the degrees of freedom.

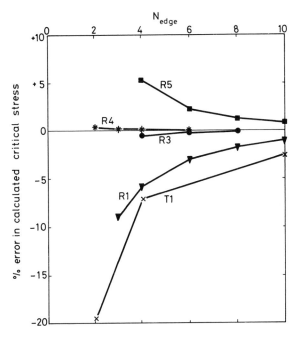

Fig. 12.11. Buckling of a simply supported square plate due to uniform σ_x: comparison of element performance. (N_{edge} is the number of elements along an edge of the complete plate).

The FEM is, of course, of greatest use in problems of a more complicated nature than that described above. Complications may be introduced through non-rectangular plate geometry and/or through the presence of non-uniform membrane stresses. In the latter case the membrane loading may be such that the distribution of membrane stresses in the plate interior has first to be calculated as a plane stress problem, using the FEM, before the element geometric stiffness matrices can be evaluated and the buckling problem itself solved. In references 42 and 43, for instance, this procedure is described with regard to rectangular elements; these are the basic rectangle for the initial plane stress analysis and the plate bending element R3 for the buckling analysis. Two particular applications, with results, are illustrated in Figs. 12.12 and 12.13. In the first of these a comparative solution is available and the FEM results show good convergence toward this solution despite the presence of large membrane stress gradients local to the applied forces P. In the second the plate is stiffened with a central stringer through which a concentrated load is applied. The stringer itself is modelled with basic bar elements for the initial in-plane analysis and with basic beam elements for the buckling analysis; its properties are that $EI_z/AD = 30$ and $A_c/Ah = 0.4$. (A_c and I_z

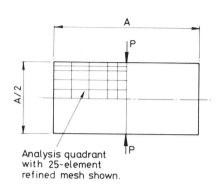

N_Q	$\dfrac{P_{cr}A}{2\pi^2 D}$
4 equal	1·466
9 equal	1·467
16 equal	1·476
25 equal	1·484
16 refined	1·496
25 refined	1·499
comparative solution	1·507

Analysis quadrant
with 25-element
refined mesh shown.

Fig. 12.12. Buckling of a simply-supported, rectangular plate due to point loads (N_Q is the number of elements in the analysis quadrant).

are the cross-sectional area and second moment of area respectively of the stringer.) No comparative solution is available but the FEM results indicate by their closeness to each other that an accurate solution has probably been attained.

Finally, it is noted that there exists a class of problem whose solution can be achieved with the use of the conventional flexural stiffness matrix and both the geometric stiffness matrix and the mass matrix introduced in this chapter. This class of problem comprises the vibration of plates (or beams) when a state of membrane (or axial) stress is present whose

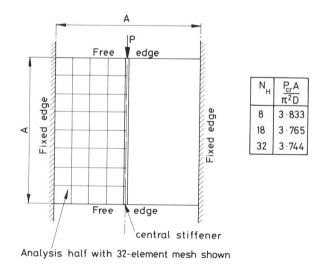

N_H	$\dfrac{P_{cr}A}{\pi^2 D}$
8	3·833
18	3·765
32	3·744

Analysis half with 32-element mesh shown

Fig. 12.13. Buckling of a stringer-stiffened plate. (N_H is the number of equal plate elements in the analysis half.)

magnitude and distribution is known and does not vary during the vibration. The equation governing the vibratory motion of the structure becomes

$$(\mathbf{K}_r - \mathbf{G}_r - \omega^2 \mathbf{M}_r)\mathbf{D}_r = \mathbf{0} \tag{12.89}$$

where the matrices $\mathbf{K}_r$, $\mathbf{G}_r$, and $\mathbf{M}_r$ are found as described earlier. This is again an eigenvalue problem. The effect of compressive stress is to lower the plate natural frequencies, that of tensile stress is to raise them.

References

1 HURTY, W. C. and RUBINSTEIN, M. F. *Dynamics of structures.* Prentice-Hall, Englewood Cliffs, NJ (1974).
2 WARBURTON, G. B. *The dynamical behaviour of structures* (2nd edn.). Pergamon, Oxford (1976).
3 PAZ, M. *Structural dynamics: theory and computation.* Van Nostrand Reinhold, New York (1980).
4 CRAIG, R. R. *Structural dynamics: an introduction to computer methods.* Wiley, New York (1981).
5 TIMOSHENKO, S. P. and GERE, J. M. *Theory of elastic stability* (2nd edn.). McGraw-Hill, New York (1961).
6 BRUSH, D. O. and ALMROTH, B. O. *Buckling of bars, plates and shells.* McGraw-Hill, New York (1975).
7 ALLEN, H. G. and BULSON, P. S. *Background to buckling.* McGraw-Hill, New York (1980).
8 WASHIZU, K. *Variational methods in elasticity and plasticity* (2nd edn.). Pergamon, Oxford (1975).
9 LECKIE, F. A. and LINDBERG, G. M. The effect of lumped parameters on beam frequencies. *Aeronaut. Q.* **14,** 224–40 (1963).
10 ARCHER, J. S. Consistent mass matrix for distributed mass systems. *Proc. Am. Soc. Civ. Eng.* **89** (ST4), 161–78 (1963).
11 ARCHER, J. S. Consistent matrix formulations for structural analysis using finite element techniques. *AIAA J.* **3,** 1910–18 (1965).
12 DAWE, D. J. A finite element for the vibration analysis of Timoshenko beams. *J. Sound Vibr.* **60,** 11–20 (1978).
13 PESTEL, E. Dynamic-stiffness matrix formulation by means of Hermitian polynomials. *Proc. Conf. on Matrix Methods in Structural Mechanics, Air Force Institute of Technology,* Wright–Patterson Air Force Base, Dayton, Ohio (1965).
14 CARNEGIE, W., THOMAS, J., and DOKUMACI, E. An improved method of matrix displacement analysis in vibration problems *Aeronaut. Q.* **20,** 321 (1969).
15 LINDBERG, G. M. Vibration of non-uniform beams. *Aeronaut. Q.* **14,** 387 (1963).
16 THOMAS, J. and DOKUMACI, E. Improved finite elements for vibration analysis of tapered beams. *Aeronaut. Q.* **24,** 39–46 (1973).
17 LEVIEN, K. W. and HARTZ, B. J. Dynamic flexibility matrix analysis of frames. *Proc. Am. Soc. Civ. Eng.* **89** (ST4), 515–36 (1963).
18 DAWE, D. J. A finite element approach to plate vibration problems. *J. Mech. Eng. Sci.* **7,** 28–32 (1965).

19 DAWE, D. J. Vibration of rectangular plates of variable thickness. *J. Mech. Eng. Sci.* **8**, 42–51 (1966).
20 DAWE, D. J. On assumed displacements for the rectangular plate bending element. *J. R. Aeronaut. Soc.* **71**, 722–4 (1967).
21 BOGNER, F. K., FOX, R. L., and SCHMIT, L. A. The generation of interelement compatible stiffness and mass matrices by the use of interpolation formulas. *Proc. Conf. on Matrix Methods in Structural Mechanics, Air Force Institute of Technology,* Wright–Patterson Air Force Base, Dayton, Ohio (1965).
22 MASON, V. Rectangular finite elements for analysis of plate vibrations. *J. Sound Vibr.* **7**, 437–48 (1968).
23 POPPLEWELL, N. and McDONALD, D. Conforming rectangular and triangular plate bending elements. *J. Sound Vibr.* **19**, 333–47 (1971).
24 ANDERSON, R. G., IRONS, B. M., and ZIENKIEWICZ, O. C. Vibration and stability of plates using finite elements *Int. J. Solids Struct.* **4**, 1031–55 (1968).
25 DICKINSON, S. M. and HENSHELL, R. D. Clough–Tocher triangular plate bending element in vibration. *AIAA J.* **7**, 560–1 (1969).
26 COWPER, G. R., KOSKO, E., LINDBERG, G. M., and OLSON, M. D. Static and dynamic applications of a high-precision triangular plate element. *AIAA J.* **7**, 1957–65 (1969).
27 DAWE, D. J. Parallelogrammic elements in the solution of rhombic cantilever plate problems. *J. Strain Anal.* **1**, 223–30 (1966).
28 ARGYRIS, J. H. Continua and discontinua. *Proc. Conf. on Matrix Methods in Structural Mechanics, Air Force Institute of Technology,* Wright–Patterson Air Force Base, Dayton, Ohio (1965).
29 HINTON, E. and BICANIC, N. A comparison of Lagrangian and serendipity Mindlin plate elements for free vibration analysis, *Comput. Struct.* **10**, 483–93 (1979).
30 ZIENKIEWICZ, O. C. The finite element method (3rd edn.), Chapter 20. McGraw-Hill, London (1977).
31 TONG, P., PIAN, T. H. H., and BUCCIARELLI, L. L. Mode shapes and frequencies by the finite element method using consistent and lump matrices. *Comput. Struct.* **1**, 623–38 (1971).
32 IRONS, B. M. Structural eigenvalue problems: elimination of unwanted variables. *AIAA J.* **3**, 961–2 (1965).
33 GUYAN, R. J. Reduction of stiffness and mass matrices. *AIAA J.* **3**, 380 (1965).
34 JENNINGS, A. Mass condensation and simultaneous iteration for vibration problems. *Int. J. Num. Meth. Eng.* **6**, 543–52 (1973).
35 RAMSDEN, J. N. and STOKER, J. R. Mass condensation; a semi-automatic method for reducing the size of vibration problems *Int. J. Num. Meth. Eng.* **1**, 333–49 (1969).
36 HENSHELL, R. D. and ONG, J. H. Automatic masters for eigenvalue economisation. *Int. J. Earthquake Struct. Dynam.* **3**, 375–83 (1975).
37 GALLAGHER, R. H. and PADLOG, J. Discrete element approach to structural instability analysis. *AIAA J.* **1**, 1437–9 (1963).
38 HARTZ, B. J. Matrix formulation of structural stability problems. *Proc. Am. Soc. Civ. Eng.* **91** (ST6), 141–57 (1965).
39 GALLAGHER, R. and LEE, B. Matrix dynamic and instability analysis with non-uniform elements. *Int. J. Num. Meth. Eng.* **2**, 265–76 (1970).

40 RODDEN, W. P., JONES, J. P., and BHUTA, P. G. A matrix formulation of the transverse structural influence coefficients of an axially loaded Timoshenko beam. *AIAA J.* **1,** 225–7 (1963).

41 KAPUR, K. K. and HARTZ, B. J. Stability of plates using the finite element method. *Proc. Am. Soc. Civ. Eng.* **92** (EM2), 177–195 (1966).

42 DAWE, D. J. Application of the discrete element method to the buckling analysis of rectangular plates under arbitrary membrane loading. *Aeronaut. Q.* **20,** 114–28 (1968).

43 DAWE, D. J. Application of the discrete element method to the buckling analysis of rectangular plates under arbitrary membrane loading. *Tech. Rep.* 68037. Royal Aircraft Establishment, Farnborough (1968).

44 CARSON, W. G. and NEWTON, R. E. Plate buckling analysis using a fully compatible finite element. *AIAA J.* **7,** 527–9 (1969).

45 WALLERSTEIN, D. V. A general linear geometric matrix for a fully compatible finite element. *AIAA J.* **10,** 545–6 (1972).

46 ABBAS, B. A. H. and THOMAS, J. Static stability of plates using fully conforming element. *Int. J. Num. Meth. Eng.* **11,** 995–1003 (1977).

Problems

12.1. Using a single basic beam element determine the fundamental natural frequency of lateral vibration of a uniform beam of length L having one end clamped and the other simply supported.

12.2. Using basic beam elements and the consistent mass matrix determine the fundamental natural frequency of vibration of a uniform clamped–clamped beam of length L with, in turn, one and two elements in a symmetric half of the beam.

12.3. Find the natural frequencies and mode shapes of lateral vibration of the uniform continuous beam shown in Fig. P12.3. Use one basic beam element in each span of the beam.

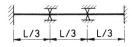

L/3 | L/3 | L/3

Fig. P12.3.

12.4. Find the lowest two natural frequencies and mode shapes of axial vibration of a uniform bar of length L held against movement at both ends. Use the consistent mass matrix and three basic elements in the full bar.

12.5. Consider a basic 4-degree-of-freedom beam element (Fig. 6.2) with non-uniform cross-section such that the width b is constant whilst the depth varies linearly from d at node 1 to $c_1 d$ at node 2. The consistent mass matrix for this element can be expressed in the form $\mathbf{m} = (\mathbf{C}^{-1})^t \mathbf{R} \mathbf{C}^{-1}$ where $\mathbf{C}^{-1}$ has its usual meaning and $\mathbf{R}$ is a 4×4 matrix, the terms of which are obtained by integrating certain quantities along the element length. Determine matrix $\mathbf{R}$ and hence show that for the particular case of $c_1 = 2$ the mass matrix becomes

$$\mathbf{m} = \frac{\rho b d l}{840} \begin{matrix} v_1 & \theta_{z1} & v_2 & \theta_{z2} \\ \begin{bmatrix} 384 & & \text{Symmetric} & \\ 58l & 11l^2 & & \\ 162 & 40l & 552 & \\ -38l & -9l^2 & -74l & 13l^2 \end{bmatrix} \end{matrix}.$$

12.6. Figure P12.6 shows a simply supported symmetric beam of length $2L$ and non-uniform rectangular cross-section. The density of the beam material is ρ, its Young's modulus is E, and the width of the cross-section is equal to $L/10$. A vertical linear spring is attached to the beam at the centre span; the spring has a stiffness equal to $E \times L \times 10^{-6}$ and the effect of its mass may be taken to be equivalent to a mass of magnitude $\rho \times L^3 \times 10^{-4}$ concentrated on the beam neutral axis at the centre span. Model the left-hand half of the beam with one beam element and determine the lowest natural frequency of lateral vibration expressed in terms of E, ρ, and L. Use the result of Problem 6.5 for the element consistent mass matrix and note that the element stiffness matrix corresponding to this mass matrix can be calculated using the result of Problem 6.3.

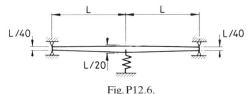

Fig. P12.6.

12.7. Derive the consistent mass matrix for axial vibration of a uniform bar finite element having three nodes and quadratic displacement (as shown in Fig. 6.8).

12.8. A uniform bar held at one end, free at the other, is modelled with a single 3-degree-of-freedom quadratic element. Find the natural frequencies and mode shapes of extensional vibration. It should be noted that the stiffness matrix for this element is given by eqn (c) of Example 6.4, and the consistent mass matrix is given by the result of Problem 12.7. It should be also noted that other solutions to this problem are included in Example 12.5.

12.9. Derive the consistent mass matrix for a uniform beam element having three nodes and quintic displacement (as shown in Fig. 6.10(a)).

12.10. A uniform simply supported beam is modelled with a single 6-degree-of-freedom quintic beam element. Find the natural frequencies and mode shapes of flexural vibration. (The frequency values are recorded in Table 12.1.) It should be noted that the element stiffness matrix is given by the result of Problem 6.12 and the consistent mass matrix is given by the result of Problem 12.9.

12.11. In Example 3.2 a 3-degree-of-freedom model of a plane frame was considered under a static loading and the structure stiffness matrix was established. Now consider the vibration of this frame in its own plane and set up the structure consistent mass matrix. Hence determine the first natural frequency of vibration corresponding to this finite-element model. Assume that $E = 200$ GN m^{-2} and $\rho = 8000$ kg m^{-3} throughout the frame.

12.12. Repeat Example 12.4 using the rectangular plate element R3 in place of element R1.

12.13. Determine the fundamental frequency of lateral vibration of a square plate having one pair of opposite edges clamped and the other pair simply supported. Use one element in a plate quadrant, hence creating a 2-degree-of-freedom model for symmetric–symmetric modes of vibration; consider the use of (a) an R1 element and (b) an R3 element in turn. Take $\nu = 0.3$.

12.14. Derive the consistent mass matrix for extensional vibration of the basic rectangular plane-stress element described in Section 8.4.1 and illustrated in Fig. 8.15.

12.15. Derive the consistent mass matrix for the basic tetrahedral element described in Section 9.3.1 and illustrated in Fig. 9.1.

12.16. Determine the fundamental natural frequency of a uniform cantilever beam when modelled with a single basic element. Use
  (a) the lumped mass matrix with only transverse inertia included
  (b) the consistent mass matrix in conjunction with the eigenvalue economizer method to eliminate the rotational degrees of freedom
(Example 12.2 details the full one-element solution.)

12.17. Repeat Problem 12.2 using the lumped mass matrix, taking account only of translational inertia.

12.18. Repeat Problem 12.4 using the lumped mass matrix.

12.19. Consider a uniform simply supported beam modelled with two basic beam elements in the symmetric half-beam. Determine the fundamental frequency of lateral vibration using
  (a) the lumped mass matrix, taking into account only translational inertia
  (b) the consistent mass matrix in conjunction with the eigenvalue economizer method to eliminate rotational degrees of freedom.
(Table 12.1 gives results for this problem when using consistent mass matrices in full solutions.)

12.20. Consider the plate problem described in Example 12.4(b) but now obtain a solution for the fundamental frequency of vibration by using
  (i) the lumped mass matrix, taking into account only translational inertia
  (ii) the consistent mass matrix in conjunction with the eigenvalue economizer method to eliminate the rotational degrees of freedom.

12.21. Determine the buckling load of a uniform clamped-free (or cantilever) strut of length L based on the use of a single basic element. (The 'exact' result is $\pi^2 EI_z/4L^2$.)

12.22. A prismatic strut of length L is clamped at both ends. Determine the buckling load corresponding to modelling a symmetric half of the strut using first one and then two equal basic cubic elements. (The 'exact' result is $4\pi^2 EI_z/L^2$.)

12.23. Derive the geometric stiffness matrix for a uniform beam element having three nodes and quintic displacement (as shown in Fig. 6.10(a)).

12.24. Find the buckling load of a uniform strut with
    (a) clamped ends
    (b) simply-supported ends
when modelled with a single quintic element of the type referred to in Problem 12.23. (The element **k** matrix is developed in Problem 6.12.)

12.25. Consider again the uniform cantilever strut of Problem 12.21. Model the strut with two basic beam elements but proceed in turn in the following two ways:
    (a) use the inconsistent element geometric stiffness matrix defined in equation (12.82),
    (b) use the consistent element geometric stiffness matrix in conjunction with the eigenvalue economizer method to eliminate rotational degrees of freedom.

12.26. The left-hand half of the stepped strut shown in Fig. P12.26 is to be modelled with two basic beam elements as indicated. Set up the structure stiffness and consistent geometric stiffness matrices (of order 4) governing buckling in a symmetric mode. Proceed to use the eigenvalue economizer method, condensing out the rotational degrees of freedom, to determine the buckling load.

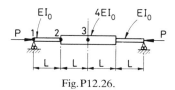

Fig. P12.26.

12.27. A square plate with all edges clamped is subjected to a uniform stress σ_x as shown in Fig. P12.27. A symmetric quadrant of the plate is modelled with one R1 plate element (**k** is given by Table 11.2 and **g** by Table 12.9). Determine the stress level σ_{xcr} at which buckling occurs. Assume $\nu = 0.3$.

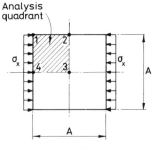

Fig. P12.27.

12.28. Repeat Problem 12.27 for a plate with the loaded edges simply supported and the other edges clamped.

APPENDIX A

MATRIX ALGEBRA AND THE
SOLUTION OF EQUATIONS

In the early part of this appendix a résumé is given of the basic definitions and operations of matrix algebra which are required for an understanding of this text. The reader requiring amplification of this material can find it in a great many texts in the realms of mathematics and numerical analysis, amongst which references 1–6 are examples. In the latter part of the appendix very brief remarks are given concerning the solution of a set of simultaneous linear equations and the evaluation of eigenvalues.

Definition of a matrix and special types of matrices

A matrix $\mathbf{A}$ of *order* $m \times n$ is a rectangular array of items set out in m rows and n columns. Thus

$$\mathbf{A} = [a_{ij}] = \begin{bmatrix} a_{11} & a_{12} & a_{13} & \cdots & a_{1n} \\ a_{21} & a_{22} & a_{23} & \cdots & a_{2n} \\ a_{31} & a_{32} & a_{33} & \cdots & a_{3n} \\ \vdots & \vdots & \vdots & & \vdots \\ a_{m1} & a_{m2} & a_{m3} & \cdots & a_{mn} \end{bmatrix}.$$

The individual items a_{ij} in the matrix are commonly referred to as *coefficients, entries, terms,* or *elements,* though the latter word is avoided, for obvious reasons, in this text.

The coefficient a_{ij} is located in the ith row and jth column of the matrix; it may be of a numeric, algebraic, differential, or integral nature.

In the particular case when $m = n$ the array is a *square matrix* of order n.

When $n = 1$ we have a *column matrix* of order $m \times 1$ (sometimes, but not in this work, termed a *vector*). A column matrix $\mathbf{B}$ is thus

$$\mathbf{B} = \{b_i\} = \begin{Bmatrix} b_1 \\ b_2 \\ b_3 \\ \vdots \\ b_m \end{Bmatrix}$$

where use of the braces { }, which is restricted to column matrices, is noted. Simply to save space a column matrix is often written across the page, as

$$\mathbf{B} = \{b_1 \quad b_2 \quad b_3 \quad \ldots \quad b_m\}$$

but no confusion should arise if the meaning of the braces is kept in mind.

When $m = 1$ we have a *row matrix* of order $1 \times n$, such as

$$\mathbf{C} = [c_i] = [c_1 \quad c_2 \quad c_3 \quad \ldots \quad c_n]$$

Some special types of matrix are now defined.

A *null matrix* is a matrix in which all the coefficients are zero, commonly denoted by $\mathbf{0}$.

A *unit* (*or identity*) *matrix* is a square matrix with unit coefficients on the *main* (or principal or leading) diagonal and all other coefficients having zero value. The main diagonal is the diagonal running from the upper left corner to the lower right corner of the array. A unit matrix may be of any order and is commonly denoted by the symbol $\mathbf{I}$, perhaps with a subscript indicating the order of the matrix; thus for unit matrix of order 4

$$\mathbf{I}_4 = \begin{bmatrix} 1 & 0 & 0 & 0 \\ 0 & 1 & 0 & 0 \\ 0 & 0 & 1 & 0 \\ 0 & 0 & 0 & 1 \end{bmatrix}.$$

The unit matrix is analogous to unity in conventional algebra.

A *diagonal matrix* is a square matrix (of any order) in which all the coefficients except those on the main diagonal are zero, e.g.

$$\mathbf{A} = \begin{bmatrix} a_{11} & 0 & 0 \\ 0 & a_{22} & 0 \\ 0 & 0 & a_{33} \end{bmatrix} = \lceil a_{11} \quad a_{22} \quad a_{33} \rfloor$$

where the form of representation on the right-hand side of this expression, with the special brackets $\lceil \ \rfloor$, is used to save space in defining a diagonal matrix. Clearly the unit matrix $\mathbf{I}$ is a particular type of diagonal matrix.

A *symmetric matrix* is a square matrix in which $a_{ij} = a_{ji}$ for all i, j.

An *upper triangular matrix* is a matrix in which non-zero coefficients only occur on or above the main diagonal, i.e. $a_{ij} = 0$ for $i > j$.

A *lower triangular matrix* is a matrix in which non-zero coefficients only occur on or below the main diagonal, i.e. $a_{ij} = 0$ for $i < j$.

A *band matrix* is a matrix in which all the non-zero coefficients are located on or near the main diagonal.

Elementary matrix operations

Equality

Two matrices $\mathbf{A}$ and $\mathbf{B}$ are equal only if all their corresponding coefficients are equal; that is $\mathbf{A} = \mathbf{B}$ if $a_{ij} = b_{ij}$ for all values of i and j.

Addition/subtraction

Two matrices can be added together only if they are of the same order. The result is obtained by adding corresponding coefficients. Thus, if $\mathbf{C} = \mathbf{A} + \mathbf{B}$ then $c_{ij} = a_{ij} + b_{ij}$.

Subtraction of matrices is achieved by subtracting corresponding coefficients. Thus if $\mathbf{C} = \mathbf{A} - \mathbf{B}$ then $c_{ij} = a_{ij} - b_{ij}$.

The sequence of additions and subtractions does not matter; for example

$$(\mathbf{A} + \mathbf{B}) - (\mathbf{C} + \mathbf{D}) = (\mathbf{A} - \mathbf{C}) + (\mathbf{B} - \mathbf{D}).$$

Multiplication by a constant

Each coefficient of the matrix is multiplied by the constant. Thus, if k is a scalar constant,

$$k\mathbf{A} = k[a_{ij}] = [ka_{ij}].$$

Multiplication of matrices

The product of two matrices can only be defined if the number of columns of the pre-multiplying matrix is equal to the number of rows of the post-multiplying matrix. Thus, if $\mathbf{A}$ is a (pre-multiplying) $m \times n$ matrix and $\mathbf{B}$ is a (post-multiplying) $p \times q$ matrix the product $\mathbf{AB} = \mathbf{C}$ exists only if $n = p$ when the matrices are said to be *conformable* for multiplication; the order of $\mathbf{C}$ will be $m \times q$.

Consider the product

$$\underset{(m \times n)}{\mathbf{A}} \quad \underset{(n \times p)}{\mathbf{B}} = \underset{(m \times p)}{\mathbf{C}}.$$

This can be generated in the following way, where it is assumed for the sake of illustration that $m = 2$, $n = 3$, $p = 2$:

$$\begin{bmatrix} b_{11} & b_{12} \\ b_{21} & b_{22} \\ b_{31} & b_{32} \end{bmatrix} \leftarrow \mathbf{B}$$

$$\mathbf{A} \rightarrow \begin{bmatrix} a_{11} & a_{12} & a_{13} \\ a_{21} & a_{22} & a_{23} \end{bmatrix} \begin{bmatrix} c_{11} & c_{12} \\ c_{21} & c_{22} \end{bmatrix} \leftarrow \mathbf{C}.$$

A coefficient c_{ij} of $\mathbf{C}$ is the summation of products of terms in row i of $\mathbf{A}$ and column j of $\mathbf{B}$. Thus

$$c_{11} = a_{11}b_{11} + a_{12}b_{21} + a_{13}b_{31}$$
$$c_{12} = a_{11}b_{12} + a_{12}b_{22} + a_{13}b_{32}$$
$$c_{21} = a_{21}b_{11} + a_{22}b_{21} + a_{23}b_{31}$$
$$c_{22} = a_{21}b_{12} + a_{22}b_{22} + a_{23}b_{32}$$

or, generally,

$$c_{ij} = \sum_{k=1}^{n} a_{ik}b_{kj}.$$

A particularly important point to note is that, unlike in ordinary algebra, the commutative property does not in general apply for products in matrix algebra, i.e.

$$\mathbf{AB} \neq \mathbf{BA}.$$

Indeed, in general, if the product $\mathbf{AB}$ exists it does not follow that the product $\mathbf{BA}$ will exist; this will be the case only if the number of rows of $\mathbf{A}$ equals the number of columns of $\mathbf{B}$.

If it is assumed that matrix products exist, matrices can be handled like ordinary algebraic quantities provided that the order of multiplication is not changed; thus

$$(\mathbf{A} + \mathbf{B})\mathbf{C} = \mathbf{AC} + \mathbf{BC}$$
$$\mathbf{C}(\mathbf{A} + \mathbf{B}) = \mathbf{CA} + \mathbf{CB}$$
$$\mathbf{ABC} = (\mathbf{AB})\mathbf{C} = \mathbf{A}(\mathbf{BC})$$

If matrix products $\mathbf{AB}$ and $\mathbf{AC}$ exist and $\mathbf{AB} = \mathbf{AC}$ it does not necessarily follow that $\mathbf{B} = \mathbf{C}$.

Some particular features of matrix products involving the special matrices $\mathbf{I}$ and $\mathbf{0}$ should be noted: these are that

$$\mathbf{AI} = \mathbf{IA} = \mathbf{A}$$
$$\mathbf{A0} = \mathbf{0A} = \mathbf{0}$$

It should also be noted that if $\mathbf{AB} = \mathbf{0}$ this does not generally mean that either $\mathbf{A} = \mathbf{0}$ or $\mathbf{B} = \mathbf{0}$.

Matrix transposition

The transpose of a matrix is obtained by interchanging the rows and columns of the original matrix. The transpose of $\mathbf{A}$ is denoted here by $\mathbf{A}^t$. If $\mathbf{A} = [a_{ij}]$ is an $m \times n$ matrix and its transpose is $\mathbf{A}^t = \mathbf{B}$, then $\mathbf{B} = [b_{ij}]$ is an $n \times m$ matrix, with $b_{ij} = a_{ji}$ for $i = 1 \ldots n$, $j = 1 \ldots m$. For example,

$$\mathbf{A} = \begin{bmatrix} a_{11} & a_{12} & a_{13} \\ a_{21} & a_{22} & a_{23} \end{bmatrix} \qquad \mathbf{A}^t = \begin{bmatrix} a_{11} & a_{21} \\ a_{12} & a_{22} \\ a_{13} & a_{23} \end{bmatrix}.$$

Obviously, if transposition is carried out twice, the original matrix is recovered, i.e. $(\mathbf{A}^t)^t = \mathbf{A}$.

If A is a symmetric matrix $\mathbf{A}^t = \mathbf{A}$.

The transpose of a product of two matrices is equal to the product of the transposed matrices in reverse order, i.e. $(\mathbf{AB})^t = \mathbf{B}^t\mathbf{A}^t$. This result can be extended to higher products, so that

$$(\mathbf{AB} \ldots \mathbf{Z})^t = \mathbf{Z}^t \ldots \mathbf{B}^t\mathbf{A}^t.$$

In structural analysis products of the form $\mathbf{B}^t\mathbf{AB}$ arise in which $\mathbf{A}$ is a symmetric matrix. Since in this circumstance $(\mathbf{B}^t\mathbf{AB})^t = \mathbf{B}^t\mathbf{AB}$ it follows that the product $\mathbf{B}^t\mathbf{AB}$ must be symmetric.

Matrix inversion

The inverse of a square matrix $\mathbf{A}$ of order n is another square matrix of order n denoted by $\mathbf{A}^{-1}$ which is defined by the fact that $\mathbf{A}^{-1}\mathbf{A} = \mathbf{A}\mathbf{A}^{-1} = \mathbf{I}$. (It should be noted that the inversion of a matrix is a process which is analogous to the division process of ordinary algebra but that $\mathbf{A}^{-1}$ is *not* $1/\mathbf{A}$). The inverse of a square matrix $\mathbf{A}$ cannot always be obtained. When this is so $\mathbf{A}$ is said to be a *singular* matrix; when the inverse can be obtained $\mathbf{A}$ is said to be a *non-singular* or *regular* matrix. One (rather inefficient) way in which the inverse of a non-singular matrix can be found is described later.

If the product $\mathbf{AB} \ldots \mathbf{Z}$ exists and is non-singular, then

$$(\mathbf{AB} \ldots \mathbf{Z})^{-1} = \mathbf{Z}^{-1} \ldots \mathbf{B}^{-1}\mathbf{A}^{-1}.$$

For any non-singular square matrix $\mathbf{A}$ the inverse of the transpose is equal to the transpose of the inverse, i.e. $(\mathbf{A}^t)^{-1} = (\mathbf{A}^{-1})^t$. Furthermore, if $\mathbf{A}$ is symmetric its inverse is also symmetric.

An *orthogonal* matrix is a square matrix whose inverse is equal to its transpose, i.e. $\mathbf{A}$ is an orthogonal matrix if $\mathbf{A}^{-1} = \mathbf{A}^t$.

Matrix partitioning

It is frequently convenient to work with the submatrices of a given matrix or matrices; individual matrices are partitioned into blocks of coefficients in any desired fashion so as to facilitate the performance of any required matrix operations. For instance, consider the matrix product $\mathbf{AB} = \mathbf{C}$

where

$$\mathbf{A} = \begin{bmatrix} a_{11} & a_{12} & \vdots & a_{13} \\ a_{21} & a_{22} & \vdots & a_{23} \\ \text{----------} & \text{----} \\ a_{31} & a_{32} & \vdots & a_{33} \\ a_{41} & a_{42} & \vdots & a_{43} \end{bmatrix} = \begin{bmatrix} \mathbf{A}_{11} & \mathbf{A}_{12} \\ \mathbf{A}_{21} & \mathbf{A}_{22} \end{bmatrix}$$

$$\mathbf{B} = \begin{bmatrix} b_{11} & b_{12} & \vdots & b_{13} \\ b_{21} & b_{22} & \vdots & b_{23} \\ \text{----------} & \text{----} \\ b_{31} & b_{32} & \vdots & b_{33} \end{bmatrix} = \begin{bmatrix} \mathbf{B}_{11} & \mathbf{B}_{12} \\ \mathbf{B}_{21} & \mathbf{B}_{22} \end{bmatrix}.$$

Here each of $\mathbf{A}$ and $\mathbf{B}$ have been partitioned into four submatrices ($\mathbf{A}_{ij}$, $\mathbf{B}_{ij}$ for $i, j = 1, 2$) which are identified by row and column indices in the usual fashion as if the submatrices are individual coefficients. The definition of the individual submatrices is indicated by the dotted lines within the statements of $\mathbf{A}$ and $\mathbf{B}$ as arrays of individual coefficients, e.g.

$$\mathbf{A}_{21} = \begin{bmatrix} a_{31} & a_{32} \\ a_{41} & a_{42} \end{bmatrix} \qquad \mathbf{B}_{11} = \begin{bmatrix} b_{11} & b_{12} \\ b_{21} & b_{22} \end{bmatrix}$$

Matrix operations can be performed with matrices expressed in partitioned form just as they are when the matrices are expressed as arrays of individual coefficients. Thus the matrix product $\mathbf{AB} = \mathbf{C}$ becomes

$$\mathbf{A} \to \begin{bmatrix} \mathbf{A}_{11} & \mathbf{A}_{12} \\ \mathbf{A}_{21} & \mathbf{A}_{22} \end{bmatrix} \begin{matrix} \begin{bmatrix} \mathbf{B}_{11} & \mathbf{B}_{12} \\ \mathbf{B}_{21} & \mathbf{B}_{22} \end{bmatrix} \leftarrow \mathbf{B} \\ \begin{bmatrix} \mathbf{C}_{11} & \mathbf{C}_{12} \\ \mathbf{C}_{21} & \mathbf{C}_{22} \end{bmatrix} \leftarrow \mathbf{C} \end{matrix}$$

where

$$\mathbf{C}_{11} = \mathbf{A}_{11}\mathbf{B}_{11} + \mathbf{A}_{12}\mathbf{B}_{21}$$
$$\mathbf{C}_{12} = \mathbf{A}_{11}\mathbf{B}_{12} + \mathbf{A}_{12}\mathbf{B}_{22}$$
$$\mathbf{C}_{21} = \mathbf{A}_{21}\mathbf{B}_{11} + \mathbf{A}_{22}\mathbf{B}_{21}$$
$$\mathbf{C}_{22} = \mathbf{A}_{21}\mathbf{B}_{12} + \mathbf{A}_{22}\mathbf{B}_{22}.$$

When the submatrices of $\mathbf{C}$ are evaluated the same result for the matrix product is obtained, of course, as would be obtained by performing the multiplication with $\mathbf{A}$ and $\mathbf{B}$ expressed as arrays of individual scalar coefficients. Clearly, when using partitioned matrices the partitions must be arranged to be compatible with the operation being carried out; thus in the above matrix multiplication the sizes of the submatrices must be such that all the products of submatrices required to define $\mathbf{C}$ actually exist.

Differentiation and integration of matrices

The differentiation (or integration) of a matrix is achieved simply by differentiating (or integrating) each individual coefficient in the conventional fashion. For example

$$\int_0^a \begin{bmatrix} 2x & 6x^3 \\ 1 & (1-x^2) \end{bmatrix} dx = \begin{bmatrix} a^2 & \frac{3}{2}a^4 \\ a & a - \dfrac{a^3}{3} \end{bmatrix}.$$

Often in this text the integration of products of matrices is required, of the form $\int \mathbf{AB}\, dx$ or $\int\int \mathbf{A^t BA}\, dx\, dy$ and so on. In these cases the matrix multiplication is performed first and then each coefficient of the resulting matrix is integrated individually in the conventional way.

Matrix statement of quadratic form

In the FEM we have seen that certain types of energy can be expressed in a quadratic form of a finite number of variables. In matrix notation we have expressions of the type

$$E = [x_1\, x_2 \ldots x_n] \begin{bmatrix} a_{11} & a_{12} & \cdots & a_{1n} \\ a_{21} & a_{22} & \cdots & a_{2n} \\ \vdots & \vdots & & \vdots \\ a_{n1} & a_{n2} & \cdots & a_{nn} \end{bmatrix} \begin{Bmatrix} x_1 \\ x_2 \\ \vdots \\ x_n \end{Bmatrix} = \mathbf{x^t Ax}$$

where $\mathbf{A}$ is a symmetric matrix. For instance, the scalar quantity E may be equated to twice the structure strain energy, $\mathbf{x}$ to the column matrix of nodal degrees of freedom, and $\mathbf{A}$ to the structure stiffness matrix. When the quadratic form has zero value only when all the variables (the x_i, $i = 1 \ldots n$) are zero and is positive for all other values of the variables, the form is called a positive definite quadratic form; correspondingly the matrix $\mathbf{A}$ is called a *positive definite matrix*. When the quadratic form is never negative but is sometimes zero, without all the variables being zero, the matrix $\mathbf{A}$ is a *positive semi-definite matrix*. Depending upon the nature of the symmetric matrix $\mathbf{A}$ other quadratic forms may be negative definite, negative semi-definite, or indefinite. The procedure adopted to obtain a solution to a structural problem may well depend on the precise nature of $\mathbf{A}$. In the displacement method described in this text the stiffness matrix (and the mass matrix) is positive definite and this is advantageous.

Determinant of a matrix and inversion by adjoint

A square matrix $\mathbf{A}$ of order n has a *determinant* associated with it which is represented symbolically as

$$\det \mathbf{A} = |\mathbf{A}| = \begin{vmatrix} a_{11} & a_{12} & a_{13} & \cdots & a_{1n} \\ a_{21} & a_{22} & a_{23} & \cdots & a_{2n} \\ a_{31} & a_{32} & a_{33} & \cdots & a_{3n} \\ \vdots & \vdots & \vdots & & \vdots \\ a_{n1} & a_{n2} & a_{n3} & \cdots & a_{nn} \end{vmatrix}$$

The parallel line symbolism implies that by performing certain operations a scalar value of the determinant can be obtained. The basic case is when $n = 2$, for which the value of the determinant is defined as follows:

$$|\mathbf{A}| = \begin{vmatrix} a_{11} & a_{12} \\ a_{21} & a_{22} \end{vmatrix} = a_{11}a_{22} - a_{12}a_{21}.$$

This procedure can be extended to deal with larger values of n through the use of the concepts of *minors* and *cofactors*.

The minor M_{ij} of $|\mathbf{A}|$ is the sub-determinant obtained by striking out the row and the column containing the coefficient a_{ij}. For instance, for the case when $n = 3$ there are nine minors of which three examples are

$$M_{11} = \begin{vmatrix} a_{22} & a_{23} \\ a_{32} & a_{33} \end{vmatrix} \qquad M_{12} = \begin{vmatrix} a_{21} & a_{23} \\ a_{31} & a_{33} \end{vmatrix}. \qquad M_{21} = \begin{vmatrix} a_{12} & a_{13} \\ a_{32} & a_{33} \end{vmatrix}.$$

The cofactor A_{ij} of coefficent a_{ij} of $|\mathbf{A}|$ is defined as

$$A_{ij} = (-1)^{i+j} M_{ij}.$$

Thus a cofactor is simply a minor with the appropriate sign; the sign will be negative when $i + j$ is odd.

The value of a determinant when n is greater than 2 can now be expressed as

$$|\mathbf{A}| = \sum_{j=1}^{n} a_{ij} A_{ij}.$$

This statement indicates that we choose a certain row i of the nth-order array, calculate the cofactors of the row, and then sum the products of the coefficients in the row with their respective cofactors. In calculating the value of the determinant by expanding by row, any row can be chosen but the work is reduced if one or more zeros is present in the row. (We could alternatively expand by column.) If the cofactors A_{ij} are themselves of order greater than 2 then the A_{ij} must be expanded in like fashion until eventually cofactors of order 2 are obtained. As an example of the procedure consider the calculation of the value of $|\mathbf{A}|$ where

$$|\mathbf{A}| = \begin{vmatrix} 1 & 2 & 3 & 4 \\ 0 & 2 & 2 & 0 \\ 3 & 0 & 1 & 2 \\ 2 & 1 & 3 & 3 \end{vmatrix}.$$

Expanding along the second row gives

$$|\mathbf{A}| = +2 \begin{vmatrix} 1 & 3 & 4 \\ 3 & 1 & 2 \\ 2 & 3 & 3 \end{vmatrix} - 2 \begin{vmatrix} 1 & 2 & 4 \\ 3 & 0 & 2 \\ 2 & 1 & 3 \end{vmatrix}.$$

Now (expanding along first row)

$$\begin{vmatrix} 1 & 3 & 4 \\ 3 & 1 & 2 \\ 2 & 3 & 3 \end{vmatrix} = +1 \begin{vmatrix} 1 & 2 \\ 3 & 3 \end{vmatrix} - 3 \begin{vmatrix} 3 & 2 \\ 2 & 3 \end{vmatrix} + 4 \begin{vmatrix} 3 & 1 \\ 2 & 3 \end{vmatrix} = -3 - 15 + 28 = +10.$$

and (expanding along second row)

$$\begin{vmatrix} 1 & 2 & 4 \\ 3 & 0 & 2 \\ 2 & 1 & 3 \end{vmatrix} = -3 \begin{vmatrix} 2 & 4 \\ 1 & 3 \end{vmatrix} - 2 \begin{vmatrix} 1 & 2 \\ 2 & 1 \end{vmatrix} = -6 + 6 = 0.$$

Thus $|\mathbf{A}| = +2(+10) - 2(0) = +20$.

There are a number of properties of determinants which can be invoked to facilitate calculation of the value of a determinant; amongst these the following are noted.

(i) If all the coefficients in any row (or any column) are zero the value of the determinant is zero.

(ii) If every coefficient in a row (or column) is multiplied by a constant, the value of the determinant is multiplied by the same constant.

(iii) Interchanging two adjacent rows (or columns) changes the sign of a determinant.

(iv) If any two rows (or columns) are identical or are in proportion, the value of the determinant is zero.

(v) The value of a determinant is not changed when a multiple of one row (or column) is added to another row (or column)

The *adjoint* of a square matrix $\mathbf{A}$ is the transpose of the matrix which is obtained by replacing each of the coefficients of $\mathbf{A}$ by its cofactor. Thus, if $\mathbf{A} = [a_{ij}]$ then adj $\mathbf{A} = [A_{ij}]^t$.

The inverse of a square matrix $\mathbf{A}$ can be obtained by dividing the adjoint matrix by the determinant of $\mathbf{A}$; thus

$$\mathbf{A}^{-1} = \frac{\mathrm{adj}\,\mathbf{A}}{|\mathbf{A}|}.$$

It is important to note that if the determinant of matrix $\mathbf{A}$ (i.e. $|\mathbf{A}|$) is zero the coefficients of $\mathbf{A}^{-1}$ cannot be found; when this condition applies the matrix is termed singular, as has been mentioned earlier.

The methods described here for evaluating a determinant and for finding the inverse of a matrix by use of the adjoint have been recorded because of their relative straightforwardness in the theoretical sense and of their historical interest. These methods are satisfactory when dealing with systems of order 3 or 4, say, but it must be realized that they become very inefficient in obtaining numerical results for matrices of any real size. Then other methods are much more appropriate [1–6], although the requirement to determine explicitly the numerical inverse of a large-order matrix can in any case usually be avoided by adopting other processes, as in the solution of a set of linear equations described in what follows.

Solution of a set of linear simultaneous equations

In the linear static analysis of structures using the MDM and the FEM we are faced with the need to solve a set of linear simultaneous equations of the form

$$\mathbf{A}\mathbf{x} = \mathbf{b} \tag{A1}$$

where $\mathbf{A}$ is a square non-singular matrix of known coefficients, $\mathbf{b}$ is a column matrix of prescribed quantities, and $\mathbf{x}$ is a column matrix of unknowns whose values are sought. (In the terminology of the displacement method, in representing the set of structure equations obviously $\mathbf{K}_r$, $\mathbf{F}_r$, and $\mathbf{D}_r$ identify with $\mathbf{A}$, $\mathbf{b}$, and $\mathbf{x}$ respectively.) If both sides of the above equations are pre-multiplied by $\mathbf{A}^{-1}$, and it is recalled that $\mathbf{A}^{-1}\mathbf{A} = \mathbf{I}$, we obtain

$$\mathbf{x} = \mathbf{A}^{-1}\mathbf{b} \tag{A2}$$

which symbolically represents the solution of the set of equations.

This implies the need to determine the inverse $\mathbf{A}$, which could be attempted through the use of the adjoint matrix as described above or by some other approach. However, other procedures are available by which $\mathbf{x}$ can be determined more efficiently without the need to invert $\mathbf{A}$ explicitly. The basic procedures really belong in the realm of numerical analysis and, since the program libraries of computer centres always have available a range of routines for equation solving, the structural analyst could be justified in treating such routines simply as 'black boxes' which will yield the correct solution (for the nodal displacements) to the set of equations that he provides. However, since the topic of equation solving is clearly of considerable importance in the overall structural analysis problem, a few general remarks are appropriate here.

The methods used for solving systems of simultaneous linear equations can be categorized into two types, direct and iterative.

The direct methods are also referred to as elimination methods and the most popular of these are Gaussian elimination and the Cholesky method. With elimination methods the aim is to transform or decompose the given set of equations, of the form of equation (A1), into an equivalent 'triangular' form which can be solved more readily. In Gaussian elimination the coefficient matrix is effectively decomposed as

$$\mathbf{A} = \mathbf{LU} \tag{A3}$$

where $\mathbf{L}$ is a lower triangular matrix and $\mathbf{U}$ is an upper triangular matrix. $\mathbf{L}$ and $\mathbf{U}$ can be generated in a number of different forms, perhaps but not necessarily with all the coefficients on the leading diagonal of one of them being unity. Equation (A1) becomes

$$\mathbf{Ax} = \mathbf{LUx} = \mathbf{b}.$$

Letting $\mathbf{Ux} = \mathbf{y}$ gives

$$\mathbf{Ly} = \mathbf{b} \tag{A4}$$

as what is termed the 'forward elimination' stage, and

$$\mathbf{Ux} = \mathbf{y} \tag{A5}$$

as the 'back substitution' stage. The complete solution can be symbolically represented as

$$\mathbf{x} = \mathbf{U}^{-1}\mathbf{L}^{-1}\mathbf{b} \tag{A6}$$

implying first the solution for $\mathbf{y}$ (equation (A4)) and then the solution for $\mathbf{x}$ (equation (A5)).

In practice the solution does not in fact require the actual calculation of the inverse of $\mathbf{U}$ and $\mathbf{L}$. The manual solution of a simple example will illustrate the basic procedure. (This procedure could be used in obtaining solutions to the small-order examples and problems given in the text.) The variant of Gaussian elimination that we choose to describe corresponds to $\mathbf{U}$ having unit terms along its leading diagonal.

Consider the solution of the following system of three equations:

$$
\begin{array}{ll}
2x_1 + 4x_2 + 2x_3 = 6 & \text{(i)} \\
4x_1 + 3x_2 + 2x_3 = 4 & \text{(ii)} \\
5x_1 + 8x_2 + 4x_3 = 10 & \text{(iii)}
\end{array}
\quad \text{or} \quad
\begin{bmatrix} 2 & 4 & 2 \\ 4 & 3 & 2 \\ 5 & 8 & 4 \end{bmatrix}
\begin{Bmatrix} x_1 \\ x_2 \\ x_3 \end{Bmatrix}
=
\begin{Bmatrix} 6 \\ 4 \\ 10 \end{Bmatrix}.
$$

Divide equation (i) by 2, solve for x_1 in terms of the remaining variables x_2

and x_3, and hence substitute for x_1 in equations (ii) and (iii):

$$x_1 + 2x_2 + x_3 = 3 \quad \text{(iv)} \qquad \text{or} \qquad x_1 = 3 - 2x_2 - x_3$$

$$\left. \begin{array}{l} -5x_2 - 2x_3 = -8 \quad \text{(v)} \\ -2x_2 - x_3 = -5 \quad \text{(vi)} \end{array} \right\} \quad \longleftarrow \quad \left\{ \begin{array}{l} \text{eliminating } x_1 \text{ from equations} \\ \text{(ii) and (iii).} \end{array} \right.$$

Retain equation (iv), renumbered as (vii) below, divide equation (v) by -5, solve for x_2 in terms of the remaining variable x_3, and hence substitute for x_2 in equation (vi):

$$x_1 + 2x_2 + x_3 = 3 \quad \text{(vii)}$$

$$x_2 + \tfrac{2}{5}x_3 = \tfrac{8}{5} \quad \text{(viii)} \qquad \text{or} \qquad x_2 = \tfrac{8}{5} - \tfrac{2}{5}x_3$$

$$-\tfrac{1}{5}x_3 = -\tfrac{9}{5} \quad \text{(ix)} \leftarrow \text{eliminating } x_2 \text{ from equation (vi)}.$$

Retain equations (vii) and (viii), renumbered as (x) and (xi) below, and divide equation (ix) by $-1/5$:

$$x_1 + 2x_2 + x_3 = 3 \quad \text{(x)}$$

$$x_2 + \tfrac{2}{5}x_3 = \tfrac{8}{5} \quad \text{(xi)} \qquad \text{or} \qquad \begin{bmatrix} 1 & 2 & 1 \\ 0 & 1 & 2/5 \\ 0 & 0 & 1 \end{bmatrix} \begin{Bmatrix} x_1 \\ x_2 \\ x_3 \end{Bmatrix} = \begin{Bmatrix} 3 \\ 8/5 \\ 9 \end{Bmatrix}, \qquad \begin{array}{l} \text{i.e.} \\ \mathbf{Ux} = \mathbf{y}. \end{array}$$

$$x_3 = 9 \quad \text{(xii)}$$

The solution for x_3 is given by equation (xii), that for x_2 then follows from equation (xi), and finally that for x_1 follows from equation (x). The result is

$$x_3 = 9 \qquad x_2 = -2 \qquad x_1 = -2.$$

In this procedure it has not been necessary to establish the lower triangular matrix $\mathbf{L}$ explicitly, though effectively equation (A4) has been used in the forward elimination stage. In fact, in the above example

$$\mathbf{L} = \begin{bmatrix} 2 & 0 & 0 \\ 4 & -5 & 0 \\ 5 & -2 & -1/5 \end{bmatrix}.$$

The Gaussian elimination procedure can be quite readily generalized for a system of simultaneous equations of any order. In general it is possible for the procedure to break down if at any stage of the elimination process a null diagonal coefficient is encountered. A solution can, however, still be obtained by rearrangement of the order of occurrence of the equations, except of course when the system of equations does not in fact have a solution. In the displacement method the possibility of a null diagonal coefficient should not arise since the structure stiffness matrix (equivalent to $\mathbf{A}$ here) is positive definite.

The Gaussian elimination procedure is applicable to any system of equations (A1) provided that the coefficient matrix $\mathbf{A}$ is square and non-singular; it applies for both non-symmetric and symmetric coefficient matrices, although it simplifies somewhat for the latter case. The alternative popular elimination method—the Cholesky method—is restricted to cases in which the coefficient matrix is symmetric and positive definite, as it is of course in the displacement method.

In the Cholesky method $\mathbf{A}$ is decomposed in the special way that the upper triangular matrix is the transpose of the lower triangular matrix, i.e.

$$\mathbf{A} = \mathbf{L}\mathbf{L}^t \tag{A7}$$

where the diagonal terms of $\mathbf{L}$ are not generally unity. The solution process is $\mathbf{A}\mathbf{x} = \mathbf{L}\mathbf{L}^t\mathbf{x} = \mathbf{b}$. The forward elimination stage is $\mathbf{L}\mathbf{y} = \mathbf{b}$ whilst $\mathbf{L}^t\mathbf{x} = \mathbf{y}$ is the back substitution stage.

As mentioned above, alternative methods to the elimination methods are iterative methods and amongst those are the Gauss–Seidel and Jacobi methods. With the iterative methods the solution procedure begins with an initial estimate of the column matrix of unknowns $\mathbf{x}$ and generates a sequence of progressively updated column matrices which hopefully converges rapidly to the correct solution. Iterative methods have advantages when dealing with very-large-order and sparse coefficient matrices but generally the elimination methods are more popular.

The above remarks have given a very brief description of possible solution procedures for sets of equations of the form of equation (A1); the reader desiring more detailed descriptions of these solution procedures is referred to texts such as references 1–6. The solution procedures discussed above are general in the sense that they are not specialized to suit the finite-element displacement method. It should be noted, though, that specialized procedures have evolved in structural analysis in which elimination methods are applied in ways which take advantage of the particular nature of the structural equations. Two kinds of procedure can be identified: these are *band solution procedures* and *front solution procedures*.

As the name suggests, a band solution procedure takes note of the banded nature of a symmetric structure stiffness matrix which generally occurs when care is taken in numbering the structure nodes and can be extreme in large-order systems. When applying elimination methods to a set of banded equations it is only necessary to work with relatively small groups, or blocks, of equations at a time and very considerable savings of both computer time and storage can result. With a front solution procedure, again only a very small part of the structure stiffness matrix is considered at any one time and the complete matrix is never assembled. Now considerations of the banding of the stiffness matrix are irrelevant. Instead the stiffness equations are considered on an element-by-element basis and the order in which elements are introduced becomes important. Front solution procedures are complicated in the sense that a considerable amount of 'bookkeeping' (programming associated with data handling) is involved but are efficient.

Amongst other sources, detailed discussion of sophisticated equation-solving procedures used in the FEM, including the band and front solution procedures, is available in references 7–14.

Solution of the eigenvalue problem

It has been made clear in Chapter 12 that the direct analytical method of solution used to determine eigenvalues in the simple worked examples of that chapter, based on the expansion of the determinant and subsequent solution of the characterstic equation, is not suitable for systems having more than a very few degrees of freedom. Consequently other methods must be used in calculating the natural frequencies of vibration, or buckling loads, and mode shapes of finite-element models of complicated practical structures which may have tens, hundreds or even thousands of degrees of freedom. It must be appreciated that the difficulty and expense involved in solving a large eigenvalue problem are of an order of magnitude greater than in a corresponding static problem. In fact not more than a couple of hundred degrees of freedom can be readily accommodated in the actual eigenvalue problem, although many more freedoms than this can be used in the initial representation of the structure provided that some scheme such as eigenvalue economization (described in Section 12.8) is used prior to the eigenvalue-solution stage to condense the effective number of degrees of freedom.

There exist many methods for the solution of the eigenvalue problem and it is outside the scope of this text to discuss these methods in any significant detail. The topic has a very considerable literature of its own within the realm of numerical methods, and is dealt with in individual chapters of many texts [1–4], or indeed in complete texts such as that of Wilkinson [15]. The program libraries of computer centres often contain a considerable variety of eigensolution procedures, from which the structural analyst can select a 'package' that seems appropriate to the task in hand and use it as a 'black box' without needing any expert knowledge of the procedures.

As far as the analysis of free vibration is concerned the structure eigenvalue problem has been presented earlier in the general forms of equations (12.23) and (12.24). The matrices $\mathbf{K}_r$ and $\mathbf{M}_r$ are normally real symmetric positive definite matrices and are often heavily banded; these properties should be borne in mind in selecting an eigenvalue-solution method. Another factor affecting choice of the solution method would be the number of eigenvalues and eigenvectors required to be calculated; in the vibration problem we may often be interested only in obtaining the first few natural frequencies whilst in the buckling problem it is usually only the lowest critical load which is of concern.

The 'determinantal' form given by equation (12.24) could be used to obtain eigenvalues. The values of the determinant could be calculated for a large number of specified values of ω^2 and hence the values of ω^2 at which the determinant becomes zero could be determined numerically. In simple terms we could plot a graph of ω^2 versus determinant value to do this; in a more sophisticated manner we could use determinant search methods based on polynomial root-finding techniques to find the eigenvalues and this can be very efficient if $\mathbf{K}_r$ and $\mathbf{M}_r$ are heavily banded.

More usually, the matrix equation (12.23) provides the basis in the free vibration problem for the calculation of eigenvalues and eigenvectors. This equation is an expression of what is termed in numerical analysis the 'generalized' eigenvalue problem represented as

$$\mathbf{A}\mathbf{x} = \lambda \mathbf{B}\mathbf{x} \tag{A8}$$

where $\mathbf{A}$ and $\mathbf{B}$ are square matrices (equivalent here to $\mathbf{K}_r$ and $\mathbf{M}_r$) and the solutions for λ and $\mathbf{x}$ are the eigenvalues and eigenvectors. Some efficient methods of solution do exist for this statement of the problem whereas other methods of solution are available for the 'standard' eigenvalue problem of the form

$$\mathbf{C}\mathbf{x} = \lambda \mathbf{x}$$

or

$$(\mathbf{C} - \lambda \mathbf{I})\mathbf{x} = \mathbf{0} \tag{A9}$$

where $\mathbf{I}$ is a unit matrix of the same order as the square matrix $\mathbf{C}$. Solutions can be obtained, i.e. the eigenvalues of $\mathbf{C}$ can be determined, if $\mathbf{C}$ is a matrix of arbitrary form, but there are considerable advantages if it is in fact symmetric and many solution procedures require this.

We could obtain equation (A9) from equation (12.23), or (A8), by either

(i) premultiplying by $\mathbf{M}_r^{-1}$ which gives the result that $\mathbf{C} \equiv \mathbf{M}_r^{-1}\mathbf{K}_r$ and $\lambda \equiv \omega^2$, together with $\mathbf{x} \equiv \mathbf{D}_r$, or

(ii) premultiplying by $\mathbf{K}_r^{-1}$ which gives the result that $\mathbf{C} \equiv \mathbf{K}_r^{-1}\mathbf{M}_r$ and $\lambda \equiv (1/\omega^2)$, together with $\mathbf{x} \equiv \mathbf{D}_r$.

Disadvantages of these direct approaches of obtaining the standard form are that they require inverting a matrix of order n and that matrix $\mathbf{C}$ becomes fully populated and unsymmetric, despite the fact that $\mathbf{K}_r$ and $\mathbf{M}_r$ are normally banded and symmetric. However, the standard eigenvalue problem (with $\mathbf{C}$ as a symmetric matrix) can be set up in either approach if extra operations are introduced to decompose (usually by Cholesky decomposition) either $\mathbf{M}_r$ or $\mathbf{K}_r$. Considering first $\mathbf{M}_r$, this matrix can be expressed as the product $\mathbf{LL}^t$ where $\mathbf{L}$ is a lower triangular matrix. Then equation (12.23) can be written $\mathbf{K}_r\mathbf{D}_r = \omega^2\mathbf{LL}^t\mathbf{D}_r$. Calling $\mathbf{x} = \mathbf{L}^t\mathbf{D}_r$ allows the equation to be expressed in the standard form of equation (A9) with $\mathbf{C} \equiv \mathbf{L}^{-1}\mathbf{K}_r(\mathbf{L}^t)^{-1}$. The eigenvalues of the original problem, i.e. the eigenvalues of $\mathbf{M}_r^{-1}\mathbf{K}_r$, are the same as the eigenvalues of $\mathbf{C}$, whilst the eigenvectors of the original problem are given by $\mathbf{D}_r = (\mathbf{L}^t)^{-1}\mathbf{x}$. It is noted that when $\mathbf{M}_r$ is a consistent mass matrix, matrix $\mathbf{C}$ is full but if $\mathbf{M}_r$ is a lumped (i.e. diagonal) mass matrix (as described in Section 12.8), matrix $\mathbf{C}$ has the same bandwidth as $\mathbf{K}_r$ and this leads to a more economical solution. In the case of the decomposition of $\mathbf{K}_r$ a similar procedure is adopted with $\mathbf{K}_r$ put equal to $\mathbf{LL}^t$. Again the standard form (equation (A9)) is produced with now $\mathbf{C} \equiv \mathbf{L}^{-1}\mathbf{M}_r(\mathbf{L}^t)^{-1}$ which is again a symmetric matrix; the eigenvalues of $\mathbf{C}$ are values of $1/\omega^2$ and are the same as those of $\mathbf{K}_r^{-1}\mathbf{M}_r$.

In general two types of method are available for solving the eigenvalue problem posed by equation (A8) or (A9): these are transformation methods and iterative methods. Transformation methods include the methods of Jacobi, Givens, Lanczos, and Householder and the QR method, and they operate on the standard form of the problem. Iterative methods include direct iteration, inverse iteration, and simultaneous iteration, and usually do not require a preliminary condensation to standard form. In general the transformation methods are preferred when all eigenvalues and eigenvectors of full matrices are required whereas the iterative methods are well suited to finding the first few highest (or first few lowest) eigenvalues when the matrices are large and heavily banded as they often are in structural problems. For further information on the features and relative advantages of available methods for determining the eigenvalues of structural problems the reader should consult references 16 and 17.

References

1 FADDEEV, D. K. and FADDEEVA, V. N. *Computational methods of linear algebra*. W. H. Freeman, San Francisco (1963).

2 FOX, L. *An introduction to numerical linear algebra*, Oxford University Press, New York (1965).

3 PIPES, L. A. and HOVANESSIAN, S. A. *Matrix computer methods in engineering*, Wiley, New York (1969).

4 FROBERG, C. E. *Introduction to numerical analysis* (2nd edn). Addison-Wesley, Reading, Mass. (1972).

5 CONTE, S. D. and DE BOOR, C. *Elementary numerical analysis* (3rd edn). McGraw-Hill Kogakusha, Tokyo (1980).

6   GERALD, C. F. *Applied numerical analysis* (2nd edn.). Addison-Wesley, Reading, Mass (1980).

7   BATHE, K. J. and WILSON, E. L. *Numerical methods in finite element analysis*. Prentice-Hall, Englewood Cliffs NJ. (1976).

8   HINTON, E. and OWEN, D. R. J. *Finite element programming*. Academic Press, New York (1977).

9   BREBBIA, C. A. and FERRANTE, A. J. *Computational methods for the solution of engineering problems*, Chapter 2. Pentech, London (1978).

10 McGUIRE, W. and GALLAGHER, R. H. *Matrix structural analysis*, Chapter 11. Wiley, New York. (1979).

11 MELOSH, R. J. and BAMFORD, R. M. Efficient solution of load-deflection equations. *J. Struct. Div. Am. Soc. Civ. Eng.* **95** (4), 661–76 (1969).

12 IRONS, B. M. A frontal solution program. *Int. J. Num. Meth. Eng.* **2,** 5–32 (1970).

13 MEYER, C. Solution of equations; state-of-the-art. *J. Struct. Div. Am. Soc. Civ. Eng.* **99** (7), 1507–26 (1973).

14 MEYER, C. Special problems related to linear equation solvers. *J. Struct. Div. Am. Soc. Civ. Eng.* **101** (4), 869–90 (1975).

15 WILKINSON, J. H. *The algebraic eigenvalue problem*. Clarendon Press, Oxford (1965).

16 BATHE, K. J. and WILSON, E. L. Solution methods for eigenvalue problems in structural mechanics. *Int. J. Num. Meth. Eng.* **6,** 213–26 (1973).

17 JENNINGS, A. Eigenvalue methods for vibration analysis. *Shock Vib. Dig.* **12,** 3–16 (1980).

APPENDIX B

USE OF THE PRINCIPLE OF VIRTUAL DISPLACEMENTS IN THE FINITE ELEMENT METHOD

IN Chapter 5 a description was given of two important principles whose application allows the generation of approximate solutions to solid mechanics problems based on the use of an assumed displacement field: these principles are the PMPE and the PVD of course. In succeeding chapters, the PMPE was employed in developing the properties of finite elements. It has been emphasized in the text that in linear elastic problems identical results are obtained if the PVD is used in place of the PMPE. The purpose here is simply to demonstrate that this is so by outlining the general finite-element formulation through use of the PVD, and hence confirm the results obtained earlier when the PMPE was used.

Consider a general linear-elastic three-dimensional body which is in a state of static equilibrium under the action of distributed body forces $\mathbf{R} = \{R_x \ R_y \ R_z\}$ and surface forces $\mathbf{T} = \{T_x \ T_y \ T_z\}$. The corresponding real displacements $\mathbf{u}$, strains $\boldsymbol{\varepsilon}$, and stresses $\boldsymbol{\sigma}$ at a general point are

$$\mathbf{u} = \{u \quad v \quad w\} \tag{B1}$$

$$\boldsymbol{\varepsilon} = \{\varepsilon_x \quad \varepsilon_y \quad \varepsilon_z \quad \gamma_{xy} \quad \gamma_{yz} \quad \gamma_{zx}\} \tag{B2}$$

and

$$\boldsymbol{\sigma} = \{\sigma_x \quad \sigma_y \quad \sigma_z \quad \tau_{xy} \quad \tau_{yz} \quad \tau_{zx}\}. \tag{B3}$$

If the presence of initial strains $\boldsymbol{\varepsilon}^0$, where

$$\boldsymbol{\varepsilon}^0 = \{\varepsilon_x^0 \quad \varepsilon_y^0 \quad \varepsilon_z^0 \quad \gamma_{xy}^0 \quad \gamma_{yz}^0 \quad \gamma_{zx}^0\}, \tag{B4}$$

is assumed, the stress–strain relationship consistent with the assumption of linear elasticity is

$$\boldsymbol{\sigma} = \mathbf{E}(\boldsymbol{\varepsilon} - \boldsymbol{\varepsilon}^0). \tag{B5}$$

Now imagine that a virtual displacement $\delta\mathbf{u} = \{\delta u \ \delta v \ \delta w\}$ is imposed on the equilibrium configuration, such a displacement being kinematically admissible but otherwise arbitrary and in general independent of the real displacement $\mathbf{u}$. The strains corresponding to the virtual displacement, in other words the virtual strains, are

$$\delta\boldsymbol{\varepsilon} = \{\delta\varepsilon_x \quad \delta\varepsilon_y \quad \delta\varepsilon_z \quad \delta\gamma_{xy} \quad \delta\gamma_{yz} \quad \delta\gamma_{zx}\}. \tag{B6}$$

The symbolic statement of the PVD is (see equation (5.22))

$$\delta U_{\mathrm p} - \delta W_{\mathrm e} = 0 \tag{B7}$$

where $\delta U_{\mathrm p}$ is the change in strain energy occurring during the virtual displacement and $\delta W_{\mathrm e}$ is the work done by the applied forces during this displacement. By direct extension of the ideas of Section 5.6

$$\delta U_{\mathrm p} = \int_{V_0} \delta\boldsymbol{\varepsilon}^{\mathrm t}\boldsymbol{\sigma}\, \mathrm d V_0 \tag{B8}$$

and

$$\delta W_{\mathrm e} = \int_{V_0} \delta\mathbf{u}^{\mathrm t}\mathbf{R}\, \mathrm d V_0 + \int_{S_1} \delta\mathbf{u}^{\mathrm t}\mathbf{T}\, \mathrm d S. \tag{B9}$$

Here V_0 is the volume of the body and S_1 is that part of the surface of the body where external tractions are prescribed.

Combining equations (B7)–(B9) and incorporating equation (B5) gives the following statement of the PVD for a general continuum:

$$\int_{V_0} \delta\boldsymbol{\varepsilon}^{\mathrm t}\mathbf{E}\boldsymbol{\varepsilon}\, \mathrm d V_0 - \int_{V_0} \delta\boldsymbol{\varepsilon}^{\mathrm t}\mathbf{E}\boldsymbol{\varepsilon}^0\, \mathrm d V_0$$
$$- \int_{V_0} \delta\mathbf{u}^{\mathrm t}\mathbf{R}\, \mathrm d V_0 - \int_{S_1} \delta\mathbf{u}^{\mathrm t}\mathbf{T}\, \mathrm d S = 0. \tag{B10}$$

Turning to the finite-element procedure, we can consider the application of the PVD at the level either of the individual element or of an assemblage of elements. Choosing the former here, we imagine a finite element in equilibrium under the action of distributed body forces $\mathbf{R}$, surface forces $\mathbf{T}$ (acting on a part of the element surface which coincides with the exterior surface of the whole structure), and nodal forces $\mathbf{P}$ which replace the effect of continuous interaction with neighbouring elements.

In the assumed equilibrium configuration the (real) displacement field of the element is represented as

$$\mathbf{u} = \mathbf{Nd} \tag{B11}$$

and the (real) strains can be expressed as

$$\boldsymbol{\varepsilon} = \mathbf{Bd} \tag{B12}$$

where $\mathbf{N}, \mathbf{B}$, and $\mathbf{d}$ have their usual meanings.

To apply the PVD we select a virtual displacement $\delta\mathbf{u}$ which has the same form as the real displacement, that is

$$\delta\mathbf{u} = \{\delta u \quad \delta v \quad \delta w\} = \mathbf{N}\,\delta\mathbf{d} \tag{B13}$$

where $\delta\mathbf{d}$ is the column matrix of virtual displacements at the element nodes. Correspondingly the virtual strains are

$$\delta\boldsymbol{\varepsilon} = \mathbf{B}\,\delta\mathbf{d}. \tag{B14}$$

As before, the element volume is denoted by V_e and that part of its surface which coincides with the loaded exterior surface of the whole structure is denoted by S_e. Bearing in mind the presence of the nodal forces $\mathbf{P}$ and using equations (B11)–(B14), we modify the statement of the PVD for a continuum given in equation (B10) to become, for the individual finite element,

$$\delta\mathbf{d}^t\left(\int_{V_e} \mathbf{B}^t\mathbf{EB}\,\mathrm{d}V_e\right)\mathbf{d} - \delta\mathbf{d}^t\int_{V_e} \mathbf{B}^t\mathbf{E}\boldsymbol{\varepsilon}^0\,\mathrm{d}V_e$$

$$- \delta\mathbf{d}^t\int_{V_e} \mathbf{N}^t\mathbf{R}\,\mathrm{d}V_e - d\mathbf{d}^t\int_{S_e} \mathbf{N}^t\mathbf{T}\,\mathrm{d}S - \delta\mathbf{d}^t\mathbf{P} = 0. \tag{B15}$$

Since the column matrix of virtual nodal displacements $\delta\mathbf{d}$, is arbitrary it follows that if we define

$$\mathbf{k} = \int_{V_e} \mathbf{B}^t\mathbf{EB}\,\mathrm{d}V_e \tag{B16}$$

$$\mathbf{Q} = \int_{V_e} \mathbf{N}^t\mathbf{R}\,\mathrm{d}V_e + \int_{S_e} \mathbf{N}^t\mathbf{T}\,\mathrm{d}S \tag{B17}$$

$$\mathbf{Q}^0 = \int_{V_e} \mathbf{B}^t\mathbf{E}\boldsymbol{\varepsilon}^0\,\mathrm{d}V_e, \tag{B18}$$

then from equation (B15)

$$\mathbf{P} + \mathbf{Q} + \mathbf{Q}^0 = \mathbf{kd}. \tag{B19}$$

Equation (B19) is a statement of the stiffness relationships for the individual element whilst equations (B16), (B17), and (B18) define the stiffness matrix, the column matrix of consistent loads corresponding to body forces and surface tractions, and the column matrix of consistent loads due to initial strain respectively. These results are, of course, precisely the same as those obtained earlier using the PMPE (see Sections 6.5, 6.12, and 7.3).

Having obtained force–displacement relationships for the individual element, those for an assemblage of elements can be established in a manner consistent with the application of the PVD to the assemblage by a procedure analogous to that described in Section 6.6 when using the PMPE; since work is a scalar quantity, the various work terms for the assemblage, whose nature is defined in equation (B10), are simply summations of element contributions whilst the work of the nodal loads $\mathbf{P}$ of the elements is replaced by that of the externally applied structure loads **F**. It is left to the reader to demonstrate that application of the PVD to an element assemblage gives precisely the same result for the structure stiffness equations as is detailed in Sections 6.6 and 6.12 when using the PMPE.

The above has been concerned only with stable structures under static loading; it should be noted, although it is not proved here, that in problems of free vibration and stability the use of the PVD again yields exactly the same results as does the use of the PMPE.

SOLUTIONS TO PROBLEMS

Chapter 2

2.1 $\{u_2 \; u_3\} = 10^{-3} \{1.0577 \; 1.1346\}$ m
423.08 kN (tensile), 23.08 kN (tensile), 226.92 kN (compressive).

2.2 $\theta_{z2} = 0.003\,75$ rad.
Reaction components are 20 kN↓ and 20 kN m ↶ at point 1, 25 kN↓ at point 2, 45 kN↑, and 60 kN m ↶ at point 3.

2.3 At load point $\{v \; L\theta_z\} = \{2 \; -1\} \times WL^3/66 \, EI_0$
5 W/11↑, 7 WL/33 ↷.

2.4 At tip $\{v \; L\theta_z\} = \{1/6 \; 1/4\} \times PL^3/EI_z$
At clamped end moment is $PL/2$.

2.5 $\{\theta_{z2} \; \theta_{z3}\} = 10^{-4} \{6 \; -18\}$ rad.
11.25 kN↓, 15 kN m ↶.

2.6 Torques in sections AB, BC, CD, and DE are 39.412 kN m, 19.412 kN m, −0.588 kN m, and −20.588 kN m.

2.7 $\{\theta_{x2} \; \theta_{x3} \; \theta_{x4}\} = 10^{-3} \{14.0562 \; 21.6868 \; 8.1325\}$ rad.
Torques in sections A, B, and C are 16.867 kN m, 6.867 kN m and −8.133 kN m.

2.8 $10^{-4}\{0.7217$ m 9.8765 m -7.4074 rad$\}$.
Bending moments at left-hand end, load point, and right-hand end are −11.111 kN m, 14.815 kN m and −22.222 kN m.

2.9 1.938 kN↑, 44.61 kN m ↷.

2.10 $\{v_2 \; L\theta_{z2} \; v_3\} = \{6 \; 8 \; 11\} \times PL^3/48EI_z$.

2.11 $\{\theta_{z1}(=-\theta_{z5}) \; \theta_{z2}(=-\theta_{z4}) \; v_3\} = \{6$ rad 48 rad 49 m$\} \times 1/90\,000$
27 kN↓, 66 kN m.

2.12 Load-point displacements are $\{7 \; 8 \; -8 \; 7\} \times WL/3AE$.
Forces are $7W/3$ (tensile), $W/3$ (tensile), $8W/3$ (compressive), $W/3$ (tensile), $7W/3$ (tensile).

2.13 $\{v_2 \; L\theta_{z2} \; v_3 \; L\theta_{z3} \; v_4 \; L\theta_{z4}\} = \{8 \; 6 \; 9 \; -3 \; 4 \; -6\} \times PL^3/72EI_0$.

2.14 $\{u_2 \; u_3\} = 10^{-3} \{1.2423 \; 1.5654\}$ m.
496.92 kN (tensile), 96.92 kN (tensile), 153.08 kN (compressive).

2.15 $\{\theta_{z1}(=-\theta_{z5}) \; \theta_{z2}(=-\theta_{z4})v_3\} = \{24$ rad 30 rad 61 m$\} \times 1/3600$.
22.5 kN↓, 75 kN m.

Chapter 3

3.1 $\bar{u}_A = 4FL/9AE$. Forces in bars AB, AC, and AD are 0.314 27F, 0.444 44F and 0.384 90F, all tensile.

3.2 $\{\bar{u}_2 \ \bar{v}_2\} = \{-16/3 \ 21\} \times WL/AE$. Forces in bars 1–2 and 2–3 are $4W/3$ (compressive) and $5W/3$ (tensile).

3.3 $\{\bar{u}_B \ \bar{v}_B\} = \{1/2 \ -1\} \times WL/AE$.
Forces in bars AB, BC, BD, and BE are $W/2$, $3W/2\sqrt{2}$, W, and $W/2\sqrt{2}$, all tensile.

3.4

	$\bar{u}_A$	$\bar{v}_A$	$\bar{u}_B$	$\bar{v}_B$	$\bar{u}_C$	$\bar{v}_C$	$\bar{u}_D$	$\bar{v}_D$	$\bar{u}_E$	$\bar{v}_E$
$\mathbf{K}_r =$	5.28									
	0	2.72								
	0	0	5.28							
	0	-2	0	2.72		Symmetric				
	-2	0	-0.64	0.48	3.28					
	0	0	0.48	-0.36	0	2.72				
	-0.64	-0.48	-2	0	0	0	4.64			
	-0.48	-0.36	0	0	0	-2	0.48	2.36		
	0	0	0	0	-0.64	-0.48	-2	0	2.64	
	0	0	0	0	-0.48	-0.36	0	0	0.48	0.36

3.5 Forces in bars 1–2, 1–3, 2–3, 2–4, and 3–4 are 84.9 kN (compressive), 80.0 kN (tensile), 40.0 kN (compressive), 60.0 kN (compressive), and 56.0 kN (tensile).

3.6 $\{\bar{u}_3 \ \bar{v}_3 \ \bar{u}_5\} = \{16.6735 \ 38.4677 \ 59.7984\} \times PL/AE$.
3.6419 P (tensile), 0.6532 P (compressive).

3.7 $\{\bar{u}_2 \ \bar{v}_2 \ \theta_{\bar{z}2}\} = 10^{-4}\{1.2655 \ \text{m} \ 1.7209 \ \text{m} \ 8.5419 \ \text{rad}\}$.
$\mathbf{P}^{1-2} = \{-33.746 \ \text{kN} \ 3.944 \ \text{kN} \ 3.638 \ \text{kN m} \ 33.746 \ \text{kN} \ -3.944 \ \text{kN} \ 8.194 \ \text{kN m}\}$
$\mathbf{P}^{2-3} = \{50.620 \ \text{kN} \ 12.315 \ \text{kN} \ 15.732 \ \text{kN m} \ -50.620 \ \text{kN} \ -12.315 \ \text{kN} \ 8.899 \ \text{kN m}\}$
$\mathbf{P}^{2-4} = \{51.628 \ \text{kN} \ 4.366 \ \text{kN} \ 6.074 \ \text{kN m} \ 51.628 \ \text{kN} \ -4.366 \ \text{kN} \ 2.657 \ \text{kN m}\}$.

3.8 $\{\bar{u}_1 \ \bar{v}_1 \ \theta_{\bar{z}1}\} = 10^{-3}\{-20.26 \ 99.36 \ 1.797\}$
$\mathbf{P}^{1-3} = \{34.73 \ 12.53 \ 927.00 \ -34.73 \ -12.53 \ 639.43\}$.

3.9 $\{\theta_{\bar{z}1} \ \bar{v}_2 \ \theta_{\bar{z}3}\} = 10^{-4}\{1.5686 \ \text{rad} \ 2.0915 \ \text{m} \ -1.5686 \ \text{rad}\}$.

3.10 $\{\bar{u}_5(=-\bar{u}_6) \ \bar{v}_5(=\bar{v}_6) \ \theta_{\bar{z}5}(=-\theta_{\bar{z}6})\} = 10^{-4}\{5.9595 \ \text{m} \ -6.2399 \ \text{m} \ 1.3499 \ \text{rad}\}$
14.162 kN, 0.148 kN, 0.451 kN m.

3.11 Tension in cables is $9.524W$, vertical reactions at A and E are $4.285W\uparrow$.

3.12 $\{\bar{u}_4 \ \bar{v}_4 \ \theta_{\bar{z}4}\} = 10^{-4}\{-10.7459 \ \text{m} \ 3.9828 \ \text{m} \ -4.0722 \ \text{rad}\}$
$\mathbf{P}^{3-4} = \{307.459 \ \text{kN} \ 1.520 \ \text{kN} \ 4.615 \ \text{kN m} \ -307.459 \ \text{kN} \ -1.520 \ \text{kN} \ 2.986 \ \text{kN m}\}$.

3.14

	u_i	θ_{zi}	u_j	θ_{zj}
$\mathbf{k} =$	AE/l			
	0	EI_z/l	Symmetric	
	$-AE/l$	0	AE/l	
	0	$-EI_z/l$	0	EI_z/l

(V_i is necessarily zero when V_j is zero)

3.15 $\{\bar{v}_2 \ \theta_{\bar{z}3}\} = \{10 \ \text{m} \ -13 \ \text{rad}\} \times 1/3600$.
Rotations at pin are $1/240$ rad in 1–2, $1/3600$ rad in 2–3.
Reaction force and moment at 1 are $25 \ \text{kN}\uparrow$, $25 \ \text{kN m} \ \text{↻}$.

3.16

$$\mathbf{K}_r = \begin{bmatrix} 2AE/L + 3EI_z/L^3 & 0 & 0 \\ 0 & AE/L + 24EI_z/L^3 & 0 \\ 0 & 0 & 8EI_z/L \end{bmatrix}.$$

where the column headings above the matrix are $\bar{u}_B$, $\bar{v}_B$, θ_{zB}.

3.17 $\{\bar{u}_B \ \bar{v}_B\} = 10^{-4}\{6.1855 \ -9.4869\}$ m
Rotations at pin are -3.558×10^{-4} rad in A–B, 3.192×10^{-4} rad in B–C.

3.18 $\{\theta_{z1} \ \theta_{z2}\} = \{102 \ 81\} \times 1/73EI_z$.
Fixing moment at point 3 is 108/73 kN m.

3.19 $\{\bar{u}_1(= \bar{u}_2) \ \theta_{z1} \ \theta_{z2}\} = \{90 \ 265 \ -505\} \times 1/21EI_z$.

3.20 $\{\bar{v}_2 \ u_3'\} = 10^{-2}\{-1.1106 \ -1.8755\}$ m.

3.21

$$\begin{Bmatrix} 0.05 \\ 0 \\ 0 \\ 0 \\ 0 \\ 0 \\ 0 \\ 0 \end{Bmatrix} = \begin{bmatrix} 206 & & & & & & & \\ 0 & 400.75 & & & & \text{Symmetric} & & \\ -6 & 1.5 & 12 & & & & & \\ -200 & 0 & 0 & 200.75 & & & & \\ 0 & -0.75 & -1.5 & 0 & 200.75 & & & \\ 0 & 1.5 & 2 & -1.5 & -1.5 & 8 & & \\ 0 & 0 & 0 & -0.64952 & -100 & 1.29904 & 50.562\,50 & \\ 0 & 0 & 0 & -1.5 & 0 & 2 & 1.299\,04 & 4 \end{bmatrix} \begin{Bmatrix} \bar{u}_1 \\ \bar{v}_1 \\ \theta_{z1} \\ \bar{u}_2 \\ \bar{v}_2 \\ \theta_{z2} \\ u_4' \\ \theta_{z'4} \end{Bmatrix}$$

Chapter 4

4.1 $pL^3/48EI_z$, $pL^2/8$

4.2 $\{\theta_{zA} \ \theta_{zB} \ \theta_{zC}\} = \{3 \ 2 \ -7\} \times WL^2/96EI_z$.
Reactions are $11W/16\uparrow$, $42W/16\uparrow$ and $11W/16\uparrow$ at A, B, and C.

4.3 $\{\theta_{zB} \ \theta_{zC}\} = \{-12 \ 16\} \times 1/EI_z$.
At A, 56 kN↑, 32 kN m ↷: at D, 16 kN↑, 20 kN m ↶.

4.4 Zero.

4.5 $\{\bar{v}_2 \ \theta_{z2}\} = 10^{-4}\{32.559 \text{ m} \ -2.2283 \text{ rad}\}$.

4.6 (a) $\{43.33 \ 17.93 \ 5.40 \ 26.67 \ 51.35 \ -30.79\}$
(b) $\{19.309 \ 11.078 \ 11.52 \ 0.691 \ 53.922 \ -7.28 \ 10 \ 15 \ -10\}$
in units of kilonewtons and metres.

4.7 $\{102.618 \text{ kN} \ 22.157 \text{ kN} \ 7.040 \text{ kN m} \ 1.382 \text{ kN} \ 62.843 \text{ kN} \ 2.940 \text{ kN m}$
$0 \ 25.0 \text{ kN} \ -37.50 \text{ kN m}\}$.
50.10 kN m.

4.8 $\mathbf{P}^{1-3} = \{28.73 \text{ kN} \ 4.53 \text{ kN} \ 677.1 \text{ kN m} \ -40.73 \text{ kN} \ -20.53 \text{ kN} \ 889.4 \text{ kN m}\}$.

4.9 $\{\theta_{zB} \ \theta_{zC}\} = \{-1 \ 4\} \times 1200/77EI_z$.
$\{M_A \ M_B \ M_{CB} \ M_{CD} \ M_{CE} \ M_D \ M_E\} = \frac{100}{77}\{74 \ -83 \ -56 \ 32 \ 24 \ 16 \ 12\}$ kN m.

4.10 $\{\theta_{z2} \ \theta_{z3}\} = 10^{-3}\{-1.375 \ 0.6875\}$ rad.
25.781 25 kN↑, 25.781 25 kN↑, 51.5625 kN↓, 206.25 kN m ↷.

4.11 $\{\bar{u}_B \ \bar{v}_B\} = \{1 \ -1\} \times \alpha LT/2$.
Forces in bars BA, BC, BD, and BE are $AE\alpha T/2$ (tensile), $AE\alpha T/\sqrt{2}$ (compressive), $AE\alpha T/2$ (tensile), and zero.

4.12 Forces in bars BA, BC, BD, and BE are $AE/1000$ (compressive), zero, $AE/1000$ (tensile) and $\sqrt{2}\,AE/1000$ (compressive).

4.13 $\{\bar{u}_2 \ \bar{v}_2 \ \theta_{\bar{z}2}\} = 10^{-4}\{9.79566 \text{ m} \ 1.79855 \text{ m} \ 4.86876 \text{ rad}\}$.
$\mathbf{P}^{1-2} = \{314.782 \text{ kN} \ 1.957 \text{ kN} \ -27.163 \text{ kN m} \ -314.782 \text{ kN} \ -1.957 \text{ kN} \ 33.034 \text{ kN m}\}$
$\mathbf{P}^{2-3} = \{391.826 \text{ kN} \ 8.001 \text{ kN} \ \ \ 9.948 \text{ kN m} \ -391.826 \text{ kN} \ -8.001 \text{ kN} \ \ \ 6.053 \text{ kN m}\}$

4.14 $\{\theta_{\bar{x}1} \ \theta_{\bar{y}1} \ \bar{w}_1\} = 10^{-3}\{22.3077 \text{ rad} \ -17.5923 \text{ rad} \ 25.5556 \text{ m}\}$
$\mathbf{P}^{1-2} = \{17.6923 \text{ kN m} \ 37.6923 \text{ kN m} \ 29.2308 \text{ kN} \ -17.6923 \text{ kN m}$
$-96.1538 \text{ kN m} \ -29.2308 \text{ kN}\}$
$\mathbf{P}^{1-3} = \{-22.3077 \text{ kN m} \ -17.6923 \text{ kN m} \ 70.7692 \text{ kN} \ 22.3077 \text{ kN m}$
$-123.8462 \text{ kN m} \ -70.7692 \text{ kN}\}$

4.15 $\{\theta_{\bar{x}1} \ \theta_{\bar{y}1} \ \bar{w}_1\} = 10^{-3}\{1.29056 \text{ rad} \ 5.03131 \text{ rad} \ 4.69186 \text{ m}\}$.
$\mathbf{P}^{1-3} = \{1.3003 \text{ kN m} \ 27.5638 \text{ kN m} \ -38.6183 \text{ kN} \ -1.3003 \text{ kN m}$
$-84.4723 \text{ kN m} \ -61.3817 \text{ kN}\}$.

4.16 $\{\theta_{\bar{x}2}(=\theta_{\bar{x}4}) \ \theta_{\bar{y}2}(=-\theta_{\bar{y}4}) \ \bar{w}_2(=\bar{w}_4)\}$
$= \{-0.056396 \text{ rad} \ -0.003589 \text{ rad} \ 0.12750 \text{ m}\}$.

4.17 $\{\bar{u}_1 \ \bar{v}_1 \ \bar{w}_1\} = 10^{-3}\{2.96204 \ -0.94785 \ 0\}$ m.
Forces in bars 1–2, 1–3, 1–4, and 1–5 are 129.253 kN (tensile), 14.361 kN (compressive), 129.253 kN (tensile), and 14.361 kN (compressive).

4.18 $\{\bar{u}_1 \ \bar{v}_1 \ \bar{w}_1\} = 10^{-3}\{-1.98796 \ -2.9279 \ 0\}$ m.
Forces in bars as for Problem 4.17.

4.19 $\{\bar{u}_A \ \bar{v}_A \ \bar{w}_A \ \bar{u}_B \ \bar{v}_B \ \bar{w}_B\} = 10^{-3}\{-0.54760 \ 3.09846 \ 0 \ 0.5 \ 1.34375 \ 0\}$ m.

Forces in bars AB, AD, AE, BC, BD, and BE are 70.7107 kN (compressive), 25.7694 kN (tensile), 25.7694 kN (tensile), 100 kN (compressive), 37.5 kN (tensile), and 37.5 kN (tensile).

Chapter 5

5.1 (a) $v_c = PL^3/48.0347EI_z$, $\Pi_p = -0.010\,4092P^2L^3/EI_z$,
(b) $v_c = PL^3/48.0149EI_z$, $\Pi_p = -0.010\,4134P^2L^3/EI_z$.

5.3 C_5 would have the value zero.

5.4 For cubic polynomial $v = qL^4/30EI_z$ at free end, $\Pi_p = -13q^2L^5/7200EI_z$.
For quartic polynomial $v = qL^4/30EI_z$ at free end, $\Pi_p = -19q^2L^5/9600EI_z$.
Exact solution corresponds to quintic polynomial field.

5.5 $v = \dfrac{pL^4}{168EI_z}\left(\dfrac{17x^2}{L^2} - \dfrac{8x^3}{L^3}\right)$, $\Pi_p = -0.010\,913p^2L^5/EI_z$.

5.6 $v = \dfrac{3pL^4}{16EI_z}\left(\dfrac{x^2}{L^2} - \dfrac{x^3}{3L^3}\right)$, $\Pi_p = -0.022\,727p^2L^5/EI_z$.

5.7 $v_c = 0.008\,3213pL^4/EI_z$, $\Pi_p = -0.001\,6863p^2L^5/EI_z$ for one-term solution.
$v_c = 0.008\,3327pL^4/EI_z$, $\Pi_p = -0.00168\,65p^2L^5/EI_z$ for two-term solution.
$(v_c = 0.008\,3333pL^4/EI_z$ in exact solution).

5.8 $v_c = 13WL^3/3072EI_z$ (which agrees with exact solution), $\Pi_p = -0.001\,7192W^2L^3/EI_z$.

5.9 (a) $u = \dfrac{2}{3}\dfrac{WL}{AE}\dfrac{x}{L}$, $\Pi_p = -0.333\,33\,W^2L/AE$.

(b) $u = \dfrac{3}{13}\dfrac{WL}{AE}\left(\dfrac{2x}{L}+\dfrac{x^2}{L^2}\right)$, $\Pi_p = -0.34615\,W^2L/AE$.

5.10 Nodal displacements are $\{0\ 5\ 8\ 9\}pL^2/18AE$.
Stresses in the three regions are $\{5\ 3\ 1\}pL/6A$.
$\Pi_p = -0.162\,04p^2L^3/AE$.

5.12 $u_2 = 0.115\,51\,WL/A_0E$, $\Pi_p = -0.057\,754W^2L/A_0E$.

5.13 $u_2 = 0.15\,WL/A_0E$, $\Pi_p = -0.075\,W^2L/A_0E$.

Chapter 6

6.1 (a) $\mathbf{Q} = \{-18\ -5l\ 18\ 3l\} \times M_0/16l$.
(b) $\mathbf{Q} = \{5\ l\ 5\ -l\} \times pl/15$.
(c) $\mathbf{Q} = \{522\ 109l\ 246\ -67l\} \times pl/3072$.

6.4 (i) $\Delta = 0.1364WL^3/EI_0$, $\phi = 0.2273WL^2/EI_0$.
(ii) $\Delta = 0.1372WL^3/EI_0$, $\phi = 0.2256WL^2/EI_0$

6.6 $u_D = 0.6857WL/AE$ for both (i) and (ii).

6.8 $N_1 = (-3+\bar{x}+3\bar{x}^2-\bar{x}^3)/48$, $N_2 = (27-27\bar{x}-3\bar{x}^2+3\bar{x}^3)/48$,
$N_3 = (27+27\bar{x}-3\bar{x}^2-3\bar{x}^3)/48$, $N_4 = (-3-\bar{x}+3\bar{x}^2+\bar{x}^3)/48$,
where $\bar{x} = 6x/l$ and l is the element length.

6.9
$$\mathbf{k} = \dfrac{AE}{40l}\begin{array}{cccc} u_1 & u_2 & u_3 & u_4 \end{array}\begin{bmatrix} 148 & & & \\ -189 & 432 & \text{Symmetric} & \\ 54 & -297 & 432 & \\ -13 & 54 & -189 & 148 \end{bmatrix}, \mathbf{Q} = \dfrac{fl}{120}\begin{Bmatrix} 13+2\lambda \\ 36+9\lambda \\ 9+36\lambda \\ 2+13\lambda \end{Bmatrix}$$

6.11 $N_{v1} = 4\bar{x}^2 - 10\bar{x}^3 - 8\bar{x}^4 + 24\bar{x}^5$, $N_{\theta z1} = l(\tfrac{1}{2}\bar{x}^2 - \bar{x}^3 - 2\bar{x}^4 + 4\bar{x}^5)$,
$N_{v2} = 1 - 8\bar{x}^2 + 16\bar{x}^4$, $N_{\theta z2} = l(\bar{x} - 8\bar{x}^3 + 16\bar{x}^5)$,
$N_{v3} = 4\bar{x}^2 + 10\bar{x}^3 - 8\bar{x}^4 - 24\bar{x}^5$, $N_{\theta z3} = l(-\tfrac{1}{2}\bar{x}^2 - \bar{x}^3 + 2\bar{x}^4 + 4\bar{x}^5)$,
where $\bar{x} = x/l$.

6.12
$$\mathbf{k} = \dfrac{EI_z}{35l^3}\begin{array}{cccccc} v_1 & \theta_{z1} & v_2 & \theta_{z2} & v_3 & \theta_{z3} \end{array}\begin{bmatrix} 5092 & & & & & \\ 1138l & 332l^2 & & & & \\ -3584 & -896l & 7168 & & \text{Symmetric} & \\ 1920l & 320l^2 & 0 & 1280l^2 & & \\ -1508 & -242l & -3584 & -1920l & 5092 & \\ 242l & 38l^2 & 896l & 320l^2 & -1138l & 332l^2 \end{bmatrix}, \mathbf{Q} = \dfrac{pl}{60}\begin{Bmatrix} 14 \\ l \\ 32 \\ 0 \\ 14 \\ -l \end{Bmatrix}$$

6.15
$$\begin{Bmatrix} V_1 \\ M_{z1} \\ V_3 \\ M_{z3} \end{Bmatrix} = \dfrac{EI_z}{l^3}\begin{bmatrix} 12 & 6l & -12 & 6l \\ 6l & 4l^2 & -6l & 2l^2 \\ -12 & -6l & 12 & -6l \\ 6l & 2l^2 & -6l & 4l^2 \end{bmatrix}\begin{Bmatrix} v_1 \\ \theta_{z1} \\ v_3 \\ \theta_{z3} \end{Bmatrix} - \dfrac{1}{8}\begin{bmatrix} 4 & -12l \\ l & -2l^2 \\ 4 & 12l \\ -l & -2l^2 \end{bmatrix}\begin{Bmatrix} P_{v2} \\ P_{\theta z2} \end{Bmatrix}$$

where P_{v2} and $P_{\theta z2}$ are the vertical force and couple, respectively, acting at node 2 due to any externally-applied loading.

6.16

$$\begin{Bmatrix} U_1 \\ U_4 \end{Bmatrix} = \frac{AE}{l} \begin{bmatrix} 1 & -1 \\ -1 & 1 \end{bmatrix} \begin{Bmatrix} u_1 \\ u_4 \end{Bmatrix} - \frac{1}{3} \begin{bmatrix} 2 & 1 \\ 1 & 2 \end{bmatrix} \begin{Bmatrix} P_2 \\ P_3 \end{Bmatrix}$$

where P_2 and P_3 are the axial forces acting at nodes 2 and 3 due to any externally-applied loading.

Chapter 8

8.1   For load case (i) the exact solution is obtained.
For load case (ii):
$\{v_1 \ u_3 \ u_4 \ v_4\} = 10^{-6}\{-21.8407 \ 24.5879 \ 85.3022 \ -11.1264\}$ m
$\{\sigma_x \ \sigma_y \ \tau_{xy}\}^{1-2-4} = \{15.75 \ 0.75 \ 0.75\}$ MN m^{-2}
$\{\sigma_x \ \sigma_y \ \tau_{xy}\}^{3-4-2} = \{4.25 \ -0.75 \ 4.25\}$ MN m^{-2}.

8.2   For load case (ii), $\{v_1 \ u_3 \ u_4 \ v_4\}$
$= 10^{-6}\{17.3329 \ -22.3496 \ 132.2397 \ -50.2999\}$ m

8.3 $\{v_1 \ u_3 \ u_4 \ v_4\} = 10^{-6}\{8.4372 \ 8.4372 \ 84.5441 \ -36.3316\}$ m
$\{\sigma_x \ \sigma_y \ \tau_{xy}\}^A = \{2.1937 \ 2.1937 \ 0\}$ MN m^{-2}
$\{\sigma_x \ \sigma_y \ \tau_{xy}\}^B = \{14.7289 \ -2.1937 \ 2.1937\}$ MN m^{-2}
For bar, $\sigma_x = 15.3870$ MN m^{-2}.

8.4

$$\begin{Bmatrix} 30 \\ 0 \\ 90 \\ 40 \\ 20 \\ 0 \\ 0 \\ 0 \end{Bmatrix} = \bar{E} \begin{bmatrix} 1.35 & & & & & & & \\ 0 & 1.35 & & & & & & \\ -1 & -0.35 & 1.7 & & & \text{Symmetric} & & \\ -0.3 & -0.35 & 0.65 & 2.35 & & & & \\ -0.35 & -0.3 & 0 & 0 & 2.35 & & & \\ -0.35 & -1 & 0 & 0 & 0.65 & 1.7 & & \\ 0 & 0.65 & -0.7 & -0.65 & -2 & -0.65 & 5.4 & \\ 0.65 & 0 & -0.65 & -2 & -0.65 & -0.7 & 1.3 & 5.4 \end{bmatrix} \begin{Bmatrix} u_1 \\ v_1 \\ u_2 \\ v_2 \\ u_3 \\ v_3 \\ u_4 \\ v_4 \end{Bmatrix}$$

where $\bar{E} = \dfrac{Eh}{2(1-\nu^2)} = 1.098\ 9011 \times 10^{-2}$ and u_1 etc. are in metres.

8.5   For load case (i):
$\{u_Q \ u_M\} = 10^{-6}\{45.5 \ 45.5\}$ m
$\{\sigma_x \ \sigma_y \ \tau_{xy}\} = \{10 \ 3 \ 0\}$ MN m^{-2} for both elements.
For load case (ii):
$\{u_Q \ u_M\} = 10^{-6}\{54.4216 \ 36.5784\}$ m
$\{\sigma_x \ \sigma_y \ \tau_{xy}\}^{LPO} = \{11.9608 \ 3.5882 \ 0\}$ MN m^{-2}
$\{\sigma_x \ \sigma_y \ \tau_{xy}\}^{LQM} = \{8.0392 \ 2.4118 \ -1.3725\}$ MN m^{-2}.

8.6   For load case (i):
$\{u_Q \ u_M\} = 10^{-6}\{45.5 \ 45.5\}$ m
$\{\sigma_x \ \sigma_y \ \tau_{xy}\} = \{10 \ 3 \ 0\}$ MN m^{-2} everywhere.
For load case (ii):
$\{u_Q \ u_M\} = 10^{-6}\{64.4583 \ 26.5417\}$ m.

8.7   For mesh (a):
$\{u_Q \ u_M \ v_R \ u_S \ v_S\} = 10^{-6}\{58.5933 \ 32.4067 \ -0.1463 \ 45.5000 \ -1.8177\}$ m
where R is the node point midway between L and P, and S is the node point midway between M and Q. Maximum stress is $\sigma_x = 12.6380$ MN m^{-2} in element PQS.

8.8 For the element shown in Fig. P8.8(a):

$$\mathbf{k} = \frac{Eh}{2(1-\nu^2)ab} \begin{bmatrix} u_1 & v_1 & u_2 & v_2 & u_3 & v_3 \\ b^2+\rho d^2 & & & & & \\ bd(\rho+\nu) & d^2+\rho b^2 & & \text{Symmetric} & & \\ \rho cd-b^2 & \rho bc-\nu bd & b^2+\rho c^2 & & & \\ \nu bc-\rho bd & cd-\rho b^2 & -bc(\rho+\nu) & c^2+\rho b^2 & & \\ -\rho ad & -\rho ab & -\rho ac & \rho ab & \rho a^2 & \\ -\nu ab & -ad & \nu ab & -ac & 0 & a^2 \end{bmatrix}$$

where $\rho = (1-\nu)/2$.

8.10 From left to right across the section AA' the consistent nodal loads are (in kN)
0.3 14.4 12.6 26.4 12.6 14.4 and 0.3
in the horizontal direction, and
15 80 50 120 70 160 and 45
in the vertical direction

8.12 (i) $(9\xi^2-1)(1-\xi)(9\eta^2-1)(1+\eta)/256$
(ii) $9(1-\xi^2)(1+3\xi)(9\eta^2-1)(1-\eta)/256$
(iii) $81(1-\xi^2)(1+3\xi)(1-\eta^2)(1-3\eta)/256$
where $\xi = x/a,\ \eta = y/b$.

8.13 $N = (\xi^2+\xi)(1+\eta)/4$ for corner node at $\xi = +1,\ \eta = +1$
$N = (1-\xi^2)(1+\eta)/2$ for side node at $\xi = 0, \eta = +1$
where $\xi = x/a,\ \eta = y/b$.

8.14 The body force provides contributions only to that part of $\mathbf{Q}$ associated with u-direction forces. These contributions, for nodes in order 1 to 9, are
$\mathbf{Q}_u = \{c\ d\ d\ c\ 4A\ 4d\ 4A\ 4c\ 16A\} \times abh/9.$
Here $c = A - aB,\ d = A + ab$, h is the thickness, and a and b are the dimensions shown in Fig. 8.15.

8.15 $\mathbf{Q} = \{15\ -17\ 45\ -54\ 45\ -81\ 15\ -28\} \times pl/120.$
(These nodal loads are listed in order of u-component and v-component at each node in turn, starting at the left-hand end (A) and running to the right-hand end (B). The positive directions are taken to be horizontal to the right for u, and vertical upwards for v.)

Chapter 9

$$\mathbf{S} = \frac{E}{6V_c(1+\nu)(1-2\nu)}[\mathbf{s}_1\ \mathbf{s}_2\ \mathbf{s}_3\ \mathbf{s}_4] \text{ where } \mathbf{s}_i = \begin{bmatrix} (1-\nu)\beta_i & 0 & 0 \\ \nu\beta_i & (1-\nu)\gamma_i & \nu\delta_i \\ \nu\beta_i & \nu\gamma_i & (1-\nu)\delta_i \\ \frac{1}{2}(1-2\nu)\gamma_i & \frac{1}{2}(1-2\nu)\beta_i & 0 \\ 0 & \frac{1}{2}(1-2\nu)\delta_i & \frac{1}{2}(1-2\nu)\gamma_i \\ \frac{1}{2}(1-2\nu)\delta_i & 0 & \frac{1}{2}(1-2\nu)\beta_i \end{bmatrix}$$

9.2 $\mathbf{B} = [\mathbf{b}_1 \ \mathbf{b}_2 \ \mathbf{b}_3 \ \mathbf{b}_4 \ \mathbf{b}_5 \ \mathbf{b}_6 \ \mathbf{b}_7 \ \mathbf{b}_8]$
where

$$\mathbf{b}_i = \frac{1}{8}\begin{bmatrix} \xi_i(1+\eta\eta_i)(1+\zeta\zeta_i)/a & 0 & 0 \\ 0 & \eta_i(1+\xi\xi_i)(1+\zeta\zeta_i)/b & 0 \\ 0 & 0 & \zeta_i(1+\xi\xi_i)(1+\eta\eta_i)/c \\ \eta_i(1+\xi\xi_i)(1+\zeta\zeta_i)/b & \xi_i(1+\eta\eta_i)(1+\zeta\zeta_i)/a & 0 \\ 0 & \zeta_i(1+\xi\xi_i)(1+\eta\eta_i)/c & \eta_i(1+\xi\xi_i)(1+\zeta\zeta_i)/b \\ \zeta_i(1+\xi\xi_i)(1+\eta\eta_i)/c & 0 & \xi_i(1+\eta\eta_i)(1+\zeta\zeta_i)/a \end{bmatrix}$$

and where $\xi = x/a$, $\eta = y/b$, $\zeta = z/c$, with ξ_i being the value of ξ at node i, etc.

9.3 $N_A = \frac{1}{8}(\xi + \xi^2)(\eta + \eta^2)(\zeta + \zeta^2)$, $N_B = \frac{1}{4}(\xi + \xi^2)(\eta + \eta^2)(1 - \zeta^2)$,
$N_C = \frac{1}{2}(1 - \xi^2)(\eta + \eta^2)(1 - \zeta^2)$.
where $\xi = x/a$, $\eta = y/b$, $\zeta = z/c$.

9.4 $N_A = \frac{1}{64}(9\xi^2 - 1)(1 + \xi)(1 + \eta)(1 + \zeta)$, $N_B = \frac{9}{64}(1 - \xi^2)(1 + 3\xi)(1 + \eta)(1 - \zeta)$
where $\xi = x/a$, $\eta = y/b$, $\zeta = z/c$.

9.5 The consistent nodal loads, acting in the direction normal to the surface are $qbc/9$ for each of the four corner-nodes, $4qbc/9$ for each of the four mid-side nodes and $16qbc/9$ for the node at the centre of the surface.

9.6 $\mathbf{B} = [\mathbf{b}_1 \ \mathbf{b}_2 \ \mathbf{b}_3 \ \mathbf{b}_4]$
where

$$\mathbf{b}_i = \frac{1}{4}\begin{bmatrix} \dfrac{\xi_i}{a}(1+\eta\eta_i) & 0 \\ 0 & \dfrac{\eta_i}{b}(1+\xi\xi_i) \\ \dfrac{1}{r}(1+\xi\xi_i)(1+\eta\eta_i) & 0 \\ \dfrac{\eta_i}{b}(1+\xi\xi_i) & \dfrac{\xi_i}{a}(1+\eta\eta_i) \end{bmatrix}$$

with $\xi = \dfrac{r - (r_a + r_b)/2}{a}$, $\eta = \dfrac{y - (y_a + y_b)/2}{b}$ and with ξ_i being the value of ξ at node i, etc.

9.7 $2\pi R_0 pc/3$, $2\pi R_0 pc$ and $\pi R_0 pc/3$ in the negative r direction.

Chapter 10

10.2 $\mathbf{N} = [N_{30} \ N_{21} \ N_{12} \ N_{03}]$

$$= \left[\frac{L_1}{2}(3L_1 - 1)(3L_1 - 2) \quad \frac{9}{2}L_1 L_2(3L_1 - 1) \right.$$

$$\left. \frac{9}{2}L_1 L_2(3L_2 - 1) \quad \frac{L_2}{2}(3L_2 - 1)(3L_2 - 2)\right].$$

10.3 See the solution to Problem 6.9.

10.7 $k_{31} = C_1(\beta_1\beta_2 + \rho\gamma_1\gamma_2)$, $k_{32} = C_1(\nu\beta_2\gamma_1 + \rho\beta_1\gamma_2)$,
       $k_{41} = C_1(\nu\beta_1\gamma_2 + \rho\beta_2\gamma_1)$, $k_{42} = C_1(\gamma_1\gamma_2 + \rho\beta_1\beta_2)$,
       where $\rho = \dfrac{1-\nu}{2}$ and $C_1 = \dfrac{-Eh}{12\Delta(1-\nu^2)}$.

10.8 $N_{300} = L_1(3L_1 - 1)(3L_1 - 2)/2$, $N_{201} = 9L_1L_3(3L_1 - 1)/2$,
       $N_{111} = 27L_1L_2L_3$, etc.

10.10 At each corner node the proportion is $-1/12$; at each mid-side node the proportion is $+1/3$.

10.11 At each corner node the proportion is $-1/8$; at each side node the proportion is $+3/16$.

10.12 (a) $N_2 = (1+\xi)(1-\eta)(9\eta^2 - 1)/32$, $N_5 = 9(1-\eta^2)(1-3\eta)(1+\xi)/32$,
         $N_6 = 9(1-\eta^2)(1+3\eta)(1+\xi)/32, N_3 = (1+\xi)(1+\eta)(9\eta^2 - 1)/32$.
       (b) $N_2 = \eta(1+\xi)(\eta - 1)/4$, $N_5 = (1+\xi)(1-\eta^2)/2$
         $N_3 = (1+\xi)(1+\eta)(\xi + \eta - 1)/4$.

10.13 At each corner node the proportion is $-1/8$; at each side node the proportion is $+1/6$.

Chapter 11

11.2 $w_3 = 0.005\,0846PA^2/D$.

11.3 $\{A\phi_2 \ \ w_3\} = \{0.0155\,36 \ \ 0.004\,0767\} \times qA^4/D$.

11.4   (i) $\mathbf{Q} = \{117 \ \ 40b \ \ 42a \ \ 153 \ \ 50b \ \ -48a \ \ 153$
                                $-50b \ \ -48a \ \ 117 \ \ -40b \ \ 42a\} \times qab/90$.
       (ii) $\mathbf{Q} = \{2 \ \ 3b \ \ 3a \ \ 18 \ \ 9b \ \ -9a \ \ 90 \ \ -27b \ \ -27a \ \ 18 \ \ -9b \ \ 9a\} \times W/128$.

11.5 $4w_{2-3} = (2 - 3Y + Y^3)w_2 + (1 - Y - Y^2 + Y^3)b\phi_2$
            $+ (2 + 3Y - Y^3)w_3 + (-1 - Y + Y^2 + Y^3)b\phi_3$
      $8a\theta_{2-3} = (Y - Y^3)w_1 + (-1 + Y + Y^2 - Y^3)b\phi_1$
            $+ (-Y + Y^3)w_2 + (1 - Y - Y^2 + Y^3)b\phi_2$
            $+ (4 - 4Y)a\theta_2 + (Y - Y^3)w_3 + (-1 - Y + Y^2 + Y^3)b\phi_3$
            $+ (4 + 4Y)a\theta_3 + (-Y + Y^3)w_4 + (1 + Y - Y^2 - Y^3)b\phi_4$.

11.6 $4w_{2-3} = (2 - 3Y + Y^3)w_2 + (1 - Y - Y^2 + Y^3)b\phi_2 + (2 + 3Y - Y^3)w_3$
            $+ (-1 - Y + Y^2 + Y^3)b\phi_3$
      $4\theta_{2-3} = (2 - 3Y + Y^3)\theta_2 + (1 - Y - Y^2 + Y^3)b\Omega_2 + (2 + 3Y - Y^3)\theta_3$
            $+ (-1 - Y + Y^2 + Y^3)b\Omega_3$.

11.7 Lateral deflection at the free corners is $0.131\,45qA^4/D$ (compared with an accurate solution of $0.126qA^4/D$): slopes at corners have magnitudes $0.186\,01qA^3/D$ and $0.064\,48qA^3/D$.

11.8 $0.002\,445qA^4/D$ for R1, $0.002\,057qA^4/D$ for R3.

11.9 $\mathbf{Q} = qab\left\{1 \ \ \dfrac{b}{3} \ \ \dfrac{a}{3} \ \ \dfrac{ab}{9} \ \ 1 \ \ \dfrac{b}{3} \ \ \dfrac{-a}{3} \ \ \dfrac{-ab}{9} \ \ 1 \ \ \dfrac{-b}{3} \ \ \dfrac{-a}{3} \ \ \dfrac{ab}{9} \ \ 1 \ \ \dfrac{-b}{3} \ \ \dfrac{a}{3} \ \ \dfrac{-ab}{9}\right\}$

corresponding to the nodal freedoms

$$\mathbf{d} = \{\mathbf{d}_1 \ \mathbf{d}_2 \ \mathbf{d}_3 \ \mathbf{d}_4\} \text{ where } d_i = \left\{ w_i \ \left(\frac{\partial w}{\partial y}\right)_i \ \left(\frac{\partial w}{\partial x}\right)_i \ \left(\frac{\partial^2 w}{\partial x \, \partial y}\right)_i \right\}.$$

Chapter 12

12.1 $\omega_1 = 20.4939\sqrt{(EI_z/\rho AL^4)}$.

12.2 $\omega_1 = 22.7359\sqrt{(EI_z/\rho AL^4)}$ for one element, $\omega_1 = 22.4030\sqrt{(EI_z/\rho AL^4)}$ for two elements.

12.3 $\omega_1 = 136.222\sqrt{(EI_z/\rho AL^4)}$, $\omega_2 = 260.845\sqrt{(EI_z/\rho AL^4)}$.

12.4 $\omega_1 = 3.2863\sqrt{(E/\rho L^2)}$, $\omega_2 = 7.3485 \sqrt{(E/\rho L^2)}$.

12.5

$$\mathbf{R} = \begin{bmatrix} l(1+c_1)/2 & & & \\ l^2(1+2c_1)/6 & l^3(1+3c_1)/12 & \text{Symmetric} & \\ l^3(1+3c_1)/12 & l^4(1+4c_1)/20 & l^5(1+5c_1)/30 & \\ l^4(1+4c_1)/20 & l^5(1+5c_1)/30 & l^6(1+6c_1)/42 & l^7(1+7c_1)/56 \end{bmatrix}$$

12.6 $\omega_1 = 0.032\,58\sqrt{(E/\rho L^2)}$.

12.7

$$\mathbf{m} = \frac{\rho A l}{30} \begin{matrix} u_1 & u_2 & u_3 \\ \begin{bmatrix} 4 & 2 & -1 \\ 2 & 16 & 2 \\ -1 & 2 & 4 \end{bmatrix} \end{matrix}$$

12.8 $\omega_1 = 1.5767\sqrt{(E/\rho L^2)}$, $\omega_2 = 5.6728\sqrt{(E/\rho L^2)}$
where L is the length of the bar.

12.9

$$\mathbf{m} = \frac{\rho A L}{13\,860} \begin{matrix} v_1 & \theta_{z1} & v_2 & \theta_{z2} & v_3 & \theta_{z3} \\ \begin{bmatrix} 2092 & & & & & \\ 114l & 8l^2 & & \text{Symmetric} & & \\ 880 & 88l & 5632 & & & \\ -160l & -12l^2 & 0 & 128l^2 & & \\ 262 & 29l & 880 & 160l & 2092 & \\ -29l & -3l^2 & -88l & -12l^2 & -114l & 8l^2 \end{bmatrix} \end{matrix}$$

where l is the element length.

12.11 $\omega_1 = 17.571$ rad s^{-1}.

12.12 $\omega_i = 37.715\sqrt{(D/\rho h A^4)}$ for clamped edges, $\omega_1 = 19.727\sqrt{(D/\rho h A^4)}$ for simply supported edges.

12.13 $\omega_1 = 25.826\sqrt{(D/\rho h A^4)}$ with R1, $\omega_1 = 28.967\sqrt{(D/\rho h A^4)}$ with R3. An accurate comparative solution is $\omega_1 = 28.946\sqrt{(D/\rho h A^4)}$.

12.14

	u_1	v_1	u_2	v_2	u_3	v_3	u_4	v_4

$$\mathbf{m} = \frac{\rho h a b}{9} \begin{bmatrix} 4 \\ 0 & 4 \\ 2 & 0 & 4 & & & \text{Symmetric} \\ 0 & 2 & 0 & 4 \\ 1 & 0 & 2 & 0 & 4 \\ 0 & 1 & 0 & 2 & 0 & 4 \\ 2 & 0 & 1 & 0 & 2 & 0 & 4 \\ 0 & 2 & 0 & 1 & 0 & 2 & 0 & 4 \end{bmatrix}$$

12.15

	u_1	v_1	w_1	u_2	v_2	w_2	u_3	v_3	w_3	u_4	v_4	w_4

$$\mathbf{m} = \frac{\rho V_e}{20} \begin{bmatrix} 2 \\ 0 & 2 \\ 0 & 0 & 2 \\ 1 & 0 & 0 & 2 \\ 0 & 1 & 0 & 0 & 2 & & & \text{Symmetric} \\ 0 & 0 & 1 & 0 & 0 & 2 \\ 1 & 0 & 0 & 1 & 0 & 0 & 2 \\ 0 & 1 & 0 & 0 & 1 & 0 & 0 & 2 \\ 0 & 0 & 1 & 0 & 0 & 1 & 0 & 0 & 2 \\ 1 & 0 & 0 & 1 & 0 & 0 & 1 & 0 & 0 & 2 \\ 0 & 1 & 0 & 0 & 1 & 0 & 0 & 1 & 0 & 0 & 2 \\ 0 & 0 & 1 & 0 & 0 & 1 & 0 & 0 & 1 & 0 & 0 & 2 \end{bmatrix}$$

The derivation of $\mathbf{m}$ is considerably simplified if volume co-ordinates L_1, L_2, L_3, and L_4 are used. Then $N_1 = L_1$, etc. for the basic tetrahedral element and the required integrals over the volume are easily established using the expression

$$\iiint_{V_e} L_1^i L_2^j L_3^k L_4^l \, dx \, dy \, dz = \frac{i!j!k!l!}{(i+j+k+l+3)!} \cdot 6V_e.$$

12.16 (a) $\omega_1 = 2.4495\sqrt{(EI_z/\rho AL^4)}$, (b) $\omega_1 = 3.5675\sqrt{(EI_z/\rho AL^4)}$.

12.17 $\omega_1 = 19.5959\sqrt{(EI_z/\rho AL^4)}$ for one element, $\omega_1 = 22.3024\sqrt{(EI_z/\rho AL^4)}$ for two elements.

12.18 $\omega_1 = 3\sqrt{(E/\rho L^2)}$, $\omega_2 = 5.1962\sqrt{(E/\rho L^2)}$.

12.19 (a) $\omega_1 = 9.8666\sqrt{(EI_z/\rho AL^4)}$, (b) $\omega_1 = 9.8726\sqrt{(EI_z/\rho AL^4)}$.

12.20 (a) $\omega_1 = 17.035\sqrt{(D/\rho hA^4)}$, (b) $\omega_1 = 18.357\sqrt{(D/\rho hA^4)}$.

12.21 $P_{cr} = 0.251\,88\pi^2 EI_z/L^2$.

12.22 $P_{cr} = 4.0528\pi^2 EI_z/L^2$ for one element, $P_{cr} = 4.0301\pi^2 EI_z/L^2$ for two elements.

12.23

$$
\mathbf{g} = \frac{P}{630l}
\begin{array}{c}
\begin{array}{cccccc}
v_1 & \theta_{z1} & v_2 & \theta_{z2} & v_3 & \theta_{z3}
\end{array} \\
\begin{bmatrix}
1668 & & & & & \\
39l & 28l^2 & & & & \\
-1536 & -48l & 3072 & \text{Symmetric} & & \\
240l & -8l^2 & 0 & 256l^2 & & \\
-132 & 9l & -1536 & -240l & 1668 & \\
-9l & -5l^2 & 48l & -8l^2 & -39l & 28l^2
\end{bmatrix}
\end{array}
$$

where l is the element length.

12.24 (a) $P_{cr} = 4.2555\pi^2 EI_z/L^2$, (b) $P_{cr} = 1.000\,56\pi^2 EI_z/L^2$ where L is the length of the beam.

12.25 (a) $P_{cr} = 0.263\,10\pi^2 EI_z/L^2$, (b) $P_{cr} = 0.250\,15\pi^2 EI_z/L^2$.

12.26 $P_{cr} = 0.1540\pi^2 EI_0/L^2$.

12.27 $\sigma_{xcr} = 9.7691\pi^2 D/A^2 h$.

12.28 $\sigma_{xcr} = 6.7854\pi^2 D/A^2 h$.

AUTHOR INDEX

Numbers in **bold type** refer to lists of publications at the ends of chapters and of Appendix 1.

SUBJECT INDEX